Smart MEMS and Sensor Systems

Smart MEMS and Sensor Systems

Elena Gaura & Robert Newman

Coventry University, UK

with contributions from

Michael Kraft

Southampton University, UK

Andrew Flewitt

Cambridge University, UK

Davies William de Lima Monteiro

Universidade Federal de Minas Gerais, Brazil

ICP

Imperial College Press

Published by

Imperial College Press
57 Shelton Street
Covent Garden
London WC2H 9HE

Distributed by

World Scientific Publishing Co. Pte. Ltd.
5 Toh Tuck Link, Singapore 596224
USA office: 27 Warren Street, Suite 401-402, Hackensack, NJ 07601
UK office: 57 Shelton Street, Covent Garden, London WC2H 9HE

British Library Cataloguing-in-Publication Data
A catalogue record for this book is available from the British Library.

Cover design by John Burns.

First published 2006
Reprinted 2007

SMART MEMS AND SENSOR SYSTEMS

ISBN-13 978-1-86094-493-2
ISBN-10 1-86094-493-0

Typeset by Stallion Press
Email: enquiries@stallionpress.com

Printed in Singapore by World Scientific Printers (S) Pte Ltd

Preface

This book has emerged as the result of the authors' research activity over the past six years. We started the journey as an Electronics Engineer and a Computer Scientist, both with working experience of sensing systems. The motivation for joint work was initially provided by the vision of a new world-changing technology put forward by vanguard researchers such as Kris Pister and Deborah Estrin. They foresaw systems composed of millions of MEMS sensors, collaborating in an 'intelligent' way to address many of the major problems of our age — environmental study and monitoring, pollution controls, transport safety and so on. These sensors would self-organise into networks, which in turn would be self-configuring, fully decentralised and would rely entirely on collaborative behaviour between sensors.

Like most researchers launching into a new venture, we started with simple theoretical and practical studies derived from our own sensing backgrounds and aimed at evolving the research towards full blown applications as above. Eventually it became clear that neither of our disciplines was sufficient to allow us to fully cover the scope of this research. Our vistas simply had to expand to allow us to make sensible choices as to direction and emphasis of research.

When approaching the detailed design of even quite small systems (in terms of sensor numbers), we found that the research questions posed required new science within our own respective disciplines, combined with a need for new science in other disciplines. To overcome the exposed hurdles, these findings had to be communicated to the scientists with the knowledge and skills to provide it. At this stage we became aware of how difficult it is for domain specialists to hold an overview of a topic as broad as that of large scale MEMS based sensing systems. As a simple example of the problem, a MEMS device designer may spend an enormous amount of effort optimising a design for those last few decibels of noise margin, when, at a system level in an intelligent sensor, that same performance increase may be obtained very simply using a slightly more sophisticated signal processing algorithm. Without at least some knowledge of what is possible in each of the domains involved here (MEMS technology and smart sensor electronics), it is impossible to make sensor system design trade-offs in a sensible way. When one moves on to the possibility of sensors collaborating and operating advanced data fusion algorithms, the design choices become even more difficult and the breadth of knowledge required tenfold larger.

A study of the many new books and papers being published in the domain of intelligent MEMS systems and wireless sensor networks reveals that they mostly take a single discipline perspective. While several of these books are excellent, a researcher embarking on a top-down sensor system design venture (which we believe to be essential to achieving success in this field) is faced with assembling a library of books, each of which books assumes prior knowledge of domains that could easily be alien to our researcher.

The aim of this book is therefore to present the leading edge of this research and indeed set the research agenda in the field of MEMS based sensing. We hope to have been able to bring in a view from all of the participatory disciplines, in an integrated way. No assumption was made of *a priori* knowledge above the level which might be reasonably expected from, say, an electronic engineer about computer science or a computer scientist about MEMS. This has proved to be a much harder task than we initially envisaged, resolved however with the help of our distinguished contributors who have risen to the challenge of stepping out of their specialist domain and presenting deep knowledge of their own field in a way accessible to the non-specialist. Hence, we are very grateful to Dr. Andrew Flewitt for a complete and clear exposition of relevant MEMS micromachining technologies; to Dr. Michael Kraft for a detailed case study on inertial sensors which reveals the merits of integrated sensor signal processing and Dr. Davies William de Lima Monteiro for a fascinating insight into the practice and potential of active optical MEMS systems.

Overall, we have tried to pay particular attention to the level of each chapter. We have not allowed ourselves to work at elementary introduction level, but have always attempted to present leading edge work in each of the three areas covered, MEMS technology, electronic system design and pervasive computer science, in a digestible fashion. The central element which facilitated this treatment was the concept of the "cogent" sensor, which we developed when it became clear how overused and overloaded the terms 'smart' and 'intelligent' are, when used to describe processor integrated sensing devices. We strongly believe that it is much more useful to define things by what they do, rather than how they are constructed, when working from a top-down systems perspective. Whereas Randy Frank proposed whimsically that "A rose with a microcontroller would be a smart rose" (Understanding Smart Sensors, pp1), we would say "a rose which provides the fragrance you need would be a cogent rose". The theme of usefulness is the essence of the top-down approach.

It is widely held that smart MEMS technology will be world-changing, which presumes it will be of real use for activities which shape the world. Rather than promoting a solution waiting for a problem, we want to enable those with problems waiting for solutions to evaluate whether smart MEMS systems will rise to that challenge. If it proves to be so, we want also to enable them to begin the process of scoping that solution.

If we have achieved this goal, we believe that the book will be uniquely useful to those faced with understanding the breadth and wealth of the opportunities that combined MEMS and pervasive computing technologies offers.

Elena Gaura,
Robert Newman.
Coventry, 2006.

CONTENTS

CHAPTER 1

MARKETS AND APPLICATIONS

1.1. Technology at Crossroads

This book concerns a particular class of Micro Electro Mechanical Systems (MEMS) device: Intelligent or smart sensors. The general area of MEMS is one that has been the subject of speculation and 'futurology' over the past few years, some of which has been quite unhelpful in providing a real appreciation of the huge potential of this technology. In the first days when micromachining became feasible, there were extreme predictions of the potential of the new technologies — micromachines that would revolutionise every aspect of daily life were predicted, often based more on science fiction that any sober assessment of the capabilities of the technology. More recently, the futurologists have turned their attention to nanotechnology, MEMS being old news. Strangely, however, the prediction that MEMS technology would affect our daily lives has turned out to be entirely true, but the effect has been in ways both more subtle and profound than envisaged by the original forecasts. MEMS has indeed proved to be a potent technology and the application of MEMS that has been most important for the realisation of this potency has been sensors: sensing, a seemingly prosaic area of technology, has been revolutionised by MEMS to the extent that the basis has been laid for completely new types of engineering systems. There is now the possibility of designing complex MEMS based systems that are sensitive and reactive to their environment and able to respond and adapt to it. In turn, this responsiveness may be used to address some of the large scale engineering problems which are crucial to the major concerns of the world today: efficiency, energy saving and environmental monitoring. This book is about the design concepts and methods that will be necessary to realise these new systems, building from the technological base provided by MEMS sensors.

One of the major problems in realising the potential of MEMS is that the fundamental characteristics of the technology, the ability to manufacture huge numbers of sophisticated electromechanical systems, gives it the power to produce systems of staggering complexity, which is reflected in the difficulty both of designing and building them. Therefore this book focuses on these systems level issues, rather than the technologies which allow fabrication of ever more capable MEMS devices — that has been extensively covered elsewhere. It charts the developments in the supporting systems technologies which enable the unique properties of MEMS sensors — mainly to do with tiny size and very low cost — to be used to build real operational systems of increasing power and capability.

The book is written at a time when there is something of a crisis of confidence in MEMS technologies, and, as suggested earlier, they are somewhat in the shade of nanotechnologies. Stephen Senturia, who has been a leading light in the field since its inception, wrote: "much of the energy and dynamism that has characterised the field for more than 20 years may flag and fail" [1]. As someone who was present at the inception of the pioneers' dreams, it is natural that Senturia should detect a waning of enthusiasm. However, those first dreams have been replaced with new, more achievable aspirations, allowing confirmation of the observation made above: that MEMS is indeed a technology which will change the world in which we live, although perhaps not in the ways which were originally put forward by the futurologists.

Crucially, for the state of the art in this field, we now have enough visibility of the world changing applications of MEMS to be able to see clearly the path towards their realisation. It is the belief of the authors that to achieve this new potential a new research agenda will be necessary, one which will include workers from other disciplines than the materials scientists and electronics engineers who have traditionally formed the mainstream of this domain. We start by surveying the importance of MEMS in the world today, the reasons for the over expectations in the past and the real potential for the future.

1.2. The Present — MEMS in the News

To understand the ways in which MEMS sensors are indeed changing the everyday world in which we live, we look at some of the recent news releases in the technical and popular press. MEMS sensors are contributing to profound developments in a number of day to day activities:

Healthcare

> *"Microelectromechanical Systems (MEMS) range from the mundane to the spectacular. At one end of the spectrum are devices such as the precisely machined nozzles used in ink jet printers. At the other extreme, MEMS are enabling the blind to see." — Semiconductor International, June 2003. [2]*
>
> *"... and then there are really esoteric MEMS devices. Researchers have created microscopic MEMS motors and minuscule mechanical manipulators that can grasp a single red blood cell.... What's not small about MEMS is their growth potential." — Design News. [4]*
>
> *"Emerging inertial sensors fashioned from microelectromechanical systems, or MEMS, promise to enhance medical equipment in ways that make them easier and safer to use. Such devices are already being aimed at defibrillators and patient-monitoring systems. MEMS technology, in the form of a DNA lab-on-chip, will also in the future play a key role as part of a set of instruments for making quick analyses of microbiological samples."*

— EE Times, August 2004. [3]

Transport

> *"Drivers in the INDY racing league have a new piece of kit this year and it isn't under the hood but in their ears. Embedded in the driver's radio earpiece is a tiny MEMS sensor system ($4.5 \times 4.5 \times 2$ mm) developed by engineers at Delphi that measures the dynamic forces applied to the drivers head during an accident. The g-force data collected will provide researchers a clearer picture of what happens in the split second of time that it takes for a crash to occur, leading to better design of the driver restraint system and safety devices."*

— Design News, May 2003. [4]

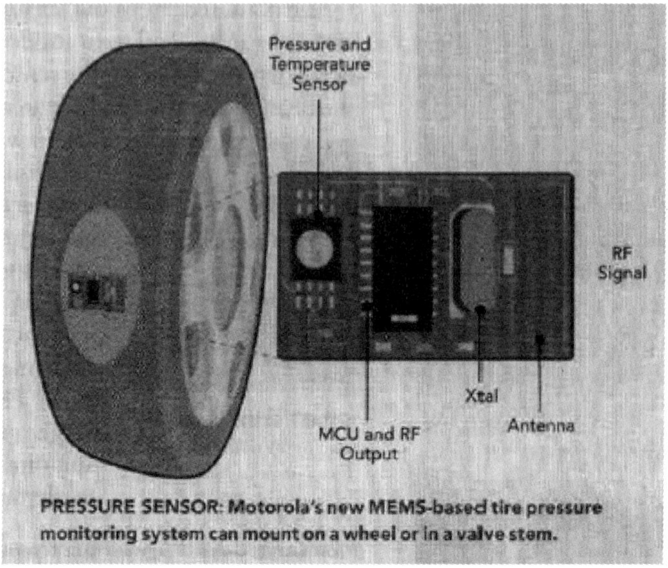

Figure 1.1: Tyre pressure sensor. From [4].

> *"From now on, even tyres will contain electronic components. Motorola's MPXY8000 tire-pressure monitoring system goes inside a tire or onto a tire's valve stem to constantly check for dangerous tire deflation (See Figure 1.1). The system wirelessly transmits information to a car's remote keyless entry receiver, where software can alert a driver to stop and add air."*

— Design News, May 2003. [4]

> *"June 10 2004 — Crossbow Technology Inc. has launched a series of systems that provide navigation, position and leveling information to air and watercraft. The NAV420 series packs MEMS-based accelerometer and gyro clusters with a global positioning satellite receiver and other sensors and software in a 3-inch cube.... The system provides information to an autopilot or display on whether the craft is straight and level, and the direction it is going, within 1 degree of accuracy. It is designed to replace larger and less reliable mechanical sensors."*

— Small Times, June 2004. [5]

> *"The accelerometers from Analog devices that deploy airbags in car crashes exhibit less than one failure per billion hours of operation."*

— Design News, May 2003. [4]

> *"The automotive industry, already the largest market for MEMS devices, will use more of them... 9.1 per vehicle in 2007... up from 5.0 in 2002."*

— Design News, May 2003. [4]

Leisure and consumer

> *"Mitsubishi Electric Corp. has designed a motion sensor by MEMSIC Inc. into a mobile phone manufactured for Vodafone Group PLC. The accelerometer enables a pedometer to measure distances and provides image orientation for the camera in either portrait or landscape modes. The sensor also allows a user to use the phone as a joystick for video games."*

— Small Times, June 2004. [6]

> *"Says Benedetto Vigna, manager of MEMS development for ST, 'I believe this will be the decade of MEMS inertial sensors for consumer applications.' Part of the impetus of new MEMS applications comes from increasingly sophisticated features. ... Performing as very sensitive motion and tilt sensors, they're starting to provide one-handed, keyless scrolling of displays on cell phones and PDAs. To scroll a mobile phone's tiny display, you just tilt the phone in the appropriate direction. ... To zoom in, you raise the phone or PDA a bit; to zoom out, you lower it."*

— Design News, May 2003. [4]

"Analog Devices Inc. has announced that its iMEMS accelerometer technology will be used in multiple platforms of IBM's ThinkPad mobile computers featuring the Active Protection System technology. An ADI accelerometer on the ThinkPad motherboard detects shocks or free-fall conditions, suggestive of an imminent impact, and within a fraction of a second signals the drive's R/W heads to temporarily park, helping prevent contact with the disk drive until the system is stabilized."

— *EE Times, January 05. [7]*

Communications

"A 80-channel optical communications switch that adopts MEMS mirrors, achieving a switching speed of 1 ms (claimed to be the fastest switch to date) was developed by Fujitsu. Measuring 150 × 400 × 300 mm, the switch offers an optical power stability within 0.5 dB. The tilt of the MEMS mirrors is precisely controlled through a feedback loop in a built-in control function which maintain the optical power at a fixed level."

— *EE Times, October 02. [8]*

Industry and construction

"Because engineering departments are run leaner and need to concentrate on core activities, design engineers are increasingly looking for complete solutions. Complete solutions are recognized as being economically attractive or perhaps more attractive because with a complete solution somebody takes overall responsibility. . . . to be effective and have wide ranging impact, smart sensors must work with a complex maze of networks and computer interfaces."

— *Industry News, December 01. [9]*

"Engineers may no longer have to struggle with wires and batteries in monitoring the structural health of buildings and bridges. A wireless, battery-free microsensor system that would enable engineers to accurately assess and monitor structural health has been developed by researchers at Sandia National Laboratories in Albuquerque, New Mexico. The energy-capturing portion of the system takes the form of a 32 by 62 by 0.5 mm strip of piezoelectric material 20 mm thick that is embedded in a concrete or steel structural element of a building or bridge along with its associated hardware."

— *Civil Engineering, August 02. [10]*

A conclusion that can clearly be drawn from these clips is that MEMS sensors already affect in a fundamental way people's day to day lives. The size of the MEMS industry today is huge (in financial terms, obviously, as the size of the physical output is tiny). The market for MEMS devices, taking a 'narrow' view of MEMS (i.e. excluding ink-jet print heads and the like) is expected by In-Stat/MDR (part of Reed Electronics Group) to grow from $3.9B in 2001 to $9.1B in 2006, with most of the growth in traditional areas, although bio-MEMS (lab-on-a-chip for DNA analysis) and RF MEMS are the fastest growing. NEXUS take a broader view, grouping MEMS with larger systems such as hearing aids and cardiac pacemakers and expects this "Microsystems" market to increase from $30B in 2002 to $68B in 2005. [11]

1.3. The Past — Great Expectations

Senturia himself gives a fascinating account of the very early days of MEMS sensor research [1]. Discussing the 1981 Materials Research Society meeting, he notes the following:

> *Eighty researchers in the field of 'Solid State Transducers' gathered from around the world to share their experiences, both technical and organisational... the only 'physical sensors' discussed during the symposium, in addition to pressure sensors, were magnetic sensors, the microdielectrometer for low frequency dielectric analysis of resins, a temperature sensor, and a dew point*

> *sensor. Chemical sensors, including both gas sensors and ion-sensitive devices, were prominent. Accelerometers, flow sensors, gyros, switches, relays and actuators of any type were nowhere in sight.*

Interestingly, the sensor types that Senturia's group was bemoaning the lack of was precisely the group that now forms the mainstay of the burgeoning MEMS industry: accelerometers, pressure sensors and gyros. The automotive industry alone consumes huge numbers of these types of sensor (primarily pressure sensors and accelerometers), and their use is growing (9.1 per vehicle in 2007... up from 5.0 in 2002) [4]. However, in 1981, one of the world's leading symposia in the field failed to consider these seriously at all. How could this be?

The answer to this question lies in the motivations and drivers of those doing the basic, underlying research in a topic. There is always pressure on researchers operating at the practical end of their domain to work in areas which are thought of as revolutionary, rather than evolutionary. The areas that are thought of as revolutionary are generally defined by the visionaries and futurologists. At this time, the ambitions of researchers were set very much higher than the mere improvement of such prosaic articles as accelerometers and pressure sensors. Some of the research agendas that were being set at the time can be judged by looking, for example, at that of the Japanese Micromachine Program I the 1990's. In a report on the Micromachine Symposium, in 1994, Kahaner summarises its goals [12]:

> *The ultimate goal of mechanical engineering is to replace human functions and labour by machines. To reach this advanced state, we must develop machines as clever as ourselves, and enable them to move according to their own decisions as our body does.*
>
> *To meet the second requirement, it is necessary to make machines much smaller, as may be realised from the fact that human movements rely on cells and their constituent substances including proteins and other biological molecules.*
>
> *Reducing machine dimensions has lagged behind the R&D of intelligent machines. However, this challenge must be faced for the progress of mechanical engineering. Developing micromachines may provide us with great innovations in industrial technologies as did the development of intelligent machines.*

Unfortunately, micromachinery has not found any definitely promising applications yet. Worse, the research investment will certainly be huge. In the private sector, therefore, research of micromachines will be too limited to achieve technological innovation.

The Industrial Science and Technology Frontier Program has been set up to develop micromachine technology to strengthen industrial technologies as well as mechanical engineering.

Later the three major goals of the program were set out.

(i) Advanced maintenance system for power plants

This is a micromachine system for the maintenance of fine tubes in power plants. The system will consist of a microcapsule, a base machine, inspection module and operation module. Necessary mechanical components (e.g. microscopic power generator and energy transmitter) of the system have been specified. The component devices are being fabricated.

(ii) Medical micromachines

Micromachines are applicable to examination and treatment inside the body cavity. A micromachine will possibly be inserted through a catheter for diagnosing and curing, for example, cerebral thrombosis and aneurysm. Component devices of such medical machines are being fabricated.

(iii) Microfactories engineering

A system for manufacturing tiny precision parts of watches, cameras, and electronic appliances with much smaller production equipment than predecessors. The system will greatly reduce energy consumption in production. The miniature equipment should be no larger than 2–10 times the size of the product. Component devices of the equipment are being fabricated.

Moreover, in a another article [13], contemporaneous with Kahaner's, Voelker states:

Micromachinery also has important potential applications in "conventional" medicine; scientists in Japan and elsewhere are working on "microrobots" designed to circulate in the bloodstream and relay temperature, pressure, pH and other conditions back to an external computer, in the manner of the miniaturized submarine in the movie "Fantastic Voyage".

In the second reference, the Science Fiction drivers are explicit. Obviously, one cannot assume that the goals of the whole research community are represented by these two statements, but it does indicate that during the 1980's and early 90's there was at least a strand of research setting near term goals for microtechnology that were visionary but ultimately unrealistic, without any clearly defined development path towards them. The actual pace of development was actually very different. Senturia wrote:

> *There has been no shortage of bright ideas in the area of microsensors, microactuators and microelectromechanical devices of all sorts. However, the track record on converting those ideas into commercially successful products has seemed uneven to some, both inside and outside the field. It has taken 15 to 20 years (or more) between early research prototypes and full commercialisation for such devices as silicon pressure sensors, accelerometers and optical displays. [1]*

At the time, the agenda for sensor researchers was being set by visions of the 'science fiction' type, which, twenty years later, have still to be achieved. Against such a background, MEMS rapidly gained a reputation as a technology that was failing to deliver. Fortunately, the stolid and reliable sensor and transducer technologists have pursued an incremental development path, resulting in technology which can deliver real benefits in real applications and which finds applications numbered in the millions. MEMS is now once again seen as a pervasive technology, affecting and enabling a host of related technologies. It will become even more so in the future.

1.4. The Future — Maturity and Pervasive Applications

The current state of the art in MEMS, particularly when linked with associated developments in VLSI and pervasive computing, has provided a basis for a new round of 'dream applications'. While still clearly visionary, these are not in quite the same grandiose league as the 90's 'fantastic voyages'. Instead they are generally designed to deliver some quite tightly specified advantages in real world applications, such as environmental monitoring, adaptive aerodynamics or scientific exploration. Unlike the previous fantastic journies, these applications highlight real research challenges, ones for which it is possible to scope and plan a line of research that can realistically deliver working technology. Below, a few such applications are surveyed, starting with near future proposals for which the technology is actively being developed, leading on to more 'blue skies' proposals, which are still at the level of feasibility studies. This is, in itself, a major qualitative difference from the proposals seen previously, which were often put forward without any well reasoned case made for feasibility.

Until recently, space research, or more colloquially 'rocket science' has been seen as one of the primary drivers of technological development, and it is to be expected that a number of visionary applications of Microsystems will apper in this domain. Some are proposed by Stenmark [14], as follows:

- *Thermal control* — using thin film technology or 'functional surfaces' (micro actuators embedded in the skin of a spacecraft — in which the thermal emittance can be changed by using an electrical control signal. It is proposed that such a system can replace mechanical louvers.
- *A Micro Propulsion Cold Gas Thruster* system — this was a multiwafer design which contained many functions in one unit (nozzle, heat exchangers, valves, pressure sensors, electronics, etc.). It is suggested that this system allows very accurate attitude control. This was achievable taking into account the state of the research art in microsystems in 2001 [15], and so is a 'safe' technology prediction.

More speculative sensor based dream applications are the Berkeley 'smart dust' proposals, based on so-called 'motes', (wireless autonomous smart sensors), which are deployed in their thousands for various environmental and battlefield sensing applications [16]. Derived from this is the GEMS proposal [17], which imagines global deployment of motes into the atmosphere for meteorological sensing. Another proposal, from NASA, is the 'ageless aircraft' in which smart materials integrate intelligent condition monitoring sensors and actuators to continuously sense and correct structural aging problems [18]. The basis for such proposals comes from the inherent, natural MEMS properties of size and potentially low cost, which encouraged the liberal usage of these devices in applications. Such usage, in turn, leads to the need to rely on and/or add efficient and clever processing of data generated by the sensing device, before such data reaches the outer world. The nature of this processing, and the design methods used to specify and code it, are rarely put forward in detail by the proposers of the application, although they are a pre-requisite for its realisation. These design methods form an important topic of this book.

Another proposal discussed here is from one of several NASA studies [19]. It proposes the use of wireless intelligent sensors (called 'tranceivers') to predict the failure of equipment (see Figure 1.2). If impending failure is predicted then a replacement, or replacement consumables can be ordered and installed 'just in time'. A block diagram of the intelligent sensor node is shown here. As well as sensors (of unspecified type) it contains a GPS system (so that it can determine its location) and radio for connection to the internet. The proposal goes on to suggest that:

> *Intelligent transceivers would have dimensions of no more than a few centimetres. They could be mass-produced relatively inexpensively by use of established integrated-circuit fabrication techniques. An intelligent transceiver would be connected with "smart-part" microchips that would be designed into major components and subassemblies of the equipment to be monitored (see figure). These microchips would contain sensors and sensor circuitry for monitoring the physical conditions and statuses of components and subassemblies.*

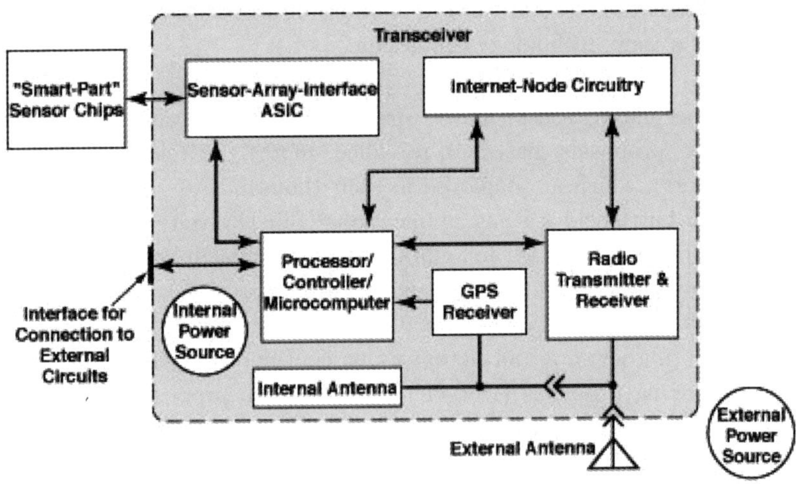

Figure 1.2: Block diagram of 'tranceiver'. From [19].

A transceiver on a damaged piece of equipment could interrogate other damaged pieces of equipment to determine what components could be salvaged and whether a replacement for a damaged component in its own damaged piece of equipment was locally available.

Each transceiver would be capable of "learning" and updating its "knowledge" of the rates of wear of critical components. An "Intelligent" Transceiver connected with "smart-part" sensor chips would provide information on the status of the equipment in which the sensor chips were embedded.

This, and the other applications put forward in this section, have a common factor that the scenarios are being detailed to the point of precise specification of the sensing devices, elaborated accounts of their function and interaction and some quite precise estimates of their size, cost and component parts. It would be possible, in most of these cases, for a suitably qualified engineer to take the scenario and start an outline design for the hardware (and this book would be a good guide for such a task). It is on this basis that these applications can be said to be feasible. However, the detail of how the interaction between the many sensing components occurs, how it is planned, designed and implemented, is much more sketchy. Research into how to undertake these tasks is a prerequisite for the realisation of all of these applications.

1.5. Drivers for Progress

The 'dream applications' for the future, presented above are all academic or agency studies. Whilst impressive, and in some cases technologically thorough and possibly feasible, none of them will be realised unless there are very sound and hard headed reasons to make the investment necessary. Some of the scenarios proposed have been detailed precisely to encourage speculation on the potential of the technology, and therefore encourage the raising of the required funding and sponsorship. Although the scenario provides some kind of 'road map' for the technological development, the precise direction in which it will proceed is, more often than not, driven by the nature of that funding, and the concerns and motivations of the sponsors. In general, these concerns and motivations derive from five high level issues: (the market, impact, competition, technology and manufacturing [2]), which are discussed below.

(i) The market

In order for a development project to go ahead, its funders need to know that the development investment will be returned, with a profit. Prediction of future markets, their possible size and growth is not an exact science, so this development funding always entails a risk. The risk is minimised if the new market is a straightforward extension or derivation of an existing market. But while such markets are more predictable, the possible returns are smaller, since the opportunity for massive growth is likely to be less than with a completely new market. A good example of such a 'developmental' market (although in the generality of MEMS, rather than sensors specifically) is that of MEMS RF filters. The MEMS products have significant size, and cost advantages over traditional filters. At the same time, given the growth of wireless technology, small, cheap RF filters are likely to be in huge demand. Thus, for instance, the development of MEMS RF filters is not a big risk.

(ii) Impact

In some cases, where the new technology offers a paradigm shift, enabling completely original products, or ones with radically different functionality from their forebears, the risks are high, and the size of the market very difficult to estimate. On the other hand, the returns can be huge. Such developments will be funded by large corporations (where the risk may be spread over a large overall development investment) or by venture capital. In either case, the sponsors' decisions are likely to be driven by the potential of the final product, rather than incremental device improvements.

(iii) Competition

In a competitive market, if one manufacturer gains an advantage by moving to a MEMS based product, then its competitors are under strong pressure to follow suit, hopefully with an improved, rather than 'me too' product. If the MEMS product radically changes the market, then manufacturers failing to follow suit are at risk.

(iv) Technology

A manufacturer's decision on whether or not to develop a product will be influenced by the investment it has to make in the technology to develop it. It is here that prior investment (either direct, or by sponsoring research organizations) in technology demonstrators (such as the 'dream applications' discussed above) may pay off. Manufacturers with existing MEMS products will be at an advantage compared to those that have to buy in (specialist MEMS foundries providing some means of 'catch-up' for these). Designs that can use or adapt existing technologies will be favoured over those requiring completely new ones.

(v) Manufacturing

Designers of technology demonstrators or research products are unlikely to consider too seriously the manufacturability of their design. On the other hand, for low cost, mass market products, issues such as device yield, plant design and calibration and configuration costs are key. Self-configuring, calibrating and fault tolerant designs will be at an advantage, so long as the "in use" cost advantages are not outweighed by higher unit costs.

In practice, it can be seen that all of these drivers are at work, pushing forward the practical limits of MEMS sensor technology. It is upon the progress currently being made that further developments will be made, which in turn will enable the 'dream applications'. Below is a survey of technological progress on three fronts, the improvement on MEMS devices themselves, the integration of those devices with other system components and functions, such as the 'intelligence' required for our 'dream applications', and finally the design methods to enable large and complex sensing systems to be realised.

1.6. Progress — Device Improvement

The maturity of the MEMS technology is shown by the subtle and sophisticated device designs now being realised. Again, looking at current press reports indicates that the ingenuity of MEMS designers has reached a very high level. Utilising tiny components to fabricate accurate and robust devices is commonplace, as indicated by the report below (see Figure 1.3).

> *"The proof mass in Analog Devices MEMS gyroscope weighs only 8 millionth of a gram. It is suspended only two microns over the device's electronic circuitry. The proof mass in accelerometers from MEMSIC beats even that. It's nothing but a gas that moves in a sealed chamber,*

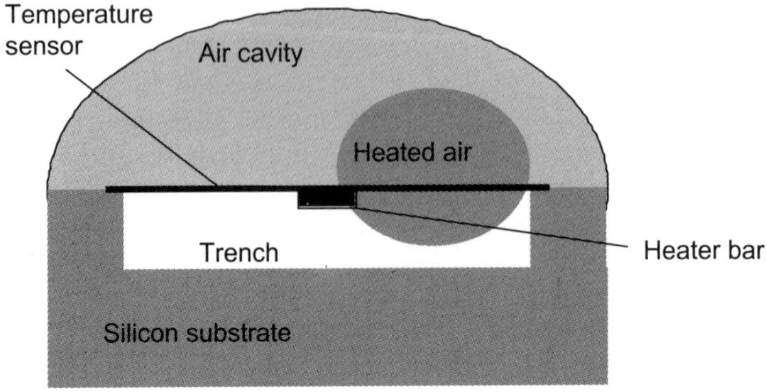

Figure 1.3: In MEMS accelerometers from MEMSIC, a heated gas changes position as the device moves. Temperature sensors measure the gas's shift to determine acceleration. From [4].

> *and because gas is the only thing that moves, there are no parts which can break. MEMSIC claims its device can withstand shocks of up to 50 000 g."*

— Design News, May 2003. [4]

However, the necessary improvement of MEMS processes to create devices such as the one presented above is not straightforward or inexpensive. We will see later the cost in dollars and time of the development of the technology on which the Analog Devices integrated MEMS systems, lauded in the quote above are based. Such technological progress is often piecemeal and can be painfully slow. For instance, Bryzek follows the development of a single MEMS sensing component, the piezoresistor — used for instance in pressure and force sensors, through more than four decades [20], from the original components fabricated by Kulite as part of a pressure sensor in 1961. The major problem that has plagued the design of this component has been stability. Successive process improvements were made by different workers to improve stability, starting with high-energy ion implantation (Honeywell, 1966), a specialised, and therefore expensive process. This process was also patented, leading other workers to look at alternative methods, ICT developing a very deep junction piezoresistor in the late 70's and NovaSensor creating a Faraday shield round the resistor using metallisation and ion-implantation in 1987, a design still in mass production. Thus, in twenty-six years, steady competitive development had

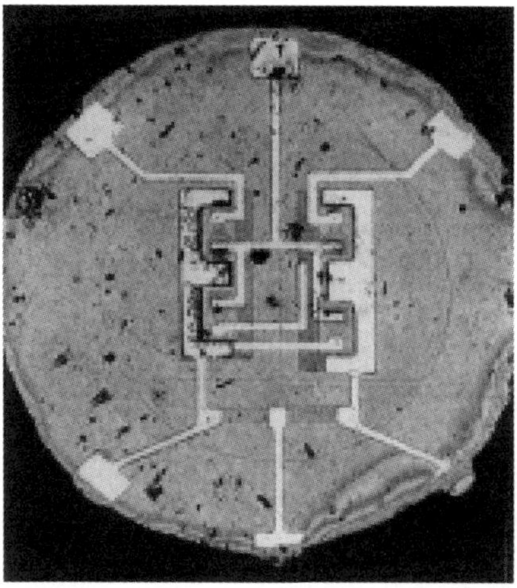

Figure 1.4: The first integrated circuit, 1961. Illustration from Fairchild Inc.

translated an effective but flawed and expensive component into one with essentially the same functionality, but without the stability problem and priced suitably for inclusion in mass market products. By any account, this is slow progress.

If, by comparison, we look at the progress of the digital VLSI industry over the same period, the slowness of progress in MEMS is seen very explicitly. In 1961 Fairchild Semiconductor (actually 'Fairchild Camera' in those days) introduced the first commercially available integrated circuit with two transistors, shown in Figure 1.4.

By 1987, 26 years later, the 'commodity' microprocessor was the Intel 386, shown in Figure 1.5, with 275 000 transistors, which outperformed $100 000 'supermini' computers of just five years before, and was produced cheaply enough to make personal computers an economic possibility for many homes. VLSI electronic process development has proceeded much faster than MEMS process development. The mass markets are obvious and established, and therefore the investment is available to fund hugely expensive process development projects. By contrast, in MEMS the mass markets are just now becoming clear, and the vast scale of investment available for new VLSI processes is still not there for MEMS. Much MEMS process research is still occurring in publicly funded university research laboratories, whereas VLSI process improvement is solely the domain of the large corporations.

Figure 1.5: Progress by 1987, Intel 386. Illustration from Intel Corp.

If MEMS process development does not proceed at the same rate as VLSI development, perhaps it is possible to use VLSI technologies, and therefore hitch a ride on that train. The use of VLSI technologies, with ultra fine scale lithography and precise control of material characteristics can make it possible to improve the performance of the transducers themselves. It is possible to use signal processing, implemented in VLSI, to ameliorate shortcomings of the MEMS transducers, and much of this book is about this very subject. The key to doing this economically is to integrate the signal processing with the MEMS sensor component. Thus device integration is a major research topic, which will be discussed in the next section.

1.7. Progress — Device Integration

This book concerns the development of intelligent sensors. We will discuss the precise meaning or interpretation of that term in Chapter 8. For the purposes of this section the salient point is that an intelligent sensor integrates electronic circuitry to enhance the performance of the sensor or to provide system level functions. Ultimately, the cost of sensor, be it intelligent or not, depends on the number of separate (as opposed to integrated) components and complexity of assembly. Given that many MEMS processes were originally derived from VLSI processes, and that typically electronic circuitry needs to be integrated with the mechanical assemblies, there is a continuing drive to integrate a 'system on a chip', although this is not always the optimal solution for a given application, as we shall see.

Some commentators see the achievement of further integration as a major constraint on progress. Ohr raises the question of why increasing levels of integration are not routinely achieved [21].

> *High on the list of obstacles slowing the proliferation of silicon micromachines is the task of integrating the sensors of microelectromechanical systems with signal-conditioning circuitry on the same CMOS chip.... "If Itanium uses 100 million transistors, why not MEMS?" asks Analog Devices' Bob Sulouff. Motorola's Dragan Mladenovic and Dave Monk say that high-density mixed-signal processes (like the company's SmartMOS7) would enable competent integration. But a single-chip MEMS controller is still a possibility for the future.*

Sometimes, integrated electronics give the only possible solution to the problems posed by the very fine scale of MEMS devices. Design News discussed this issue. [4]

> *The changes in capacitance that a MEMS device detects are just as tiny as MEMS mechanical components... (at the order of zeptofarads)... so small that having on-chip circuitry to measure it and process the reading is preferable to off-chip circuitry, which can affect readings.*

Just as device integration can make some sensors possible to produce, it can also be used to improve the performance of a device. To illustrate this, we return to Bryzek's tale of the development of the piezoresistor [20]. In the previous section, we saw how process improvement had eliminated the stability problems over a 26 year period. A parallel process, at least to the latter part of that time, has been the use of integrated signal processing to eliminate these same stability problems. ICT started such a development in the 1970's but it was shelved due to lack of development capital (we can speculate that had the process improvement been essential to the development of commodity semiconductors, the funding would have been found). The baton was taken up by National Semiconductor, firstly with integration of active temperature stabilisation circuitry, and secondly with integrated on-chip bipolar analogue signal processing circuitry for a mass-market automotive application. Both were ultimately unsuccessful. Finally, Honeywell succeeded in the early 1980's, on the back of funding for a high value aerospace application. Motorola achieved a low cost integrated solution in the 1990's. Subsequently other manufacturers have produced integrated MEMS with on chip analogue and digital electronics, but no 'standard' process has become established — a testament to the difficulties of integration. Bryzek summarises the existing approaches to integration [20].

> *The first is integration of the MEMS components on top of a fabricated IC. The major constraint is the need for strict process compatibility, which rules out many common MEMS process steps, such as LPCVD polysilicon and silicon fusion bonding. Another restriction is that the exposed materials, typically a low-temperature oxide and aluminum, limit the chemistries available for processing.*

One of the most successful products built with this approach is Texas Instrument's optical display chip, the Digital Light Processor DLP. A complete account of this technology, and the process required to produce it is given by Hornbeck [22]. The DLP chip is an array of articulated mirrors, each of which can be individually swivelled using electrostatic forces. A visualisation of an individual mirror is shown in Figure 1.6.

The complex micromachining process required to build this structure on top of a CMOS chip is shown in Figure 1.7.

The length of development (17 years) and the amount of money spent ($1B) is a good indicator of the difficulty of using this approach. Hornbeck presents a chart (Figure 1.8), showing the incremental design improvements that have been made during those years.

> *A second approach is lateral or side-by-side MEMS and IC Integration. Here, any CMOS incompatible processes are fabricated first,*

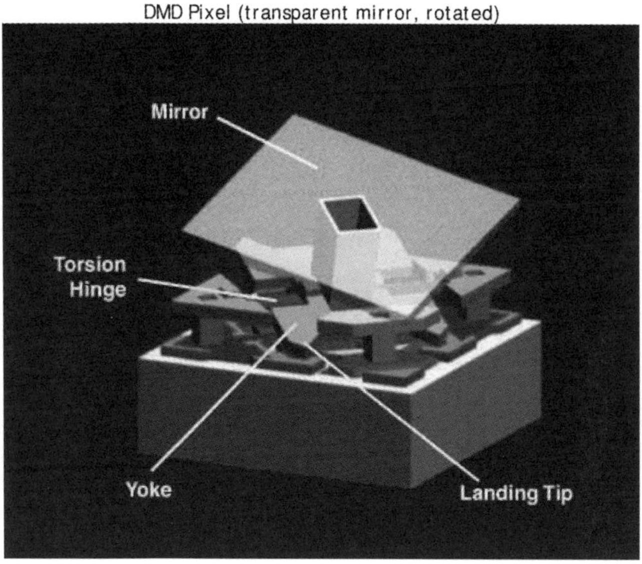

Figure 1.6: Digital light processor mirror assembly. From [22].

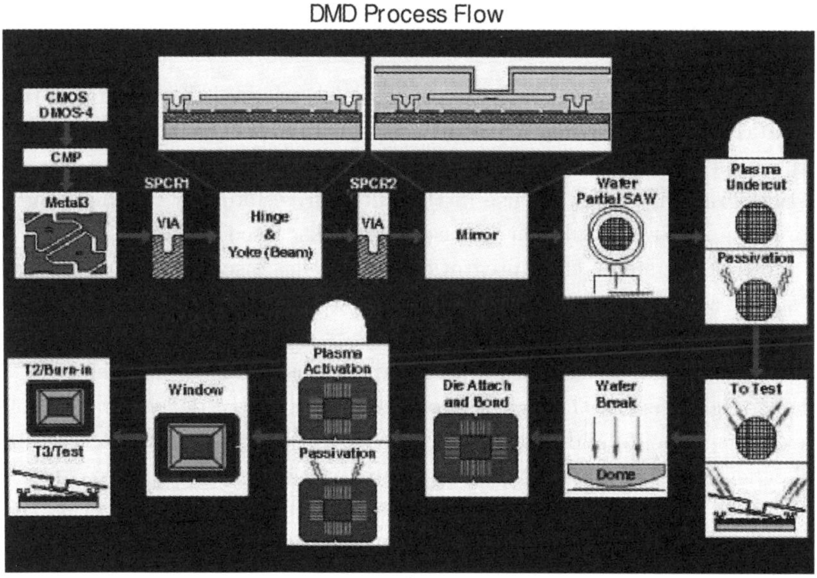

Figure 1.7: Process flow for Digital Mirror Device. From [22].

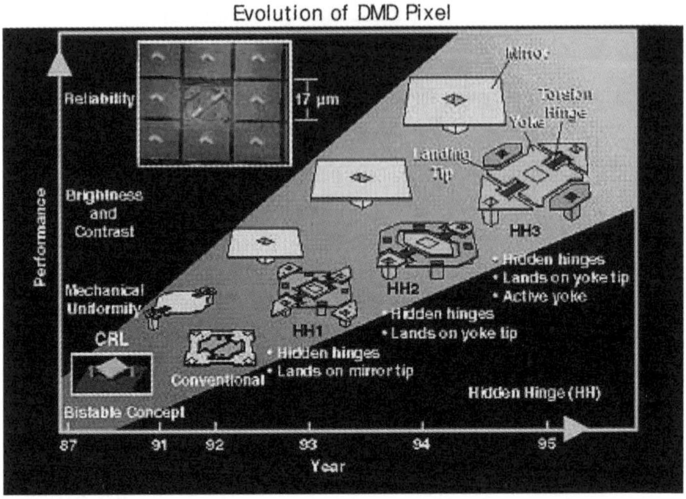

Figure 1.8: Evolution of DMD over nine years. From [22].

and the CMOS circuitry fabricated second, along with process compatible MEMS steps. However, this approach is does not appear to be a great deal more straightforward to develop.

Analog Device's acceleration sensors, which adopted this approach, took about 10 years to debug MEMS-IC integration process and design. The prize which kept such a development on hand was the supply of millions of acceleration sensors for use in the automotive industry, in the kinds of application (airbags and active suspension) discussed earlier.

Both of the single chip integration approaches discussed above required a significant process development effort, stretching across multiple years and costing hundreds of millions of dollars. However, in both cases the market potential of the resultant product was sufficient for the companies to persevere with such a protracted and costly development. Both Texas instruments and Analog Devices are large corporations. It is doubtful whether small or start-up enterprises could have completed successfully such a technological development. The end result is, however, that both approaches to device integration have been proven to yield viable products.

1.8. Smart MEMS — The Research Agenda

The conclusion to be drawn from the digest of the 'news' given in Section 1.2 is that integrated 'smart' MEMS is now a technology which is mature enough, at least at the device level, to begin to plan some of the 'dream applications' that the new generation of visionaries in the field have proposed. It is now possible to fabricate an integrated assembly of MEMS based transducers and the required ancillary analog and digital electronics, to combine MEMS and VLSI to produce a systems component with the required sensory and processing capabilities, to enable at least some of these applications. Moreover, it appears that it should be possible to produce these devices at a cost which does not render the proposed scenarios infeasible — particularly if the volumes of device required are really huge.

The authors of this book have been active in the development and detailing of a scenario for the use of a multiplicity of MEMS based intelligent sensor devices for planetary exploration, which is detailed in the final chapter. When one comes to look closely at the way such an enterprise would actually work, it becomes apparent that the necessary hardware technologies are very nearly there. Those which have not already been developed can be seen as the likely end-point of current research programmes. However, this does not mean that trying to make the scenario a reality is straightforward. As the detail of the operation is studied with more precision, it is clear that there is a great deal of research and design elaboration to be done, but little of it is in the areas that have traditionally been the preserve of MEMS researchers. While continued improvements in the technologies discussed above will be welcome, and will increase the possible scale of the scenario, or decrease the possible costs (which will be enormous), or give the designers more degrees of freedom in planning the envisaged system, none of them, at the current state of development, or one that could be extrapolated for the very near future, is actually a necessary precondition as a part of the basic research to enable the scenario to go ahead. Rather, there are a number of problems with its design for which no simple solutions are available in the research literature. These broadly fall into two areas:

(i) If MEMS sensors are to be built into systems expressly designed around them, how does this affect the design of the sensor devices themselves? Are there constraints put on to the design of MEMS systems by current applications, which could be removed, thereby making new design solutions possible at the device level? Another possibility, which is being explored by several researchers is that of integrated intelligent multi-sensor devices. Such devices would make little sense in the context of traditional applications, but can provide an attractive building block for some of the large sensory systems being proposed, at the same time making best use of the scale and integration potential of MEMS technologies.

(ii) Does the integration of processing capability onto a MEMS sensor affect either the design of the sensor itself, or the system into which it is designed? As we have seen above, the original purpose of integration of signal processing into MEMS devices was to overcome shortcomings of the devices themselves. The integration of a great deal more processing power may open up new possibilities in sensor design, by using advanced calibration and linearisation techniques, such as the Artificial Intelligence based techniques described later in the book. Another possibility is to enable new types of sensor, by using computational power to derive the desired quantity from other sensory data. A third is to use computational power local to the sensor to ameliorate resource shortages in other parts of the system, perhaps reducing data to minimise communication bandwidth, or offloading computation from a central processor.

Both of these questions are essentially positioned around the design of sensor components themselves. They pose the possibility of a change in design motivation for these devices from one that is essentially 'bottom-up', working from an isolated specification of the sensing device itself, and then subsequently working that into a system design. Often this bottom-up motivation is based around the prior development of a set of technological solutions to the design of certain types of sensor. Sometimes there are completely new ideas for sensor operating principles, such as, for instance, the MEMSIC accelerometer, but the end result is still designed towards a relatively fixed end point, defined by the existing applications and usually defined by older technologies. The presentation of the development of MEMS technology in Sections 1.6 and 1.7, suggests that the developments that have been successful (usually those for which the organisations responsible have had very deep pockets) have been those in which the resultant device forms a part of some existing system, or linear development of it. For instance, the Texas Instruments DLP does not (currently) enable new types of system, it enhances existing ones (although TI has an active programme to encourage third parties to think of entirely new applications). DLP's are used to build data projectors, televisions and digital image printers. For each application there are alternative technologies available. Looking at the other MEMS sensor development discussed extensively here, the MEMS accelerometer, we find here these were initially designed to replace macro-machined accelerometers, and designed to have similar characteristics. Later examples have been designed to be a part of various automotive systems, particularly air-bag deployment systems. The salient point here is that within the system design, the accelerometers are acting entirely as sensors, the design of the system has taken no account of the potential of intelligent MEMS, and conceptually could be built just as easily with non-MEMS accelerometers (except, of course, it would be too expensive to be viable).

By contrast the future systems we are looking at are precisely those enabled only by intelligent MEMS. For these systems, the design goals of the intelligent sensor components will probably be quite different from those designed to fit into existing systems. The new design motivation will be 'top-down', working from a starting point of problem solving in large scale application domains, through solutions which make use of the intrinsic properties of intelligent MEMS devices, namely tiny size, large scale reproducibility, low cost and inbuilt intelligence and autonomy. In turn, the specification of the devices themselves will be derived from the overall systems design. This book includes several chapters intended to open up discussion and lead to research and development in the design methods appropriate to this way of working.

Consideration of the second question above leads on to another question. Given that under consideration is the building of systems, which explicitly use the potential of intelligent MEMS devices, these systems are likely to consist of very many autonomous devices (as can be seen from several of the 'dream applications'). Thus the third question is:

(iii) Can we solve the design and organisational issues involved in the realisation of well-defined and reliable systems composed of a multiplicity of autonomous intelligent devices? The sole example we have presently of a system of this type, of the scale envisaged is the Internet. This cannot be an exemplar of future practice for at least three reasons. Firstly, the Internet depends on a huge infrastructure of communications lines and intelligent switches and routers, that would not be practical in a sensor system; secondly, the resources needed by the nodes are at a level above those likely to be found in an intelligent sensor node, even given a period of further development; thirdly, the levels of reliability achieved by the Internet, while impressive in absolute terms, are below what will be required for some of the applications being proposed.

A complete work addressing the full potential of MEMS based sensing systems must therefore address these issues of large scale organisation as well. Without an understanding of the overall function and organisation of the system, it is impossible to take the top-down view on which successful design of sensing systems such as those envisaged in the 'dream applications' is predicated.

1.9. Structure of the Book

Working from the discussion above, this book needs to have a scope which proceeds from the basic, enabling MEMS technologies, through to the design principles on which are based the ability to produce operational examples of systems using thousands or millions of MEMS devices. In between those end points, the various ancilliary technologies, necessary to produce a complete system design, must be covered, along with examples and case studies of particular areas of interest which are met along the way. The book is structured as follows:

Firstly, in this chapter, the overall area has been surveyed, with a perspective of gaining an understanding of the fundamental nature of intelligent MEMS sensor devices (that is, a view unconstrained by reference to previous technologies).

Chapter 2 concerns itself with MEMS design and fabrication techniques, to gain a view of the potential and constraints imposed by the methods of design and fabrication which have been developed for implementing MEMS devices and the underlying concepts of MEMS sensor systems, particularly concerning itself with the physical, mechanical and electrical operating principles, and how they utilise the mechanical components that can be fabricated using the techniques described in the chapter.

The next five chapters cover the design of the anciliary systems which are necessary to use the underlying MEMS technology to its best advantage, and will be included in any complete MEMS based sensing system.

Chapter 3 examines the priciples and design of the anciliary electronic systems required to make the electro-mechanical part of a MEMS sensor operative.

Chapter 4 studies the use of integrated electronics to allow calibration and compensation of the intrinsic faults of transducers. The next three chapters present case studies which explore aspects of integrated MEMS/electronic systems.

Chapter 5 shows how sigma-delta modulation techniques can be used to provide an accelerometer with enhanced performance characteristics.

Chapter 6 is concerned with the design and implementation of advanced optical MEMS systems.

Chapter 7 examines how integrated signal processing using artificial intelligence can be used to ameliorate the basic shortcomings of a sensor device.

In the following chapters the discussion moves to the wider systems level and begins to look at the issues surrounding the third question posed in Section 1.8, looking specifically at how large systems may be built from autonomous intelligent MEMS sensors.

Chapter 8 surveys the potential of integrated artificial intelligence to provide a higher level of functionality, a level dubbed the 'cogent sensor' by the authors. Here, the sensor device not only collects data, it also provides a level of interpretation, transforming the data to information, thereby simplifying the design of applications systems, by freeing them from the need to collect and interpret large amounts of raw data from a multiplicity of sensors.

Chapter 9 considers the scope, applications and device design (at the integration, not hardware level) of advanced sensor systems of this type.

Chapter 10 discusses how systems may be designed and built using very large arrays of intelligent sensors networked together to form a single integrated system.

Finally, Chapter 11 looks forward, detailing one of the 'dream applications' that are being proposed and relating the way such an application could use the technologies and methods explored in the rest of the book, and how ongoing and future research might inform their design.

Taken together, the book is intended to provide a newcomer to the field, or a worker in one of the sub-disciplines involved in MEMS based system design, with a broad enough scope to appreciate the requirements, and also opportunities, which follow as the result of a top-down view of the design of the kind of future-shaping applications currently being proposed by visionaries in the field. It is to be hoped and expected that several of these visions are sufficiently practical that at least some readers of this book will be involved in their development in the near future.

References

1. Senturia, S.D. (2003) Perspectives on MEMS, past and future: the tortuous pathway from bright ideas to real products, *Proc. of the 12th Int. Conf. on Solid State Sensors, Actuators and Microsystems Transducers'03*, Boston, MA, pp. 10–15.
2. Device Basics, (June 2003) *Semiconductor International*, pp. 128, www.semiconductor.net.
3. Grieco, B. L. (August 2004) S MEMS advance medical electronics, *EE Times.*
4. Legg, G. (2003) Cover story, *Design News*, pp. 73, www.designnews.com.
5. Crossbow integrates GPS with MEMS, *Small Times*, http://www.smalltimes.com/document_brief.cfm#brief_4.
6. MEMSIC sensors in Mitsubishi mobile phone, *Small Times*, http://www.smalltimes.com/document_brief.cfm#brief_4.
7. New Products (2004) ADI MEMS accelerometers prevent information loss, *EE Times*, http://www.eetasia.com.
8. New Products (2003) Fujitsu optical switch based on MEMS mirrors, *EE Times*, http://www.eetasia.com.
9. Industry news (December 2001) *Standard Challenge For Sensor Vendors*, p. 5.
10. Harris, C. A. (August 2002) Structural health monitors go wireless, *Civil Engineering*, Vol. 72, Issue 8, p. 19, 1/3p, http://search.epnet.com/direct.asp?an=7278400&db=buh.
11. Device Basics (June 2003) Semiconductor International, p. 128, www.semiconductor.net.
12. Kahaner, D. K. (November 1994) Micromachine Center Progress, *Micromachine Symposium*, Tokyo, http://www.atip.org/public/atip.reports.94/micromac.94.html.
13. Voelker, M. A. (1992) Synchrotron etches microstructures for tiny 3d components, *SPIE*, OE Reports, p. 104.
14. Stenmark, L. (2001) Microsystems in Space exploration, *Proc. of Transducers'01*, paper no. 3A1.01.

15. Kohler, J. *et al.* (2001) A hybrid cold gas microthruster system for spececraft, *Proc. of Transducers'01*, paper no. 3A1.02.
16. Pister, K. S. J., Kahn, J. M. and Boser, B. E. (1999) Smart dust: wireless networks of millimeter-scale sensor nodes, Highlight Article in *1999 Electronics Research Laboratory Research Summary*.
17. Manobianco J., Evans, R. J., Pister, K. S. J. and Manobianco, D. M., GEMS: A revolutionary system for environmental monitoring, *Proc Nanotech 04*, pp. 422–425.
18. Abbott, D. *et al.*, Development and evaluation of sensor concepts for ageless aerospace vehicles, *Threats and Measurands, NASA*, http://techreports.larc.nasa.gov/ltrs/PDF/2002/cr/NASA-2002-cr211772.pdf.
19. http://www.nasatech.com/Briefs/June00/NPO20699.html.
20. Bryzek (October 2003) MEMS-IC integration remains a challenge, *PlanetAnalog*. http://www.planetanalog.com/showArticle.jhtml?articleID=15800098.
21. Ohr, S. (October 2003) MEMS, EE Times, http://www.eetimes.com/article/showArticle.jhtml?articleId=18309906.
22. Hornbeck, L. J., *Digital Light Processing: A New MEMS-Based Display Technology*, Texas Instruments White Paper. http://www.dlp.com/dlp_technology/dlp_technology_white_papers.asp.

CHAPTER 2

MICROFABRICATION TECHNOLOGIES

by *Andrew Flewitt*

In developing a new MEMS device, it is clearly essential for the designer to consider the process flow by which the device can be fabricated during the initial design phase. Failure to do so runs the risk of the final design being impossible to fabricate or simply uneconomical even if fabrication is physically possible. It is therefore necessary for the designer to have some understanding of the basic MEMS fabrication technologies and materials sets at their disposal together with some of the fundamental processing 'rules'. The aim of this chapter is not to provide the reader with detailed information about specific processes, as many excellent reference texts already exist for this purpose, but to provide an overview of microfabrication to allow the designer to make a rule of thumb decision on the feasibility of a particular device.

To this end, a variety of MEMS devices will be examined as case studies of microfabrication. Initially, some simple passive devices will be considered as a means of introducing the basic micromachining technologies by example. Both actuating and sensing device will then by discussed together with their related materials systems. Finally, we will see how these elements can be integrated using a larger process flow to produce complete microsystems.

2.1. Introduction

MicroElectroMechanical Systems (MEMS) are devices which contain mechanical components with a length scale of the order of micrometres, and that employ both an electronic movement of charge and mechanical movement for operation. The magnetic disk drive [1], inkjet print head and scanning probe microscope [2] can all be considered to be forms of MEMS devices.

MEMS device development can be traced back to the 1970's. However, it is in the period since 1995 that there has been a significant proliferation of new device applications as a variety of new materials and bulk micromachining technologies have become available. Despite this broadening of the field, MEMS devices may still fall into three classes:

Actuators
Devices whose principal aim is to produce mechanical motion (e.g. micro-tweezers [3] and micro-engines [4]);

Sensors
Devices whose principal aim is to produce an output which is dependent upon their environment (e.g. pressure sensors, accelerometers);

Passive devices
Devices whose principal aim is to respond passively to their environment without producing an output signal (e.g. hinges, micro-lenses, linkages).

These MEMS device technologies have grown out from the microelectronic device fabrication industry which itself has shown remarkable development since the invention of the integrated circuit by Jack Kilby at Texas Instruments in 1958. The concept of the IC was formulated to meet the need to reduce the physical size of electronic systems by integrating many devices onto a single semiconductor substrate. The devices are connected together by depositing a layer of metal interconnects onto the surface of the system. The increased processing power achieved since the 1960's has essentially been reached through continuing miniaturisation of the physical size of electronic devices, as shown in Figure 2.1 [5]. It was hence required to develop a technology for actually fabricating these small structures, which is now known as micromachining.

In comparison with other material processing technologies, micromachining is characterised by the extremely high tolerance ($\sim$1 μm) and low surface roughness ($\sim$10 nm) with which physical structures can be produced (Figure 2.2) [6]. Silicon has been dominant as the material of choice for semiconductor devices and consequently, it is for this material system that micromachining technology has been best developed.

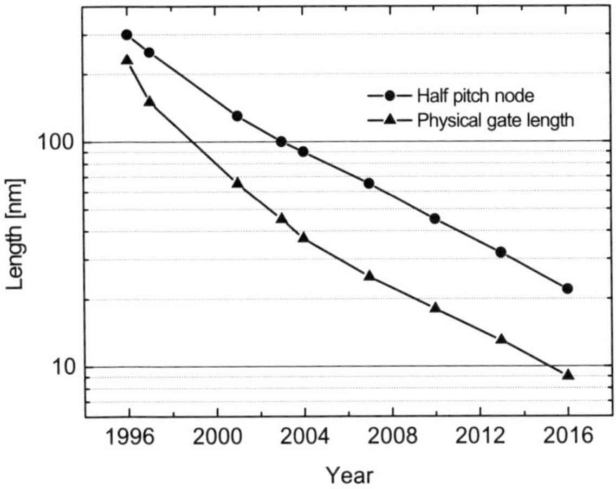

Figure 2.1: There is a constant driving force in the semiconductor industry for reducing feature size in CMOS devices. Feature sizes after 2003 are those predicted by the 2003 International Technology Roadmap for Semiconductors 2003. From [5].

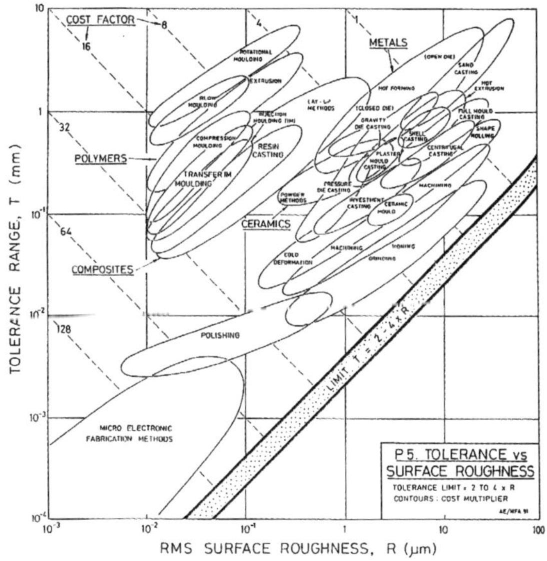

Figure 2.2: Micromachining is unique amongst material processing technologies in terms of both the tolerance and surface roughness that can be achieved. Consequentially, this is a high cost technology. From [6].

Micromachining has permitted the development of a large variety of MEMS products such as the Texas Instruments' 'Digital Micromirror Display' (DMD), for example. The DMD is a projection display in which light from a suitable source is shone upon an array of small, square aluminium mirrors, each with a side length of 4 μm. Each mirror is hinged and can be rotated through an angle of $\pm 12°$ using electrostatic actuation. This allows light to be deflected either to a light stop or through a set of projection optics to produce an image on the screen. Switching occurs in under 20 μs, allowing pulse width modulation to be used to achieve greyscale control. The device is fabricated by micromachining layers of (generally) either silicon oxide or aluminium on top of a silicon substrate that incorporates the drive electronics for the display [7].

High precision micromachining has an associated high cost factor. In addition, the cost of traditional complementary metal oxide semiconductor (CMOS) fabrication is essentially a function of the area of silicon being processed and the number of process steps employed. The same expense is therefore required to produce a single transistor on an 8″ silicon wafer as to produce millions of the same device on an 8″ wafer.

Therefore, although the cost per unit area to the end user of a CMOS processor, such as Intel's Pentium 4®, is in excess of €100 000 m^{-2}, as the density of transistors is over $10^{10}\,\mathrm{m}^{-2}$ only a small area of silicon is used for an individual processor, and so the cost to the end user is reasonable (Figure 2.3). Some MEMS devices, such as the DMD in which the density of devices is high and the cost of the finished product is also high, fit into this economic model. However, many MEMS devices, such as a sensor for measuring tyre pressures on a car, must be manufactured for a few Euros as discrete components with a low surface density, and so do not fit into the same economic structure as CMOS devices.

Large area electronic (LAE) devices, such as thin film transistor (TFT) active matrix liquid crystal displays (AMLCDs) can be manufactured economically at low surface densities. In order to reduce the cost to the end user per unit area to below €10 000 m^{-2}, manufacturing tolerances are relaxed and a cheaper material set employed (for example the use of glass substrates instead of silicon). Therefore, despite the fact that the density of devices is below $10^{8}\,\mathrm{m}^{-2}$, the finished display is affordable (Figure 2.3).

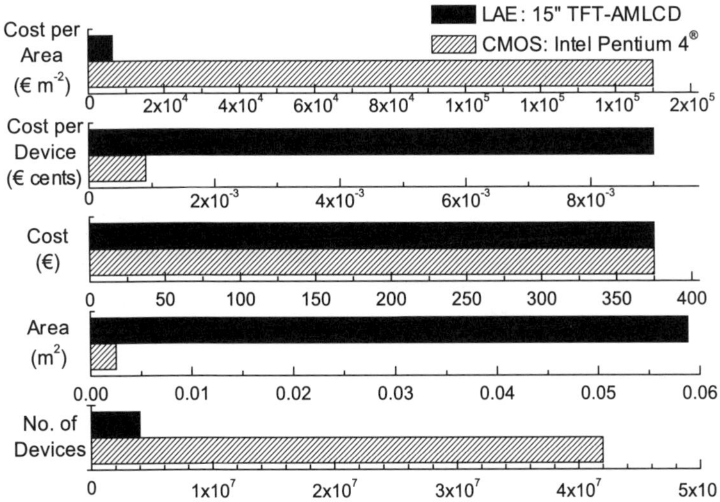

Figure 2.3: The relative cost in 2002 to the end user of a 15″ TFT-AMLCD display and an Intel Pentium 4® processor as examples of LAE and CMOS devices respectively.

Most MEMS devices, and in particular sensors, are more akin to LAE devices than CMOS devices. Hence, the driving forces in the development of MEMS micromachining technology is the reduction in process cost through the use of new, cheap material sets (e.g. polymers, glasses) and low cost processes (e.g. electroplating, embossing).

The importance of processing cost on the practical development of MEMS devices is exemplified by the implementation of MEMS accelerometers in the automotive industry during the 1990's. The MEMS accelerometer works on the principle that a small proof mass will experience a force, when accelerated, that will cause the mass to be deflected. By measuring this deflection, the acceleration in the axis of motion may be determined. During the 1990's, the cost of fabricating one accelerometer fell from ~€100 to below €10. As the cost fell to an acceptable level given the 'value added' nature of the product, the market for these devices increased rapidly (Figure 2.4). However, market saturation was reached by the late 1990's, and so further significant fabrication cost reduction or performance enhancement will be required for these devices to penetrate new markets.

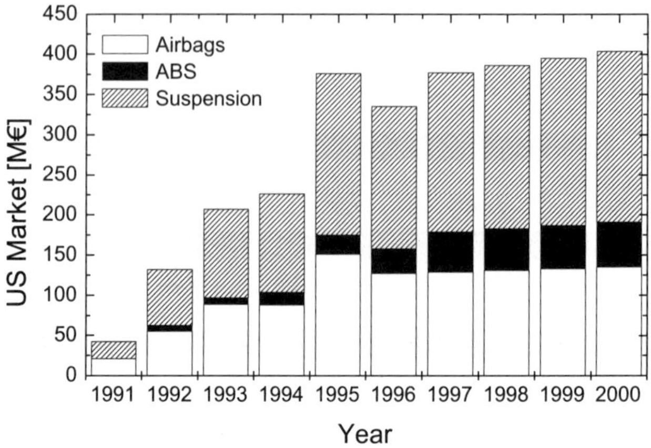

Figure 2.4: The market for automotive MEMS accelerometers per year during the 1990's separated by application: air bags, antilock braking system (ABS) and active suspension.

2.2. Passive Components

2.2.1. *Introduction to Passive Components and Their Fabrication*

Passive MEMS devices are those components whose principal aim is neither actuation nor sensing, but which respond passively to their environment.

An example of such a passive device is a simple MEMS hinge that allows two planes of material to be moved relative to each other about a common line. This either requires the two planes to be glued to each other using a compliant material as the hinge [8], or for the two planes to be physically interlocked [9], as illustrated in Figure 2.5. In either case, an externally applied force is required to cause the hinge to move. Surface tension has been shown to provide a suitable force for the self-assembly of simple three-dimensional microstructures with hinged components [10].

More complicated mechanical passive components include MEMS linkages and gears [11]. Once more, these components do not actually *generate* any motion in their own right, but instead can be used to mechanically transfer motion from an actuator to an object that needs to be moved. Such a system can also transform the direction in which a force is applied from that in which it is generated, as well as changing the force-displacement characteristics to match those required to drive the load.

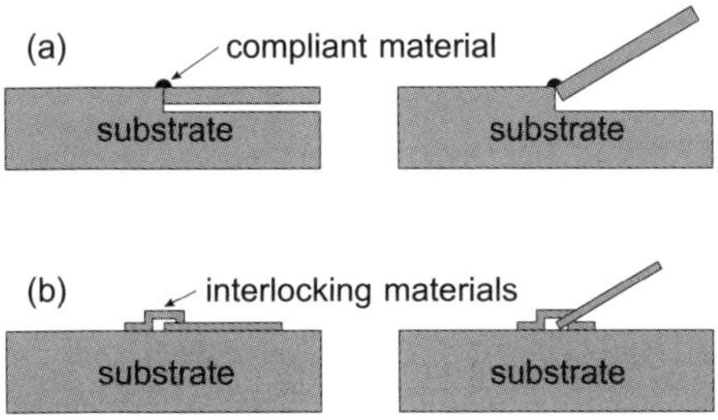

Figure 2.5: A passive MEMS hinge can be created by (a) using a compliant material or (b) interlocking materials.

Passive mechanical MEMS components allow flat structures that are easily fabricated in the plane of a surface to be pushed out of the plane and locked in place. This has allowed, for example, the development of the micro-optical table. Passive optical MEMS mirrors, gratings, lenses and beam-splitters can be arranged above the surface of a silicon wafer to control the optical path of a beam of light. The micro-lens consists of a number of concentric rings of silicon which are arranged to produce a Fresnel zone plate that acts to focus light passing through it to a point [12]. Simple passive MEMS cantilever clips can then be used to align an optical fibre to the micro-optical table, allowing external optical systems to be efficiently coupled to the micro-optical table [13].

These passive devices are all fabricated in essentially the same way (Figure 2.6). A surface layer of a silicon-based material is produced on top of a sacrificial layer of material which itself is supported on a crystalline silicon substrate. The top layer is then patterned using photolithography and unwanted material is etched away. This exposes some of the underlying sacrificial material that can then be selectively etched without affecting the top structure, which is left suspended above the substrate to produce the required structure, such as a simple cantilever clip.

The reminder of this section, will discuss in some detail several basic passive MEMS components and the factors that must be considered in their design and fabrication.

2.2.2. *Beams and Cantilevers*

Beams and cantilevers represent perhaps the simplest of all microstructures, requiring only the ability to create a released structure for their

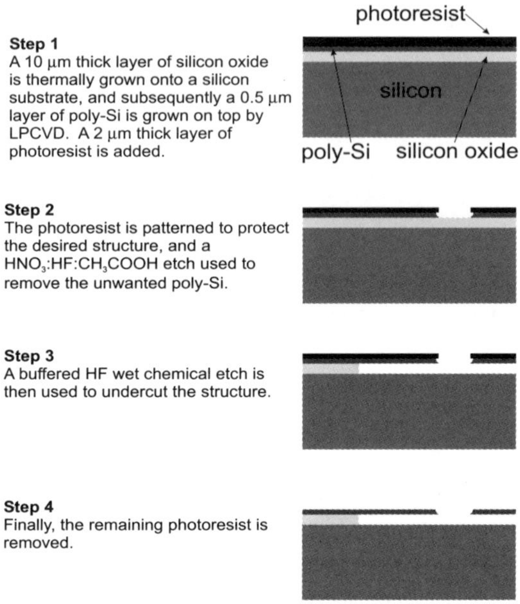

Figure 2.6: Free-standing structures can be produced by sacrificially etching a layer of silicon dioxide lying underneath the structure to be released which can be made from poly-Si.

manufacture, and they are widely used in microsystems. In some cases, the free standing structure is designed simply to hold another component in place. Other applications include devices such as the fibre-optic clip and interlocking clamp which have already been mentioned [13].

Another rapidly growing application of passive beams and cantilevers is in compliant systems for converting motion, either by varying the force-displacement characteristics, or by changing the type of motion, for example from a linear translation to a curved translation [14]. Passive devices, such as torsion rings [15] and beams [16, 17], are also finding applications as components for materials characterisation in their own right.

Several factors must be considered in the design of any passive components. As with all microsystems devices, the operation of the device itself will place limitations on the selection of materials and on the geometry of the device. For example, the clips for fibre-optic cables must have a high fracture strength to avoid failure as the cable is inserted, whilst a compliant system must be resistant to fatigue cracking, requiring the elimination of points of high stress concentration. In many ways, the design criteria are similar to those of macroscopic components (which will be considered in Section 2.5.7).

However, microfabrication presents the designer with significant additional constraints, of which, perhaps the most important is stress. There are essentially two sources of stress in released thin film materials: externally applied stresses and *intrinsic stress* (in thin film materials that are not released, but still adhered to a substrate, stress due to thermal mismatch must also be considered). If we assume that externally applied stresses are known and accounted for in the design process, then it is the intrinsic stress that we must additionally consider. The component of intrinsic stress that is homogeneously distributed throughout the material is known as *residual stress*, and this can be caused by the presence of impurities, vacancies, microvoids or grain boundaries. Additionally, some materials, most notably polymers, are susceptible to water absorption and desorption, and the consequent expansion or contraction of the material can also lead to a residual stress. Structures that are clamped at two points, such as beams, are particularly affected by residual stress: compressive stress will cause a beam to bow out of plane, as shown in Figure 2.7(a), and may ultimately result in buckling (the formation on a kink in the structure that will be a point of mechanical weakness), whereas tensile stress can lead to the formation of microcracks in the structure as well as deformation perpendicular to the length of the beam. Cantilever structures, on the other hand, which are clamped at one point only, will simply change their length upon release to accommodate the residual stress. However, these structures will deform in the presence of a *stress gradient*, where the intrinsic stress varies as a function of depth in the thin film. This will cause the released cantilever to bend out of plane along its length, as shown in Figure 2.7(b). The degree of intrinsic stress that can be accepted in a particular passive device structure will depend on its geometry and the degree of deformation of the structure that can be tolerated, but in practice an intrinsic stress that is less than a few hundred MPa is normally required in a suspended structure.

The fabrication technology to be employed also has a profound influence on device design. Techniques for surface micromachining will be considered in Section 2.6. However, in general, release of a microstructure will require the removal of a sacrificial material from underneath the structure using

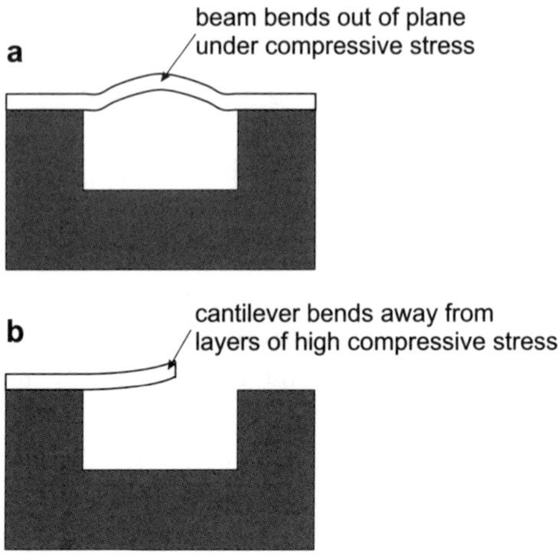

Figure 2.7: (a) A simple residual compressive stress will cause a beam which is clamped at two points to bow out of the surface plane, whilst (b) a stress gradient is required to cause a cantilever which is clamped at only one point to bend out of the surface plane.

either a wet or dry etch. The etchant must be able to physically reach the sacrificial material. Release structures with a large surface area therefore present a problem, as the etchant has to diffuse a long distance underneath the structure in order to reach all of the sacrificial material. If a solid structure is required, this can limit the spatial extent of the structure in one dimension (for example the width of a beam). However, in many cases, it is acceptable to fabricate the structure with an array of holes in the surface to allow local ingress on the chemical etchant, as illustrated in Figure 2.8. Furthermore, if a wet chemical etchant is to be used, then surface tension during the drying of the structure will tend to cause the free standing structure to collapse onto the underlying substrate. Various techniques, such a freeze drying or substrate texturing have been developed to mitigate this effect, but ultimately large free standing structures will always be prone to collapse due to this effect, which is commonly called *stiction*. Dry etching is therefore preferred for the fabrication of such structures.

2.2.3. *Hinges, Gears and Rotors*

A logical extension of the MEMS beams are the microhinges. These allow two components to rotate about a line with respect to each other

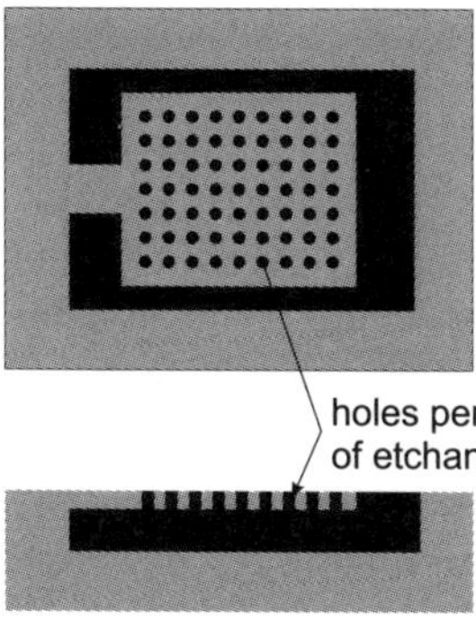

Figure 2.8: Release of a structure with a large area requires a pattern of holes to be designed into the structure to permit the etchant to reach the underlying sacrificial release layer without the need for diffusion over long distances which would take an unacceptably long time.

(Figure 2.5). In this case, there are two interlocking structures, both of which possess underlying sacrificial materials that must be removed, normally by employing the same material in both cases, and removing both layers in a single etch step. Alternatively, a compliant structure is used to allow a degree of rotation to be achieved, as found in the Digital Micromirror Display.

Gears and rotors both rotate about an axis, and are an essential element in microturbines [18] and micromotors [19]. The bearings are most frequently formed using bulk micromachining to release a rotating element from a wafer and then using wafer bonding to ensure capture of the released element [20], although surface micromachining also allows for simple bearing formation [21].

In all of these cases, the new issue of friction arises, as two surfaces are moving against each other. Friction is a major issue on the microscale. Whereas volume scales as the cube of dimension, surface area only scales as the dimension squared, and so the ratio of surface area to volume increases substantially at the microscale. Friction is essentially a surface area effect, and so is relatively of greater importance than might be expected from our 'common sense' experience of the macroscale world. Friction not only is a problem in terms of energy loss, but also in terms of the wear that will result on the two surfaces in contact, where wear rate, W, is given by $W = K_W p v$, where p is the pressure, v is the sliding velocity and K_W is the dimensional wear coefficient [22]. Therefore, for a given bearing, the wear rate will be a function of the rotation speed, the geometry of the bearing and the wear coefficient for the system. In addition, great care must be taken during fabrication to avoid surface asperities that might locally increase the pressure applied, resulting in an enhanced and uneven wear rate that could subsequently lead to device failure. Carbon nanotubes are known to have a well-defined surface structure, have a high fracture strength, but also have a low coefficient of friction, and for this reason, there have been recent attempts to employ these as bearing elements [23].

2.3. Sensing Components

2.3.1. *Introduction to Sensing Components*

Microsensors offer three major advantages over their macroscale counterparts, all of which are associated with their size. Firstly, the high surface area to volume ratio at the microscale means that microsensors tend to have unprecedented levels of sensitivity. For example, in gas sensing, the gas to be detected interacts with a surface to produce a signal. The amplitude of the signal is therefore proportional to the active area of the device. The underlying substrate serves no purpose in detection, and is only a source of noise. Therefore, sensors with a high surface to volume ratio will also have a good signal to noise ratio, and many high sensitivity gas sensors have been demonstrated using this technology [24]. Secondly, the small size of microsensors makes it possible to obtain local measurements of a particular effect in remote locations, such as the use of flow meters to measure fluid flows in microfluidic systems [25] or the proposed use of sensors in biomedical applications [26]. Thirdly, microsystems allow the integration of a degree of signal processing into a sensor with a resulting improvement in signal to noise ratio [27]. The unique opportunities offered by microsensors are making this an area of rapid development. However, most of theses sensors use one of a limited number of sensing principles, and these will be briefly reviewed in this section.

2.3.2. *Mechanical Sensors*

Mechanical sensors have arguably become the most commercially successful of all the microsystem devices with applications such as pressure sensors,

Table 2.1: Classification of common mechanical MEMS sensors.

Detection method	Constant mass or stress	Variable mass or stress
Non-resonant detection	Pressure sensors, flexure accellerometers	Stress change chemical sensors
Resonant detection	Gyroscopes	Cantilever chemical sensors

accelerometers, gyroscopes and chemical sensors. They generally fall into one of two categories: non-resonant devices and resonant devices, and within each of these categories a division may be drawn between constant mass or stress and variable mass or stress devices, as summarised in Table 2.1.

In the case of non-resonant devices, the deflection of an object under the application of a force is normally being measured. For example, a flexure with a proof mass attached forms a simple accelerometer. Under the action of an acceleration, the proof mass will be deflected by the accelerating force, and this deflection may then be detected. Alternatively, a pressure sensor may be formed by monitoring the deflection of a thin membrane that has been formed over a sealed cavity which has a fluid at a known pressure contained within it.

In either case, it is necessary to be able to detect the small deflection of a plane, and this may be successfully achieved in several ways:

Optical detection:
One of the best developed and most sensitive methods for detecting small deflections is to reflect a beam of laser light from the moving surface. Any deflection in the surface will cause the path of the reflected laser beam to change. Using an oblique angle of incidence will result in a significant change in the optical path for very small deflections of the surface and this may be accurately measured using a charged coupled device (CCD) detector. Sub-nanometre movements of a surface may be accurately detected using this system, however there is significant complexity in aligning the laser light source, the deflecting surface and the CCD. Alternatively, for semi-transparent planar diaphragms, shining laser light through the diaphragm so that it reaches a fully reflective, stationary, parallel mirror may form a simple Fabry-Perot etalon interferometer [28]. The normal separation between the diaphragm and mirror should be greater than one quarter of

the wavelength of the laser light. Interference will occur between the light reflected off the diaphragm and the mirror, which will be dependent on the separation. Measuring this reflected light therefore allows any deflection in the diaphragm to be determined.

Capacitive detection:
The capacitance between two parallel conducting plates is a function of both the area of the plates, A, and their separation, x. Therefore, if one conducting plate is rigid, then the deflection of the second plate may be determined by measuring the change in capacitance that will result from the relative movement of the second plate. Such a system is easily realised through the fabrication of a conducting cantilever beam over a conducting substrate surface. The system will produce a change in capacitance when the separation of the plates, x, changes. However, the resolution of such detection systems tends to be low, as the size of the movement is typically much smaller than the distance between the plates. Devices in which the area overlap of the plates varies with movement (rather than the distance between them) have a higher resolution and produce a signal in which the capacitance is linearly dependent on movement.

Piezoresistive detection:
Materials whose resistance changes as a function of mechanical stress are known as piezoresistors. In these materials, Ohm's law of conduction becomes:

$$\mathbf{E} = (\boldsymbol{\rho} + \mathbf{\Pi} \cdot \boldsymbol{\sigma}) \cdot \mathbf{J}, \tag{2.1}$$

where $\mathbf{E}$ is the electric field matrix, $\mathbf{J}$ is the current density matrix, $\boldsymbol{\rho}$ is the resistivity matrix, $\mathbf{\Pi}$ is the piezoresistivity tensor and $\boldsymbol{\sigma}$ is the stress matrix. A track of a thin film piezoresistive material may therefore be fabricated over a deflecting membrane. Any change in the curvature of the membrane will induce a strain in the piezoresistor and therefore a stress will result. This will cause a measurable change in the resistance of the piezoresistive track, allowing the deflection of the membrane to be determined.

Piezoelectric detection:
Piezoelectric materials possess a coupling between electrical polarisation and mechanical stress or strain. Although they are commonly used as high frequency actuators, they may also be used as a mechanical sensor in which a mechanical deformation of the piezoelectric will produce an electric field

that can be measured. However, they tend to only be suitable for high frequency applications, as the polarisation charge produced will tend to dissipate through a parallel leakage resistance with time. The RC time constant of the piezoelectric detector will define the lowest frequency of signals that can be reliably measured.

In many mechanical detection systems, it is the application of a force to a flexure that produces motion and hence a signal. However, in some biological and chemical sensing applications, it is a change in the mass and/or surface stress of the flexure that produces a signal. This is achieved by coating the flexure, for example, a cantilever beam, in a sensitising agent that will selectively react with a particular chemical species. The attachment of the detected species to the flexure will cause a change in the mechanical properties of the structure [29]. In many cases, a simple change in deflection results, due to the intrinsic stress of the newly added layer of material. However, highly sensitive detection may also be achieved by monitoring the resonant frequency of the flexure, which will change as the mass and intrinsic stress of the flexure varies. This has the added complexity, however, of the need to integrate an actuating system to mechanically oscillate the flexure at high frequencies. The mechanical response of the flexure may then be measured optically or capacitively, as described previously.

Gyroscopes also rely on mechanical resonance to detect motion. This uses the Coriolis force, F_C, which acts on a body of mass m moving in a rotating frame of reference of angular velocity $\mathbf{\Omega}$ with a radial velocity $\mathbf{v}$, where:

$$F_C = 2m(\mathbf{v} \times \mathbf{\Omega}). \tag{2.2}$$

Therefore, if a proof mass is made to resonate by two independent oscillators acting at 90° to each other in an $x - y$ plane, there will be no coupling between the oscillators unless the mass is rotated out of the $x - y$ plane, in which case a coupling will occur that may be sensed. Several possible practical geometries exist for achieving this, including tuning forks [30] and ring oscillators [31].

2.3.3. *Acoustic Sensors*

The development of piezoelectrics has enabled the production of sensors that are based on detecting changes in the velocity or amplitude of acoustic

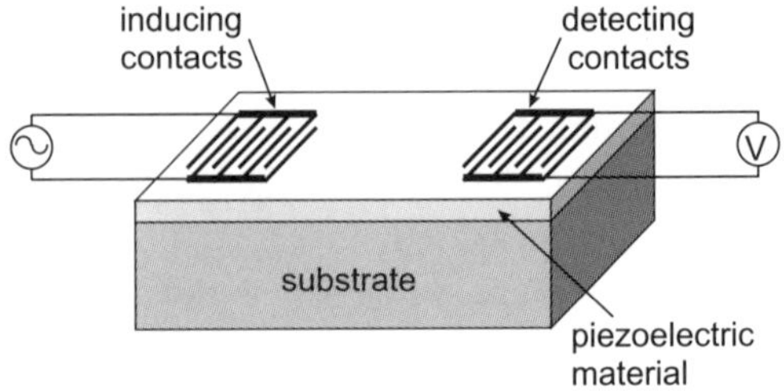

Figure 2.9: Schematic diagram of the basic structure of a surface acoustic wave device. The area between the inducing and receiving contacts will normally be coated in a sensitising agent to which the substance to be detected will attach. A thin barrier layer is normally placed between the sensitising agent and the piezoelectric to prevent undesired interactions between these materials.

waves travelling at or near the surface of a solid. The structure of these *surface acoustic wave* (SAW) devices is shown schematically in Figure 2.9. They use an interdigitated electrode structure on the surface of a piezoelectric to generate a mechanical vibration that passes through a transmission medium which is in intimate contact with the piezoelectric. A second piezoelectric layer is also fabricated a short distance from the first. Mechanical vibrations, which are transmitted to this second piezoelectric from the first, will generate an electric field that may be detected using a second interdigitated pair of electrodes.

The transmission of these acoustic waves between the generator and receiver electrodes will be strongly influenced by the surface of the transmission medium. Therefore, by making this surface chemically active (often by coating it in a sensitising agent), if a particular chemical species is present, then it will react with the surface causing an associated change in the surface properties. This will affect either (or both) the velocity and amplitude of the acoustic waves, and the receiver will measure this change [32].

2.3.4. *Optical Sensors*

Photodiodes offer the simplest means of detecting an optical signal in the visible range. The photodiode comprises a simple pn junction that has been fabricated from a direct gap semiconductor with a bandgap that is smaller than the energy of the photons that are to be detected. A depletion region forms at the interface between the p and the n-type semiconductor. When the device is placed under a reverse bias (with a positive voltage applied to the n-type semiconductor with respect to the p-type semiconductor), no current flows across the junction. However, when a photon of light hits the depletion region of the junction, it may be absorbed to form an electron-hole pair. The electron and hole are rapidly swept out of the depletion region by the electric field that exists in the depletion region, and a current flows. This current is largely independent of the applied bias, but is proportional to the light flux. Sensitivity may be improved by the addition of an intrinsic (undoped) region of semiconductor between the p-type and n-type regions to form a pin junction, as this increases the volume of material in which photon absorption will lead to carrier generation.

2.3.5. *Thermal Sensors*

The ability to locally measure the temperature of a system is an essential element of many microsystems. The method employed will normally depend on the temperature range that is to be detected and the resolution that is required, and several options are available:

Microbolometers:
These devices are designed to remotely measure temperature by detecting the infrared radiation emitted. The principle of operation is that infrared radiation falling on a thin film material will be absorbed and this will lead to a change in temperature. This change can lead to a well-defined change in the resistance of the thin film material, and it is this that is directly measured [33]. Several materials are suitable as the absorbing layer. These include metals, metal oxides and traditional semiconductors, such as silicon and germanium. Such devices may be fabricated in an array to produce a temperature map of an area, which makes these devices uniquely suitable for temperature imaging applications, although this is balanced by the disadvantage that temperature is being measured indirectly, and so other techniques can yield a more accurate absolute temperature measurement.

Thermocouples:
A difference in thermoelectric temperature coefficient (sometimes called the *Seebeck coefficient*) between two conducting materials in intimate contact at a junction results in the production of a potential difference that is

dependent on the temperature of the junction. Thermocouples are widely employed for temperature sensing on the macroscale, over temperature ranges of many hundreds of Kelvin. However, a combination of surface and bulk micromachining has allowed the fabrication of microthermocouples which are formed by producing a small junction — normally at the end of a tip — with reported diameters as small as ~100 nm [34, 35]. These microthermocouples not only permit localised measurements of temperature, but their small volume means that they have a small thermal mass and hence a fast response time.

Resistors and thermistors:
Most metals undergo a well-defined (although generally non-linear) increase in resistivity with increasing temperature. This effect may be employed for determining temperature by microfabricating a small metallic structure with two contacts of known dimensions. The resistivity of the metal may therefore be determined by measuring the resistance between the two contacts, and hence the temperature can be inferred. Platinum has an almost linear relation between resistivity an temperature close to room temperature, and so this material is frequently used for macroscopic temperature sensors. This principle is also applicable on the microscale, and several micro-resistive temperature sensors have been reported using a variety of metals, including tungsten, nickel and chromium [36, 37]. Thermistors, on the other hand, are made from semiconductors. Whilst highly pure, intrinsic semiconductors show an Arrhenius dependence of resistivity with temperature, this is not the case for most doped semiconductors where a more complex, but well-defined variation in resistivity with temperature is observed. Microfabrication is well suited to the formation of semiconductor devices, and therefore the production of semiconductor thermistor sensors is routine.

Semiconductor devices:
Many semiconductor devices, such as the pn junction diode, the bipolar junction transistor and the metal oxide semiconductor field effect transistor, have current-voltage characteristics that depend on temperature. In most applications, this effect is undesired — indeed the strong temperature dependence of germanium-based semiconductors places a severe limit on their general application. However, this effect can be put to good use in temperature sensing devices. The simplest of these devices is the thermodiode.

2.4. Actuating Components

2.4.1. *Introduction to Actuating Components*

An important aspect of MEMS devices is the ability to generate mechanical motion — the application of a force over a distance. There are several 'classes' of actuators, which are grouped according to the means by which motion is generated, amongst which the following are the most common:

Electrostatic actuators:
The attraction or repulsion between charges generates a force;

Shape memory alloy actuators:
A phase change of a material produces a physical change in local atomic bonding that produces a change in volume of the material and hence a force;

Thermal actuators:
The expansion of a material when heated produces a force;

Piezoelectrics:
The application of an electric field to a piezoelectric material causes a deformation of the crystal lattice and hence a change in dimension which generates a force;

Magnetic actuators:
The application of a magnetic field produces a force on a magnetic material or moving charge.

The critical parameters to consider in choosing a method of actuation for a particular system are:

- the magnitude of the force that can be produced compared with the distance over which it can be applied — or more fundamentally, the maximum stress and strain;
- the minimum step change in displacement that can be produced, known as the strain resolution;
- the maximum frequency of operation;
- the temperature of operation;
- the volume and density of the actuator relative to the power that can be produced and
- the efficiency of the device (power out relative to power in) [38].

The principle of operation of the above classes of actuators will be considered in the following sections.

2.4.2. *Electrostatic Actuators*

Although there are numerous geometries of electrostatic actuation systems, including commonplace parallel plate [39–42] and comb structures [3, 43], the underlying principle of operation is the same in all cases. Fundamentally, two conductors are required with an insulating region between them. Such a system will have a capacitance, C. When a potential difference, V, is applied between the two conductors, a net charge will be induced on each of the conductors of magnitude:

$$Q = CV, \tag{2.3}$$

with one conductor being positively charged and one negatively charged. An electric field is produced between the conductors and the opposite charges cause the two conductors to experience an attractive force, F. If the conductors are held at a constant potential difference, then any movement, δx, of the conductors relative to each other will cause a change in charge, δQ, and using the principle of virtual work:

$$\delta W_E - F\delta x = V\delta Q, \tag{2.4}$$

where δW_E is the change in electrostatic energy stored in the system. The change in charge and electrostatic energy is due to the change in capacitance of the system,

$$\delta Q = V\delta C, \tag{2.5a}$$

$$\delta W_E = \frac{V^2 \delta C}{2}. \tag{2.5b}$$

Substitution into Equation (2.4) gives:

$$-F\delta x = \frac{V^2 \delta C}{2}. \tag{2.6}$$

For example, for two parallel conducting plates of area A that are separated by an air gap x whose capacitance is

$$C = \frac{\varepsilon_0 A}{x}, \tag{2.7}$$

where ε_0 is the permittivity of free space, it is clear that

$$\delta C = -\frac{\varepsilon_0 A}{x^2}, \tag{2.8}$$

and hence

$$F = \frac{V^2 \varepsilon_0 A}{2x^2}. \tag{2.9}$$

The displacement produced by this force will then depend upon the spring constant, k, of the mechanical system which will produce a restoring force, F_r, given by

$$F_r = k(x_0 - x), \tag{2.10}$$

where x_0 is the air gap with no applied potential difference. The system will be at its equilibrium displacement when $F = F_r$. An important consequence of this is the phenomenon of *pull-in*, which occurs because, as the air gap decreases, the electrostatic force increases with an x^2 dependence, whereas the restoring force is linearly dependent on x. Consequently, once x has decreased to a critical 'pull-in gap', x_{pi}, an equilibrium point will no longer exist, and the two conductors will snap together. This critical gap may be shown to be:

$$x_{pi} = \frac{2x_0}{3}. \tag{2.11}$$

Electrostatic actuators have become a mainstay of MEMS devices, allowing high force actuation at room temperatures in an easily controllable fashion through the application of a potential difference between the conductors. The maximum frequency of operation is normally limited by the mechanical resonant frequency of the system. Design and fabrication of electrostatic actuators is also straightforward, requiring the deposition of two conducting regions, which are sometimes separated by a sacrificial layer that can be easily removed to produce an air gap.

2.4.3. *Shape Memory Alloy Actuators*

Shape memory actuators utilise the phenomenon that a small number of metallic alloys, such as TiNi [44], exist in one of two phases: a low temperature phase in which the alloy can be easily deformed, and a high temperature phase in which the alloy is rigid. The alloy will change between the two phases at a well defined transition temperature, T_t. When the alloy is heated above this temperature it will exert a force as it attempts to take up the rigid structure that it has previously been formed in. The shape of the rigid structure is defined by forcing the alloy into the desired shape whilst at a temperature below T_t. The alloy is then heated above this transition temperature whilst being held in the desired shape before being cooled once more. The alloy will then always return to this 'fixed' shape when heated above T_t.

Shape memory alloys can undergo changes in length of over 10% in this transition, and can exert very high forces. However, the need to heat the sample in order to induce the phase change means that the maximum frequency of operation is limited by the heat capacity of the system, and so is very low [38]. However, this type of actuation has been successfully used in microgripper devices.

2.4.4. *Thermal Actuators*

Thermal actuators use the fact that almost all materials suffer a volume expansion when heated. Quantitatively, the change in strain, $\Delta\varepsilon$, induced by a change in temperature, ΔT, is given by:

$$\Delta\varepsilon = \alpha_T \Delta T, \tag{2.12}$$

where α_T is the linear coefficient of thermal expansion of the material. This expansion can be used to generate a force, F, which is related to the change in strain by the Young modulus, E, according to

$$F = AE\Delta\varepsilon = AE\alpha_T \Delta T, \tag{2.13}$$

where A is the cross sectional area of the plane perpendicular to the force. It is clear that materials which have both a high linear coefficient of thermal expansion and a high Young modulus are required to generate appreciable forces. Therefore, metals are best suited for thermal actuator devices. They also have the advantage that they can carry a current and so may be resistively heated.

Two types of thermal actuation are possible in the case of thin film technologies: in-plane actuation in which a force is generated in the plane of the material, and out-of-plane actuation in which a force in generated in the direction perpendicular to the plane of the surface, as shown in Figure 2.10. In the case of in-plane actuation, a single layer of material is used and the

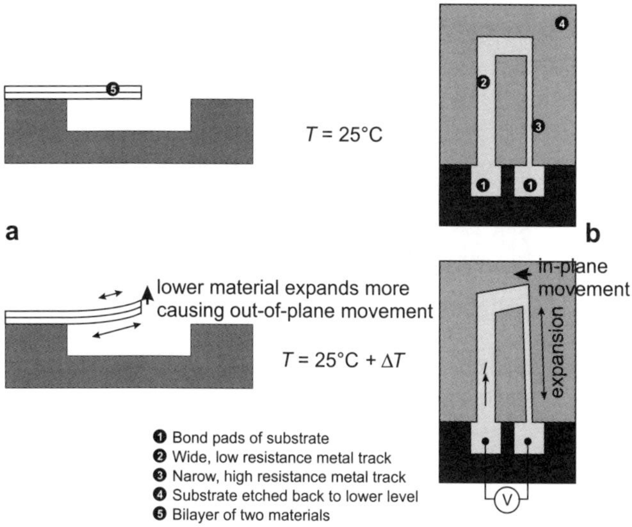

Figure 2.10: (a) A metal bilayer structure in which one material has a higher thermal expansion coefficient will produce out-of-plane actuation when heated. (b) A single metal track of varying width will rise to a higher temperature where the track is thinnest and the resistance is highest when a current is passed through it, leading to greater expansion on one side and in-plane actuation.

local temperature controlled to give an appreciable expansion in a long direction. Two layers of material, called a thermal bimorph, are required to achieve out-of-plane actuation. In this case, the metal is in intimate contact with a second material that has a very low linear coefficient of thermal expansion. When the two layers are heated, the metal will expand more than the second material. This will cause the bilayer to curl with the metal layer on the convex face. Figure 2.11 shows a microgripper that has been fabricated from a bilayer made from nickel and diamond-like carbon (DLC) [45]. DLC has a very low linear coefficient of thermal expansion, allowing the bimorph structure to curl through 180° with a change in temperature of only a few hundred Kelvin.

2.4.5. *Piezoelectric Actuators*

Dielectric materials can be polarised by the application of an electric field with the result that a charge will build up on the surface of the dielectric. Therefore, the electric flux density, **D**, in these materials is related to the

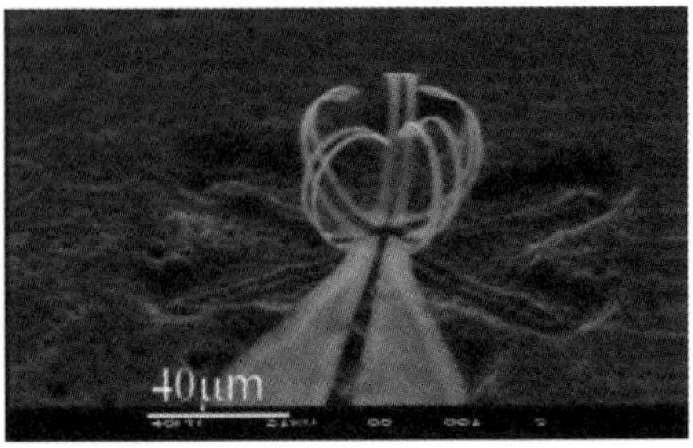

Figure 2.11: Thermally actuated microgripper consisting of a bilayer of nickel on diamond-like carbon. The nickel has a higher thermal expansion coefficient, causing the microgripper to uncurl when heated by passing a current through the device using the contacts at the bottom of the image [45].

electric field strength, **E**, and polarisation, **P**, according to:

$$\mathbf{D} = \varepsilon_0 \varepsilon_r \mathbf{E} = \varepsilon_0 \mathbf{E} + \mathbf{P}, \tag{2.14}$$

where ε_0 is the permittivity of free space and ε_r is the relative permittivity of the dielectric. However, if the dielectric has a non-centrosymmetric crystal structure, then there will be a relative shift in the position of the positively charge ions in each unit cell relative to the negatively charged ions when the system is under strain. As a consequence, the application of a uniaxial stress to a piezoelectric will also produce a polarisation. Furthermore, the application of an electric field to the piezoelectric will cause a strain to develop, as each unit cell will be distorted. Equation (2.14) therefore becomes:

$$\mathbf{D} = d\sigma + \varepsilon_0 \varepsilon_r|_\sigma \mathbf{E} = e\varepsilon + \varepsilon_0 \varepsilon_r|_\varepsilon \mathbf{E}, \tag{2.15}$$

where d is the piezoelectric constant under conditions of an applied stress, e is the piezoelectric constant under conditions of applied strain, σ is the stress and ε is the strain.

Piezoelectrics may both be used as actuating elements (as the application of an electric field will produce a force), and as sensing elements (as the application of a force will produce an electric field). The electromechanical coupling coefficient of a piezoelectric is defined as:

$$k = \sqrt{\frac{E_s}{E_a}} = \sqrt{\frac{de}{\varepsilon_0 \varepsilon_r|_\sigma}}, \tag{2.16}$$

where E_s is the electrical or mechanical energy stored and E_a is the mechanical or electrical energy input.

An early example of the use of piezoelectrics for actuation can be found in the scanning tunnelling microscope, in which they are used to accurately control the position of a tip in relation to a surface with accuracy better than 0.1 nm [2, 46]. Piezoelectric materials have found applications as actuators for a variety of microsystems due to the high force that can be developed and the very high frequency at which they can operate (which is a consequence of the direct link between the electrostatic and mechanical properties of piezoelectrics).

2.4.6. *Magnetic Actuators*

Of the actuation methods discussed in this section, magnetic actuators are the least common. This is because the magnetic force does not scale down well to small dimensions, the fabrication of small structures made from magnetic materials is complicated and a three dimensional current carrying structure is required to generate a magnetic field using moving charges, whereas most micromachined structures are two dimensional with a simple extrusion into a third dimension (so called 2.5D structures).

However, magnetic actuators do have a significant advantage in that levitation of structures is readily achieved. Frictional forces on bearings are a major problem on the microscale, due to the high surface area to volume ratio at this length scale.

Bearings made from permanent magnets may be levitated to eliminate frictional forces.

Magnetic actuators may also be fabricated from materials that exhibit a piezomagnetic effect in which the material suffers a change in strain upon magnetisation. This effect may either me used to create simple in-plane motion, or may be combined with a non-piezomagnetic material in a bilayer structure to produce an actuator that will give out-of-plane motion (in a similar fashion to the thermal bimorph considered in Section 2.4.4).

2.5. Materials and Growth

2.5.1. *Introduction to Materials for Microsystems*

A good understanding of the physical properties of materials and the ways in which materials can be grown is an integral part of designing and fabricating a microsystem, and being able to predict how it will behave. Although a range of novel materials are being developed for microsystems applications, such as diamond-like carbon, gallium arsenide and low temperature polymers or resins (most notably SU8 and polyimide) the traditional CMOS materials — silicon, silicon dioxide, silicon nitride and a range of metals — still dominate. They are likely to continue to do so for the foreseeable future, because the foundry processes for these materials are well developed.

Crystalline silicon wafers form the backbone of most MEMS devices, allowing integration of CMOS devices with the microsystem on the same substrate. The wafers themselves are usually produced from purified molten silicon by introducing a small crystalline silicon seed into the melt. In what is known as the *Czochralski process*, the silicon seed is rotated with respect to the melt, and slowly withdrawn from the solution. Silicon atoms solidify on the seed in the same crystal orientation as the seed, to produce a large ingot of single crystal silicon. This ingot is then cut, polished and lapped to produce the silicon wafers that we are familiar with, and upon which devices can be built.

This section looks at how to produce other materials on the surface of the silicon wafer. In the case of silicon dioxide, a conversion process can be employed in which the surface of the silicon wafer is oxidised. For other silicon based materials, a chemical vapour deposition is usually employed. Metals, on the other hand, are normally deposited either by physical vapour deposition or by electroplating, and these techniques are also discussed here.

Finally, one requires some method for determining whether a particular material is suitable for a given application. This process is known as *material selection*, and the methodology for achieving this will be briefly considered at the end of the section.

2.5.2. Silicon Oxidation

Silicon dioxide (SiO_2) is one of the most commonly used sacrificial materials in the fabrication of crystalline silicon (c-Si) based MEMS due to the existence of both wet and dry etch chemistries which efficiently remove SiO_2 whilst leaving c-Si largely intact. In addition, c-Si can be easily oxidised to form a high quality surface layer of SiO_2.

Oxidation may be most simply carried out by heating c-Si to temperatures between 1200 K and 1500 K in a dry gas mixture of oxygen and nitrogen at a pressure of one atmosphere. Small impurity concentrations of H_2O in the gas stream are thought to catalyse the creation of oxygen ions which diffuse through the surface oxide layer to the c-Si/SiO_2 interface where they then further oxidise the c-Si [47]. Oxidation can be enhanced by the deliberate addition of H_2O to the gas stream by bubbling the O_2/N_2 gas mixture through a simmering water bath before contact with the silicon in a process known as *wet oxidation.* This not only significantly enhances the growth rate, but also reduces the defect density in the SiO_2 as dangling bonds in the material can be passivated by hydrogen atoms from the water.

Deal and Grove [48] showed that the time t required to produce a silicon oxide layer of thickness x is given by:

$$t = \frac{x^2}{B} + \frac{Ax}{B} - \tau, \tag{2.17}$$

where A and B are constants and τ is an incubation time for oxidation, all of which are dependent on the oxidation temperature, method, and surface orientation of the c-Si lattice.

It is important for the process designer to remember that the oxidation of silicon is a material transformation, and silicon is *consumed* in the process. Approximately a 460 nm thick layer of c-Si will be required to produce a 1000 nm thick layer of SiO_2.

2.5.3. *Chemical Vapour Deposition*

It is frequently the case that we wish to produce a thin silicon layer, which will actually form the moving part of the device, on top of a sacrificial SiO_2 layer on a silicon wafer (Figure 2.12). A common way of achieving this is to flow a gaseous silicon-bearing precursor over the surface to be coated. The silicon precursor chemically reacts with the exposed surface to form a thin film of silicon. This is known as chemical vapour deposition (CVD).

There are several methods by which a chemically active silicon precursor can be produced. One of the simplest methods for producing a high quality silicon thin film is low pressure (LP) CVD. A schematic diagram of an LPCVD reactor is shown in Figure 2.13. The samples to be coated stand vertically in a quartz tube which is kept under vacuum and heated to between 800 and 900 K. Silane (SiH_4) gas (which is normally diluted in an inert nitrogen carrier gas) is flowed into the chamber at a partial pressure

Figure 2.12: SEM image of a suspended amorphous silicon cantilever over a silicon substrate.

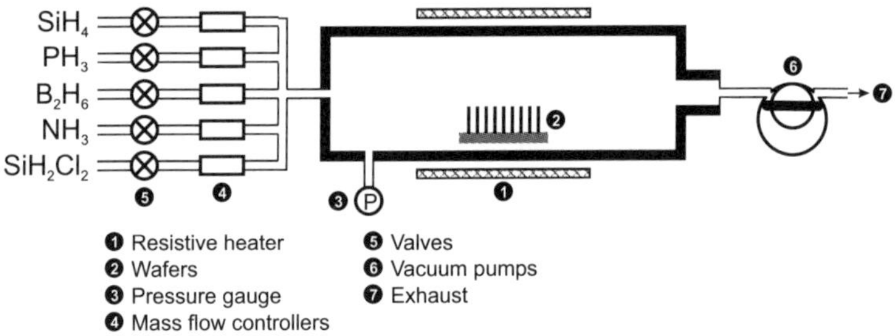

Figure 2.13: Schematic diagram of a low pressure chemical vapour depoisition system.

of between 10 and 30 Pa. Silane is a highly unstable gas and so is readily dissociated at the hot surfaces into active SiH_x radicals which can diffuse over exposed surfaces and bond at favourable locations. Excess hydrogen is driven out of the growing film at the high temperature to leave a polycrystalline silicon (poly-Si) thin film consisting of crystalline regions of silicon with a preferred $\langle 110 \rangle$ surface orientation, typically between 100 and

500 nm in diameter, separated by well-defined grain boundaries [49–52]. Growth rates up to $10\,\text{nm}\,\text{min}^{-1}$ can be readily achieved. Dopants can be introduced into the poly-Si through the addition of impurity gases, such as phosphine (PH_3) and diborane (B_2H_6).

Amorphous silicon nitride (a-SiN) is also preferentially produced by LPCVD. In this case, a feedstock gas mixture of dichlorosilane (SiH_2Cl_2) and ammonia (NH_3) is employed [53]. At temperatures above 800°C, a chemical reaction is thermally induced between these gases,

$$3SiH_2Cl_2 + 4NH_3, \quad Si_3N_4 + 6HCl + 6H_2, \tag{2.18}$$

yielding a silicon nitride thin film on exposed substrates and releasing HCl and H_2 as waste gas byproducts. Stoichiometric silicon nitride has a Si:N atomic ratio of 3:4, however such material is usually unsuitable for most microsystems applications as it has a high intrinsic tensile stress, but this can be reduced by increasing the silicon content of the material and this allows the production of very low stress material [54].

Accurate control of material properties is possible by LPCVD through the adjustment of the deposition process parameters, namely: temperature; gas mixture; total gas flow and gas pressure.

As most chemical reactions in LPCVD are thermally activated, temperature acts to control rate of reaction and hence the rate of deposition. Gas mixture, on the other hand, is the primary method by which stoichiometry or dopant concentration adjusted. For example, silicon rich a-SiN is produced by allowing an excess of SiH_2Cl_2 to flow into the reaction chamber relative to NH_3. Meanwhile, the total gas flow and gas pressure together determine the residence time of species in the reaction chamber and hence which chemical reactions will take place in the gas phase. This high level of control allows the deposition of material with excellent physical properties. Furthermore, as the deposition rate is limited solely by the thermally activated chemical reaction at the growing surface, step edges are conformally coated.

Whilst LPCVD produces excellent thin film materials, the high deposition temperature severely limits the application of this technique, as many materials cannot withstand such heat. Therefore, LPCVD tends to be limited to the coating of (oxidised) crystalline silicon wafers.

Catalytic CVD (cat-CVD) — sometimes called hot wire CVD — has been developed in recent years in an attempt to overcome this problem

[55–57]. In this case, the source gases at low pressures (<50 Pa) are passed over a tungsten wire which is heated to between 600°C and 1800°C. Silane gas (which is normally diluted in hydrogen) dissociates catalytically in the presence of tungsten at these temperatures before passing over the substrate to be coated, which is located downstream of the heated wire. As the source gas is already chemically highly reactive, the substrate only needs to be heated to between 300°C and 500°C in order to yield a high quality poly-Si deposition. As with LPCVD, doping may be achieved through the addition of PH_3 or B_2H_6 to the feedstock gases, whilst the addition of ammonia leads to the formation of silicon nitride.

Cat-CVD is a relatively new technology, and as such is rarely found in fabrication facilities. Plasma enhanced chemical vapour deposition (PECVD) on the other hand is widely available, and provides a means for depositing a diverse range of thin film materials, including hydrogenated amorphous silicon (a-Si:H), a-SiN and amorphous silicon oxide (a-SiO) at temperatures between 80°C and 350°C. Although several varieties of PECVD exist, they all use the principle that high quality thin film materials can be deposited at relatively low substrate temperatures if energy is first put into the feedstock gas to produce a plasma containing a mixture of ions, electrons and radicals [58]. The different forms of PECVD then differ only in the method by which the plasma is generated.

Radio frequency (rf-) PECVD is the most common of the low temperature thin film deposition techniques. A schematic diagram of a typical rf-PECVD system is shown in Figure 2.14. The sample to be coated sits on a heated, earthed plate inside an evacuated chamber. A second plate sits parallel to the sample plate with a gap of a few centimetres between the two. This plate is designed with an array of holes to form a showerhead through which the source gases are passed to create a uniform gas flow through the chamber with a pressure of between 10 and 100 Pa. An rf electric field is applied between the two plates to create a glow discharge plasma [59]. Pure silane gas is used to deposit a-Si:H [60–62], whilst the addition of a mixture of NH_3 and N_2 produces a-SiN or alternatively the addition of N_2O with helium results in a-SiO [63].

The properties of materials deposited by rf-PECVD are determined by the following factors: substrate temperature; power density of the rf radiation; frequency of the rf radiation; gas pressure; gas flow ratio and total gas flow.

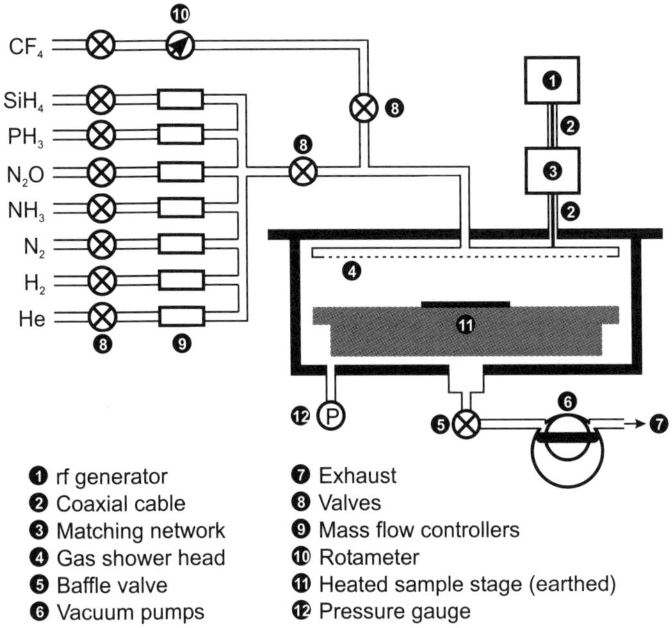

Figure 2.14: Schematic diagram of a rf plasma enhanced chemical vapour deposition system.

The power density of the rf radiation affects how the gases which are injected into the deposition system are dissociated. The gas pressure, P, and total gas flow, f, then determine the residence time of species in the vacuum chamber, t_r, according to:

$$t_r = \frac{pVL_A}{RTf}, \tag{2.19}$$

where V is the chamber volume, L_A is the Avogadro constant, R is the molar gas constant, T is the gas temperature and f has units of molecules per second. Given this residence time and the dissociation of the injected gases, a number of gas phase reactions may take place to produce the precursor species that will actually lead to deposition. For this reason, considerable effort has been expended over many years to quantify the reaction rates for many of the common rf-PECVD gas mixtures [64–67]. Once the deposition precursors reach the growing material surface, they will diffuse over the growing surface or into the material bulk until they either chemically bond to the material or desorb from the surface back into the gas phase (which may also include the abstraction of material from the growing surface — an etching process). These processes are all thermally activated, and as such are strongly affected by the substrate temperature.

The higher mobility of electrons in the plasma, relative to ions, leads to the formation of a 'plasma sheath' near the two electrodes. These sheaths are regions that are depleted of electrons, and so they have a net positive charge, whereas the bulk of the plasma has no net charge. The plasma bulk therefore has a positive potential, V_p, relative to the earthed substrate electrode, and ions are accelerated across the plasma sheath towards the substrate. Ion bombardment tends to produce material with a high compressive intrinsic stress that is normally undesirable in microsystems structures. Reducing the plasma potential reduces the ion bombardment energy. This may be achieved by reducing the rf power density, although this will also adversely decrease the dissociation of the gas. Alternatively, increasing the frequency of the rf radiation also reduces the plasma potential without the need to change the power applied, but tends to lead to less uniform growth over large areas. Therefore, material deposited at low frequencies tends to have a compressive intrinsic stress whilst high frequencies often produce material under tensile stress [62]. Finally, ion bombardment may also be reduced by increasing the gas pressure so that the mean free path of species in the gas phase is less than the width of the plasma sheath. In this circumstance, ions crossing the sheath undergo multiple collisions and lose kinetic energy in the process.

Ion bombardment also affects the conformality of rf-PECVD. In general, CVD produces excellent step edge coverage of patterned surfaces. However, if ions form a significant fraction of the deposition precursors, then coverage of step edges is reduced if ions are accelerated vertically across the sheath so that their angle of incidence with the growing surface is close to the normal [68, 69]. Side-walls therefore see a reduced flux of film forming species and the deposition rate on these surfaces is reduced, as indicated in Figure 2.15.

It is clear that rf-PECVD is a complex system and that each of the factors controlling deposition have a complex effect upon the properties of the material produced. Therefore, deposition conditions for a particular material are normally optimised empirically. However, the low substrate temperature coupled with the ability to deposit high quality materials uniformly

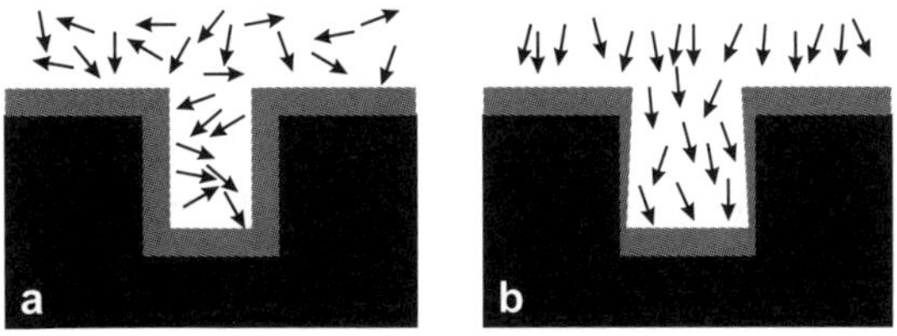

Figure 2.15: (a) Conformal (uniform) coating of step edges occurs when incoming precursors have a broad angular distribution of incidence and/or there is a high degree of surface diffusion and attachment is limited by chemical reaction with the surface, as found in LPCVD. (b) A non-conformal coating results when the precursor species have a narrow angular distribution of incidence, as found in rf-PECVD because of the acceleration of ions across the plasma sheath, and surface diffusion is limited.

over very large areas (over $4\,m^2$) makes this an attractive thin film deposition technique.

In certain circumstances a greater degree of control over deposition is required than that offered by rf-PECVD. This can be achieved by generating the plasma remotely from the sample, which is located so that chemically active species generated in the plasma subsequently pass over the sample. The growth of polycrystalline diamond, for example, is achieved by microwave CVD [70]. A highly diluted gas mixture of CH_4 in H_2 is activated using microwave radiation to provide a high degree of dissociation. The CH_3 and H radicals produced then pass at high pressure ($\sim$5000 Pa) over the substrate to be coated which is heated to around 850°C. Although the growth rate of polycrystalline diamond by this technique is relatively low ($\sim$10 nm min^{-1}), the average grain size is frequently in excess of 100 μm. Polycrystalline diamond films have been successfully employed as passive microsystem components, such as gears, where the low coefficient of friction and high wear resistance of this material makes it ideally suited for this application.

Plasma dissociation and ionisation may be further enhanced through the application of a magnetic field to a microwave generated plasma, as found in electron cyclotron resonance (ECR-) PECVD. The extremely dense plasma produced can be used to successfully deposit silicon based materials at very low temperatures below 100°C onto plastic substrates [71–73]. In this case, the need for a high substrate temperature is removed by supplying energy to the growing surface from a high flux of low energy ions.

The flexibility of PECVD means that this generic technique is used to deposit thin films of a uniquely diverse range of materials. Consequently there is a continuing research effort to develop this technology to give greater control of deposited material properties with recent variations including the expanding thermal plasma [74], helicon plasma sources [75] and electron cyclotron wave resonance PECVD [76].

2.5.4. *Sputter Coating of Metallic Thin Films*

Sputter coating is a widely used technique for producing metallic thin films uniformly over large areas. Although there are several variants, all sputter coating techniques essentially use the principle that firing heavy atoms of an inert gas, such as argon, with a high kinetic energy, at a solid material (called the *target*) will cause atoms in the target to be physically knocked out of the solid and into the gas phase. These gas phase atoms will condense on any exposed surfaces to form a thin film of the original target material. The source of energetic atoms is usually a low pressure plasma of the inert gas. This is formed by the application of an electric field between the target, which acts as the cathode, and an anode upon which the surface to be coated is placed. The positively charged ions in the plasma are then accelerated towards the cathode by the electric field.

A useful measure of the efficiency of the sputtering process is the *sputtering yield*, S, which is defined as the average number of atoms that are removed from the target by one incident inert gas ion. If the kinetic energy transferred by the incident ion to the target material is insufficient to overcome the binding energy of atoms in the target material, then no sputtering will take place. However, above this threshold, which is typically less than 50 eV, the sputtering yield is given by

$$S = \frac{\varepsilon E_i}{H f_m}, \tag{2.20}$$

where E_i is the kinetic energy of the incident ion, H is the heat of sublimation of the target material, f_m is a nearly linear function of the ratio of the mass of the target atom to that of the incident ion, m_t/m_i, and ε is the proportion of momentum transferred to a target atom in an elastic collision which is given by [77]

$$\varepsilon = \frac{4 m_i m_t}{(m_i + m_t)^2}. \tag{2.21}$$

The various sputtering techniques that are available only differ in the method by which the inert gas plasma is generated. The simplest technique is *DC sputtering* in which a DC bias is applied between the cathode and anode to create a glow discharge plasma. This allows successful sputtering of metallic thin films [78]. The addition of other reactive gases to the plasma, such as oxygen, allows the deposition of more complex materials, such as metal oxides [79]. This technique is known as *reactive sputtering*. Alternatively, in *rf sputtering*, an rf electric field is applied between the two electrodes to produce a plasma. Whilst this technique can be used in the same way as a DC sputtering system, the fact that an rf electric field is not screened by insulators, as is the case for DC electric field, allows the use of insulating target materials, such as silicon dioxide, to produce insulating thin films. This technique has an advantage over reactive sputtering in that it is easier to produce thin films of the correct stoichiometry. This is because although the different elements in the target will have varying sputtering yields, during a short equilibration period the surface layer of the target will be depleted of the element with the higher sputtering yield. In this way the flux of this species is reduced to compensate for the different sputtering yields and so the stoichiometry of the deposited thin film will almost exactly match that of the target. Finally, the sputtering rate of both DC and rf systems may be enhanced by the application of a magnetic field across the target which increases the ionisation of the plasma in this region, and this is called *magnetron sputtering*.

Despite this, a major limitation of sputtering is that growth rates tend to be very low — typically no more than $1\,\mathrm{nm\,s^{-1}}$. Furthermore, some ion bombardment of the growing thin film is inevitable, and this leads to the creation of defects and intrinsic compressive stress in the material, excessive heating of the sample after significant sputtering times and the inclusion of atoms of the inert gas in the thin film bulk leading to porosity. However, if the area of the target is greater than that of the sample, then this technique may be effectively used to coat step edges on a sample surface.

2.5.5. *Evaporation of Metallic Thin Films*

One of the simplest means of producing thin metallic films is by evaporation under vacuum. In this technique, energy is supplied to cause a small quantity of the desired metal to melt inside a vacuum system, as shown in

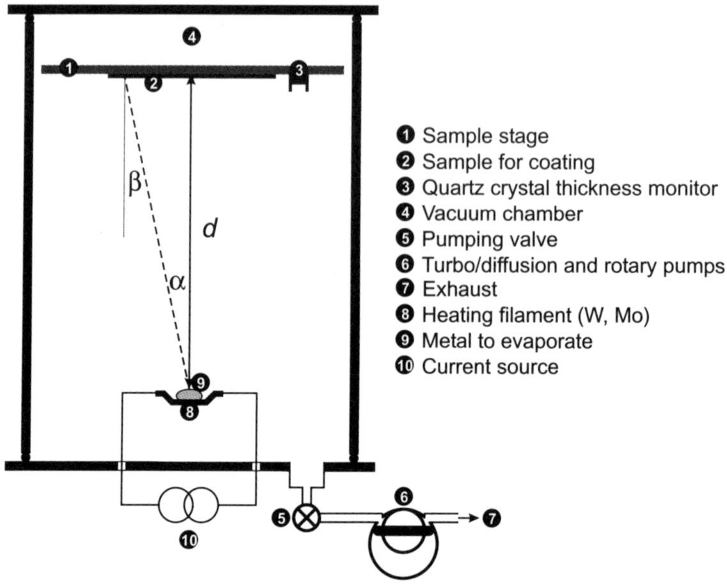

Figure 2.16: Schematic diagram of a thermal evaporation system.

Figure 2.16. This source of metal is normally in the form of small pellets or short lengths of a thin wire of high purity. The metal atoms readily evaporate at these low pressures. The gaseous metal atoms will then travel in a straight line until they reach a cool surface where they will condense. Therefore, a thin metal film, normally with an amorphous structure, can be produced on a substrate that has been placed a short distance from the evaporation source.

It is clearly important that the evaporating metal atom have a clear path from the source to the substrate. Therefore, the background pressure in the vacuum chamber must be sufficiently low that the mean free path of atoms (the average distance travelled between collisions with other gas phase species) is longer than the distance between the substrate and the source. In practice, for most evaporation systems where this distance is less than one metre, a background pressure lower than 10^{-4} Pa will suffice. Other background gases may be intentionally introduced to the system during evaporation, such as oxygen, to enable the growth of metal oxide thin films, and this is known as *reactive evaporation*.

Another consequence of the line-of-sight nature of evaporative growth is that the growth rate, R, will depend on both the angles of the source and substrate relative to each other and the distance between the two as:

$$R \propto \frac{\cos\alpha\cos\beta}{d^2}, \tag{2.22}$$

where all symbols are defined in Figure 2.16. Consequently, uniformity of deposition will be improved by increasing the distance between the source and the substrate, but this will also cause a significant decrease in the deposition rate. Furthermore, the thickness of the metal film will be highly non-conformal, and step edges (where β is locally 90°) will be uncoated. This is in stark contrast to sputtered metal coatings. In addition to long-range variations in thickness, evaporated metal layers may also possess short-range topographic features with some evaporated metals — most notably gold — producing an island-like surface structure, as shown in Figure 2.17.

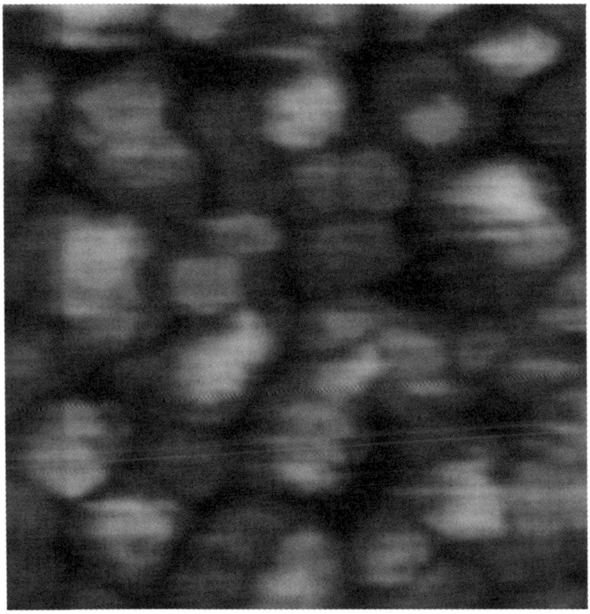

Figure 2.17: Scanning tunnelling microscopy image of a 2 µm×2 µm area of thermally evaporated gold on a mica substrate. The height variation in the image is 40 nm.

Several methods exist for actually providing energy to cause the metal source material to evaporate:

Thermal evaporation:
The metal source material is placed in intimate contact with a resistive heating element that is made from a metal which has a much higher melting point than the metal to be evaporated, such as tungsten or molybdenum. A current of the order of tens of amps is passed through the filament, which then heats up, causing the source metal to evaporate over a short period of time. Whilst being the simplest of the evaporation techniques, accurate control of film thickness is difficult, as growth rate varies with time due to the changing quantity of source material on the filament. Consequently, placing the correct 'dose' of source metal on the filament provides the best means of controlling the film thickness.

Electron beam evaporation:
A crucible made from a material with a high melting point, such as quartz, is filled with the source metal. An electron beam (e-beam) is directed at the source material, leading to localised melting and evaporation. By ensuring that an excess of metal is placed in the crucible, and with careful control of the electron beam, it is possible to ensure than the molten metal does not come into contact directly with the crucible, and it is therefore possible to produce very high purity metal thin films. Furthermore, the deposition rate can be accurately controlled by varying the electron energy and beam current. Therefore, a quartz crystal microbalance (QCM) is frequently employed to allow measurement of deposited film thickness in real time. However, care must be taken to ensure that X-rays are not produced when the e-beam interacts with the metal, as this can be harmful both to the sample and the operator!

Laser heating:
Alternatively, pulses from a high power laser may be used to sublime a metal source in a crucible. This technique has many of the advantages of e-beam evaporation, but without the risk of X-ray production. However, it is extremely expensive to set up.

Radio frequency induction heating:
Finally, an rf induction heater may be used to melt the metal source material in a crucible. In this case, however, the molten metal is in contact with

the crucible and a higher level of contamination may be incorporated into the deposited film.

2.5.6. *Plated Metallic Layers*

Plating is a cheap and simple alternative to vacuum-based thin film coating techniques in which the sample to be coated is dipped into a solution containing metal ions. These metal ions are then electrochemically attached to the substrate surface using one of the following three methods:

Immersion plating:
This is the simplest of the plating methods. A galvanic displacement is used to produce a thin metal coating. For immersion plating to work, the metal from which the thin film is to be composed must be more electropositive than the element on the surface of the substrate to be coated. A galvanic reaction will then cause the surface layer of the substrate to go into solution, and a layer of the metal from the solution will replace it. If this new metal layer is impermeable, then the reaction will automatically stop, and a conformal metal coating will be the result.

Electroless plating:
Although electroless plating also only requires dipping the substrate to be coated into a solution containing the metal ions to be deposited, as a displacement reaction is not involved, thick metal layers may be produced. In this case, the solution contains negatively charged ions that are easily oxidised by a metal-coated substrate surface. The metal ions in solution are therefore reduced to form a coating on the substrate which allows the deposition reaction to continue.

Electroplating:
A conducting substrate surface is connected as the cathode to a DC voltage or current source whilst an inert conducting plate is connected as the anode. The two plates are then placed in an aqueous solution of a metal salt. The power supply is then switched on and the positively charged metal ions will be attracted to the substrate (cathode) where they will be discharged and form a deposited metal layer. The metal ions in solution will be depleted with increasing time, resulting in a reduced deposition rate. This may be countered by replacing the inert anode with one made from the metal to be coated. Metal ions will be removed from the anode during plating and

the concentration of metal ions in the solution will be maintained. If this method is adopted, then the deposition rate, R, is given by:

$$R = \frac{JA}{\rho z F}, \tag{2.23}$$

where J is the current density due to metal ions, A is the atomic weight of the metal, z is the valence number of the metal ions, ρ is the density of the metal coating and F is the Faraday constant, which is 96 500 C.

2.5.7. *Material Selection*

In designing a microsystem, it is important to consider whether a particular material is suitable for the task to which is will be assigned, and the aim should be to adopt a quantitative methodology for achieving this. Methods for material selection are well covered by Ashby [6], and the reader should consult this work for a detailed understanding of the selection process. However, at its heart lies the general need to design a device to perform a particular function given certain constraints. The key to material selection, therefore, it to generate a simple analytical model that explicitly shows how the particular function depends on the geometry of the system and the properties of the material.

This concept is best explained by example. Let us consider a situation where one wishes to measure an acceleration in one dimension. One could attempt to do this by making a cantilever proof mass of width w, height h and length l. It is desired to maximise the deflection of the beam under the application of a particular acceleration, as shown in Figure 2.18. From simple beam theory, it is known that the beam will deflect by a distance δ at its end under the application of a force F applied uniformly across its area according to:

$$\delta = \frac{3Fl^3}{2Ewh^3}, \tag{2.24}$$

where E is the Young modulus of the cantilever material. The force is related to acceleration by the Newton Law:

$$F = ma, \tag{2.25}$$

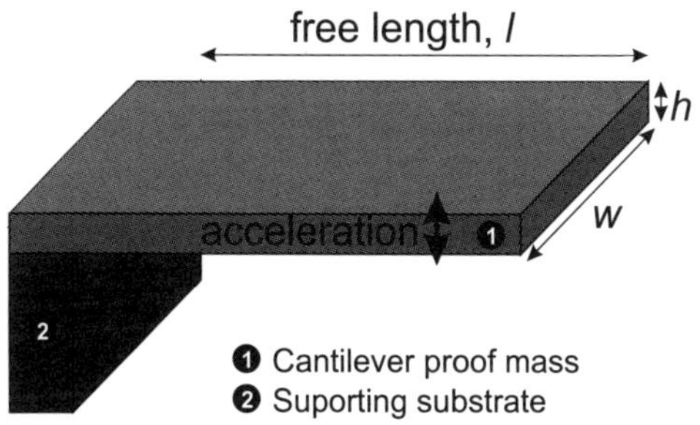

Figure 2.18: Schematic diagram of the geometry of a simple proof mass cantilever accelerometer.

and the mass is related to the density ρ of the material by:

$$m = \rho l w h. \tag{2.26}$$

Hence, the deflection of the cantilever is given by:

$$\delta = \frac{3}{2} \cdot \frac{l^4}{h^2} \cdot \frac{\rho}{E} \cdot a. \tag{2.27}$$

Therefore, it is clear that a material with a high ratio of density to Young modulus would be most appropriate for this application. Table 2.2 gives this ratio for some common microsystems materials. It is apparent that metals, such as aluminium or nickel, are most appropriate for this application. A further design metric might then be that we wish the cantilever to maximise deflection without fracture, and the ratio of fracture strength to Young modulus then becomes significant. Equation (2.27) also gives us an indication of how to optimise the geometry of the cantilever for this application, suggesting that the ratio l^4/h^2 is significant — in other words, the beam should be long and thin. However, such cantilevers are prone to failure due to stiction effects during processing, and this will place an upper limit on this ratio.

This design methodology may also be applied to actuating systems. One could consider the thermal actuator shown in Figure 2.19 which is heated to 900 K by passing a current through the u-shaped device. This causes the material to expand, generating a force in the direction shown. The design metric is that the force should be maximised at this temperature. It is known that the force generated when the system expands by a distance Δl

Table 2.2: The ratio of density and Young modulus for some common MEMS materials.

Material	Density, ρ ($\text{kg}\,\text{m}^{-3}$)	Young modulus, E (GPa)	ρ/E ($\text{kg}\,\text{m}^{-3}\,\text{GPa}^{-1}$)
Aluminium	2710	69	39.3
Crystalline silicon	2400	190	12.6
Nickel	8900	207	43.0
Polycrystalline diamond	3500	1035	3.4
Polycrystalline silicon	2320	160	14.5
Silicon dioxide	2200	70	31.4
Silicon nitride	3440	380	9.1

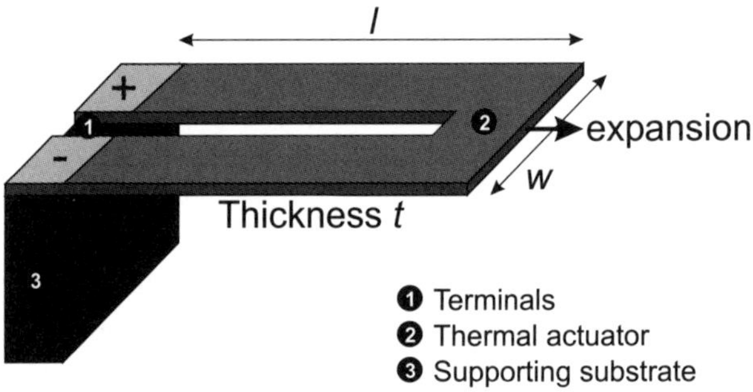

Figure 2.19: Geometry of a simple thermal actuator.

is related to the Young modulus, E, by

$$F = \frac{Ewh\Delta l}{l}, \tag{2.28}$$

where w and h are the width and thickness of the cantilever beam. However, the expansion generated when a material is increased in temperature by ΔT is related to the thermal expansion coefficient, α, of the material by:

$$\Delta l = l\alpha\Delta T. \tag{2.29}$$

By eliminating Δl, it can be shown that the force generated is related to the temperature rise of the system according to:

$$F = wh \cdot E\alpha \cdot \Delta T. \tag{2.30}$$

Therefore, a material with a high product of Eα is required to maximise the force developed at a given temperature. In terms of geometry, it is also clear that a device with a high cross sectional area will lead to an increased force.

The materials selection charts produced by Ashby [6] provide an easy graphical means of making sensible materials selections based on this design methodology by grouping materials together by class (metals, polymers, etc.). It should be remembered, however, during material selection, that the chosen material must not only be suitable for the task to which it will be employed, but also that it can be fabricated. For example, many polymers cannot withstand the elevated temperatures associated with microsystem fabrication. Therefore, selection must be made in the context of designing a full process flow for fabrication, and this is the subject of Section 2.6.7.

2.6. Fabrication Techniques

2.6.1. *Introduction to Fabrication Techniques*

So far, this chapter has considered a range of elements found in microsystems, and grouped these into passive, sensing and actuating components. It was also shown how different materials can be produced on a substrate and a methodology for selecting materials appropriately for a given application was discussed. The process by which these microsystem elements can be fabricated from the basic materials is considered next.

The process by which structures are manufactured from a material is known as *micromachining*. This is further subdivided into *surface micromachining* in which only the surface layers of material are acted upon, resulting in devices which are no more than a few tens of micrometres deep, and *bulk micromachining* in which large quantities of the substrate material, perhaps hundreds of micrometres in depth, are removed.

Light sensitive polymers, called photoresists, are normally employed for initially creating a pattern on the surface of a sample to be micromachined, and the process by which this is achieved is discussed before the principal micromachining technologies — etching, bonding and planarisation — are described.

2.6.2. *Photolithography and Cleanliness*

Having produced a series of material layers that are to be converted into a structure, it is necessary to be able to actually produce a pattern in the material layers. Photolithography is the most common means of achieving this at the micrometre length scale by coating the surface of the material to be patterned in a light sensitive polymer. The polymer is then exposed to a predefined pattern of light, allowing certain areas of polymer to be selectively removed. The regions of material that are not now covered by the polymer may be removed by etching.

Of all the process steps employed in the fabrication of microsystems, photolithography is perhaps most sensitive to particulate contamination, and so should always be performed in a dust free environment, called a Clean Room. A clean environment is achieved by limiting the dust that can enter the environment. Therefore, those working in a Clean Room should wear appropriate coveralls and a facemask. Also, the Clean Room should be air conditioned and kept under a positive air pressure relative to atmosphere to ensure that dust cannot be blown into the clean environment. Furthermore, any dust that inevitably does enter the room is removed by frequently circulating the air through high efficient particle arrestor (HEPA) filters. Where a particularly high level of cleanliness is required, a laminar airflow is employed in which air enters the Clean Room through the ceiling and is removed through holes in the floor so that there is a constant down-flow of air acting to carry away dust particles. The cleanliness of a Clean Room is then rated by the ISO14644 classification. In this system, a clean room is given a Class Number, N, where the concentration per cubic metre, C, of particles with a diameter greater than D (measured in micrometres) is given by:

$$C = 10^N \left(\frac{0.1}{D}\right)^{2.08}. \tag{2.31}$$

A Class 5 Clean Room is sufficient for most photolithographic process down to feature sizes of 0.5 µm.*

It is clearly not only important that the environment is clean if photolithographic patterning is to be attempted, but also that the surface to be patterned is also free from debris. There are several common recipes for producing a clean surface using wet chemistry. Which recipe is employed

*The reader should note that the ISO14644 nomenclature is relatively recent, and most Clean Rooms are still classified according to the US Federal Standard 209b where the Class of a room is the number of particles greater than 0.5 µm in diameter per cubic foot. ISO14644 Class 5 is approximately equivalent to 209b Class 100.

depends on the level of cleanliness required and the reactivity of materials that are exposed to the cleaning solution — it is clearly important that the cleaning solution should not itself remove the material to be patterned. Some of the most common cleaning chemistries are considered below.

RCA Clean:
The RCA clean has become a standard method for removing contaminants from silicon wafers, and is a two-stage process. In stage one (RCA1), one part of hydrogen peroxide (H_2O_2) and the sample to be cleaned are added to an already boiling solution of one part 25% aqueous ammonia diluted in five parts of deionised (DI) water. After ten minutes, almost all organic contaminants on the wafer surface will have been removed (this process is not, therefore, suitable for polymer containing samples). The second stage of the RCA clean (RCA2) is designed to remove metal ion contaminants. One part of H_2O_2 and the sample added to an already boiling solution of one part of HCl diluted in six parts of DI water. This cleaning step may be unsuitable for samples where metal is exposed.

HF Cleaning:
A thirty second dip in buffered hydrofluoric acid (HF) is used to remove the thin native silicon oxide layer on silicon wafers and in doing so any surface contaminants. However, HF is an extremely hazardous acid. Furthermore, HF will also attack glass samples and many metals, thereby limiting its applicability.

Fuming Nitric Acid:
A wider range of materials may be cleaned by immersion in a bath of fuming nitric acid for five minutes. This will efficiently remove organic contaminants, particularly if ultrasonic agitation is also employed. This should be followed by a thorough rinse in DI water.

Solvent Clean:
The least aggressive of the wet chemical cleaning recipes — and hence the most widely applicable — is the solvent clean. In this process, the sample is first cleaned for ten minutes in an aqueous solution of a proprietary degreasing agent whist being ultrasonically agitated. The sample should then be placed in an ultrasonically agitated bath of DI water for thirty minutes. The DI water should be changed every ten minutes during this time to ensure that the degreasing agent is completely removed. The sample should then

be placed in an ultrasonically agitated bath of acetone (propanone) for ten minutes followed by an ultrasonically agitated bath of isopropanol for ten minutes. Finally, the sample undergoes a ten minute clean in an ultrasonically agitated bath of DI water before being spin dried under a flow of dry nitrogen gas and baked in an oven for thirty minutes at 125°C. In practice, if the sample is already nominally clean, then the degreasing agent and thirty minute DI water bath cleaning steps may be omitted. It should also be noted that acetone attacks some plastics, and so this process is not suitable for samples where such materials are exposed.

Once the sample surface is clean, then the photoresist may be applied. Photoresists are solutions of an organic polymer resin in a liquid solvent together with a sensitiser. A dose of the photoresist solution is applied to the sample surface which is then spun at several thousand revolutions per minute so that an even coating is produced that is normally a few micrometres thick. The thickness of photoresist actually used is dependent on the size of the smallest features to be produced and the thickness and nature of the underlying material to be etched. In general, the thinner the photoresist layer, the smaller the minimum feature size that can be produced. However, the photoresist is present to protect the underlying material from subsequent etching. Therefore, the photoresist layer must not be completely removed by the etching process to be subsequently employed, and this will set a minimum required thickness for the photoresist layer. The coated sample is then baked to cause the solvent to evaporate, leaving a solid polymer layer. If a sample surface is particularly hydrophobic then the photoresist may not bind to the sample. This may be countered by applying an adhesion promoter, such as hexamethyldisilazane (HMDS) to the sample surface before the photoresist.

The photoresist coating must now be selectively exposed to ultraviolet (UV) light. This is achieved using a UV transparent plate, normally made from quartz or soda lime glass, on one side of which there is an opaque pattern of metal, called a *mask*. The mask is used in one of three ways to selectively expose the photoresist to the UV light:

Proximity Printing:
In this method, the mask is held parallel to the surface of the sample with a small *print gap* between the two which is normally less than 30 μm. UV light is shone through the metal, and the patterned mask casts a shadow

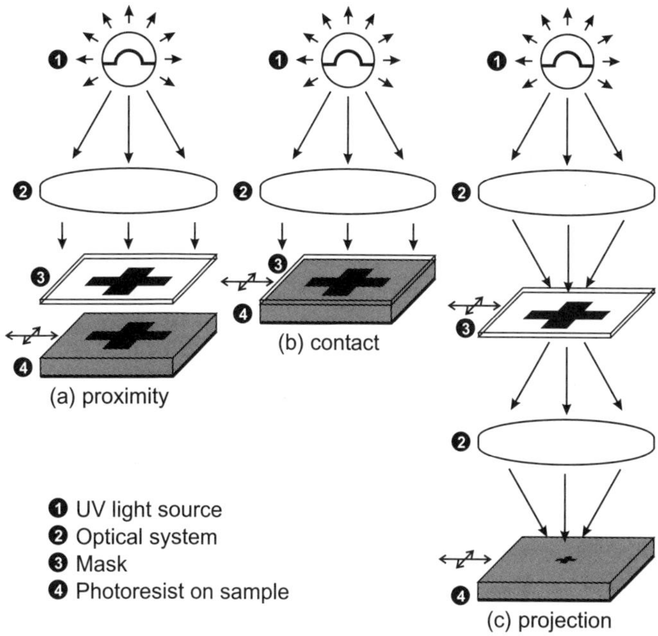

Figure 2.20: Schematic diagram comparing (a) proximity, (b) contact and (c) projection photolithographic printing techniques.

pattern on the photoresist, resulting in a selective exposure, as desired (Figure 2.20a). The critical resolution, R, (the size of the smallest features that can be produced) is dependent on the diffraction of light around the mask, and so is controlled by the wavelength of the light used, λ, the printing gap, g, and the thickness of the photoresist layer, t, according to

$$R = \frac{3}{2}\sqrt{\lambda\left(g + \frac{t}{2}\right)}. \tag{2.32}$$

It is clear that proximity printing will lead to the exposed pattern on the photoresist to be a 1:1 copy of the mask pattern.

Contact printing:

It is clear from Equation (2.32) that resolution can be improved by reducing the printing gap between the sample and the mask to zero. This mode of operation is known as contact printing (Figure 2.20b). Some pressure must be exerted to hold the sample and mask together in contact printing. If gravity alone is used to hold the mask on the sample, then this is called *soft contact*. In *hard contact*, additional mechanical pressure is applied, whilst in *vacuum contact*, the region between the mask and the sample is evacuated so that 1 atmosphere of pressure holds the two together. Vacuum and hard contact may be used together to generate up to 3 atmospheres of pressure. Whilst resolution improves with increasing pressure, it is also the case that damage to the metal coating on the mask will increase, and so the lifetime of the mask (which can cost thousands of euros) is greatly reduced. As with proximity printing, contact printing will yield a 1:1 copy of the mask pattern.

Projection printing:
Projection printing avoids any contact between the mask and the sample by shining collimated UV light through a large scale mask and then focussing the shadow pattern onto a small area of the sample some distance below using a lens. The sample is then *stepped* underneath the mask and exposure repeated to generate the pattern required over the whole sample (Figure 2.20c). In this case, the resulting pattern is a scaled down copy of the mask. The critical resolution of projection printing is given by

$$R = \frac{k\lambda}{N}, \tag{2.33}$$

where λ is the wavelength of the UV light, N is the numerical aperture of the focusing optics and k is an experimental constant that is normally around 0.3.

There are two classes of photoresist: positive and negative. In the case of positive resists, development in a weak base solution causes the areas of resist which were exposed to UV light to be removed, leaving a copy of the mask pattern in the photoresist. Development of negative resists, however, leads to removal of those areas which were not exposed to UV light, leaving an inverted copy of the mask pattern in the photoresist.

Both positive and negative photoresists require a certain 'dose' of UV light to have an effect. As the UV light passes through the photoresist, it is absorbed, and so the photoresist closest to the sample receives a lower dose that at the top surface. Therefore, underexposure of a positive resist will lead to only the top part of the photoresist layer being removed by the developer with an unwanted layer of photoresist remaining at the bottom of the exposed areas, as shown in Figure 2.21(a). Underexposure of

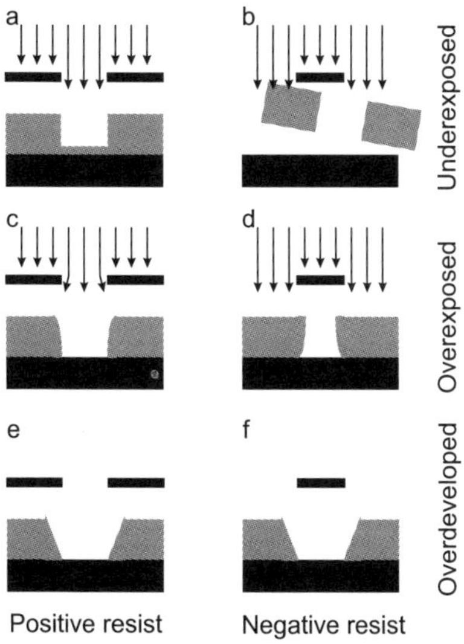

Figure 2.21: The effect of exposure and development on the shape of features.

a negative photoresist, on the other hand will leave the bottom layer of resist open to attack by the developer, allowing the whole pattern to be lifted away from the substrate (Figure 2.21b). Overexposure of photoresist is also undesirable. Light has a tendency to diffract as it passes the metal layer in the mask. Therefore, some light is diffracted into the shadow under the metallised regions. In the case of positive photoresists, overexposure will tend to allow some of the top layers of photoresist under the masked region to be sufficiently exposed, as shown in Figure 2.21(c), resulting in a narrower structure than intended which tapers towards the top. On the other hand, overexposure of a negative resist will tend to cause the opposite effect, with such photoresist layers tending to have an upper overhanging region (Figure 2.21d).

Development of photoresists must also be correctly timed. Whilst there should be a high ratio between the etch rate of exposed and unexposed photoresist in a developer solution, this is not infinite. Therefore, excessive development will tend to lead to the top layers of resist being removed, resulting in structures which are tapered towards the top (Figures 2.21e and f). Too short a developing time will leave unwanted resist at the bottom of structures. It should also be noted that negative photoresists have a tendency to swell in developer solutions — an effect that is less common in positive resists. Therefore, negative photoresists are not normally suitable for producing structures with features smaller than 2 μm.

Whilst such 'shape' effects caused by photoresist exposure and development are normally undesirable, this is not always the case. For example, undercut structures, as in Figure 2.21(d), are often very good for producing lift-off structures, as deposition on the sidewalls is very difficult (Section 2.5.5). Furthermore, controlling the local UV dose received by a photoresist has been successfully employed to created complex 3D structures using 'thick' photoresists, such as SU-8, which can either be used as mechanical structures in their own right [80, 81], or as moulds for producing structures in other materials [82].

It is frequently the case that a full device requires the patterning of many layers, and each new layer must be correctly *aligned* with previous layers. This is achieved through the use of alignment marks on each mask. These normally consist of a set of inverted and non-inverted crosses that are regularly repeated over the mask and also over the sample from previous patterning stages. The sample is moved relative to the mask in a mask aligner prior to exposure until two spatially separated sets of crosses overlap, as shown in Figure 2.22. If these two points are correctly aligned, then it must be the case that the whole system is correctly aligned and exposure may proceed. In practice, alignment of two layers is never perfect, and some tolerance must be built into the design of any device to allow for the

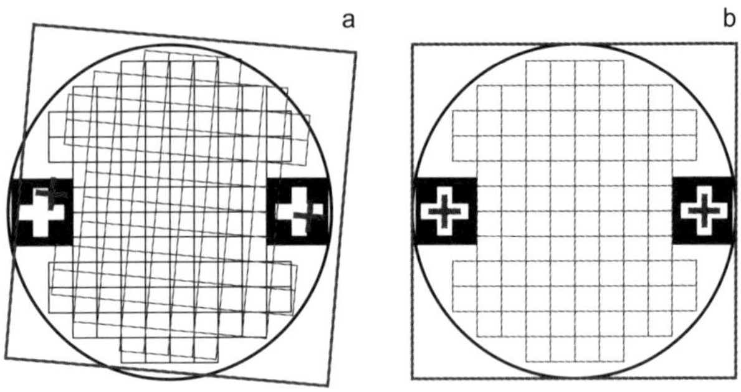

Figure 2.22: (a) When the two alignment marks on opposite sides of the sample are out of line with those on the mask, then the whole pattern is misaligned. (b) However, when the two alignment mark do overlap, the whole sample is correctly aligned to the mask.

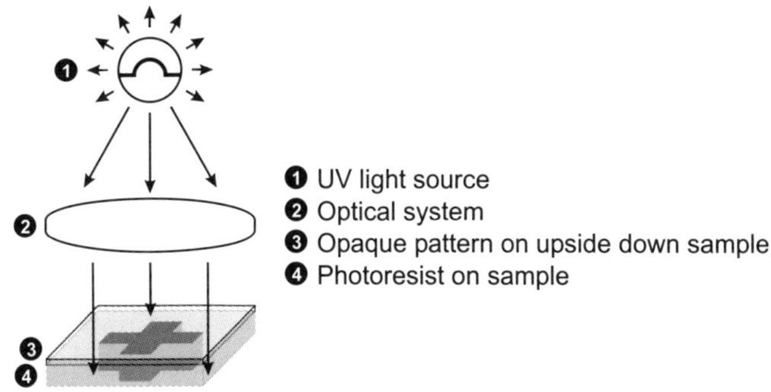

Figure 2.23: Schematic diagram of a self-alignment where an opaque pattern is produced from a previous process step on an optically transparent sample. The sample is coated with photoresist and turned upside down before being exposed to UV light. The opaque pattern shadows the photoresist producing almost perfect alignment.

tolerance of the fabrication apparatus to be used. In particular, photoresist swelling, development errors, resolution of the printing method used and the accuracy with which a particular instrument can align two patterns must all be considered in calculating the alignment tolerance of the process as a whole.

Alignment errors can be mitigated through the use of a self-alignment process. In this case, an opaque pattern from a previous processing step is used as the mask for a subsequent step, as in Figure 2.23. However, as this requires backside illumination of the sample, this method can only be used for structures on UV transparent substrates, such as quartz.

2.6.3. *Wet Etching*

All processes which result in the removal of solid material from a system by immersion in a liquid are collectively known as wet etches. Wet etches fall into two broad categories: isotropic etches and anisotropic etches.

Isotropic wet etches remove material at the same rate in all directions, and most etchants fall into this category. As a consquence, isotropic wet etches will tend to undercut a mask pattern to leave a larger cavity underneath. They are therefore a common means of producing suspended structures. For example, buffered hydrofluoric acid (bHF) acts as an isotropic etchant of silicon oxide, but does not etch poly-Si significantly. Hence, it

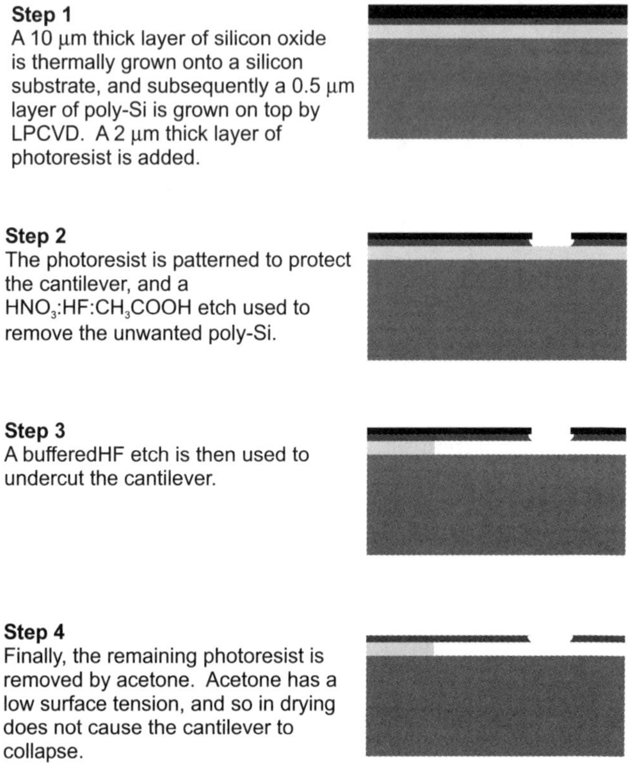

Figure 2.24: Process flow demonstrating the use of a wet etch to remove a sacrificial layer of material leaving behind a free-standing structure.

can be used to produce suspended poly-Si structures as demonstrated in Figure 2.24.

In such cases, it is important that the wet etchant removes all of the sacrificial underlayer before the top masking layer is itself etched significantly. This property is known as the selectivity of the etch, and determines the minimum thickness of the masking layer. For example, 50 μm of silicon oxide may have to be etched in bHF while a photoresist coating protects areas which must not be removed. If the etch rate in bHF of the silicon dioxide compared with the photoresist is in the ratio of 100:1, then the photoresist layer must be at least 0.5 μm thick to provide sufficient protection. Table 2.3 details suitable masking materials for particular wet etchants.

Table 2.3: Compatibility of etches for some common microsystems materials.

	Al	Cr	DLC	Photoresist	Si	Si_3N_4	SiO_2	Ti
Acetone	M	M	M	–	M	M	M	M
bHF	–	–	M	M	M	M	–	–
bHF/HNO_3	–	–	M	M	M	M	–	–
HNO_3	M	M	M	–	M	M	M	M
H_3PO_4	–	–	–	–	–	–	M	–
KOH	–	–	M	–	–	M	–	M
–		Material is etched by this etchant						
M		Material acts as a suitable hard mask for this etchant.						

Etchants which remove material at different rates depending on the crystal plane which is exposed are called *anisotropic*. For example, the (111) surface of crystalline silicon is particularly stable in aqueous potassium hydroxide relative to the other planes, and this allows the production of cavities with a well-defined shape as the etch will stop wherever a (111) surface is exposed. Therefore, if a masking pattern is produced on the surface of a Si(100) wafer with sides in the $\langle 110 \rangle$ crystal direction, then a groove will be etched in the silicon whose sides will be at an angle of 54.7° with respect to the plane of the surface (Figure 2.25a). Alternatively, if the sides of the masking pattern are aligned in the $\langle 100 \rangle$ direction, then only (100) surfaces will be exposed and the surface pattern will be undercut to produce a cavity, as shown in Figure 2.25(b). It should be noted that convex edges — even between two (111) planes — are always unstable, and this allows the production of fully released structures on silicon. Figure 2.26 shows an optical microscope image of a set of silicon nitride cantilevers on a silicon (100) wafer. Whilst the shortest cantilevers have been completely released, the longer cantilevers (to the left of the image) have only been partially etched, and the etching of the convex edges can be clearly observed.

2.6.4. *Dry Etching*

It has already been shown in the previous section that wet chemicals may be effectively used to remove unwanted material during microsystem fabrication, and that both isotropic and anisotropic etching is possible

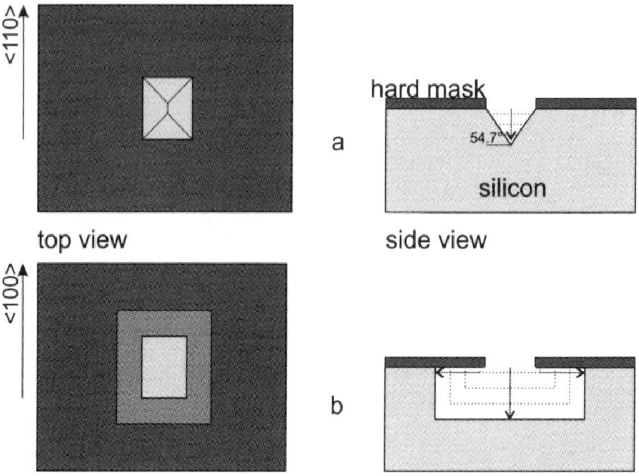

Figure 2.25: (a) A KOH etch will produce a grooved structure in silicon when the edges of the hard mask are aligned with the ⟨110⟩ direction. (b) An undercut cavity will result when the edges of the hard mask are aligned with the ⟨100⟩ direction.

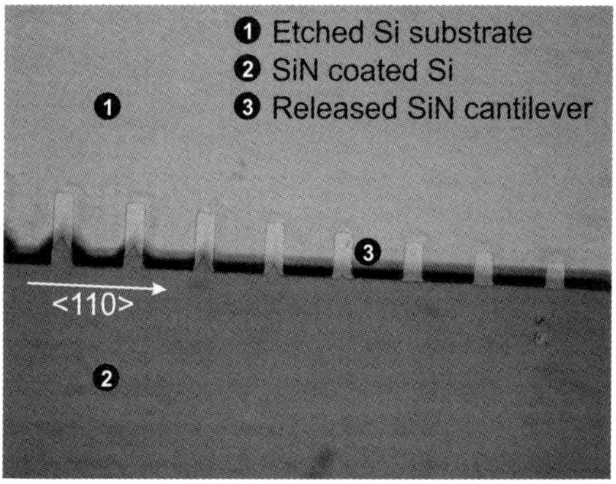

Figure 2.26: Optical micrograph showing silicon nitride cantilever beams on a silicon substrate. The short beams have been fully etched, and the etch has terminated on a (111) crystal plane. The long beams continue to be etched as convex edges which are always unstable, even between two (111) planes.

in crystalline materials. However, wet processing has several associated problems. Firstly, there is a finite supply of the wet chemical in any container, and so the concentration of etchant varies with time during processing, and constantly replenishing the etch solution through the use of a flowing system is very expensive and environmentally unfriendly. Therefore, the ability to predict the amount of material that will be removed is limited, unless an etch stop is employed. Secondly, it can be very difficult to remove all of the etch chemical from the substrate being processed at the end of the etch, and this can lead both to over-etching and long-term contamination. Finally, and perhaps most seriously, a substrate that has been wet etched must eventually be dried. During the drying process, the surface tension of the liquid will tend to cause free standing structures to collapse and stick to the underlying substrate. This effect is known as *stiction*. Therefore, the substrate must either be dried using a liquid with a low surface tension, such as acetone, or a technique such as freeze drying must be used.

In dry etching, a low pressure gas is used to chemically remove exposed, unwanted material from a substrate. The constant flow of gas over the substrate means that the etching rate does not vary as function of time due to the changing concentration of the etchant chemical. Furthermore, vacuum pumps can be used to efficiently remove the etchant gas together with the gas phase products of the etching reaction, leaving behind an uncontaminated substrate. In this way, there is a high level of control over the quantity of material removed, reducing the need for an etch stop. Residual gas analysis of the gas flow leaving the reaction vessel also allows investigation of the products of the reaction. Changes in the composition of the outflow gases may be used to determine when a particular layer of material has been completely removed. Alternatively, spectroscopic ellipsometry can be used to optically study the thickness of any surface layers that are being etched in real time, providing further control over the quantity of material being removed. Such techniques for determining when an etch is completed are known as *end point detection*. Finally, stiction is not an issue during the release of free standing structures. There is, therefore, a preference for using dry etching over wet etching whenever possible (although is should be noted that the costs associated with setting up and running a dry etching system are orders of magnitude greater than the cost of a wet etching system).

A few gases are sufficiently reactive that simply passing them over a solid material will result in a chemical reaction causing removal of the solid. For example, *vapour etching* of silicon is efficiently performed using xenon difluoride gas (XeF_2) [83, 84]. The etch is usually performed at room temperature in a vacuum system with a gas pressure if ~400 Pa and results in a net chemical reaction,

$$\mathrm{Si_{(s)}} + 2\mathrm{XeF_{2(g)}}, \quad \mathrm{SiF_{4(g)}} + 2\mathrm{Xe_{(g)}}, \tag{2.34}$$

that yields an isotropic etch of silicon.

Reactive ion etching (RIE) provides a wider range of etching possibilities by exciting an otherwise stable gas to produce chemically active species that will perform an etch. Normally, the excited gas is in the form of a low pressure plasma, and this is frequently generated by applying rf radiation at 13.56 MHz between two plates in a vacuum system. Rf-RIE therefore has similarities with rf-PECVD, and the system used for etching is essentially that shown in Figure 2.14 with the exception that, for etching, the top electrode is earthed and the rf signal is applied to the substrate electrode. Whereas vapour etching is an entirely chemical process, bombardment of the substrate by energetic ions in the plasma causes some material to be removed by sputtering, which is a physical process. The combination of chemical and physical etching leads to etch rates ~100 nm min^{-1}. Furthermore, at low pressures, ions will not suffer collisions in crossing the plasma sheath, and so an anisotropic etch will result, whereas at high pressures, collisions in the sheath cause ions arriving at the substrate to have a broad range of angles of incidence, leading to a more isotropic etch (see Figure 2.15 for the equivalent situation in rf-PECVD). As a result, it is possible to undercut structures using dry etching to leave free-standing components. However, for this to work, there needs to be a high degree of selectivity between the etch rate of the sacrificial material and that of the structural material.

Whilst rf-RIE is sufficient for performing surface micromachining, the etch rate is too slow to perform bulk etching of materials, and truly anisotropic etching is difficult to achieve. For this reason, deep reactive ion etching (DRIE) has been developed, allowing anisotropic etch rates into bulk silicon and silicon dioxide of ~10 μm min^{-1}. This is achieved using a cyclic process developed at Robert Bosch GmbH [85, 86].

In this technique, a high density plasma consisting of low energy ions is generated at low pressure by inductively coupling rf radiation into a vacuum chamber using a single turn coil electrode [87]. The plasma passes into a main etching chamber where the sample to be etched sits on a platen which is kept cool — typically below 50°C — to avoid plasma heating of the sample. Pulsed rf radiation may also be applied to the platen to change the energy of the incident ions. Initially, sulphur hexafluoride (SF_6) gas is introduced into the DRIE system for a few seconds. The negative fluorine ions in the plasma efficiently etch silicon, resulting in the removal of material. After between 10 and 20 seconds, the SF_6 gas flow is stopped and is replaced by a flow of C_4F_8 gas. The plasma decomposition of this gas leads to the deposition of a Teflon-like fluorocarbon material on the sample surface. When the C_4F_8 gas is once again replaced by SF_6, the greater degree of ion bombardment experienced by surfaces that are parallel to the sample surface means that this Teflon-like coating is selectively removed from the bottom of the area to be etched whilst sidewalls remain coated. Therefore, the SF_6 only etches in a vertical direction, as shown in Figure 2.27 [87]. In this way, very deep trenches may be produced that are only a few micrometres wide, but have aspect ratios up to 100:1.

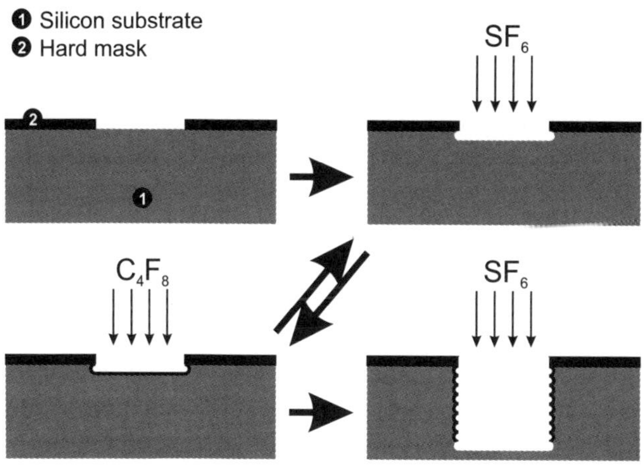

Figure 2.27: Flow diagram of the deep reactive ion etching process. A patterned substrate is exposed cyclically to an SF_6 plasma which anisotropically etches silicon and a C_4F_8 plasma which isotropically passivates exposed surfaces.

A consequence of the cyclic etching process is that sidewalls tend to gain a scalloped structure, with a roughness of a few hundred nanometres. This can be overcome using *cryogenic plasma etching* [86]. In this case, only SF_6 gas is used and the substrate is cooled down to cryogenic temperatures. At these low temperatures, gaseous species containing fluorine condense on the substrate. The ion bombardment of horizontal surfaces by the SF_6 plasma means that these condensates are efficiently removed, allowing etching to proceed in a vertical direction. However, vertical surfaces do not suffer significant ion bombardment, and so remain protected by the condensate. A highly anisotropic etch results, allowing the production of high aspect ratio structures without and scalloping of the sidewalls. Despite this advantage, DRIE tends to be favoured over cryogenic plasma etching due to the higher etching rate of the former, and the increased complexity of having to cryogenically cool the substrate.

In recent years, DRIE has become a mainstay of the microsystem processing industry, allowing the production of high aspect ratio structures in three dimensions with micrometer-scale accuracy that would have been impossible to fabricate using the traditional etch technologies borrowed from the microelectronics industry. DRIE allows entire silicon wafers to be etched through in under an hour, allowing the production of completely free structures, such as microbearings which have been effectively employed in devices such as the microturbine developed at MIT for high density power generation [18, 88, 89].

However, great care has to be taken to avoid contamination of the etch chamber by other materials, as this can have a profound effect upon the sensitive etching chemistry. Furthermore, careful process optimisation is required to ensure that features of different widths etch at the same rate — a phenomenon known as *lagging* caused by variations in the ratio of ion and neutral plasma species arriving at the sample surface and the local depletion of reactant precursors. This localised gas *loading* can also cause regions with a high etch surface area density to be etched more slowly than regions where little material is to be removed. These effects, together with localised variations in the charge density at exposed surfaces can result in the production of a wide variety of non-ideal trench shapes if etching conditions are not carefully controlled.

The production of many microsystems devices now relies on DRIE. The excellent selectivity of the process is due to the ratio of the etch rate of other materials relative to silicon which can be as high as 300:1 relative to some resists. As a consequence, DRIE has been very effectively used in conjunction with silicon-on-insulator (SOI) wafers, as the buried silicon dioxide layer has a very low etch rate and effectively acts as an etch stop. This allows the simple formation of suspended structures [90, 91] for both sensor and actuator applications. For example, out-of plane actuation of several micrometres can be achieved by coating a membrane in a deposited piezoelectric layer [92], whilst in-plane actuation can be achieved through a combination of back and front side etching of SOI wafers to form electrostatic comb actuators [93, 94]. Although originally developed for the etching of bulk silicon, DRIE etch chemistries are also now appearing for a variety of material systems, such as silicon dioxide [95, 96].

2.6.5. *Bonding*

Although developments in bulk micromachining have allowed the formation of deep structures in bulk materials, such as silicon, in many respects these structures cannot really be described as truly three-dimensional (3D). Instead, these structures are better described as being 2.5D, as they are simply vertical projections of a 2D pattern into a bulk material. Bonding refers to the process by which several substrates of material are patterned individually before being brought into intimate contact to create a complex structure that can properly be said to be 3D in nature. A good example of what can be achieved using this technique is the micro-engine developed at MIT. This is a micromachined, fuel burning turbine for power generation. It requires many components, such as a rotor, stator and fuel injection system, and is fabricated by processing six individual silicon wafers that are then brought together into intimate contact to make the final device, as shown in Figure 2.28 [18].

Perhaps the most common application of bonding is in the production of silicon-on-insulator (SOI) wafers that have a layer of silicon dioxide buried inside a silicon wafer [97]. SOI wafers are most commonly produced by oxidising the surface of a crystalline silicon wafer and bonding a clean silicon wafer on to top of the silicon dioxide layer produced.

Two silicon wafers may be directly joined together in a process known as *direct bonding* (sometimes also called *fusion bonding*) [98]. In this method, the surfaces of the two wafers to be joined are thoroughly cleaned to

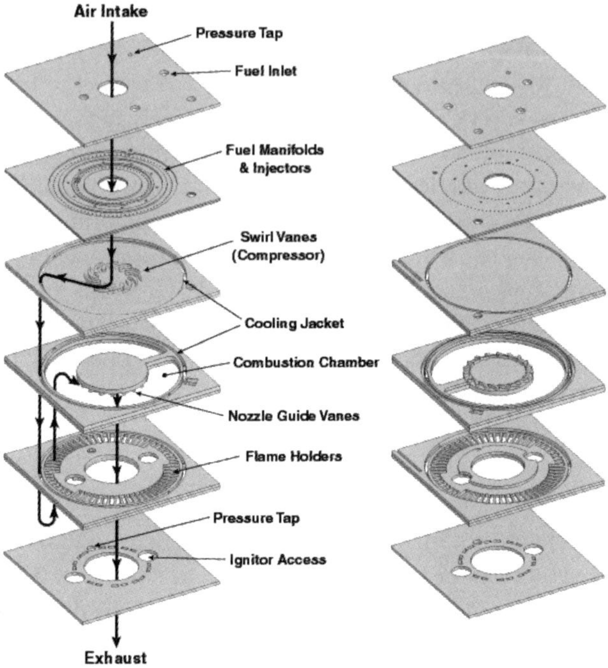

Figure 2.28: Six wafers are individually micromachined before being bonded together to form a microengine developed at MIT by Mehra *et al.* [18].

remove any surface particulates that would inhibit the bonding process, and are hydrophilised to ensure that the surfaces are terminated in a layer of chemisorbed water molecules. The two surfaces are then brought into intimate contact, and pressure applied at a point to locally reduce the distance between the two surfaces. Hydrogen bridge bonds form at this contact point, resulting in a net energy gain through the elimination of the two wafer surfaces. This energy gain then causes the contact point to spread over the surface of the two wafers in a 'zip up' process, resulting in the two wafers being in intimate contact over their entire surface. Finally, the wafers are heated to drive out the hydrogen and excess oxygen from the system, resulting in the formation of strong Si–O–Si bonds. The strongest bonding requires temperature of $\sim$1000°C, but other limitations (such as the presence of dopants or metals) often means that lower temperatures have to be used. Temperatures as low as $\sim$100°C can be usefully employed, but there is a significant reduction in the strength of the resulting bond [99]. A further limitation on the strength of the resulting bond is the presence of patterns on the wafer itself. For the 'zip up' process to be effective, there must be a continuous silicon-silicon surface to join. However, the presence of microstructures on the two wafers means that this is rarely the case, and this can result in the formation of microvoids at the bonding interface [100]. Such microvoids can lead to subsequent device failure. Thought must be given during device design to ensure that the likelihood of microvoid formation is minimised, however, infrared transmission imaging of the bonded silicon wafers can be used to detect if such microvoids are present during the fabrication process [101].

Whilst direct bonding provides a method for joining two wafers together, *anodic bonding* (sometimes called *Mallory bonding*) provides a low temperature alternative for joining a glass substrate such as Corning 7740, which has a similar thermal expansion coefficient to silicon (or a silicon wafer that has been coated in a spin-on or sputter deposited glass [102]) to a silicon wafer or a metal substrate [103]. In this process, the glass (which must contain sodium ions) and the wafer surfaces are brought into intimate contact. The system is put under vacuum and heated to between 300°C and 400°C. A negative bias of several hundred volts is then applied to the glass relative to the silicon wafer [104]. The electric field causes sodium ions in the glass to drift to the silicon wafer, resulting in a negative space charge region in the glass. This enhances the local electric field, and the resulting strong electrostatic force between the glass and the wafer produces a bond after a few tens of minutes [100]. It should be noted that the resulting bond cannot subsequently withstand temperatures approaching the bonding temperature. Furthermore, sodium ions are highly mobile [105] and can easily diffuse through silicon where they act as a strong p-type dopant. Therefore, this process can have a serious impact upon the electronic properties of devices on the silicon wafer. Despite these constraints, anodic bonding has found particular relevance in microsystems device packaging, allowing capping layers of glass to be formed on either side of a device. Deep reactive ion etching allows the production of sampling channels in these glass layers. This is put to good effect in microfluidic [106] and gas sensing systems [107].

In a few situations, the intimate bonding of two layers is not possible, and in these cases *intermediate layer bonding* is required in which another thin film of material is introduced between the two layers to be bonded, such as a soft metal, glass or polymer [100].

2.6.6. *Planarisation*

A frequent difficulty with bulk micromachining is that rough surfaces are frequently created. The surface roughness of a deposited material tends to increase with deposition thickness, and so films of material several micrometres thick frequently possess high surface roughness [108–110]. Subsequent etching can then produce surface features that are tens of micrometres deep. This presents difficulties for subsequent photolithography, and layers of photoresist on such structures will vary considerably in depth as a function of position, changing both the exposure time and proximity gap required in a mask aligner as well as the developing time. The deposition of structural materials on topographically complex surfaces will also be unpredictable. Therefore, a variety of techniques have been developed to allow the planarisation of rough surfaces to produce a smooth surface for subsequent processing:

Chemical mechanical polishing (CMP):
This is the most powerful planarisation technique. Planarisation is achieved by pressing a rotating platen against the surface to be smoothed. A slurry containing both a chemical etchant and an abrasive is introduced between the sample and the rotating platen, resulting in the removal of surface asperities and the production of a smooth surface. Care must be taken, however, if different materials are exposed to the slurry at any one time, as each material will tend to have a different etch rate. This can result in localised variations in etch rate with pattern density and material, and dummy structures are often employed to ensure a uniform etch rate over the entire wafer surface. CMP also has the disadvantage of being an expensive technology as large quantities of slurry have to continuously flow into the system to ensure constant and uniform etch results.

Resist etchback:
An alternative subtractive technique for planarising a rough surface is to coat the sample in a polymer layer. When in solution, surface tension will cause the polymer to flow over the sample until a flat surface is produced, which will be maintained when the polymer is dried and cured. The sample

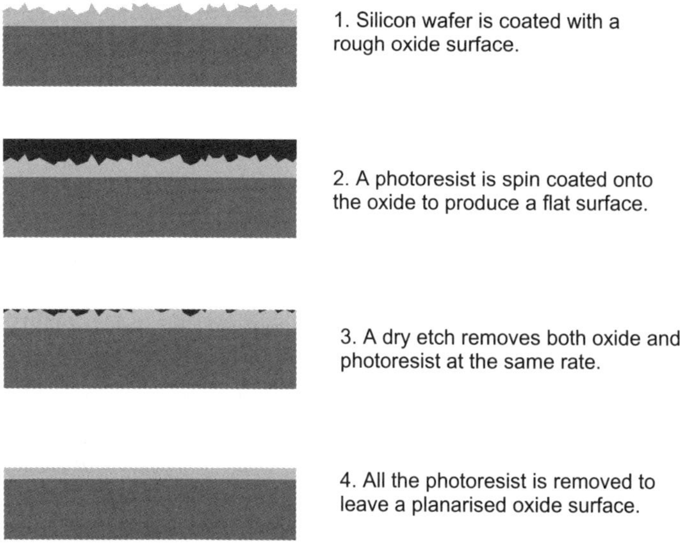

Figure 2.29: Step by step process flow of the resist etchback planarisation technique.

may then be exposed to a dry etch. If the polymer and the underlying rough material have similar etch rates, then a planarised surface will be produced, as shown in Figure 2.29.

Polymer planarisation:
If a flat surface only is required, rather than the smoothing of a particular material surface, then polymer planarisation provides an additive means of achieving this. In a similar fashion to the resist etchback technique described previously, a polymer in solution is allowed to flow over the surface of a rough sample. Surface tension will tend to cause the polymer to produce a flat surface, which will be largely retained upon drying and curing, and this may be used as a structural platform for further processing. The application of several polymer layers can be used to further reduce the surface roughness.

2.6.7. *Yield and Process Flow*

The practical world of device fabrication is not ideal. Inaccuracies can occur in the alignment, exposure and development of photoresists, wet chemical etches can vary with concentration as a function of time, samples can become contaminated with particulates, physical effects such as stiction and human error can all lead to failure of the finished device. The percentage of devices that have been successfully fabricated of the total number processed is known as the *yield*. In the introduction to this chapter, microfabrication was presented as being an expensive technology, both due to the installation and running costs of the equipment required and the cost of the time spent by the skilled technician undertaking the work. Therefore, a low device yield can lead to commercial disaster in the industrial environment. However, even in a research and development setting, a low yield can hold back work and prove costly. Care must obviously be taken during device fabrication to ensure that all process steps are carefully carried out and that the environment of the Clean Room in which the work is being performed is maintained. Furthermore, good design of the fabrication *process flow* can also help to significantly improve yield.

The process flow for fabrication of a device is a detailed, step-by-step specification of everything that has to be done to a substrate from taking it out of its box from the supplier to handing over a finished, packaged device. By way of an example, Table 2.4 shows a process flow for the simple, single mask fabrication of an array of silicon nitride cantilevers of a silicon wafer. Each step in the process flow is clearly identified by a line in the process flow, and all processing conditions, such as wet etch concentration, substrate temperature and processing times, are included explicitly. A schematic diagram of the masks used and any plan and cross section views of the devices at critical points in the process that would assist the fabrication technician to understand the process usually accompany the process flow.

Construction of the process flow requires careful thought, and several software packages now exist to assist the engineer in this process. In particular, attention must be paid to the effect that any one step will have on the process as a whole. On one scale, it should be remembered that most processes, and in particular wet etching, affect both sides of a substrate. Therefore, passivation of the backside of a substrate is often necessary to avoid undesired thinning of the substrate material. Etches will attack all exposed materials, and will frequently undercut thin, exposed layers of material at their edges. Several helpful studies have been carried out which comprehensively review the effect of different common etch

Table 2.4: Process flow for the production of silicon nitride cantilevers on a silicon substrate.

Step no.	Step Code	Descritpion
1	WAF1	Start: Silicon (100) substrate [100 mm diameter, n-type]
2	RCA1	RCA Clean 1
3	RCA2	RCA Clean 2
4	NIT1	LPCVD of 250 nm thick silicon nitride [800°C, 70 Pa, 5 sccm SiH_2Cl_2, 30 sccm NH_3, 50 min]
5	PHO1	Spin on AZ5214E photoresist [4000 rpm, 30 s]
6	BAK1	Photoresist pre-bake on hotplate [100°C, 60 s]
7	EXP1	Exposure through mask [15 μm alignment gap, 10 μm print gap, 5 s exposure]
8	DEV1	Develop in 1:1 solution of AZ developer with DI water for 60 s
9	BAK2	Photoresist post-bake on hotplate [120°C, 120 s]
10	RIE1	Reactive ion etch of silicon nitride [100 W rf power, 150 mTorr, 40 sccm CF_4, 5 sccm O_2, 150 s]
11	KOH1	KOH etch [25% KOH in DI water, 85°C, 1.5 hrs]
12	DIW1	Finish: DI water rinse

chemistries upon a wide range of materials to allow compatibility issues to be checked [94, 111].

The substrate temperature of any processing step must also be carefully reviewed, as all materials present will be heated to this degree, and polymers (in particular photoresist) can rarely withstand much over 100°C. Glasses deform under their own weight at their softening point, which is frequently well below their melting temperature. Thin metal layers, on the other hand, tend to agglomerate into small islands of material due to surface tension effects at temperatures as low as ~500°C. Diffusion is a thermally activated process and increases exponentially with temperature, resulting in impurity migration between layers at elevated temperatures.

A particular process flow consideration for microsystems devices is that the device is structurally viable at all stages during the process, and will not structurally fail during processing because, for example, a structural support is added later on, or a structure, such as a bearing, is fully released before a capturing structure is added. It should also be noted that most processes will have some spatial variation across a substrate. For example, in reactive ion etching the presence of a substrate will affect both the flow of gas and the distribution of the plasma, which can lead to the centre of the substrate being etched at a different rate to the edges. Finally, the packaging of the device must be carefully designed. Many microsystems devices fail because the packaging process is not compatible with the device itself or does not effectively protect the device from its operation environment.

The materials employed in the fabrication of a particular device must be carefully considered with the process flow in mind, and this will form part of the material selection process described in Section 2.5.7. It is frequently the case that the 'best' material to use based on the quantitative material selection cannot be used because of processing difficulties, and a compromise is usually necessary.

Yield and process flow are therefore inextricably linked, and the process engineer is advised to identify those steps in the process that are most likely to lead to a reduction in yield. Once identified, measures can be taken to mitigate the effect that these steps will have on the total yield. If possible, steps that are likely to reduce overall yield should be included near the start of the process before significant time and effort has been expended rather than at the end of the process once a large investment has been made in the device. The inclusion of test structures should be considered to allow individual process steps to be critically assessed to ensure that they have been completed successfully. For example, a simple cantilever or beam structure at the edge of a substrate can be used to assess if stiction effects have had an impact after a wet etching step. Changing the process to improve yield is also a possibility. For example, wet etches are notoriously difficult to control, and precise etching of a particular thickness of material is difficult. The inclusion of an etch stop layer — a layer of material that is not affected by the wet etch at the required depth to terminate an etch — can improve the yield of that step. However, there is a possibility that adding steps to create the etch stop layer will have a more detrimental affect on the overall yield of the process than the problem that it was trying to cure, and a quantitative assessment is necessary. If a step with a low yield is unavoidable well into the process flow, then it is sensible to consider a contingency plan that would allow the affected part of the structure to be removed and replace so that the earlier processing is not lost.

Good design of the whole process flow, taking serious consideration of each step in the process can therefore have a profound impact upon the

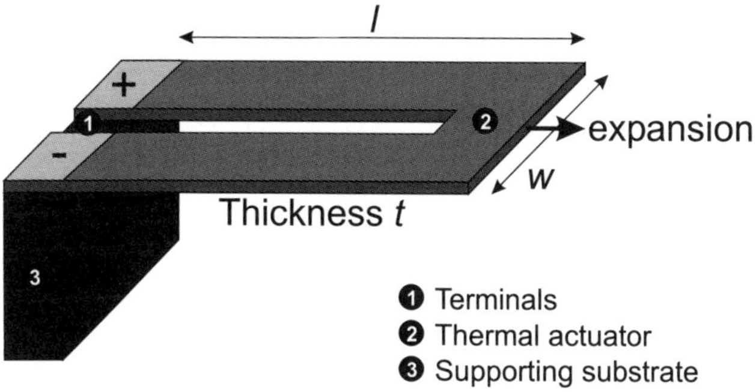

Figure 2.30: Geometry of a simple thermal actuator.

manufacturability of the final device and therefore the success of a given project.

2.7. Conclusions

It has been shown in this chapter that the microfabrication technologies born out of the microelectronics industry have provided a means of producing mechanical structures with dimensions and tolerances that are far smaller than those offered by traditional machining techniques. This has consequently permitted the development of a range of MEMS devices that are now widely employed in everyday life. However, as MEMS manufacturing has matured, a broad spectrum of materials that lie within the CMOS materials set have been incorporated into devices. This has required the development of a variety of tools for producing these materials on substrates and subsequently patterning them into the three dimensional structures desired. This chapter has reviewed both the principle technologies for the growth of these materials and the methodology for selecting the appropriate material for a given application before discussing micromachining techniques for producing structures and the integration of these principles into a process flow.

The cost of microfabrication is high, and therefore it is essential that the development of the microfabrication process associated with producing a new MEMS device is as straightforward as possible. This is possible through careful design and consideration of the issues discussed in this Chapter.

Fortunately, there is a wealth of public-domain information and computer aided design software to assist the Process Engineer further in this task, and time well spent in the design phase of a new project will greatly enhance the chances of final success!

References

1. Grundy, P. J. (1998) Thin film magnetic recording media, *Journal of Physics D — Applied Physics* **31**, 2975–2990.
2. Binning, G., Rohrer, M., Gerber, C. and Weibel, E. (1982) Surface studies by scanning tunneling microscopy, *Phys. Rev. Lett.* **49**, 57–61.
3. Volland, B. E., Heerlein, H. and Rangelow, I. W. (2002) Electrostatically driven microgripper, *Microelectronic Engineering* **61**(2), 1015–1023.
4. Epstein, A. and Senturia, S. (1997) Macro power from micro machinery, *Science* **276**, 1211–1211.
5. *International Technology Roadmap for Semiconductors* (2003).
6. Ashby, M. (1992) *Materials Selection in Mechanical Design* (Pergamon, Oxford).
7. Van Kessel, P. F., Hornbeck, L. J., Meier, R. E. and Douglass, M. R. (1998) MEMS-based projection display, *Proceedings of the IEEE* **86**, 1687–1704.
8. Ishimori, M., Song, J., Sasaki, M. and Hane, K. (2003) Si-wafer bending technique for a three-dimensional microoptical bench, *Japanese Journal of Applied Physics Part 1 — Regular Papers Short Notes & Review Papers* **42**, 4063–4066.
9. Kolesar, E. S. *et al.* (2001) Three-dimensional structures assembled from polysilicon surface micromachined components containing continuous hinges and microrivets, *Thin Solid Films* **398**, 566–571.
10. Syms, R. R. A., Yeatman, E. M., Bright, V. M. and Whitesides, G. M. (2003) Surface tension-powered self-assembly of micro structures — The state-of-the-art, *Journal of Microelectromechanical Systems* **12**, 387–417.
11. Sniegowski, J. J. and de Boer, M. P. (2000) IC-compatible polysilicon surface micromachining, *Annual Review of Materials Science* **30**, 299–333.
12. Lin, L. Y., Lee, S. S., Pister, K. S. J. and Wu, M. C. (1994) 3-dimensional micro-fresnel optical-elements fabricated by micromachining technique, *Electronics Letters* **30**, 448–449.
13. Lu, T. J., Moore, D. F. and Chia, M. H. (2002) Mechanics of micromechanical clips for optical fibers, *Journal of Micromechanics and Microengineering* **12**, 168–176.
14. Mankame, N. D. and Ananthasuresh, G. K. (2004) A novel compliant mechanism for converting reciprocating translation into enclosing curved paths, *Journal of Mechanical Design* **126**, 667–672.
15. Chua, H. C. (2005) In *Engineering*, Cambridge University, Cambridge.

16. Denhoff, M. (2003) A measurement of Young's modulus and residual stress in MEMS bridges using a surface profilometer, *J. Micromech. Microeng.* **13**, 686–692.
17. Chen, S., Baughn, T., Yao, Z. and Goldsmith, C. (2002) A new *in situ* residual stress measurement method for a MEMS thin fixed-fixed beam structure, *J. Microelectromech. Syst.* **11**, 309–316.
18. Mehra, A. *et al.* (2000) A six-wafer combustion system for a silicon micro gas turbine engine, *Journal of Microelectromechanical Systems* **9**, 517–527.
19. Livermore, C. *et al.* (2004) A high-power MEMS electric induction motor, *Journal of Microelectromechanical Systems* **13**, 465–471.
20. Hara, M., Tanaka, S. and Esashi, M. (2003) Rotational infrared polarization modulator using a MEMS-based air AMP turbine with different types of journal bearing, *Journal of Micromechanics and Microengineering* **13**, 223–228.
21. Deng, K. R. and Mehregany, M. (1998) Outer-rotor polysilicon wobble micromotors, *Sensors and Actuators A — Physical* **64**, 265–271.
22. Williams, J. A. (2001) Friction and wear of rotating pivots in MEMS and other small scale devices, *Wear* **250**, 965–972.
23. Bourlon, B., Glattli, D. C., Miko, C., Forro, L. and Bachtold, A. (2004) Carbon nanotube based bearing for rotational motions, *Nano Letters* **4**, 709–712.
24. Bang, Y. I. *et al.* (2004) Thin film micro carbon dioxide sensor using MEMS process, *Sensors and Actuators B-Chemical* **102**, 20–26.
25. Rasmussen, A., Mavriplis, C., Zaghloul, M. E., Mikulchenko, O. and Mayaram, K. (2001) Simulation and optimization of a microfluidic flow sensor, *Sensors and Actuators A-Physical* **88**, 121–132.
26. Richards Grayson, A. C. *et al.* (2004) A BioMEMS review: MEMS technology for physiologically integrated devices, *Proceedings of the IEEE* **92**, 6–21.
27. Noda, T. *et al.* (2004) Characteristics of high-resolution hemoglobin measurement microchip integrated with signal processing circuit, *Japanese Journal of Applied Physics Part 1 — Regular Papers Short Notes & Review Papers* **43**, 2392–2396.
28. Wang, W. J., Guo, D. G., Lin, R. M. and Wang, X. W. (2004) A single-chip diaphragm-type miniature Fabry-Perot pressure sensor with improved cross-sensitivity to temperature, *Measurement Science & Technology* **15**, 905–910.
29. McKendry, R. *et al.* (2002) Multiple label-free biodetection and quantitative DNA-binding assays on a nanomechanical cantilever array, *Proceecings of the National Academy of Sciences of the United States of America* **99**, 9783–9788.
30. Seshia, A. A. *et al.* (2002) A vacuum packaged surface micromachined resonant accelerometer, *Journal of Microelectromechanical Systems* **11**, 784–793.

31. Ayazi, F. and Najafi, K. (2001) A HARPSS polysilicon vibrating ring gyroscope, *Journal of Microelectromechanical Systems* **10**, 169–179.
32. Gizeli, E. (2000) Study of the sensitivity of the acoustic waveguide sensor, *Analytical Chemistry* **72**, 5967–5972.
33. Zerov, V. Y. and Malyarov, V. G. (2001) Heat-sensitive materials for uncooled microbolometer arrays, *Journal of Optical Technology* **68**, 939–948.
34. Lee, D. W., Ono, T. and Esashi, M. (2002) Fabrication of thermal microprobes with a sub-100 nm metal-to-metal junction, *Nanotechnology* **13**, 29–32.
35. Suzuki, Y. (1996) Novel microcantilever for scanning thermal imaging microscopy, *Japanese Journal of Applied Physics Part 2 — Letters* **35**, L352–L354.
36. Li, M. H. and Gianchandani, Y. B. (2000) Microcalorimetry applications of a surface micromachined bolometer-type thermal probe, *Journal of Vacuum Science & Technology B* **18**, 3600–3603.
37. Li, M. H. and Gianchandani, Y. B. (2003) Applications of a low contact force polyimide shank bolometer probe for chemical and biological diagnostics, *Sensors and Actuators A — Physical* **104**, 236–245.
38. Huber, J. E., Fleck, N. A. and Ashby, M. F. (1997) The selection of mechanical actuators based on performance indices, *Proceedings of the Royal Society of London Series A — Mathematical Physical and Engineering Sciences* **453**, 2185–2205.
39. Dai, C.-L. (2003) *In situ* electrostatic microactuators for measuring the Young's modulus of CMOS thin films, *J. Micromech. Microeng.* **13**, 563–567.
40. Gaspar, J., Chu, V. and Conde, J. (2003) Electrostatic actuation of thin-film microelectromechanical structures, *J. Appl. Phys.* **93**, 10018–10029.
41. Srikar, V. T. and Spearing, S. M. (2003) Materials selection for microfabricated electrostatic actuators, *Sensors and Actuators A — Physical* **102**, 279–285.
42. Knapp, J. and de Boer, M. (2002) Mechanics of microcantilever beams subject to combined electrostatic and adhesive forces, *J. Microelectromech. Syst.* **11**, 754–764.
43. Grade, J. and Jerman, H. A. K. (2003) Design of large deflection electrostatic actuators, *J. Microelectromech. Syst.* **12**, 335–343.
44. Roch, I., Bidaud, P., Collard, D. and Buchaillot, L. (2003) Fabrication and characterization of an SU-8 gripper actuated by a shape memory alloy thin film, *Journal of Micromechanics and Microengineering* **13**, 330–336.
45. Luo, J. K., Flewitt, A. J., Spearing, S. M., Fleck, N. A. and Milne, W. I. (2004) Normally closed microgrippers using a highly stressed diamond-like carbon and Ni bimorph structure, *Appl. Phys. Lett.* **85**, 5748–5750.
46. Binnig, G. and Rohrer, H. (1982) Scanning tunneling microscopy, *Helvetica Physica Acta* **55**, 726–735.

47. Verway, J., Amerasekera, E. A. and Bisschop, J. (1990) The physics of SiO2 layers, *Rep. Prog. Phys.* **53**, 1297–1331.
48. Deal, B. E. and Grove, A. S. (1965) General relationship for the thermal oxidation of silicon, *Journal of Applied Physics* **36**, 3770–3778.
49. Duffy, M. *et al.* (1983) LPCVD polycrystalline silicon: growth and physical properties of diffusion-doped, ion-implanted, and undoped films, *RCA Review* **44**, 313–325.
50. Harbeke, G., Krausbauer, L., Steigmeier, E., Widmer, A., Kappert, H. F. and Neugebauer, G. (1983) LPCVD polycrystalline silicon: growth and physical properties of *in situ* phosphorus doped and undoped films, *RCA Review* **44**, 287–312.
51. Harbeke, G., Krausbauer, L., Steigmeier, E., Widmer, A., Kappert, H. F. and Neugebauer, G. (1984) Growth and physical properties of LPCVD polycrystalline silicon films, *J. Electrochem. Soc.* **131**, 675–682.
52. Brotherton, S., Ayres, J. and Young, N. (1991) Characterisation of low temperature poly-Si thin film transistors, *Solid-State Electronics* **34**, 671–679.
53. Olson, J. M. (2002) Analysis of LPCVD process conditions for the deposition of low stress silicon nitride. Part I: preliminary LPCVD experiments, *Materials Science in Semiconductor Processing* **5**, 51–60.
54. French, P. J., Sarro, P. M., Mallee, R., Fakkeldij, E. J. M. and Wolffenbuttel, R. F. (1997) Optimization of a low-stress silicon nitride process for surface-micromachining applications, *Sensors and Actuators A — Physical* **58**, 149–157.
55. Heintze, M., Zedlitz, R., Wanka, H. and Schubert, M. (1996) Amorphous and microcrystalline silicon by hot wire chemical vapor deposition, *J. Appl. Phys.* **79**, 2699–2706.
56. Matsumura, H., Umemoto, H. and Masuda, A. (2004) Cat-CVD (hot-wire CVD): how different from PECVD in preparing amorphous silicon, *Journal of Non-Crystalline Solids* **338–340**, 19–26.
57. Matsumura, H. (1998) Formation of silicon-based thin films prepared by catalytic chemical vapor deposition (Cat-CVD) method, *Japanese Journal of Applied Physics, Part 1 — Regular Papers Short Notes & Review Papers* **37**, 3175–3187.
58. Chen, F. F. (1984) *Introduction to Plasma Physics and Controlled Fusion*, (Plenum, New York).
59. Kohler, K., Coburn, J. W., Horne, D. E., Kay, E. and Keller, J. H. (1985) Plasma potentials of 13.56 MHz RF argon glow discharges in a planar system, *J. Appl. Phys.* **57**, 59–66.
60. Perrin, J. (1995) *Plasma Deposition of Amorphous Silicon-Based Materials*, (eds.) Bruno, G., Capezzuto, P. and Madan, A. 177–241 (Academic Press, London).
61. Perrin, J. (1991) Plasma and surface reactions during a-Si:H films growth, *J. Non-Cryst. Solids* **137** and **138**, 639–644.

62. Flewitt, A. J., Robertson, J. and Milne, W. I. (1999) Growth mechanism of hydrogenated amorphous silicon studied by *in situ* scanning tunneling microscopy, *Journal of Applied Physics* **85**, 8032–8039.
63. Hess, D. W. (1984) Plasma-enhanced CVD: oxides, nitrides, transition metals, and transition metal silicides, *Journal of Vacuum Science & Technology A: Vacuum, Surfaces, and Films* **2**, 244–252.
64. Sommer, T. and Kushner, M. (1992) Numerical investigation of the kinetics and chemistry of rf glow discharge plasmas sustained in He, N_2, O_2, $He/N_2/O_2$, $He/CF_4/O_2$ and SiH_4/NH_3 using Monte Carlo-fluid hybrid model, *J. Appl. Phys.* **74**, 1654–1673.
65. Kushner, M. (1999) Plasma chemistry of $He/O_2/SiH_4$ and $He/N_2O/SiH_4$ mixtures for remote plasma-activated chemical-vapor deposition of silicon dioxide, *J. Appl. Phys.* **74**, 6538–6553.
66. Kushner, M. (1992) Simulation of the gas-phase processes in remote-plasma-activated chemical-vapor deposition of silicon dielectrics using rare gas-silane-ammonia mixures, *J. Appl. Phys.* **71**, 4173–4189.
67. Kushner, M. (1988) A model for the discharge kinetics and plasma chemistry during plasma enhanced chemical vapor deposition of amorphous silicon, *J. Appl. Phys.* **63**, 2532–2551.
68. Hamers, E., Bezemer, J. and van der Weg, W. (1999) Positive ions as growth precursors in plasma enhanced chemical vapor deposition of hydrogenated amorphous silicon, *Appl. Phys. Lett.* **75**, 609–611.
69. Hamers, E., Fontcuberta I. M. A., Niikura, C., Brenot, R. and Roca I. C. P. (2000) Contribution of ions to the growth of amorphous, polymorphous, and microcrystalline silicon thin films, *J. Appl. Phys.* **88**, 3674–3688.
70. Celii, F. G. and Butler, J. E. (1991) Diamond chemical vapor deposition, *Annual Review of Physical Chemistry* **42**, 643–684.
71. Flewitt, A., Dyson, A., Robertson, J. and Milne, W. (2000) Hydrogenated amorphous silicon and silicon nitride deposited at less than 100°C by ECR-PECVD for thin film transistors, *Mat. Res. Soc. Symp. Proc.* **609**, A28.2.1.
72. Flewitt, A., Dyson, A., Robertson, J. and Milne, W. (2000) Low temperature growth of silicon nitride by electron cyclotron resonance plasma enhanced chemical vapour deposition, *Thin Solid Films* **383**, 172–177.
73. Rashid, R., Flewitt, A. J. and Robertson, J. (2003) Physical and electrical properties of low temperature (<100°C) SiO_2 films deposited by electron cyclotron resonance plasmas, *Journal of Vacuum Science & Technology A* **21**, 728–739.
74. van de Samden, M., Severens, R., Kessels, W., Meulenbroeks, R. and Schram, D. (1998) Plasma chemistry aspects of a-Si:H deposition using an expanding thermal plasma, *J. Appl. Phys.* **84**, 2426–2435.
75. Shinohara, S. and Tanikawa, T. (2004) Development of very large helicon plasma source, *Review of Scientific Instruments* **75**, 1941–1946.

76. Weiler, M., Lang, K., Li, E. and Robertson, J. (1998) Deposition of tetrahedral hydrogenated amorphous carbon using a novel electron cyclotron wave resonance reactor, *Appl. Phys. Lett.* **72**, 1314–1316.
77. Sigmund, P. (1969) Theory of sputtering. I. Sputtering yield of amorphous and polycrystalline targets, *Phys. Rev.* **184**, 383–416.
78. Gautier, C., Moulard, G., Chatelon, J. and Motyl, G. (2001) Influence of substrate bias voltage on the *in situ* stress measured by an improved optical cantilever techniques of sputtered chromium films, *Thin Solid Films* **384**, 102–108.
79. Westliner, J. *et al.* (2002) Simulation and dielectric characterization of reactive DC magnetron cosputtered $(Ta_2O_5)_{1-x}(TiO_2)_x$ thin films, *J. Vac. Sci. Technol. B* **20**, 855–861.
80. Nicolas, S. *et al.* (1998) Fabrication of a gray-tone mask and pattern transfer in thick photoresists, *Journal of Micromechanics and Microengineering* **8**, 95–98.
81. Yao, J. *et al.* (2001) Refractive micro lens array made of dichromate gelatin with gray-tone photolithography, *Microelectronic Engineering* **57–58**, 729–735.
82. Coutrot, A. L. *et al.* (2002) Copper micromoulding process for NMR microinductors realization, *Sensors and Actuators A — Physical* **99**, 49–54.
83. Vugts, M. J. M., Verschueren, G. L. J., Eurlings, M. F. A., Hermans, L. J. F. and Beijerinck, H. C. W. (1996) Si/XeF_2 etching: temperature dependence, *Journal of Vacuum Science & Technology A — Vacuum Surfaces and Films* **14**, 2766–2774.
84. Bahreyni, B. and Shafai, C. (2002) Investigation and simulation of XeF_2 isotropic etching of silicon, *Journal of Vacuum Science & Technology A — Vacuum Surfaces and Films* **20**, 1850–1854.
85. Douglas, M. A. (1987) **9** (Texas Instruments Inc., USA).
86. Douglas, M. A. (1988) **13** (Texas Instruments Inc., USA).
87. Rangelow, I. W. (2003) Critical tasks in high aspect ratio silicon dry etching for microelectromechanical systems, *Journal of Vacuum Science & Technology A* **21**, 1550–1562.
88. Ayon, A. A., Braff, R., Lin, C. C., Sawin, H. H. and Schmidt, M. A. (1999) Characterization of a time multiplexed inductively coupled plasma etcher, *Journal of the Electrochemical Society* **146**, 339–349.
89. Khanna, R. (2003) MEMS fabrication perspectives from the MIT Microengine Project, *Surface & Coatings Technology* **163**, 273–280.
90. Docker, P. T., Kinnell, P. K. and Ward, M. C. L. (2004) Development of the one-step DRIE dry process for unconstrained fabrication of released MEMS devices, *Journal of Micromechanics and Microengineering* **14**, 941–944.
91. Milanovic, V. (2004) Multilevel beam SOI-MEMS fabrication and applications, *Journal of Microelectromechanical Systems* **13**, 19–30.

92. Zinck, C., Pinceau, D., Defay, E., Delevoye, E. and Barbier, D. B. (2004) Development and characterization of membranes actuated by a PZT thin film for MEMS applications, *Sensors and Actuators A — Physical* **115**, 483–489.
93. Hah, D., Choi, C. A., Kim, C. K. and Jun, C. H. (2004) A self-aligned vertical comb-drive actuator on an SOI wafer for a 2D scanning micromirror, *Journal of Micromechanics and Microengineering* **14**, 1148–1156.
94. Lee, D., Krishnamoorthy, U., Yu, K. and Solgaard, O. (2004) Single-crystalline silicon micromirrors actuated by self-aligned vertical electrostatic combdrives with piston-motion and rotation capability, *Sensors and Actuators A — Physical* **114**, 423–428.
95. Williams, K. R., Gupta, K. and Wasilik, M. (2003) Etch rates for micromachining processing — Part II, *Journal of Microelectromechanical Systems* **12**, 761–778.
96. Donohue, L. A., Hopkins, J., Barnett, R., Newton, A. and Barker, A. (2004) *Micromachining Technology for Micro-Optics and Nano-Optics*, pp. 44–53.
97. Celler, G. K. and Cristoloveanu, S. (2003) Frontiers of silicon-on-insulator, *Journal of Applied Physics* **93**, 4955–4978.
98. Turner, K. T., Thouless, M. D. and Spearing, S. M. (2004) Mechanics of wafer bonding: effect of clamping, *Journal of Applied Physics* **95**, 349–355.
99. Cho, Y. and Cheung, N. W. (2003) Low temperature Si layer transfer by direct bonding and mechanical ion cut, *Applied Physics Letters* **83**, 3827–3829.
100. Turner, K. T. and Spearing, S. M. (2002) Modeling of direct wafer bonding: effect of wafer bow and etch patterns, *Journal of Applied Physics* **92**, 7658–7666.
101. Schmidt, M. A. (1998) Wafer-to-wafer bonding for microstructure formation, *Proceedings of the IEEE* **86**, 1575–1585.
102. Chung, G. S. and Kim, J. M. (2004) Anodic bonding characteristics of MLCA/Si-wafer using a sputtered Pyrex #7740 glass layer for MEMS applications, *Sensors and Actuators A — Physical* **116**, 352–356.
103. Wallis, G. and Pomerantz, D. I. (1969) Field assisted glass-metal sealing, *Journal of Applied Physics* **40**, 3946–3949.
104. Schjolberg-Henriksen, K., Jensen, G. U., Hanneborg, A. and Jakobsen, H. (2004) Anodic bonding for monolithically integrated MEMS, *Sensors and Actuators A — Physical* **114**, 332–339.
105. Schjolberg-Henriksen, K., Jensen, G. U., Hanneborg, A. and Jakobsen, H. (2003) Sodium contamination of SiO2 caused by anodic bonding, *Journal of Micromechanics and Microengineering* **13**, 845–852.
106. Roberts, D. C. *et al.* (2003) A piezoelectric microvalve for compact high-frequency, high-differential pressure hydraulic micropumping systems, *Journal of Microelectromechanical Systems* **12**, 81–92.

107. Tian, W. C., Pang, S. W., Lu, C. J. and Zellers, E. T. (2003) Microfabricated preconcentrator-focuser for a microscale gas chromatograph, *Journal of Microelectromechanical Systems* **12**, 264–272.
108. Tong, W. and Williams, R. (1994) Kinetics of surface growth: phenomenology, scaling, and mechanisms of smoothing and roughening, *Annu. Rev. Phys. Chem.* **45**, 401–438.
109. Mullins, W. (1959) Flattening of a nearly plane solid surface due to capillary, *J. Appl. Phys.* **30**, 77–83.
110. Herring, C. (1950) Effect of change of scale on sintering phenomena, *J. Appl. Phys.* **21**, 301–303.
111. Williams, K. R. and Muller, R. S. (1996) Etch rates for micromachining processing, *Journal of Microelectromechanical Systems* **5**, 256–269.

CHAPTER 3

SENSOR ELECTRONICS

by *Elena Gaura and Robert Newman*

3.1. Introduction

The sensor read-out circuit (sometimes called sensor interface or sensor pick-off) enables a MEMS sensing element to become a fully functional MEMS sensor or transducer. Whilst two decades ago describing the read-out/interfacing techniques for sensors would have been a relatively straightforward matter, this is no longer the case, as interface circuits of today are no longer just an essential block of electronics to be placed behind the sensing device in order for the sensing task to be accomplished. As new generations of cheap and 'smart' sensors (meaning 'easy-to-use' for most people) are being developed and finding their way into applications, the interface circuits have evolved to encompass signal conditioning circuits, ADCs, bus interfaces and sometimes data (or even information) communication circuitry and sensor networking facilities. This, added to the sensor specific design techniques being developed for increased accuracy, low power consumption and the speculative integrated decision making capabilities foreseen for deployed sensors, makes the task of this chapter a rather difficult one.

Separating the 'sensor interface electronics' from the 'sensor electronics' in terms of current designs and implementations on one hand and tracing the particulars of various interface function implementations (filtering, signal separation, signal conversion) in sensing applications lead designs, on the other hand, is almost impossible. Consequently, the approach to discussing the 'sensor electronics', in this chapter, is by pointing out the interface requirements and characteristics of MEMS sensors in the context of the sensor as a digital or mixed signal processing system.

The functions of sensor systems (sensing, transduction/pick-off, analogue signal processing, digitization, specific digital signal processing functions) will be briefly introduced. An examination of the specific requirements of MEMS sensors electronics, which needs to match both the characteristics of the sensor and that of the eventually integrated system will be provided. The challenges raised, particularly relating to scale and low level signals to be manipulated, noise and signal quality improvement will be highlighted. Several successful sensor system designs and design trends will be presented, mostly from a systems perspective but with some sample implementations.

3.2. Functions of a Sensor System

This book addresses the design of sensor systems at a number of different levels, from the fabrication steps and resultant mechanical assemblies covered in Chapter 2 through to networked sensor systems comprising thousands of sensors in Chapter 10. At each level of design, the designer will tend to think of the word 'system' as referring to the set of components which are the subject of that particular design exercise. This chapter is concerned with the design of a sensor and the electronic components required to interface the sensor to whatever or whoever is the ultimate user of the data provided by the sensor, and it is in this context that the term 'sensor system' is used, notwithstanding the fact that at a different level of abstraction, the designer of some larger system of which the sensor and its ancillary electronics are a part, may use the term 'sensor system' to refer to this larger system. Likewise, the location of the boundary between the 'sensor electronics' and the electronics which are part of the higher level system depends very much on one's point of view as a designer. For the purposes of this chapter, the concern is with the functions required to process the data from an individual sensor, be they implemented using analogue hardware, digital hardware or software, and be they physically located adjacent to the sensing device itself, or at a distance.

The diagram in Figure 3.1 shows a 'generic' sensor system processing the measurand from a single sensor device. The role of the system is to read the data from the sensing element itself, and deliver it in the form required by

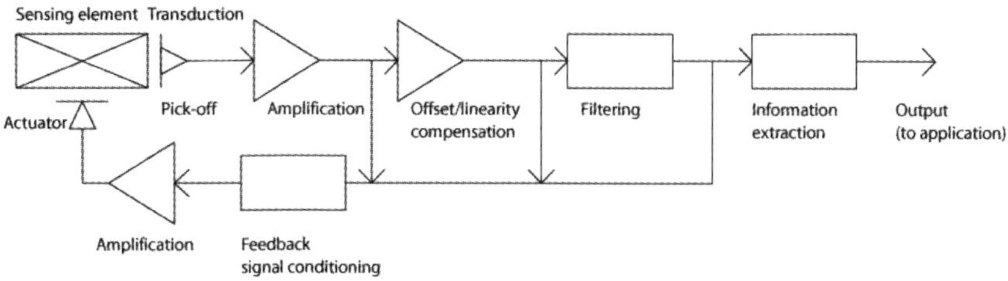

Figure 3.1: Functions of a sensor system.

the user — be it a human being or an information system. Generically, the user of the sensor's output (data or information) is called the 'application'.

The diagram in Figure 3.1 represents system functions, rather than components. In different system designs, the functions may be realised as separate components and subsystems. Alternatively, a function may be spread across several system components or subcomponents. The functions in Figure 3.1 may be either implemented locally to the sensor or centrally with the application.* The decisions as to which functions are implemented where form an important discussion in modern sensor systems design, and are dealt with fully in Chapters 8, 9 and 10.

Also, the ordering of functions will not be invariably as illustrated here, sometimes being determined by higher level design and integration decisions. The forward chain in the diagram represents the signal processing stages which must occur in practically every sensor system. The feedback chain is used in 'closed loop' sensors, a category discussed later in this chapter and in Chapters 4 and 5.

As has been hinted at in the discussion above on the use of the term 'system', one of the issues that complicates discussions on sensor systems is the variable use of terminology in different contexts and by different authors. For clarity, the terms as used in our diagram are defined below. The definitions are at an elementary level. Many readers of this book (those with an electronics background) will be quite familiar with these terms, although their precise usage may need clarification; others, maybe from a mechanical of computing background, may be less familiar with this terminology.

*If, indeed, the application is centralised. Currently most are, but distributed sensing systems open up the possibility that the 'application' (or the software that serves it) may be decentralised, being distributed over a number of processors.

Sensing element:
All sensors sense some form of energy. The sensing element is the component which receives the energy from the physical entity being sensed. The types of sensing element used in MEMS devices and their principles of operation were discussed in Chapter 2, which surveyed MEMS fabrication technologies and the physical properties of the sensing devices that can be fabricated using them. Those physical properties produce characteristic behaviour of those devices when used as a sensing element. This chapter is concerned with that behaviour and the consequent requirements which are placed on the interface electronics.

Transduction/pick-off:
Transduction is the process of translation between one form of energy and another, for instance between mechanical energy and electrical energy. The pick-off is the means by which this is done. Generally the concern here is with picks-off which produce an electrical output.

Amplification:
Amplification translates an electrical signal to one of a different (generally higher) amplitude. The signal may be expressed as a voltage (voltage amplification), a current (current amplification) or a charge (charge amplification). The factor of amplification is the gain. The term 'amplification' will also be used to refer to some slightly more generalised processes. The first generalisation is conversion between two signal forms, for instance a current signal to a voltage one (transresistance amplifier) or a voltage input to a current (transconductance amplifier). The second generalisation is level translation within digital systems, in which a digitally encoded signal is 'amplified' simply by multiplying the samples by the 'gain'. Since, in the functional sense that our diagram represents, this is equivalent to amplification, it will be called amplification here.

Offset/linearity compensation:
This is the process of eliminating the offset and nonlinearity from the sensor signal/transfer characteristic. Offsets and linearity errors originate from both the characteristics of the sensing device itself and from imperfections of the interface electronics. The correction of the characteristic offsets and nonlinearities of the sensing device is a sufficiently important topic to

warrant its own chapter (Chapter 4). This chapter deals with correction of imperfections raising from the interface electronics only. This separation is slightly artificial — in many designs, both sets of errors are dealt with together. However, in others they are not, necessitating the separate discussions, as proposed here.

Filtering:
Many sensors sense dynamic quantities, producing an AC electrical signal. The quality of the part of that signal that is required (or useful) for the particular application may often be improved by frequency selection. The term 'filtering' is used to denote any modification of the frequency or phase response of an AC signal, including processes such as time integration and differentiation.

Information extraction:
The sensor itself provides *data*, whilst what is ultimately required by the user of application system is *information*, which is selective and context dependent.† For the purposes of the present discussion, it is sufficient that it is identified that somewhere in the processing chain, the required information must be extracted. This can occur at several places in the processing chain, including in the brain of the (human) user. For instance, the plotting of sensor data onto a graph, which can be more easily interpreted by a human being than can raw numerical data, is one step in a process of information extraction, which will finish with a human user making some judgement as a result of visualising that graph. The presence of the information extraction in the per-sensor processing chain reflects the fact that it will be necessary to derive some information from an individual sensor and pass that information onward for further processing in the application system. A simple example would be a case where the sensor was required just to provide an alert as to whether or not the quantity it was sensing was above some predetermined limit. This is a piece of information that can easily be extracted from the sensor data by comparing its data with the limit value. In some application system designs this information extraction will

†The differentiation between the concepts 'data' and 'information' is discussed fully in Chapter 8. In this chapter we are considering only the electronics adjacent to the sensor. In the context of these discussions, there is no sense of what the data will actually be used for, thus we are generally discussing 'data' rather than 'information'.

take place centrally, as part of a program which inputs the raw data from the sensor and performs the comparison, in others there may be an analogue comparator located with the sensor, which simply delivers a Boolean value to the application system. As technology advances, it is feasible to implement ever more sophisticated information extraction functions using per-sensor hardware, thus opening up the design choices available to the applications system designer.

Feedback signal conditioning:
As will be discussed further in this chapter, some sensors use a 'closed loop' topology, feeding back a signal to the sensing element. The term 'feedback signal conditioning' is used here to denote the process of producing the required feedback signal from the forward signal. In practice, this could include similar stages to any or all of those found in the forward path.

Actuator:
In a closed loop topology, energy is fed back to the sensing element, in an appropriate form. For example, light would be fed back to an optical sensor, mechanical energy to a mechanical one. Closed loop topologies are discussed extensively in Chapter 5, and an account of how feedback operates can be found there. Most transduction processes are reversible, so it is often the case that actuators in a sensor system take the same form as the pick-off. For instance, a variable capacitance structure can be used as a capacitive pick-off and also as an electrostatic actuator. Generally, however, the optimum design of the electromechanical system will be not be the same for an actuator and a pick-off.

The seasoned sensors literature reader must have noted by now that, the analogue to digital converter (ADC), which generally appears somewhere in a block diagram of a sensor system, is nowhere to be found in the diagram in Figure 3.1. This is because, from a functional viewpoint, the analogue to digital conversion forms no part in the process of conditioning and processing data. The data remains the same (except for artefacts produced by the conversion process), what has changed is the way that the data is represented, the conversion is simply a transform from the analogue to the digital domain. The conversion process itself may appear as a part of one of several of the 'blocks' illustrated in Figure 3.1. Indeed, it is not always the case that analogue to digital conversion is exclusively identified with a single component or subsystem. Its functions may be distributed between several of them, as shall be seen in Chapter 5.

If the data from the sensor is ultimately required in digital form, it would normally be expected that a single conversion from analogue to digital representation will be made, at some point in the chain shown in Figure 3.1. Selection of the precise point in the chain where the conversion happens is a design choice which is influenced by a number of factors, some of which are discussed in Section 3.3.

The diagram in Figure 3.1 will form a reference point through the rest of this chapter. Over the years, designers of sensor systems have used a bewildering variety of configurations and technologies for the various functions. From a top level, systems design viewpoint, the two design choices mentioned above — the position of the analogue to digital boundary and distribution of functions — have proved to be the ones that determine the overall 'architecture' of the system. A system designer might make these choices based on a number of different considerations, such as existing practice, availability of off-the-shelf subsystems that suit a particular configuration, or local technological capabilities. As the examination of the electronics associated with a sensor proceeds, it will be seen that many different solutions and variations have been developed in the continual quest for improved sensing systems. The diagram will be used to provide context and draw together commonalities and differences between the various approaches.

3.2.1. Transduction and Pick-off

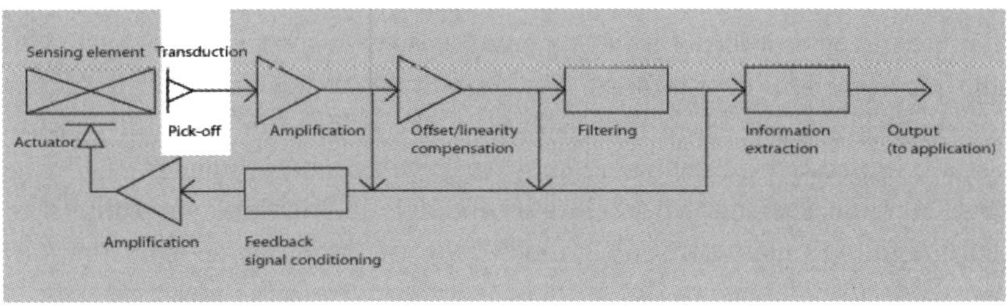

The various transduction mechanisms found in MEMS sensors were discussed in detail in Chapter 2. What we are concerned with in this chapter is the electronic requirements posed by the different types of transducer.

The most prominent MEMS sensing techniques are based on the following devices [1]:

- Strain gauges, using piezoresistive or piezoelectric detection
- Capacitive detection using mechanically variable capacitors
- Thermal sensing using microbolometers, thermocouples, thermoresistors or semiconductor devices
- Tunnelling effect
- Optical sensing, using photodiodes as the detection element (which may be integrated into a photodiode array of charge coupled device (CCD)

Strain gauges

Strain gauges transduce a mechanical signal into an electrical one by measuring the change in resistance of a strained metallic conductor. A semiconductor can produce a much larger effect (10–100 times larger) and integrate particularly well with semiconductor based MEMS technologies. Miniature silicon piezoresistors may therefore be integrated into sensors which are fabricated from metals, or semiconductor compounds, such as silicon. The measurement of mechanical strain may be used to provide some other data, such as the deformation of some mechanical component. This effect is used, for example in pressure sensors, where the deformation of a membrane is measured, and also in acceleration sensors where the deflexion of a flexure attached to a proof mass is measured. Two examples (a vacuum sensor and an accelerometer) are given below.

Figure 3.2 shows the configuration of piezoresistors implanted on the diaphragm of a Micro Pirani vacuum sensor (a specialist pressure sensor) [2]. Here four resistors are used to allow a Wheatstone bridge input configuration. Since all of the resistors in the bridge are identical piezoresistors, implanted on the same chip, they will tend to undergo the same characteristic changes with temperature. R1 and R3 are not subject to diaphragm strains, whereas R2 and R4 are. A bridge made from the four resistors will balance out any drift or instability from the piezoresistors.

In an accelerometer piezoresistors are placed on the supporting beams of the seismic mass. Stress in the beam generated by movement of the seismic mass causes them to change their resistance. In a silicon based technology, the piezoresistors can be directly diffused into the supporting beams which makes the manufacturing process fairly simple and cheap. Different

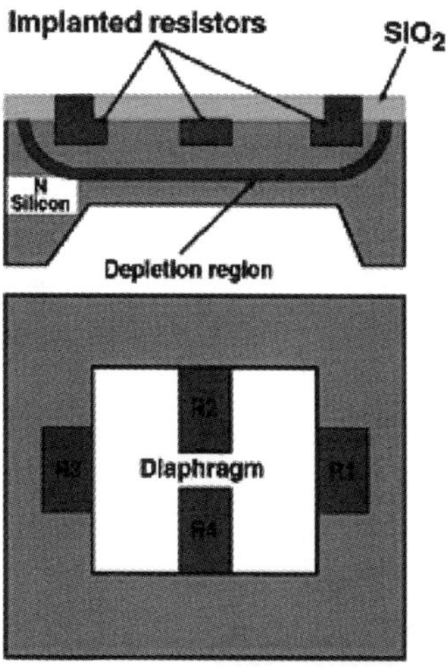

Figure 3.2: Micro Pirani vacuum sensor, showing implanted piezoresistor bridge. From [2].

pick-off designs are possible, for example half-bridge, full bridge configurations, and additional piezoresistors may be used for temperature compensation. Figure 3.3 shows the structure of a piezoresistive hinge assembly in an accelerometer reported by Warneke *et al.* [3]. The required pick-off electronics is relatively straightforward, consequently this technique used to be attractive for low cost applications [4] and has some advantages for stand-alone sensors.

However, piezoresistors have a number of disadvantages: they have a low level output (typically 10–100 mV full-scale output), have a large sensitivity drift with temperature, and self-test features and closed loop operation can only be realized through an additional electrode and increased complexity of electronics. As a consequence, piezoresistive sensors are usually open-loop, low cost devices for low performance applications.

Capacitive pick-off

The capacitive position sensing method uses the variation in capacitance caused by electrode motion in order to transduce a position to a mechanical signal; common designs used are:

- Parallel plates, as shown in Figures 3.4(a) and 3.4(b), for vertical motion detection [5];

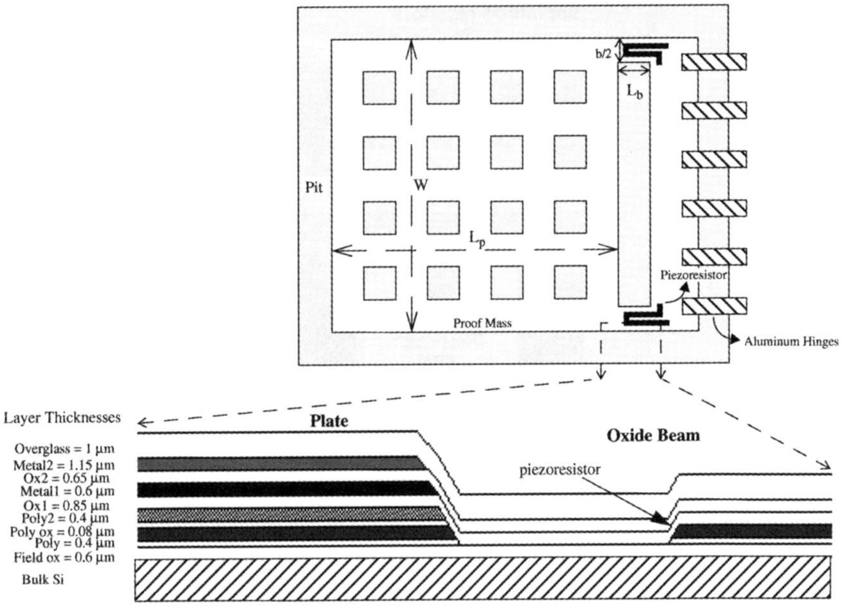

Figure 3.3: Piezoresistive hinge assembly. From [3].

- Arrays of interdigiated electrodes, as shown in Figures 3.4(c) and 3.5, for sensing changes in lateral position (in-plane motion for example) [6].

With applications to a variety of sensors such as proximity detectors, micrometers, motion detectors, pressure sensors, touch pads and microphones, capacitive pick-off transducers have become very popular. Capacitive sensing has some advantages over the piezorezistive methods, particularly when sophisticated electronics is closely integrated with the sensor, including small size, high sensitivity, low power consumption and closed loop design opportunities. Generally, capacitive sensors (with their associated pick-off electronics) have good linearity, higher output levels and very low sensitivity to temperature drift. The drawback lays with the challenge of measuring very small differential capacitances (down to fF in some cases), meaning that they are very sensitive to parasitic capacitances and electromagnetic interference, and that the pick-off electronics may be complex (accomplishable however for digital, integrated sensors).

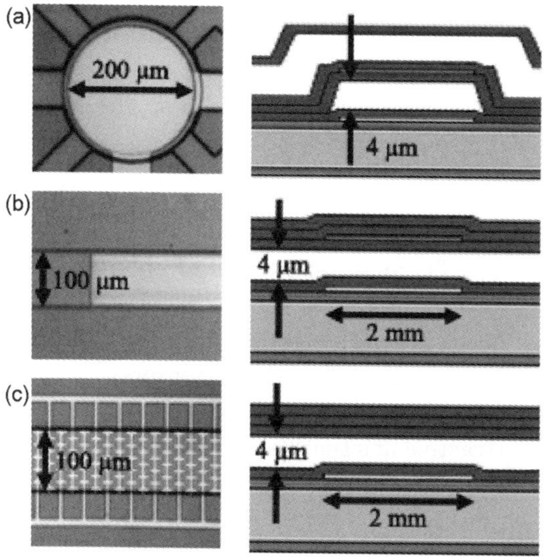

Figure 3.4: Capacitive pick-off in a pressure sensor. From [5].

Figure 3.5: Interdigitated capacitive pick off. From [6].

Thermal pick-off

Thermal sensing is generally used to sense properties of fluids, either using the change in thermal capacity of a gas as its pressure changes, or heat loss caused by flow of a fluid. The sensing element itself is often an electrically heated 'hot wire'. Its temperature may be sensed by means of measuring its resistance, or a closed loop sensor may measure the amount of energy required to heat it to a given temperature. Temperature sensors may be semiconductor junctions, resistive or thermocouples, all of which can be fabricated using MEMS technologies.

The illustration in Figure 3.6 shows the operating principle of a thermal accelerometer produced by Memsic Inc. [7]. A heater heats a gas bubble held above the chip. Thermal sensors placed on either side of the heater sense the temperature of the gas. At zero acceleration the temperature is the same either side. Acceleration changes the gas dynamics, and a temperature difference occurs between the temperature sensors.

Thermal pick-off (when used for sensing quantities other than direct temperature) is attractive in situations in which direct measurement of the thermodynamic properties of fluids is required, and allows the design of sensors which sense these properties, using no moving parts. This, in turn can allow the design of sensors which are more robust, or have greater ability to operate in extreme environments. The disadvantages are that complex pick-off electronics are required and that the output is intrinsically subject to thermal noise. The requirement for thermal actuation (heaters) means that thermal pick-off is unlikely to be considered for very low-power systems.

Tunnelling effect based pick-off

This category of micromachined sensors emerged in the mid-1990's. They utilise a constant tunnelling current between one tunnelling tip (attached to a movable structure) and its counter electrode to sense deflection. Figure 3.7 [8] shows a micromachined tunnelling accelerometer. When the tip is within a few Å of the counter-electrode, the potential between the tip and the electrode generates a current due to the quantum tunnelling effect, which will remain constant so long as the distance between the tip and electrode, and the bias voltage remain constant.

The equation for the resulting current is of the form:

$$I_{\text{tun}} = I_0 e^{-\alpha\sqrt{\phi x}}, \tag{3.1}$$

where I_{tun} is the tunnelling current, I_0 is a constant, α is the tunnelling constant, x the separation gap and Φ the tunnelling barrier.

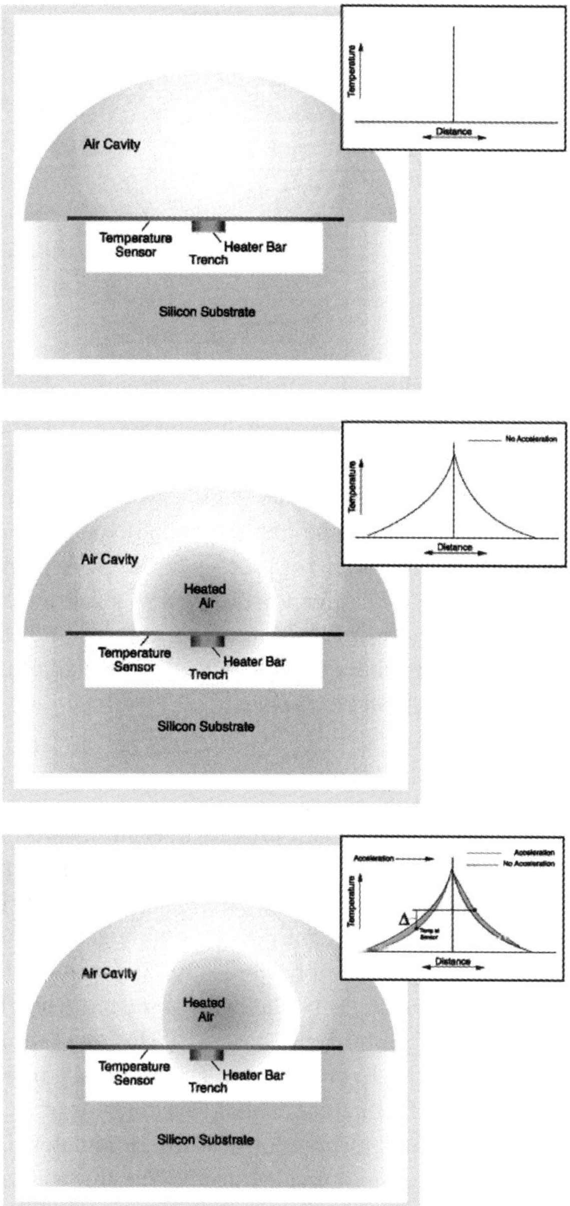

Figure 3.6: MEMSIC thermal pick-off. From [7].

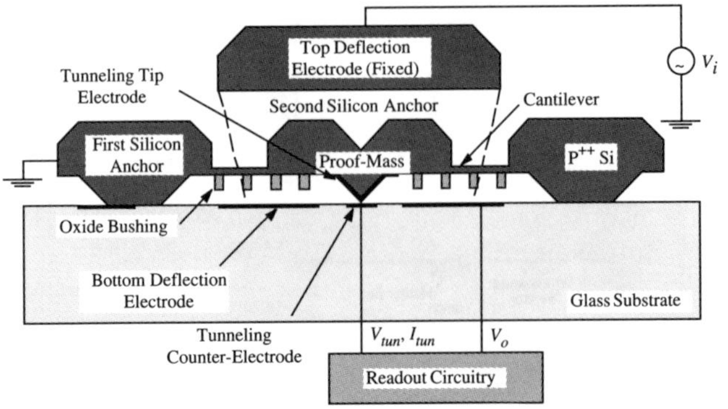

Figure 3.7: Tunnelling effect pick-off accelerometer. From [8].

If the proof mass and hence the tip moves, the current will change. Tunnelling accelerometers can achieve very high sensitivity with a small size of seismic mass since the tunnelling current is highly sensitive to displacement, typically changing by a factor of two for each Å of displacement. However, these devices have larger low-frequency noise levels. Given the exponential relationship between the tunnelling current and the tunnelling distance, these microsensors have a small device area, high sensitivity, wide bandwidth and simple pick-off electronics.

Optical pick-off

Light (either infra-red or visible) can provide a useful means of signal pick-off from a moving mechanical element, but only if opto-electronic technologies are available. For optical pick-off, a *pick-off mirror* is fabricated on the moving assembly. As it moves, the mirror deflects a light beam. This deflection is sensed either by an array of photo-receptors, or the sensor is included in a closed-loop system, which keeps the beam aimed at a single sensor. An alternative mechanism, for rotation sensing, is to couple an *interruptor wheel*, which interrupts the light beam as the assembly rotates. A third mechanism, unique to MEMS, is the MEMS Fabry-Perot interferometer. Two parallel reflective plates form an optical interferometer. Displacement of one plate creates a change in optical transmission, due to the interference effects between the incident and reflected light. Lee *et al.* [9] have built and characterised a Fabry-Perot displacement sensor, and showed that it can detect a movement of 7 nm using an optical power of 1 μW.

Optical pick-off is widely used in macro scale electro-mechanical devices (for instance, computer mice), but infrequently used in MEMS, since it requires a light source and detectors. This entails the use of opto-electronic technologies, which are not commonly integrated with MEMS systems, except perhaps in highly integrated, multi-sensor devices.

3.2.2. *Analogue Signal Processing: Front-end Amplification*

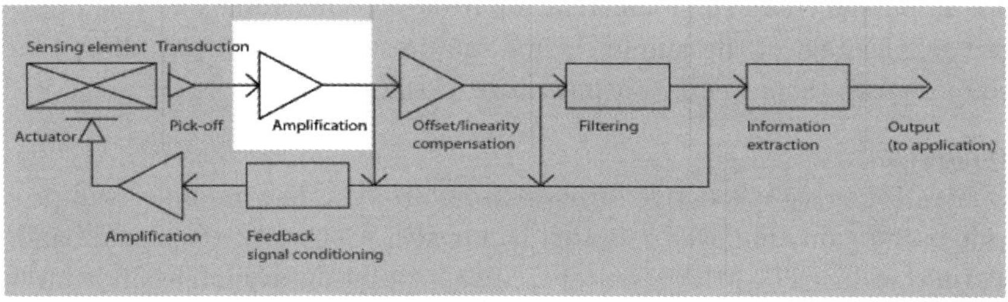

In this section the common concerns which must be addressed in the design of the 'front-end' are discussed (that is, the immediate electronic interface with the transducer). The discussion is aimed at those readers who are not familiar with analogue electronics — those who are could skip the section. For a more detailed treatment, suitable for those with an electronics background, the reader is referred to Brignell and White [10].

Most if not all front-end circuits for MEMS sensors receive low-level electrical signals from the sensing elements. Whilst handling low-level signals is not unusual in microelectronics, specific sensors introduce additional constraints which require dedicated solutions and careful consideration of the technology to be used. The interface requirements of various transducer types will be dealt with in Section 3.5. Here we look at the general issue of low noise amplification, particularly with respect to integrated transducer systems. The figures of merit for front-end circuits are the following:

Noise:
All amplifiers introduce noise to the signal from the transducer. The ratio of the amplitude of the noise signal to the amplitude of the signal from the

transducer determines the analogue resolution of the available signal. The level of noise determines directly the sensitivity of the amplifier.

Linearity:
Its opposite is known as 'distortion' or 'nonlinearity'. Ideally, the transfer function between input and output of an amplifier should be linear. Generally, within the range of measurement and precision required of sensor systems, amplifier linearity is not a problem, and it will not be considered further here.

Offset:
All amplifiers entail some inherent input offset. This input offset is amplified by the gain of the amplifier, and may result in an output offset that compromises the available output range of the signal.

Bandwidth:
Generally the signals derived from sensors are AC. The amplifier will provide useful gain only over a limited bandwidth. Generally, the signal bandwidths produced by MEMS sensors are well within the available bandwidths of modern integrated semiconductor systems.

Noise

Figure 3.8 shows a circuit model for a transducer pick-off element connected to an amplifier.

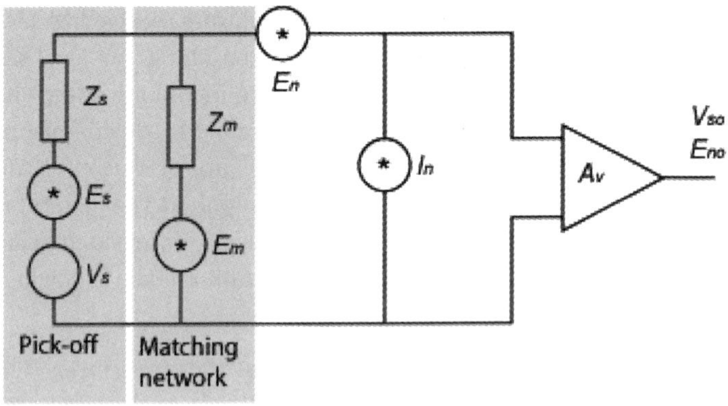

Figure 3.8: Front-end amplifier noise sources.

Here the pick-off generates a signal V_s, and a noise voltage E_s, and has a source impedance Z_s. An impedance matching network connects the pick-off to the amplifier. The impedance matching network has an impedance Z_m and produces its own noise E_m. The amplifier itself produces its own input noise: voltage noise E_n and current noise I_n. The output from the amplifier is V_{so} (the required signal), along with the unwanted noise E_{no}.

The effective input noise E_{ni} has the form:

$$E_{ni}^2 = A^2E_s^2 + B^2E_n^2 + C^2I_n^2Z_s^2 + D^2E_m^2, \tag{3.2}$$

where A, B, C and D, are scale coefficients for each noise component. For a current output pick-off the effective input noise, I_{ni} could be expressed as:

$$I_{ni}^2 = J^2I_{ns}^2 + \frac{K^2E_n^2}{Z_s^2} + L^2I_n^2 + \frac{M^2E_m^2}{Z_m^2}, \tag{3.3}$$

with

$$I_{ns} = E_s/Z_s, \tag{3.4}$$

where J, K, L and L are once again scale coefficients. The input noise cannot be removed later in the signal chain, so the designer's aim must be to minimise it.

For the designer of the electronic part of a MEMS system, the noise produced by the pick-off is a given, so the design efforts must concentrate on reducing the matching network noise and the amplifier input noise. By clever design, it may be possible to match the input of the amplifier to the source without need for a matching network, in which case this additional source of noise can be eliminated.

The following stages will see not only the transducer and matching network noise, but also the noise produced by the input amplifier, so this must also be minimised. Noise in semiconductor amplifiers is caused by a number of effects. The first is resistor or Johnson noise, caused by the thermal excitation of the electrons carrying the current. The root-mean-square noise voltage of Johnson noise is $v_n = 2\sqrt{keTR\Delta f}$, where k is the Boltzmann constant, e is the charge on an electron, Δf is the signal bandwidth, T is the absolute temperature and R is the resistance. A 1-kΩ resistor has noise of approximately $4\,\text{nVHz}^{-1/2}$ at room temperature. Obviously, Johnson noise may be minimised by reducing any or all of the operating temperature, input resistance or bandwidth.

The next source of noise is shot or Schottky noise. This is caused by the fact that the current is carried in discrete quanta of charge. The noise current is given by $I_n = \sqrt{2eI_q\Delta f}$, where I_q is the quiescent current in the device and e and Δf are as above. Schottky noise is minimised by minimising the quiescent (bias) currents. The Schottky noise of a typical bipolar operational amplifier is of the order of $250\,\mathrm{fAHz}^{-1/2}$, for $I_b = 200\,\mathrm{nA}$, where I_b is the bias current, whereas for JFET input operational amplifiers the noise is of the order of $4\,\mathrm{fAHz}^{-1/2}$ at $I_b = 50\,\mathrm{pA}$. The later will however increase as temperature increases, since the bias current is temperature dependent.

In addition, FETs produce other sources of noise: there is a small component from the (tiny) gate current and also a $1/f$ flicker effect. For this reason, FETs are often optimum at high frequencies, where they can be superior in noise performance to bipolar transistors. Generally bipolar transistor circuits have lower voltage noise (due to the lower input resistance) and FETs lower current noise (due to the very low input current). Overall, the noise performance of bipolar circuitry is, at typical sensor frequencies, superior.

Offset

The input transistors of an amplifier require to be biased into their linear region for analogue operation. A two input operational amplifier will require a bias current for each input device. Unless the two bias currents are precisely matched, the difference will be amplified and occur as a DC offset voltage at the output.

In some applications, an offset is not a problem. If the system is to be used in a frequency domain monitoring application, such as vibration monitoring, the offset may be removed simply by AC coupling the stages of the amplification chain. In systems where the absolute quantity of the measurand is important, however, the offset must be eliminated.

At an overall system level, the amplifier offset simply adds to any transducer offset, and may be removed along with that offset, using techniques that will be explained in Chapter 4. However, in many cases this offset must be handled within the amplification chain, for several reasons. Firstly, amplifier offset is usually temperature dependent, so it cannot be handled by a simple subtraction of a fixed quantity from the measured signal. Moreover, if the input offset and gain are sufficiently large, the output offset may take the signal outside of the operating range of subsequent stages. Particularly in systems which operate with a low supply voltage (as will be typical of low power, integrated sensors) there may not be sufficient amplification headroom to allow for a large offset. It must either be designed out of the amplification chain, or some other means used to eliminate it.

Bandwidth

Most sensor systems operate with a signal bandwidth in the kHz or tens of kHz range, at most. With modern process technologies, which can produce amplifiers with bandwidths in the MHz, lack of bandwidth is hardly a problem. Moreover, amplifier bandwidth is a commodity that can be traded for other desirable properties. Particularly, it provides the basis for performance improvement using feedback techniques, both within the electronic part of the system and, using the 'hardware in the loop' techniques to be presented in Chapter 5, for the mechanical part as well. The disadvantage of high speed, and therefore high bandwidth circuits, is that they consume more power, which may be critical in many applications.

3.2.3. *Design of Integrated Front-end Amplifiers*

Generally, it is only at the front-end that the electronic circuitry must be designed particularly for low noise, since the front-end is uniquely in a position where its noise will be amplified by the full gain of the amplification system. Thus, it is in this part of the circuit that special low noise devices are likely to be selected. However, freedoms in the past afforded to electronic designers are not always present for the designers of modern integrated sensing systems. In an integrated system, if only one part of the chain requires a low noise technology, this technology choice dictates that used for the whole of the integrated subsystem, and selecting a process that can provide suitable low-noise input devices may adversely affect cost of compatibility with MEMS device fabrication. As an example, minimisation of voltage noise requires low transistor resistances, which may be achieved either by use of large geometries or by heavy doping. The technology of choice would be bipolar, if there were no other constraints. On the other hand, very low current inputs would suggest the choice of a FET input. Optimally, the electronics integrated with the transducer must have the appropriate low noise devices available (the question of integration strategies is dealt with more fully in Section 3.6). For the purposes of the present discussion, the designer of a fully integrated monolithic MEMS/ electronics system‡ has the following three choices, in respect with front-end amplifier design:

- Limit the level of integration to those possible with appropriate low noise technologies. This is an approach often used in traditional sensor systems, but in practice it usually means that the level of integration is limited to an amplified transducer. Alternatively, multi-chip hybrid integration technologies can be used, but these will result in higher price and larger size than monolithic integration of the sensor electronics.
- Find a technology capable of low noise and high levels of integration. Generally, the 'high integration' processes are digital, and digital processes (especially at very small geometries) have poor analogue performance, since they are optimised to have the minimum possible characteristics to provide two distinct binary signal levels, whereas analogue circuits require a range of signal levels. Similarly, high integration digital processes are designed to use very low supply voltages, so as to conserve power (which is proportional to the square of the supply voltage), and thereby, reduce the heat dissipated by the circuit. By contrast, to preserve good noise headroom, analogue circuits require relatively higher supply voltages. Furthermore, in some input configurations, bipolar transistor circuits are preferable to FETS. Also, analogue circuits require resistors (at least) as well as transistors. Pure digital technologies, which nowadays are almost exclusively CMOS, are generally not optimised to produce efficient resistors. Processes which attempt to combine good digital and analogue characteristics are known as 'mixed signal technologies'. Generally, there are two approaches to mixed signal technologies. The first extends the digital CMOS technology, to allow higher supply voltages and variation of gate doping to give a range of threshold voltages. This allows the flexibility to make high quality CMOS operational amplifiers, at the same time maintaining the digital capabilities of the process. The second class of processes are mixed bipolar/CMOS or BiCMOS,

‡Generally in this chapter monolithic integration is not assumed. The additional constraints imposed by such a design are discussed fully in Section 3.6. The design choices discussed here apply also, however, to the designer of a single chip interface to a sensor device, even if that chip does not include the MEMS components. The requirements of cost and space dictate increasingly that these (monolithic or two chip assemblies) are the choice for new sensor electronc systems.

adding additional process steps to a CMOS process to allow fabrication of bipolar transistors. Such processes will be more expensive than straightforward CMOS, due to the extra steps.

- Find ways of ameliorating the poor performance of high integration technologies. The high integration of these processes may allow additional signal processing, which can negate (at least to some extent) the shortcomings of the analogue input section. Although input noise can never be removed once added to a signal, it is possible to extract the required information from a noisy signal, using sophisticated processing, often called 'noise reduction'. For instance, bandpass filtering can pull a narrow band signal out of wideband noise. Such techniques require some prior knowledge of the context in which the device will be used. Similarly, temperature sensitivity and nonlinearity may be compensated using analogue or digital post-processing techniques. The enhanced functionality offered by increased integration provides other means of handling symptoms such as input offset. The availability of clock oscillators and MOS signal switches allows use of switched capacitors, which, as will be seen in the next section, can be used to provide a substitute for precision resistors, and chopper stabilisation which can eliminate amplifier input offset. An internally-nonlinear/externally-linear approach was proposed by Carro *et al.* [11], suggesting the use of non-linear MOS gate capacitors and MOS transistors as a linear resistor. All known non-linearities are thereafter compensated using adaptive filters. An area saving of 50% is claimed compared to the analogue area in a process supporting linear devices.

3.2.4. *Performance Enhancement Opportunities Using Integrated Electronics*

In this section we discuss techniques available to the designer of the front-end electronics which allow for some compensation for inadequacies of the semiconductor processes, that may be available in high integration technologies. Essentially the tasks that must be undertaken here are the elimination of temperature dependent amplifier derived offsets and filtering to eliminate out of band noise. Elimination of offset, along with linearity correction, will also often be required to correct these and other faults inherent to the transducer itself, and in that context, compensation issues are covered in Chapter 4.

Autoadjustment and correction of amplifier offset

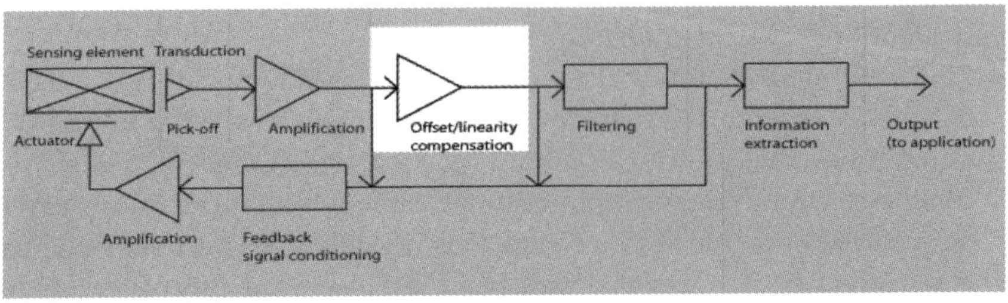

In a conventional precision electronic circuit, input offset would be trimmed out, to adjust the DC level of the reference input of the input operational amplifier so as to produce a zero output offset. This is obviously not possible in mass produced highly integrated circuits (although trimming is possible, using either laser trimmed resistors or digitally switched resistor trees). An alternative technique is to build auto correction into the circuit. This technique provides a servo loop to feed in an input current to set the DC output level at zero. The servo loop can either be analogue, in which case a low pass filter is used to remove the AC components of the output and this signal fed back to the input, or it can be digital, in which case the digitised output signal is monitored, and if an offset is detected, a correction is fed back via a digital to analogue converter.

Switched capacitor techniques

The idea behind a switched capacitor (SC) circuit is that a capacitor presents an impedance to an AC signal. In the circuit in Figure 3.9, if S_1 and S_2 are controlled by alternate phases of a clock, then for each cycle — S_2 closed, S_1 open, followed by S_2 open, S_1 closed — a charge q is transferred from the input to the output. The charge transferred is $\Delta q = C_1(v_2 - v_1)$. If the frequency of the clock is f, then the amount of charge transferred in unit time (otherwise known as the current) is $i = C_1(v_2 - v_1)f$. Rearranging this gives $\frac{v_2 - v_1}{i} = \frac{1}{C_1 f}$. Thus, there is effectively a resistance $1/C_1 f$ between the input and the output. This resistance may be varied by changing the clock frequency. Furthermore, because the geometries of capacitors may be precisely controlled, potentiometers with precise ratios may be constructed. The technique provides a means of producing resistors in a technology that does not readily provide them.

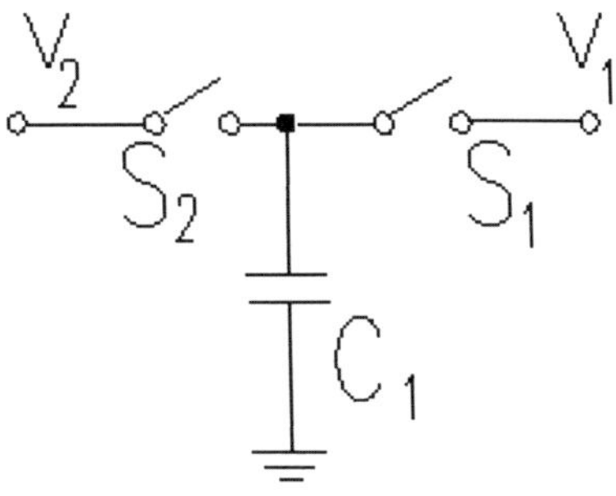

Figure 3.9: Switched capacitor.

Switched capacitor circuits, since they provide 'programmable' resistors, may be used for a variety of auto trimming and filter circuits. They have been used extensively in integrated sensor front-end circuits. Senturia describes a number of SC input configurations for a capacitive pick-off front-end in [12]. Zhang *et al.* describe a SC input as the capacitive pick-off element of a universal micro-sensor interface chip in [13]. Chavan *et al.* describe a sophisticated multi-function programmable front-end based on switched capacitor techniques [14]. This circuit has 15 programmable gain settings, provided by varying the capacitor switching waveform. Gola *et al.* [15] describe a fully differential switched capacitor charge amplifier chain for amplifying the signal derived from a capacitive pick-off in a rotational accelerometer.

Switched capacitor circuits may be used as another means of auto-zeroing correction. Here, the offset voltage captured during one phase of the clock, is fed back to compensate the input signal in the second phase. Such an arrangement is described by Maloberti *et al.* [16].

Chopper stabilisation

A chopper stabilised amplifier eliminates the input offset by switching the input of an amplifier alternately between the input signal and ground. In the grounded phase, the amplified offset appears at the output of the amplifier. In the signal phase, the amplified input signal plus amplified offset is present. Thus the output is a DC offset plus a square wave, the amplitude of which is the required output signal. This is synchronously demodulated and low pass filtered to produce an output free of the offset signal.

3.2.5. *Filtering*

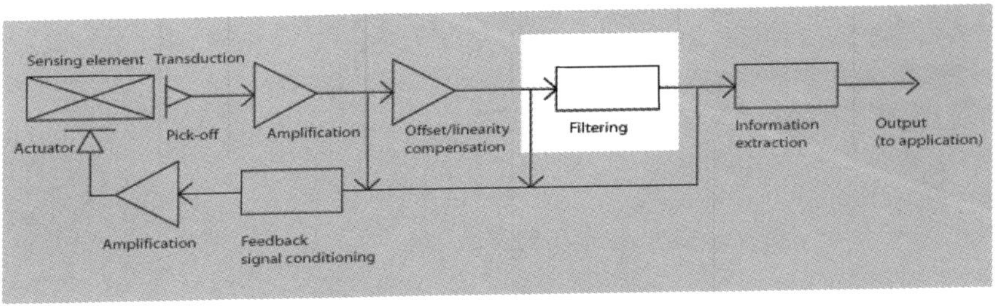

By contrast with the previous functions, the fourth function, filtering, operates in the frequency domain. It is used in signal conditioning for one of two reasons: to correct frequency dependent errors or to restrict the bandwidth to the part of the signal spectrum of interest, thereby filtering out broad-band noise or noise at specific frequencies (such as mains noise). Filters may be considered as a transfer function from an input, continuous time-based function $f(t)$, to another, output function, $g(t)$. In the frequency domain, filters are specified by their frequency response. However, they must be designed in the time domain, as transfer functions of the Laplace variable $s(= j\omega)$. In this form, the transfer function of any filter can be described by polynomials:

$$H_s = H_0 \frac{(s - z_1)(s - z_2) \cdots (s - z_n)}{(s - p_1)(s - p_2) \cdots (s - p_n)}, \tag{3.5}$$

where the roots of the numerator are 'zeros' and the roots of the denominator 'poles' and n is the 'order' of the filter.

In theory, the required frequency response can be obtained by producing a circuit configuration which generates the required time delays to model the given transfer function. In practice, for filters above the second order, this is difficult. However, the transfer function can be rewritten using second order polynomials:

$$H_s = H_0 \frac{(s^2 + b_{11}s + b_{10}) \cdots}{(s^2 + a_{11}s + a_{10}) \cdots}, \tag{3.6}$$

suggesting that a higher order filter can be fabricated using a series of second order filters. In integrated applications, filters are invariably active, that is they use the gain of an operational amplifier to provide the transfer

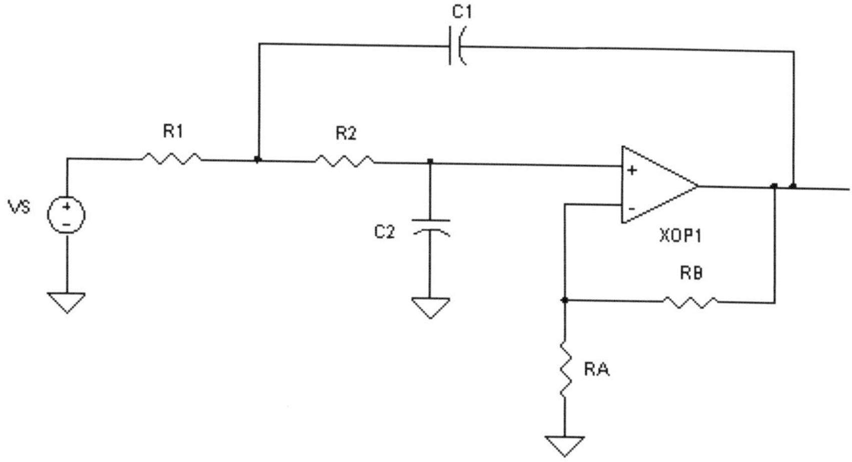

Figure 3.10: Sallen-Key low-pass filter.

function. The time delay may be produced using resistor capacitor combinations, or switched capacitors. A popular filter configuration is the Sallen-Key filter (Figure 3.10), which produces a second order filter using a single operational amplifier. Strings of such filters may be used to generate the five basic filter functions: high-pass, low-pass, band-pass, notch (band-reject) and all pass (phase shift). The precise frequency response required depends on the correct alignment of the filter parameters. Commonly used alignments are based on particular polynomials, including Butterworth (which produces a maximally flat passband), Chebychev (which produces the steepest cut-off slope) and Bessel (which is a compromise between the two extremes). Achieving the precise alignment requires accurate component selection. Switch capacitor filters can produce precisely aligned filters, which are tunable by varying the frequency of the capacitor switching. The disadvantage is that they are more noisy than active filters.

3.3. Analogue and Digital Design Options

Figure 3.11 shows a generic diagram of a mixed signal processing front-end sensing sub-system as viewed by Carro *et al.* [11]. The context of the discussion around this diagram is the use of digital signal processing to correct non-linearities in the analogue signal processing chain, which chain,

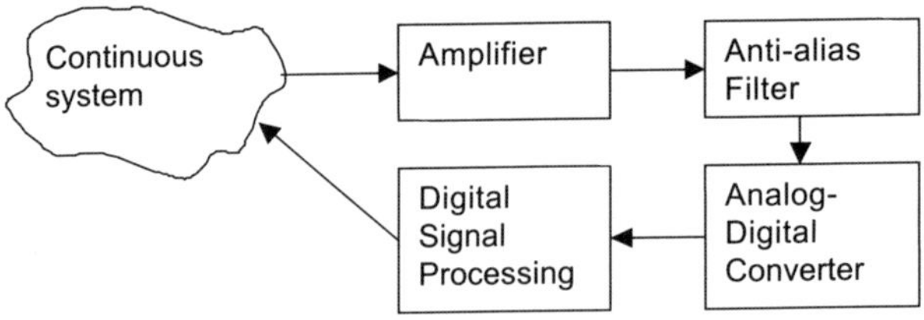

Figure 3.11: Front-end subsystem, according to Carro *et al.* From [11].

here, is limited simply to amplification. In this diagram all of the substantive signal processing occurs within the box labelled 'Digital Signal Processing'. The analogue to digital conversion occurs at the earliest stage possible in the signal chain, as soon as the signal has been amplified sufficiently for conversion. This view represents one extreme in the range of choices to be made concerning the stage at which to 'go digital' in the chain shown in Figure 3.1, and is suitable in the case when the analogue circuits available are non-linear, therefore their use should be as limited as possible within the sensor system. Other choices may be appropriate in particular contexts, for instance, if high performance analogue technology is available, the designer may choose to perform a great deal of the signal processing using analogue circuitry, with the conversion to digital representation occurring only at the output to the application system (to the right hand end of Figure 3.1).

In this chapter the primary concern is the techniques which are used to correct offset and linearity generated within the amplifier chain. Sometimes the same techniques may also be used to correct the offsets and non-linearities produced by the transducer itself (we shall see such an example in the circuit configurations used to correct the inherently non-linear response of a tunnelling effect pick-off in Section 3.5.4). The options for the design of these functions are discussed more fully in Chapter 4, which is concerned with the correction of the faults in transducers.

The designer's choice with respect to changing the balance between analogue and digital circuitry is likely to revolve around the correction requirements for the particular sensor and the relative costs, in the context of the overall design, and also whether it is possible, by making a particular design choice, to correct both types of fault (transducer and amplifier chain) with a single subsystem.

Before discussing the exact stage at which the conversion to digital might take place, we should observe that these functions — the offset/linearity correction and filtering stages — are not always placed in the order shown in Figure 3.1, and in some designs partial functions may be interleaved or integrated. We will assume for the sake of this discussion that the order shown in Figure 3.1 is the norm. In this case, there are three possible stages at which to make the conversion:[§]

Before the offset/linearity function:
Both offset/linearity correction and filtering are performed digitally. This choice may be made for a number of reasons, namely:

- the pick-off system may be one that lends itself particularly to a digital output (such as an AC capacitive pick-off);
- the offset or linearisation correction may be one very difficult to perform using analogue circuitry;
- the transducer may need extensive correction, which justifies digital signal processing, and signal chain error correction may be easily integrated with that function;
- there may already be a powerful digital signal processor integrated with the system, in which case, calibration/compensation come 'for free' — they are simply additional software tasks;
- the integrated electronics use a process with poor analogue characteristics (for instance, CMOS). In this case, it makes sense to perform as many functions using digital circuitry as possible.

Between offset/linearity correction and filtering:
This choice is likely to be made if there is sufficient digital signal processing capacity only for one of these tasks. Depending on the precise nature of the functions required, the two may be reversed, with the signal conditioning

[§]This is in itself a simplification. Some designers split functions between the 'analogue' and 'digital' sides, so that, for instance, part of the calibration/compensation (maybe simple, linear scaling) is performed using analogue circuitry, while the rest (more complex, non-linear functions) are performed digitally. It has even been known for analogue calibrator/compensators to be implemented using digital circuitry, with an ADC at the front, and a DAC at the back to produce the analogue output. With so many options open to the designer, and so many engineers ready and willing to use any or all of them, a coherent and inclusive classification is almost impossible.

occurring before the calibration/compensation, and thereby using analogue circuitry. Another reason for this to be necessary would be if the analogue amplification and analogue to digital chain had insufficient signal headroom, given the gain required and offset errors needed. In this case the offset needs to be corrected before digitisation.

After both functions:
This option is likely to be taken when there is limited digital capability, or integration has been made with a technology highly optimised for analogue circuitry. The functions in both blocks are likely to be simpler than those achieved using digital circuitry.

3.3.1. *Analogue to Digital Conversion*

There are many different configurations of analogue to digital converter available to the designer, but almost all of them fall into one of two broad categories: sampling converters or integrating converters.

Sampling Converters:
They sample the level of the analogue signal at an instant. One type of sampling converter is a 'flash converter' in which there is a comparator circuit for each detectable input level. Thus, a four bit flash converter would have $2^4 = 16$ comparators, and an encoder, which produces the required output code. Flash converters are the fastest type of converter, but are also the most inaccurate, and use the most resource (power and chip real estate). The most commonly used type of sampling converter is a parallel feedback converter, often of the 'successive approximation' type, generically shown in Figure 3.12. Here there is a single comparator, which compares the input signal with the output from a digital to analogue converter (DAC). The DAC is placed within a digital feedback loop, the function of which is to equalise the ADC input and the DAC output. When equality is achieved, the digital value feeding the DAC is taken as the digital value corresponding to the input. The DAC can use a resistor ladder network, or can be capacitive.

All sampling converters require the input signal to be held constant while the conversion takes place. This is achieved using a 'sample and hold' circuit, typically a transistor feeding a capacitor.

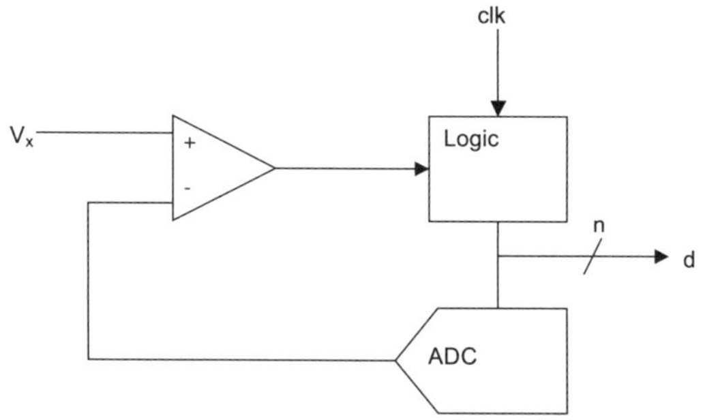

Figure 3.12: Parallel conversion converter.

Sampling converters suffer from aliasing, in which high frequency components of the input waveform produce spurious low frequency components in the output. They therefore require that high frequency components be removed with a low pass filter, the *anti-aliasing filter*, which must be implemented using analogue circuitry, using the techniques discussed in Section 3.2.5.

Integrating converters:
These naturally produce a single bit, frequency modulated output, using a single comparator. There are several types of integrating converters, as discussed below.

In a *dual slope* converter a capacitor is charged by the input, then discharged at a constant rate. The discharge time (and hence frequency) gives a measure of the level of the input signal. A *charge balancing* converter uses a similar mechanism, but removes the charge in discrete packets, if enough charge is present on the capacitor. Thus the pulse rate of the output is proportional to the input.

The most sophisticated form of integrating converter is a *sigma-delta* converter. This works in the same way as a one bit successive approximation converter, but the sampling frequency is many times higher than the highest input frequency. The one bit digital output is processed by a digital filter which integrates many one bit samples to produce a single multi bit sample.

Integrating converters do not require sample and hold or anti-aliasing filters, since the inherent integration performs these functions. The disadvantage is that they are inherently slow. However, digital circuitry can now operate at very much higher frequencies than typical analogue processes. Integrating converters, particularly sigma-delta converters are almost entirely digital (apart from a single comparator), and so can be fabricated using modern digital technologies with a conversion rate adequate for most sensor applications. As a result, the sigma-delta converter has become the most widely used form of converter, particularly in integrated applications utilising digital processes.

Matching the converter to the application

Given the characteristics of a particular technology, it is possible to optimise a converter to suit, and a fair amount of research has been undertaken on such optimisation. A number of variations on the converter types discussed above can be found in the literature, optimised for an individual or a class of applications. Some considerations to be taken account of in selecting an optimal converter type are discussed below.

The narrow bandwidth of most sensors (at least the physical ones) compared with the clock rates of modern digital circuitry make oversampled converters (incremental ADCs and sigma-delta modulators) a popular choice for digitisation since they allow high resolution to be achieved. However, when low power consumption and fast conversion time are most important, successive approximation ADCs are preferred [16].

In many cases, however, the choice of the digitisation method stems from system level considerations. A good example here is the capacitive accelerometer presented in Chapter 5, where signal quality enhancement through closed loop operation led to the use of sigma-delta modulation schemes, which in turn performed the digitisation task as well. Another example is the capacitive humidity sensor presented by Maloberti *et al.* [16], where variation in capacitance at the level of a few tens of aF are measured by using either a switched capacitor read-out, if analogue output is desired, or a fully differential switched capacitor second order sigma delta modulator configuration for digital output. The sigma-delta converter is an extreme example of how digital post processing can overcome inherent weaknesses of a digital technology used for analogue purposes.

All this having been said, it is often the case that the selection of digitisation method is made not as a result of application level considerations, but as a result of a previous decision to use a particular VLSI technology.

3.4. Digital Signal Processing

Digital signal processing provides a much wider range of signal conditioning capabilities than analogue signal processing, which may be applied with advantage to the signal processing functions already discussed. For a start, digital signal processors may effortlessly perform most compensation tasks, provided they are endowed with sufficient resources. The simplest technique, but one which requires the largest memory usage, is a simple look-up table. The table is indexed by the ADC output values, (one entry per sample level), and provides the corrected output value. If compensation is required for more than one variable (a secondary variable is usually temperature), then multiple look-up tables may be used, indexed on the secondary variable value. Memory space can be saved by using look-up tables which provide only a subset of the input values. Intermediate outputs are calculated using interpolation, either linear, or some more sophisticated interpolation functions, such as a cubic or a spline function. These provide higher accuracy at the cost of increased processing power required. Alternatively, a transfer function can be computed from scratch, providing the minimum memory requirement but needing the largest computational resource. A detailed account of the construction and use of look-up tables is given in Chapter 4.

Another function in which digital signal processing may provide advantages over its analogue equivalent is filtering. Whereas analogue filters may be considered as a transfer function from a continuous time based function to another, output, function, a digital filter, may, by analogy be considered to be a transformation from one input set of discrete values, or samples, $[f_m]$, to another, output set of values $[g_m]$. Again, the required frequency response is transformed into a time domain transfer function, this time a discrete function. The transfer function is described by:

$$\bar{H}(j\omega) = \sum_{n=-N}^{N} b_n e^{-jn\omega T}. \tag{3.7}$$

This is simply a discrete Fourier transform, and thus the coefficients b_n can be determined using a filter synthesis formula:

$$b_n = \frac{T}{2\pi}\int_{-\pi/T}^{\pi/T} \bar{H}(j\omega)e^{jn\omega T}d\omega \quad -N \le n \le N. \tag{3.8}$$

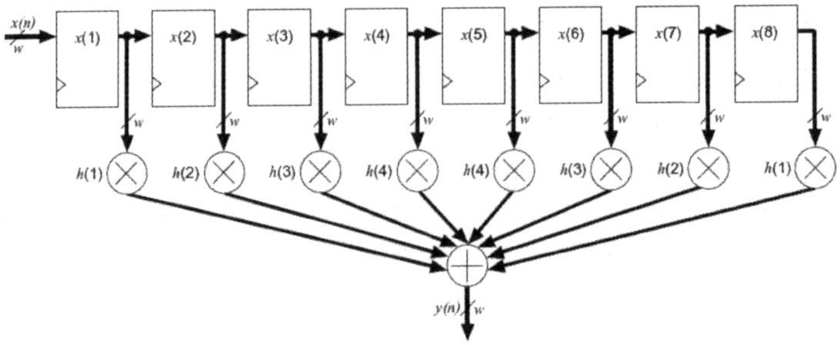

Figure 3.13: Digital filter configuration.

The ability to directly approximate virtually any required transfer function is a huge advantage of digital filters. Practically, the time delay required is achieved simply by clocking sample values along a shift register. A bespoke digital filter therefore consists of a tapped register feeding a series of multipliers (which multiply each sample by the coefficient) and adders (which successively accumulate the output value), as shown in the Figure 3.13. A more practical way of providing the function is to calculate the filter function algorithmically, using a suitably fast processor. Typical processors for this type of use (termed digital signal processors, DSP) will have parallel multiplier accumulators to allow single cycle calculation of the filter terms. Filters built using the principles above are called Finite Impulse Response (FIR) filters, since for practical reasons the length of the coefficient sequence must be truncated. Using DSPs, it is also possible to build Infinite Impulse Response (IIR) or recursive filters, in which the coefficients act on the output of the filter as well as the input. IIRs can produce better approximations of a required transfer function for a given number of coefficients, at the cost of more complex hardware or greater processing overhead.

Huddleston [17] identifies several ways of improving the accuracy of measurement using digital filtering. The context here is thermocouples, but the principles can be applied to other types of sensor. The five techniques suggested are:

Oversampling:
Uses sampling at rates well above the Nyquist rate. By doing so, we can spread the replicated spectra further apart. This minimises leakage between adjacent spectra and allows the use of powerful algorithms without incurring significant signal delays.

Removing Power-Line Interference:
The application of a low-pass filter with a very low cut-off frequency can significantly improve system accuracy by removing much of the power line noise. Even though the power of the AC noise is usually much higher than that of the sensor signal itself, because the noise spectrum is separated from the signal of interest, we can apply a unity-gain low-pass filter to kill the noise very effectively. The implementation of the low-pass filter can be in the form of either an infinite impulse response (IIR) or a finite impulse response (FIR) filter, depending on the amount of memory and computational resources available. IIR filters usually consume fewer resources, but FIR filters allow the designer to tailor the filter characteristics more precisely.

Median Filtering for the Removal of Shot Noise:
Noise that occurs sporadically, also known as shot noise, may have a variety of electrical or mechanical sources. The applicable technique here is median filtering. A group of samples is sorted in ascending or descending order. The median value is picked as the filter output. In effect, a median filter is another form of low-pass filter but tends to do a better job than an averaging filter in the presence of shot noise because it rejects non-representative samples rather than merely attenuating their effect.

Multichannel Averaging:
Multichannel averaging offers a way to remove noise that is induced in a particular path, should additional, redundant sensors be available. Simply averaging readings across multiple sensors measuring the same (or nearly the same) location, significantly diminishes the effects of non-common-mode noise. Median filtering can also be applied across sensor readings (assuming that we're measuring the same spot) to further reduce the effects of shot noise.

The balance between analogue and digital

For the designer, the questions raised by the hybrid analogue/digital functions described above are as follows:

- Given the use of a large amount of digital circuitry within the system whether it is or not advisable to 'go the whole hog' and use digital functions for calibration and compensation;

- If the signal is to be digitised in any case, may not the memory used to control the calibration resistance networks, for example, be used simply for a look-up table to perform the same function digitally?

The 'best' answer (in the absence of any other constraints) is almost totally dependent on the relative performance of the available analogue and digital technologies. For a given semiconductor technology analogue computation can provide more 'computational power' than digital computation, because analogue computers model the required functions directly using varying voltages or currents, whereas digital computers must perform many thousands or millions of individual calculations to produce the same result. Hybrid solutions seek to exploit the power of analogue computation with the flexibility, and some of the algorithmic sophistication available to digital computing. This allows a small microcontroller with limited memory resources to be used, essentially, to sequence and control a powerful analogue computational system. Such a mix is typical of mixed signal technologies, which do not have the capability, generally, to build very complex digital systems. However, by and large, we live nowadays in a digital world. A consequence of this is that digital semiconductor processes have seen huge investment and development, and are now significantly more refined and capable than their analogue counterparts.

Selection of digital processor

When selecting a suitable processor to support integrated digital functions for a sensor, the designer is usually offered a choice between one of four, generic types of processor device.

Microprocessor:
This is a generic term for a single chip computer processor. From the original microprocessor (the Intel 4004) it has been expected that a microprocessor will integrate the ALU, sequencer and register set for the processor, but will not include memory or peripherals. Sometimes the term 'microprocessor' is used simply to denote a processor aimed at the general purpose computer market — some of these may be suitable for use in embedded systems such as 'intelligent' sensors and will tend to provide more processing power for a lower cost than specialised processors, due to the economies of scale made

possible by the large consumer markets. Microprocessors will tend to be used within sensor systems when there is a need for a high level of processing power and large amounts of memory, such that sufficient resource cannot be packaged within a microcontroller (see below), or where the microcontroller I/O hardware does not suit the particular sensor application. Another factor may be the size of the software system envisaged. Microprocessors will tend to have easier to use instruction sets and a higher level of software development tools available than will some of the alternative choices given below, and if the software system is to be large, this may be an important consideration.

Microcontroller:
These were developed specifically to provide a single chip solution for the processing part of small embedded systems. Typically a microcontroller integrates a processor with sufficient data and program memory (program memory usually in the form of ROM), to allow a simple control program to run, and a number of generic peripheral device interfaces. The microcontroller provides a good solution if the on-chip memory and peripheral devices are sufficient or suitable for running the programs envisaged for the digital part of the system, and if its processing power is sufficient to provide those functions in real time.

Digital Signal Processor (or DSP):
This is a term for a microprocessor with an architecture optimised for signal processing requirements. Usually these optimisations consist of an ability to perform fast multiplication operations and a single clock execution cycle using a Harvard architecture (separate memory ports for instructions and data, allowing both to be transferred simultaneously). A DSP is a good solution when sophisticated digital signal processing, such as digital filtering, is to be used.

Digital Signal Controller (DSC):
This bears the same relationship to a DSP as does a microcontroller to a microprocessor, namely it is a DSP with integrated memory and peripheral interfaces. A DSC will be a good choice when a highly integrated solution is required, along with the level of processing power provided by a DSP.

An engineer's view of the choices available and the factors which drive the decision is given by Huddleston [17] as follows:

> *The ability to perform single-cycle high-precision mathematical operations is essential to implementing digital signal processing algorithms. Not only do the mathematical operations themselves need to execute rapidly, but the accumulator (register) that holds the results must be able to store the results from many operations. Usually, we're talking about accumulating 16-bit by 16-bit multiplications (32-bit results), which requires a 40-bit-or-better register that can quickly identify and recover from arithmetic overflows. Microcontrollers can't do this whereas the wide, single-cycle accumulators found in digital signal controllers easily support high-speed, high-precision data flow for implementing digital filters and other signal processing algorithms.*
>
> *Another requirement for maximum signal processing throughput is the ability to read from two separate areas of memory in a single cycle (for example, to get the filter coefficient and the associated data sample). Very few microcontrollers support this capability, known as a Harvard or modified Harvard architecture. Without it, the time to retrieve sample data and filter coefficients is doubled, effectively halving the processing throughput. Digital signal controllers all support either a Harvard or modified Harvard architecture, maximizing their ability to deliver data to the mathematical engine without delay.*
>
> *Finally, most microcontrollers do not support an internal bus wide enough to move 24- to 32-bit data efficiently. The overhead required to transfer high-precision data severely limits the amount of data the microcontroller can handle and unnecessarily complicates the associated code. With their wider internal data paths, digital signal controllers eliminate this overhead from both a processing and coding perspective.*

Huddleston's clear-cut views as to the suitability of DSP's over general purpose controllers are rooted in a fixed view of the division between the different categories of processing device available. However, as technology improves, the clear differentiation has become blurred. 'Microprocessors' have gained single cycle arithmetic operations and have Harvard architectures, and can perform signal processing computation at equivalent rates to specialised digital signal processors. The market for highly integrated computing devices such as PDAs and mobile phones has spawned powerful microprocessors with integrated memory and peripherals. Although these are not marketed as 'microcontrollers' they have precisely the same characteristics, apart from being endowed with much more processing power than is traditional for a microcontroller. Essentially, they have the same resources as a digital signal controller, but with the more easily programmed and more flexible architecture of the general purpose microprocessor. For example, probably the most ubiquitous 'microcontroller' in use today is the ARM, which is in fact a general purpose microprocessor. The ARM processor core exists as an 'IPR' (intellectual property rights) design, implemented as part of very many ASICs in a wide field of use. Most tellingly, the ARM is the processor of choice in mobile phones and many PDAs, so global volumes are enormous. The ARM has gained this position because the design was optimised from the start for low power and low chip real estate usage. Nonetheless, it is a powerful, 32 bit RISC design. As such, it includes many of the characteristics that Huddleston ascribes to DSPs, such as single cycle instruction execution, fast multiply/accumulate function, modified Harvard architecture and 32 bit wide data paths. At the same time, it is an orthogonal, easy to program processor, with a wealth of high level programming tools, ability to address a large linear address space and general processing capability. For many designs it will be a more tractable alternative to 'traditional' DSPs. Other modern 32-bit microcontrollers, such as those based around the MIPS and Motorola 68000 instruction sets, have similar characteristics.

3.5. Interface Configurations for Different Transducer Types

Having presented the purpose of the different functions in the signal processing chain, and their implementation requirements, the discussion will now return to the source of the signal processed: the transducer. The major transducer types will be examined again and the way in which the techniques presented in this chapter may be used for each of them will be explored. Some examples of the design choices made by researchers using each of these transducer types will be presented.

3.5.1. *Piezoresistors*

Peizoresistors require considerable amplification and other signal conditioning such as temperature correction. Due to the relatively poor performance of piezoresistors as a transduction mechanism, they are generally used in applications where high resolution is not required, hence the input amplifier does not need exceptionally low noise performance, and standard operational amplifier circuits suffice. Offset can, however, be a problem, since the magnitude of the offset can be comparable with the piezoresistor output. Figure 3.14 shows the input configuration of a Motorola pressure sensor [12]. Three operational amplifiers are used. The circuit uses laser trimmed resistors to adjust the temperature compensation of both span (R_s) and offset (R_{TO}) and the offset and the gain (R_g).

An example of a low cost 3-axis accelerometer based on the piezoresistive behaviour of polysilicon in standard CMOS is given by Kruglick *et al.* [18]. In this design, the resistance of the piezoresistor is $4\,k\Omega$, and the sensor sensitivity is $17.5\,mV/g$. The thermal noise in the resistor is $16\,nVHz^{-1/2}$. Using a standard CMOS process, a CMOS amplifier can be made with a noise figure of $38\,nVHz^{-1/2}$. This is sufficient to give a minimum detection threshold of 0.3 mg over a bandwidth of 500 Hz. The design can be fabricated without the cost of a custom process, and can be integrated monolithically with electronics, thus reducing chip count and wire bonding.

3.5.2. *Capacitive Pick-off*

If the capacitor is operated in a constant charge regime, then changes in capacitance will cause a change in voltage across the capacitor. This regime

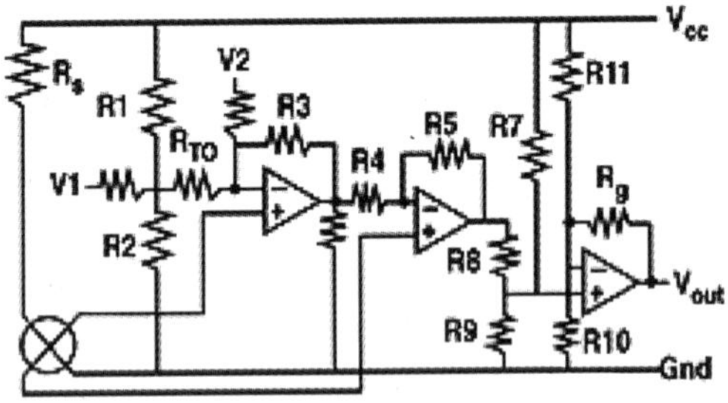

Figure 3.14: Integrated input amplifier for Motorola pressure sensor. From Senturia [12].

requires a very high impedance input amplifier (typically FET or MOS input), so as not to discharge the capacitor. Alternatively, the capacitor may be kept at a constant voltage, in which case capacitance changes will cause charge (and hence a current) to flow to and from the capacitor.

Given the very small capacitance changes, the capacitors are likely to be configured as a bridge with capacitors sensing the opposed motion or with identical non-variable capacitors for reference. In the case of accelerometers, the sensing element would consist of a seismic mass which can move freely between two fixed electrodes, each forming a capacitor with the seismic mass used as the common centre electrode. The differential change in capacitance (generated by the movement of the seismic mass as a result of an applied acceleration) is proportional with the deflection of the seismic mass from the centre position. An example of the input circuitry for a capacitive bridge is given by Luo *et al.* [19], in Figure 3.15. In this application, the high input impedance of the CMOS amplifier is a good match to the requirements of the capacitive pick-off. In this design the input offset is eliminated using an AC modulated mechanism, which exploits the AC impedance of the variable capacitor. By measuring the AC current flow through the bridge, the DC offsets are eliminated after the AC signal is demodulated. Figure 3.16 shows the input configuration for Luo's design.

The 2 MHz differential drive signal feeds both the bridge and a synchronous demodulator. The input amplifier (Figure 3.15) provides an amplified differential AC signal from the bridge, which is demodulated into a differential DC signal. Residual AC components are removed by the low pass filter. This particular accelerometer is a closed-loop design, with the force feedback being provided through the tap off R1 and R2.

Lee and Park [20] describe an input for an accelerometer using capacitive pick-off. This example uses a capacitor bridge, with each limb driven by opposite phases of an AC excitation signal. The front-end amplifier uses

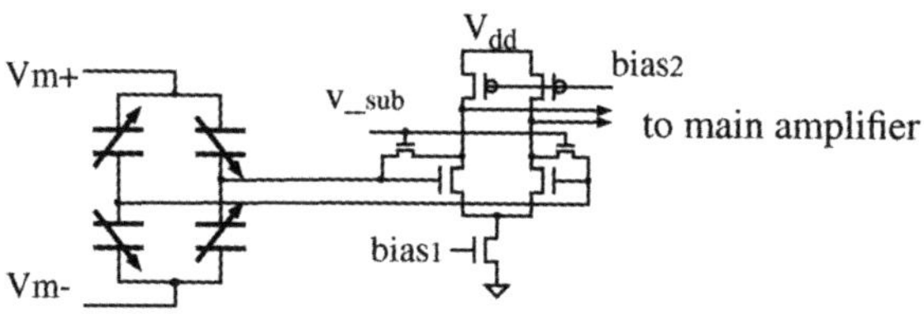

Figure 3.15: CMOS capacitive bridge amplifier. From [19].

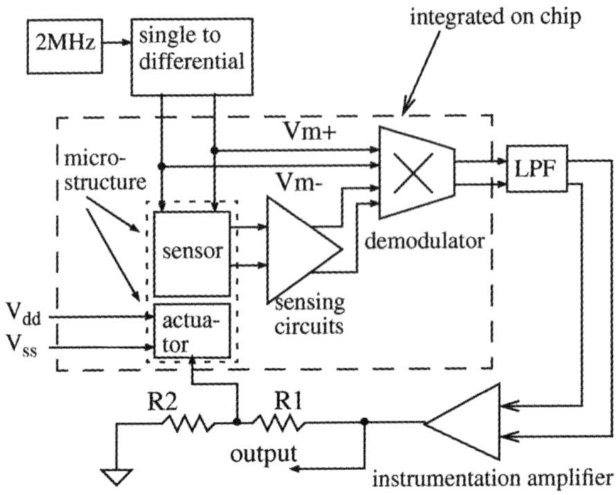

Figure 3.16: AC drive of capacitive bridge pick-off. From [19].

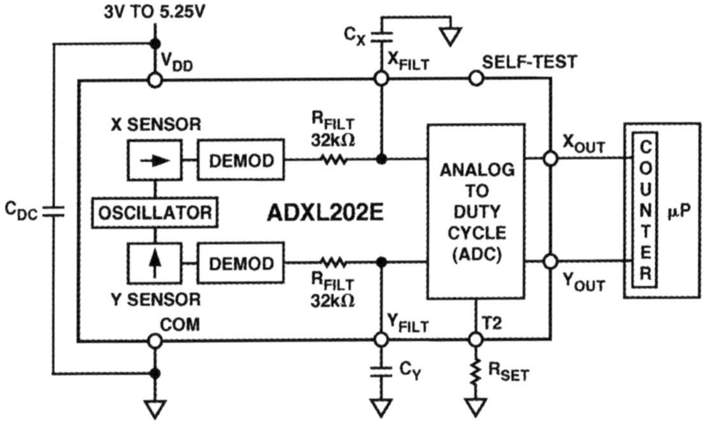

Figure 3.17: AC capacitive pick-off and signal processing used in Analog Devices accelerometer. From [21].

three CMOS operational amplifiers, with gain and feedback controlled by switched capacitors.

From the commercial range, the ADXL202 is a good example of integrated capacitive accelerometer which can measure both dynamic accelerations and static ones (Figure 3.17). Acceleration can be measured both on

X and Y axis and the differential capacitance is measured using synchronous modulation/demodulation techniques [21]. The user may limit the bandwidth and thereby lower the floor noise by adding a capacitor at Cx and Cy pins. The device resolves signals down to 20 zF, on top of common mode signals several order of amplitude larger. Analogue Devices took full advantage of the potential of full integration to provide a simply packaged and simple to use device in this successful product.

3.5.3. *Thermal Sensing*

An example is presented by Haberli *et al.* [22] which describes an integrated CMOS thermal pressure sensor system for the range 10^2–10^6 Pa. The operating principle of the sensor is based on the pressure dependent heat transfer across the air gap separating a heat source from a heat sink, known as Pirani effect, and realised here by means of a resistively heated wire with a high temperature coefficient of resistance. The filament is part of a Wheatstone bridge. Other mechanisms are also available for thermal sensing, including thermocouples and semiconductor junction sensors.

Another example is an IR flow sensor based on on-chip temperature detection using a thermopile [23], whose output signals could be in the μV range. Low noise and low offset amplification are major constraints for the electronic interface. CMOS technology can achieve this if the signal bandwidth is assumed to be low. The required sensitivity can be achieved by implementing autozero techniques and successive low frequency filtering of out of band noise.

Makinwa and Huijsing [24] describe a thermal wind sensor, implemented in CMOS. The principle of operation is shown in Figure 3.18. Thermopiles detect heat carried from heaters as the wind flows over the CMOS chip. Amplifiers A1 and A2 (Figure 3.19) receive a differential input and provide differential output. The CMOS technology used has a high and thermally dependant offset, which is removed using chopper stabilisation.

3.5.4. *Tunnel Effect*

Initial designs for tunnel effect pick-off were operated at high voltages (several tens of Volts or higher), limiting therefore the range of applications for these devices. Yeh and Najafi proposed a 10 V operated tunnelling

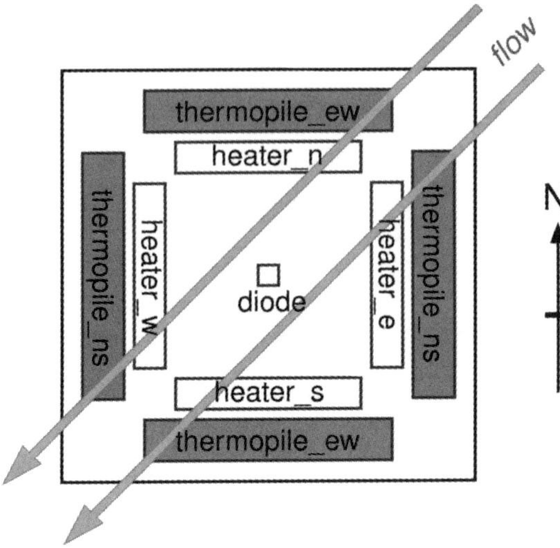

Figure 3.18: Thermal wind sensor. From [24].

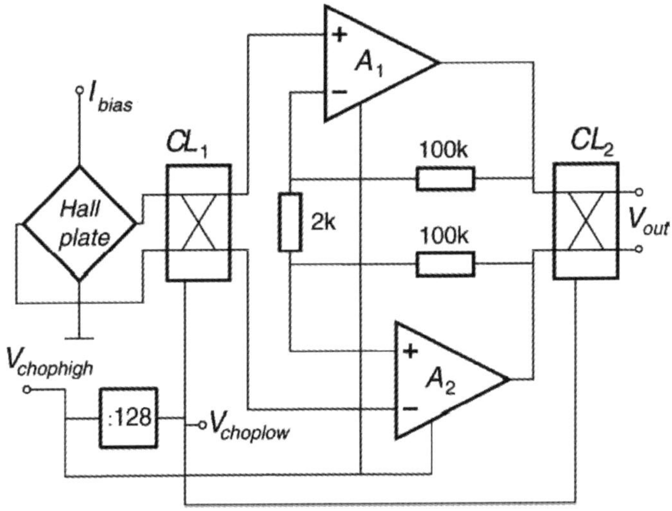

Figure 3.19: Input amplifier for thermal wind sensor. From [24].

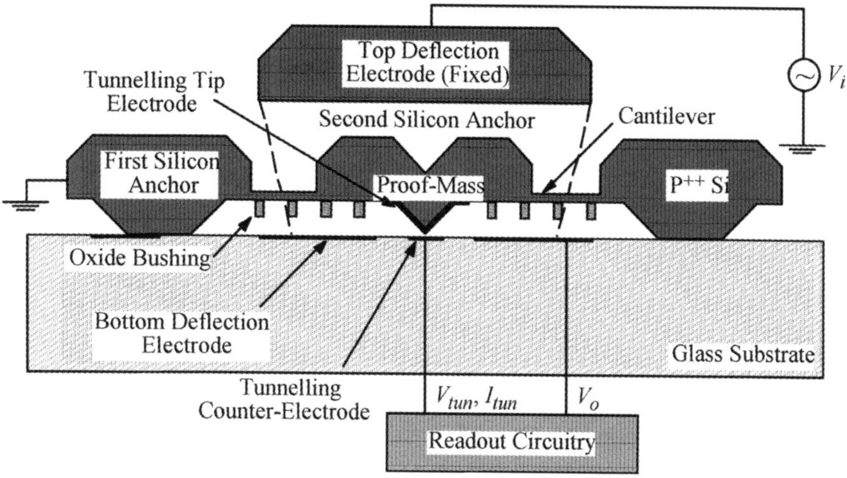

Figure 3.20: Structure of tunnelling accelerometer. From [8].

accelerometer [8], implemented using bulk silicon micromachining and the boron-etch-stop-dissolved wafer process, allowing the readout electronics to be CMOS compatible (Figure 3.20).

The interface circuitry uses a pn-junction diode as a logarithmic current to voltage converter, to eliminate the exponential non-lineality between the current and distance. A linear relationship between the acceleration and the output voltage can be obtained. The device is incorporated in a closed loop system, therefore the distance to current gain does not limit the measurement range and the device sensitivity is solely determined by the electrostatic feedback force rather than the tunnelling barrier height. The device produced by Yeh and Najafi, dissipates 2.5 mW, has a sensitivity of 125 mV/g, a bandwidth of 2.5 kHz, and a measurement range of 30 g. The long term output offset voltage and sensitivity drift are 0.125% and 0.15% respectively. Figure 3.21 shows the sensor's interface circuitry.

This design tackles the non-linearity of the tunnelling current by using a closed loop system. Force feedback is supplied to the tip to keep the tunnelling current constant. The output from the tip is amplified and compared with a reference. At the same time an AC 'dither' signal is added. After 'compensation' (filtering) the error signal is fed back to the control electrode via a high voltage amplifier.

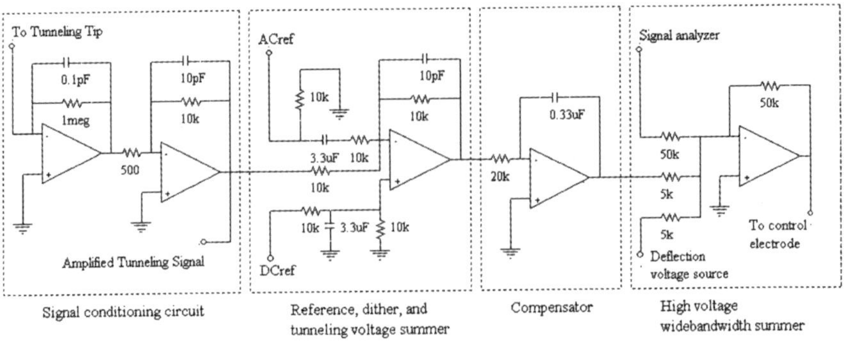

Figure 3.21: Input circuitry for a tunnelling accelerometer. From [8].

3.5.5. *Optical Pick-off*

In optical pick-off, the sensing devices are usually photodiodes. These may either be used in 'photovoltaic' mode, in which the electro motive force (e.m.f) across the diode junction is amplified using a very high input impedance amplifier. Alternatively, the diode may be used in 'photoconductive' mode, in which the diode is reversed biased, and the photo-current passing through the junction is measured. In this case, a transimpedance amplifier with a very low input offset current is needed. Generally, photovoltaic mode gives higher sensitivity, and photoconductive mode faster response. The requirements of both modes are large current gain and low noise [16] and are best met with a precision JFET input amplifier. Burr Brown Inc. give some examples of precision amplifiers in their application notes [25], which are reproduced in Figures 3.22 (photovoltaic mode) and 3.23 (photoconductive mode).

However, JFET input amplifiers are not always available in processes which are commonly used for electronics integrated with MEMS systems, which are generally CMOS or bipolar. Maloberti *et al.* discuss the design of an interface for a UV sensor based on a UV photodiode which does not make use of JFETs [16]. In this application bipolar transistors are used (which are deemed to be more suitable than CMOS based implementations). Indeed, a bipolar transimpedance amplifier (gain of 10^9 V/A) was designed and integrated with a UV diode to form a UV detector in a modified bipolar process, with good results.

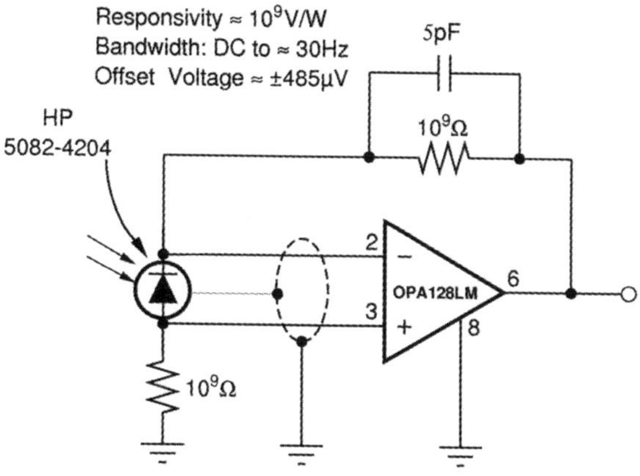

Figure 3.22: Photovoltaic mode photodiode amplifier. From [25].

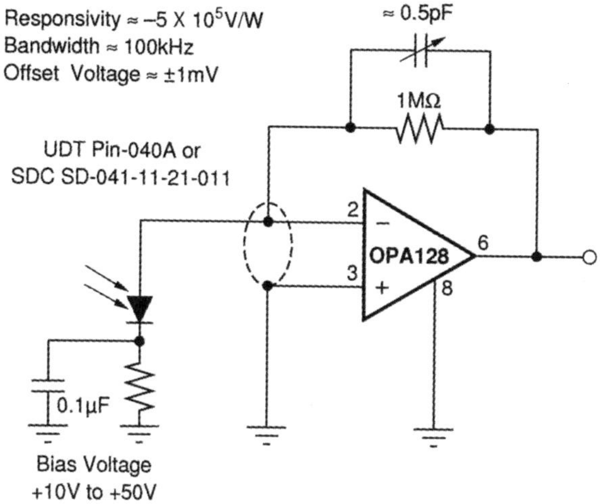

Figure 3.23: Photoconductive mode input amplifier. From [25].

3.5.6. *Design Options with a View to Integration*

The examples in this section showed how the different types of sensor pick-off place different requirements on the interface circuitry, particularly on the input amplifier. Ideally, the requirements for high sensitivity, low noise, and an input impedance well matched to the requirements of the pick-off dictate specialist input devices such as low-noise JFETs and bipolar transistors. In selecting a VLSI technology, such requirements severely limit the choice of the designer, who is likely to find that the most readily available and lowest cost processes are CMOS. Nonetheless, as we have seen, a number of designers have succeeded in producing input circuitry in CMOS, by using techniques such as chopper stabilisation and switched capacitors to deal with input offsets and precision setting of gain. Additional signal processing techniques may be available in a highly integrated system, such as filtering and the various techniques put forward by Huddleston [17]. Using these, the required information may be extracted from the input signal, even if the input amplifier is less than optimal.

3.6. Integration

The discussions in this chapter have alluded several times to the design constraints inherent to the integration of MEMS and sophisticated electronic circuitry to produce integrated sensor systems (sometimes denoted as 'smart' by vendors and designers alike). When integration is a product requirement, the goal is to select a means of integration which leads to meeting the constraints for the design in hand, a major such constraint being cost.

3.6.1. *Integration Options*

In his review of progress in MEMS integration [26], Bryzek presents the following options:

Integration of MEMS on Top of IC:
The most straightforward method of integration is to build the MEMS device directly on top of the CMOS wafers. This has the severe disadvantage of requiring strict process compatibility. MEMS structures are limited to those which can be surface-micromachined within the CMOS thermal budget (500°C). Another restriction is that the exposed materials, typically a low-temperature oxide and aluminium, limit the chemistries available for processing.

Lateral (Side-by-side) MEMS and IC Integration:
This approach overcomes some process incompatibilities between MEMS and CMOS. In this method, any CMOS incompatible processes are fabricated first. It is used for both bulk and surface micromachining.

Vertical Wafer-level MEMS-IC Integration:
This approach is based on wafer bonding of two or more wafers, at least one MEMS and at least one CMOS, each fabricated in a dedicated foundry. Compared to the effort of integrating MEMS and active circuitry at the process level, vertical wafer level integration of circuits and MEMS has the advantages that it lacks the restriction of process compatibility and it does not suffer the real estate penalty and inefficiency of lateral integration techniques. Because the MEMS and CMOS are fabricated separately prior to integration, it affords the designer absolute flexibility in the choice of active circuit process options. High-density digital, mixed signal, high voltage, BiCMOS, RF etc. can all be integrated using the same process steps.

In addition to the options above, there have been examples of *lateral, chip level integration*, in which electronics and mechanical chips are bonded side by side on a substrate (hybrid IC) and connected using bond wires. The illustrations in Figure 3.24 show an accelerometer by Silicon Designs Inc. [27], using this technique.

The first two options result in 'monolithic' or single chip implementations, which will lead to lower unit costs, all other things being equal. However, other things are rarely equal. Both processes are extremely difficult to master. The former (MEMS on top of IC), was used by Texas Instruments for their Digital Light Processor chips. The length of development (17 years) and the amount of money spent ($1B) is a good indicator of the difficulty of using this approach. Analog Devices used the second approach (Lateral) for their range of 'smart' MEMS accelerometers. The process took over ten years to perfect. One of the reasons that the processes are so hard to make economic is to do with process yield in LSI. The yields of each process step combine geometrically. Since device cost is determined largely by yield, the unit cost will also multiply with the yield reduction caused by each step. Unless individual step yields are very high, the total yield in a process with many steps will be very low, and the unit cost high.

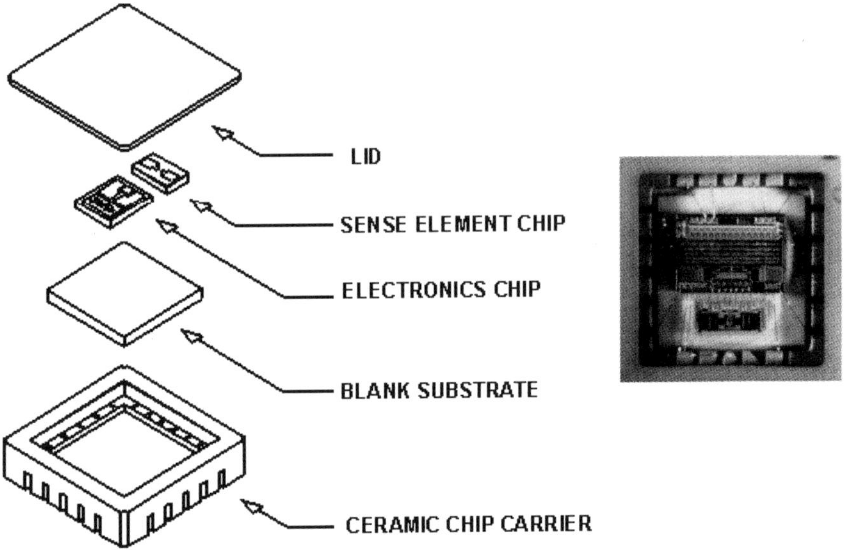

Figure 3.24: Multi-chip side by side integration packaging (Silicon Designs Inc.). From [27].

Both of the monolithic options add additional steps to a process which is likely to be already on the point of economic yields. The additional steps often involve aggressive micromachining, which may in itself have a very low step yield. Thus, unless the processes are very carefully refined, unit costs may well be higher than with the third option, vertical integration.

Wafer or chip level integration has the added advantage of allowing process optimisation for each chip. Thus, while most integrated single chip MEMS/VLSI processes are based on low cost CMOS LSI, esoteric mixed signal or analogue processes may be used with vertical integration. Similarly, mechanical components in the monolithic MEMS/VLSI processes are likely to be fabricated as silicon, silicon dioxide or connect metal (usually aluminium), since these are the materials available in the LSI process. By contrast, the mechanical component of a vertically integrated system may be fabricated using a range of materials, allowing optimisation for the required mechanical properties. The Silicon Designs mechanical chip in Figure 3.24, for example, is fabricated from nickel, which was found to have a very good set of mechanical properties and to be easy to pattern.

For vertical, wafer level integration, the issue is how to connect the electronics to the pick-off. The easiest integration can be achieved for electrostatically coupled devices, such as capacitive sensors, as no conductive electrical connection between IC wafer and MEMS wafer is necessary. However, for consistent performance, the placement and separation of the wafers must be very precise, and configurations such as interdigiated capacitors are impossible. This is one area in which sophisticated integrated electronics can aid process design. If the electronics allows self calibration and compensation of the device, then it may be possible to relax assembly tolerances. For pick-off technologies requiring electrical connection, conductive bonding or through wafer vias may be employed. According to Bryzek, there are currently three main technologies for CMOS to MEMS wafer level bonding: organic, anodic and metal-metal.

Organic wafer bonding has been demonstrated with such materials as benzocyclobutene (BCB). The advantages of using adhesives to bond wafers are that they can planarise over otherwise difficult topology. The disadvantages include their lack of long-term reliability and their questionable ability to form a long-term hermetic seal.

CMOS compatible anodic wafer bonding has been demonstrated, but requires high voltages (up to 700 V) which are potentially damaging to many sensitive circuits.

In respect with metal-metal bonding, there are several possible choices for suitable metals. In-Au, In-Sn, Solder-Au, and Al-Ge have all been demonstrated. Bonding temperatures for these different metal systems range from 140°C to 575°C. An example of process steps for solder bonding a CMOS wafer to a MEMS wafer is shown in Figure 3.25. The two wafers are pre-processed to fabricate the bond metallisation. The bonding recipe requires a precise bonding pressure and temperature profile.

Of special note is the fact that it is possible to create bonds which are purely structural, or, which can also conduct signals between the MEMS and the CMOS, thus functioning as a part of the electrical circuit.

3.6.2. *The Effect of Integration on Fundamental Design Choices*

In Section 3.4 we discussed some of the options available concerning the choice of a digital processor. This was, however, from the point of view of selection of a suitable part 'off the shelf'. Seemingly, the designer of a

Ti/Ni/Au are patterned in the appropriate locations on the MEMS wafer. *MEMS wafer is ready for bonding.*	
A seed layer of Ti/Cu is deposited on the CMOS wafer.	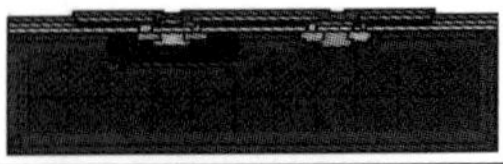
A photoresist mold is patterned for electroplating of a Cu spacer layer and a solder bonding layer. By patterning these bonding pillars over electrical contacts in the CMOS, vias are created that route signals up to the MEMS device.	
Cu and solder are electroplated to the desired thickness. The photoresist is removed. A short etch is used to remove the plating seed layer. *CMOS wafer is ready for bonding.*	
Cross-section of a MEMS and CMOS wafer bonding with a solder-Au bond. The outside bonds make contact to metal of the CMOS. The inside bond is electrically isolated	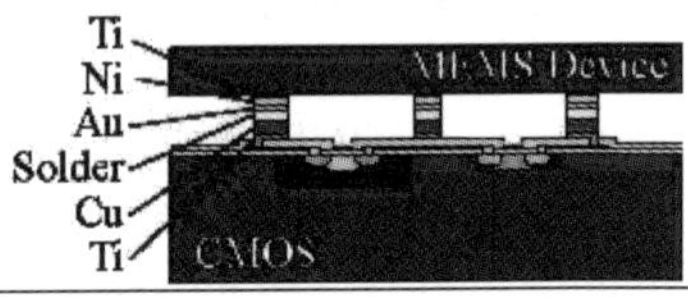

Figure 3.25: Vertical wafer-level MEMS-IC integration. From [26].

new, integrated, single chip MEMS/VLSI system would have total freedom in the design of an appropriate or optimum processor and custom signal processing circuitry. In practice, the most sensible choice may be the use of an existing processor design, realising the custom functions as 'firmware', even though such a choice may be less than optimum in terms of chip real estate usage, total data throughput or power usage. The reason for this arises from the economics of VLSI system development.

In the design of complex systems, in which each design iteration requires a large investment, minimising iterations by eliminating design faults is vital. Thus it is necessary to find development methods which cut out the most common reasons for failure. Figure 3.26, by Miller [28], shows the reason for failure of the *first silicon* wafers produced for a number of different new designs (notice that the bars add up to more than 100% — many designs have multiple faults). VLSI development is fully representative of the type of process discussed above, in that by the time first silicon is produced, a very large level of expenditure in the design has already been made. Furthermore, every iteration of the design reticules for the chip involves considerably more expense if it requires a substantial redesign.

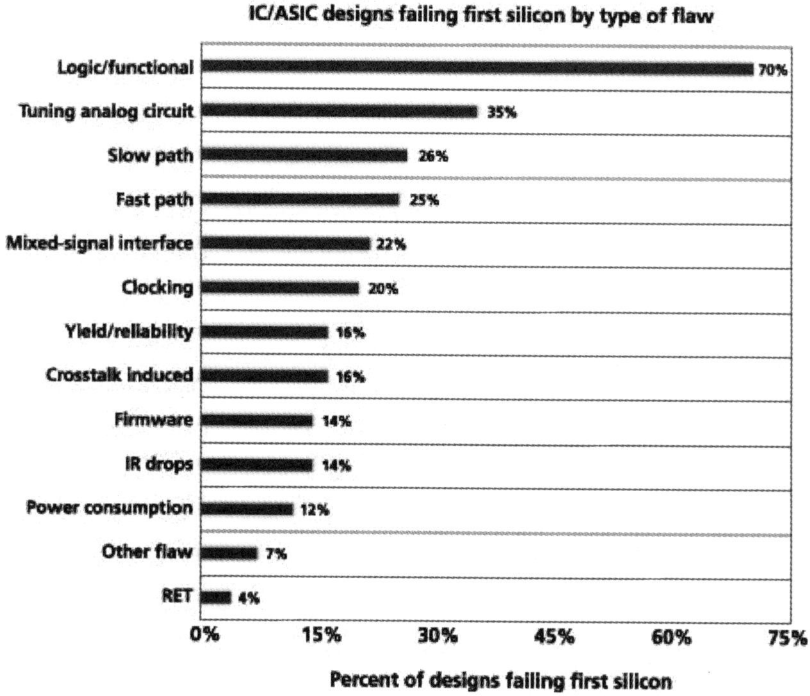

Figure 3.26: Reason for failure of ASIC designs. From [28].

From the chart, it can be seen that the tuning of analogue circuits accounts for 35% of failed designs, with the mixed signal interface accounting for 22%. Minimisation of analogue circuitry, and adoption of simple, standardised mixed signal interfaces is one way round this. This seems to be at odds with the observation that digital, i.e. 'logic/functional' failures, are the greatest cause of failure, at 70%. However, this refers to new digital designs. If the design is based around an existing, proven processor core, then the associated error is the firmware problem, at 14%. Even this can be removed if EEPROM or RAM is used for program storage, allowing the chip to be reprogrammed after production (however, chip programming is a major production cost, and is unlikely to be acceptable in large volume applications). As was discussed above, yield maximisation at each stage is fundamental to the achievement of economic integrated microsystems. Thus, these factors will tend to favour digital, programmable solutions.

To cater for the demand for these solutions, in the field of modern ASIC development, subsystems such as digital processors (including general purpose and specialist signal processors) are available as 'IPR' (Intellectual Property Rights — the term misused by the industry to denote pre-designed and validated subsystems which a chip designer might integrate into a larger system, so long as the property rights are paid for). Once again, the programmability of digital processors allows such a subsystem to be precisely configured to the job in hand without making any structural changes which might invalidate the IPR vendor's warranty.

The net result of these tendencies — the post production correctability of digital processor based designs and increasing availability of pre-designed and tested digital IPR — is that digital signal processing circuitry is used increasingly, even when its use would seem to be an over-sophisticated option, when viewed 'bottom-up', by a design engineer.

3.7. Design for Power Awareness

The per-sensor electronic systems discussed in the preceding sections of this chapter require a power supply to work. This is not usually a problem when the electronics is housed in a central unit, and the remote sensor devices are purely passive transducers. However, many types of transducer need a power supply to work at all, and, as has been seen, require integrated electronics to operate well. If such sensors are wired to a centralised application system, then power can be provided via the wires. However, not all applications system will use wired sensors. In Chapters 9 and 10 we will examine a new class of sensor system, composed of *ad hoc*, wireless connected and networked intelligent sensors. We will see that to operate in these systems sensors need to be equipped with considerable amounts of processing capability, as well as the ability to communicate using radio, optical or other wireless transmission medium. Since they have no physical connection with their controller (centralised application system), they need to be self-powered, typically from batteries. Applications of this type have meant that design for low power usage is a topical research subject, since innovative system-level techniques are required to eliminate energy inefficiencies (which would have been overlooked in the past) if sensors are to be part of distributed networks and operate within long missions on tiny batteries. (Whilst the transistor densities in digital chips double every

18 months, the energy density of batteries doubled only every 5–20 years. It is therefore necessary to adopt a global system-level perspective on the node when designing the sensor component [29].

3.7.1. *Design for Low Power Consumption*

Small, deployable sensors (nodes) imply limited physical space for batteries, with most applications making periodic battery replacement inconvenient or impossible. A state-of-the-art lithium battery offers an energy density of $2\,\mathrm{KJ\,cm^{-3}}$. Assuming that $1\,\mathrm{cm}^3$ is the battery space available within the node and that the desired device lifetime is one year, the average power dissipation must be less than 63.4 uW. This value exceeds the stand-by power of most digital systems [30].

Successful implementation and deployment of intelligent and autonomous sensors heavily depends on the ability to resolve in a cost and size-effective manner the powering of such devices for long periods of time. If the most 'attractive' and adventurous sensor projects are considered, such as the embedding of these sensors by the thousands in materials and spreading them by the millions for environmental monitoring (throw away sensors), it is clear that the life-time of the sensors has to be long enough to justify their deployment, cheap enough to account for redundancy, loss at deployment and finally loss at end-of-life and also small enough to match the micro scale of the rest of the system.

Various research groups working in the area of low-power circuit design have developed several approaches to tackle this challenge, of which, some are:

Hardware Level Efforts:
Focused around novel circuit design optimised for low-power consumption. Two strands here are:

- New architectures
  Examples include mixed-signal microcontroller architectures, as proposed by Ravindan *et al.* [31] which uses a 900 mV analogue front-end (AFE) for single-chip instrumentation and an on-chip distributed capacitance LC clock generator (chip to follow migration into SOI to explore relevant power saving optimisations); the AFE contains input voltage buffers, a programmable gain amplifier, a second order sigma delta modulator and

Figure 3.27: WIMS mixed signal microcontroller. From [31].

a third order digital comb filter; switched capacitor circuits are employed throughout due to their superior low voltage, low power performance and robustness. The resultant chip is shown in Figure 3.27.

- New device level power saving strategies
  Particularly in CMOS, exploiting sub-threshold transistor functionality. One of the challenges in designing robust low power circuits comes from the fact that traditionally, supply voltage has been scaled along with technology to keep power density within manageable limits. The threshold voltage has been simultaneously reduced to maintain speed. However, because of the exponential relationship between leakage current and threshold voltage in the weak inversion mode of the transistor, leakage power becomes significant at low threshold voltages and needs to be effectively controlled. Gate-to-body tunnelling leakage is also cause for concern when the gate oxide thickness is reduced to a few nanometers [31].

System level:
Development of efficient strategies for sharing power (on demand, time shared) between components within the sensing system. This applies equally to a single sensor system and multiple-sensor systems.

Software level:
Development of retargetable power-aware compilers that enable efficient power management in the system; the aim is to reduce the demand for

power-critical computations and enable the compiler to explicitly manage the processor resources (such as the memory system) in a fine-grained manner [31]. For example, data with high access affinities will be grouped together in the memory layout and therefore the amount of memory active at any one time can be minimized allowing the reminder to be powered down. Another line of work coming from Ravindram *et al.* [31] involves customising the instruction set architecture for the microcontroller.

Of these approaches, the second is probably the most represented, with a host of groups continuously proposing novel power efficient analogue or digital designs. The mixed-signal circuitry area also receives a great deal of attention with VLSI designers striving to develop mixed-signal micro controllers which can be appropriately interfaced with solid state sensors [31].

The success could be in addressing the power considerations over the full range of design levels, from architecture to interconnect, which reinforces the imperative for the top-down microsystems design approach supported by the authors.

Interestingly, in parallel with the above 'old-electronics'/software optimisation efforts, a new line of thought appears to be blooming, which suggest replacing whole electronic complicated functional subsystems with their micro-mechanical functional equivalent. Two examples here are the mechanical amplifier [32] and the DC-to-DC converter proposed in [31]. Traditional step-up DC-to-DC converters are hybrid devices consuming a large amount of area, typically implemented by storing energy in an inductor, breaking the current flow by opening a switch, and pulling the energy off the inductor. The inductor will have a high voltage because of the inductive kick. A scalable such converter was designed and implemented with 3–10 V and output 300 V at 1 W, with a target area less than 1 cm^2, based on a MEMS switch which minimises the parasitic losses and improves efficiency.

Whilst the power consumption concern is relatively new to sensing devices designers and MEMS technologists, the large majority of researchers and programmes concerned with systems of microsensors (i.e sensor arrays and sensor networks in all their variety) make this issue the only technological one remaining to be solved (that is apart from all software ones) in order for large scale MEMS applications to become a reality.

It is for this reason that most of the 'low-power' sensor designs and related publications are from the domain of distributed microsensor systems. Here, one of the main design requirements is to optimise across all levels of abstraction, with the goal of minimising energy dissipation. As a consequence, protocols required for collaborative sensing and information distribution, system partitioning considering computation and communications costs, low energy electronics, low power system design and energy harvesting techniques are all playing a role in the bigger picture of creating efficient large scale systems of microsystems. The discussion on low power strategies and circuitry goes beyond the scope of this chapter and refers to the whole of electronics associate with a sensor, be it for pick-off, signal conditioning, compensations and other computations and ultimately for communications and network infrastructure support. Some brief considerations are given in the next section.

3.7.2. *Power Aware Systems*

Whilst traditionally one would assume, when designing a sensor system that the requirements are static as such, when referring to applications such as Wireless Intelligent Networked Sensors discussed in Chapter 10, the amount of resources available (e.g. battery lifetime), the quality requirements (e.g. accuracy of the sensing results) and the latency requirements can vary over time. Scaling quality or latency with respect to energy dissipation can be done by exploiting system-level power down. At circuit level, dynamic voltage scaling allows the energy dissipation of a processor to be scaled with computation latency or Quality of Service [33].

Moreover, many sensor systems need to be designed to deal with low duty cycles. Many sensing applications involve the sensor remaining in a sleep-mode until some interesting event happens, detected by some simple front-end electronics. Maximising battery lifetime means the stand-by power of the computation (and communication) circuitry must be minimised. In digital electronics this is a major problem since low voltage technology (i.e. low-threshold devices) results in significant sub-threshold leakage. Several technologies such as *multiple threshold CMOS* and *substrate controlled variable threshold CMOS* are emerging and are promising in controlling the leakage current in stand-by mode. Low threshold devices may not be necessary in many sensor applications, since the throughput requirements are low, but they are a major concern for high performance applications such as image sensors [33].

Power saving and digital processors

One of the most successful initiatives in the area of power aware design for sensors and sensor networks was the μAMPS (μ-Adaptive Multi-Domain Power Aware Sensors) project at MIT. The aim in the μAMPS project is to achieve an average power dissipation of 100 μW per sensor node, in order to take advantage of ambient energy sources using 'energy harvesting' [34]. One of the techniques put forward was that of Dynamic Voltage Scaling on the sensor node's processor (implementation and experiments with off-the-shelf components by Min *et al.* [35]. The proposed technique, applied to sensor nodes built with off-the-shelf components and computation supported by a StrongARM SA-1100 processor showed a 53% reduction in the node energy consumption, from 1 W to 450 mW, whilst a strategy of 'sleep' for the sensing device and the ADC lead to 4% savings.

Apart from those described here, there may be opportunities to save power by optimising the detailed design at different parts of the signal processing and communications chain. Min, R. *et al.* [35] give a detailed discussion on digital processing circuitry typically used for digital signal processing of gathered data and implementation of the protocol stack. Wang and Chandakasan [36] take as an example a source tracking and classification application and a detailed analysis of power consumption and strategies for reducing it are presented, with calculations and concerns separated by levels of abstraction.

System level power saving

Most work seems to be concentrating on power saving at processor level and the communications block. The sensing circuitry, which requires power for bias voltages/currents, A/D conversion, amplification and filtering is seen as having a relatively constant power dissipation while on and improvements to their energy efficiency depends on increasing integration and skilled analogue circuit design.

More or less most avenues for power saving and management both at sub-system level and overall sensor node level in a wireless sensor network were explored and examples were treated in Chandrakasan *et al.* [33]. One idea is to use hard-wired (or application specific) processors rather than programmable solutions as it is claimed they offer more than three orders of magnitude reduction in energy dissipation to implement a function. Of course there are disadvantages associated with the use of hardwired solutions. The reader is referred to the section on design choices and integration in this chapter.

Power can be aggressively managed by designing the hardware to anticipate the application requirements of the microsensor domain. To address the low duty-cycle encountered in many applications, fine grained sub system shutdown can be devised to minimise energy dissipation in idle nodes. Energy consumption can be scaled gracefully with performance (both for computation and communications). The challenges identified by Wang and Chandrakasan [36] for next generation sensor nodes are as follows:

Low Voltage Design:
Sub-threshold/leakage reduction by use of fine grained threshold voltage control or Multi-threshold CMOS (MTCMOS), which reduces stand-by leakage power by severing a circuit from the power rails. Any unused circuit regions can enter sleep mode while surrounding circuits remain active.

Stand-by Voltage Scaling for Leakage Reduction:
Lowering V_{DD} (limited by the requirement to preserve circuit state) during stand-by mode will reduce power by decresing V_{DD} along with subthreshold current and gate leakage. An open-loop approach pinches in the rail voltages during stand-by using diode stacks and a power gating MOSFET, reducing V_{DD} by approximately 40%. Closed loop approaches are even more effective.

Energy Scalable Computing:
Energy aware design is in contrast with low power design which targets the worst case scenario and may not be globally optimal for systems with varying conditions, such as many sensor applications. A new metric for design, proposed by Bhardwaj, Min and Chandrakasan [37] is to maximise energy awareness. The above authors are of the opinion that the energy awareness of a system can be increased by adding hardware to cover functionality over many scenarios of interest and by tuning the hardware such that the system is energy efficient over a range of scenarios.

Technology advances and circuit and architectural optimisation allow for supply voltages reductions to 1 V and below. Such techniques need to be leveraged in sensor systems to minimise energy dissipation. Designing electronics that provides a knob to trade-off energy and quality/latency is one of the future challenges [33].

Power Harvesting

An alternative or complementary approach to extreme power conservation is the use of novel power sources. Particularly attractive is the harvesting of power from the general environment of the systems, which might come in several different forms. The most familiar sources of ambient energy (which is to be converted into electrical form) are solar power, thermal gradients, RF and mechanical vibration. Given that the power consumption of low to medium throughput DSPs is projected to be scaled to 10's to 100's of microVolts, energy harvesting is an attractive option, explored in many research papers (see for example Chandrakasan *et al.* [33]). The following account by Williams [38] discusses current research in this area:

> *Advances in microfabrication techniques have allowed the development of MEMS that integrate intelligent electronic control systems with a mechanical system on the micro scale. MEMS researchers have developed all types of diagnostic and measuring instrumentation, as well as miniature actuators and other sensors. The next step in making MEMS truly autonomous is to make them independent of classic macroscale power sources. The advantages of using alternative energy power systems to supply micro power to MEMS, making them truly self-sufficient, include abundance, simplicity, reliability and low cost.*
>
> *Solar Power*
>
> *MEMS tend to have a very low power requirement, usually less than 1 Watt. At its surface, the sun has an energy density of approximately 62.5 MW/m^2. By the time it has reached a satellite orbiting the earth, the radiation output is an abundant 1340 W/m^2. A crystalline solar panel, with an efficiency of 34% and measuring only 5 cm × 5 cm (or a thin film panel measuring 7 cm × 7 cm, only 17% efficient) could potentially supply 1 W of power to MEMS.*

Because they are made with the same micromachining techniques, thin-film cells are easily incorporated into existing MEMS designs. Their disadvantage is a very poor energy density.

Newer research is being done on multi-junction Monolithically Interconnected Modules (MIM). MIMs layer several materials of different bandgaps to maximize the radiation captured. MIMs will potentially become the most efficient solar panels.

Many MEMS researchers already have an appreciation for solar energy as a power source. Companies and laboratories such as the Interuniversity MicroElectronics Center, NASA's Jet Propulsion Laboratory, and the Department of Energy's Sandia and Oak Ridge National Laboratories are researching solar-powered MEMS in conjunction with thin-film microbatteries.

For MEMS with a low power requirement used in space applications, the panel efficiency and supplied radiation are both high enough to justify use of solar panels in conjunction with a lithium micro-battery. Both the battery and the thin-film panel can be directly integrated into the MEMS. Used in this specific application, solar panels are a very viable alternative energy source for MEMS.

Thermoelectric Power

A thermocouple is made from two dissimilar semi-conductors materials connected electrically in series. Electricity is generated by pumping heat into the 'hot junction' of a thermocouple and having it rejected from the thermally parallel 'cold junction'. Macroscale thermoelectric generators like the Stirling R-55 have most commonly been used in deep space power systems.

Development of thermoelectric power generators on the microscale has begun in the past couple of years. It has not yet moved from research to industry, so manufacturing and testing is not yet standardized. Each μ-TEG typically consisted of an array of about 100 thermocouple elements electrically connected in series and sandwiched between the hot nd cold junctions. Each element was anywhere from 20–600 μm tall with a 60–80 μm diameter or thickness.

Thermoelectric power has the potential to be a viable alternative-energy source for MEMS when used in an environment with a large constant temperature gradient, such as an engine interior, or kitchen stove. The disadvantage of a small efficiency (typically 3%) is minimised by a reliable power source that is modular, easy to fabricate, and can be used almost anywhere a heat sink is used. Imagine waste heat from a laptop battery recharging an additional battery, or, running a microsensor that monitors the battery-life!

Other Power Sources

There are many other interesting alternative energy power sources useful for MEMS applications, several are briefly summarised below:

Scavenged Microwaves

At the University of California, Berkeley, a power density of $70\,\mu\,Wcm^{-3}$ *was demonstrated using microwave vibrations of 100 Hz to drive an off-the-shelf* $1\,cm^3$ *piezoelectric membrane in 2003. The key advantage of this device was its ability to scavenge waste power from a variety of sources eliminating the need for storage.*

Micro Fuel Cells

At Stanford University, a power density of $40\,mWcm^{-2}$ *was demonstrated in a four-cell micro hydrogen fuel cell prototype. The unique benefit of this 2002 design was the large energy density with a non-polluting hydrogen fuel.*

Nuclear Batteries

In 2002, at the University of Wisconsin-Madison, a radioactive nickel thin-film was used to actuate a copper cantilever for an average power of 0.4 pW. The main advantage of this device is its life-cycle — a reliable battery that lasts 100 years.

Biomolecular Motor

At Cornell University, the enzyme F-ATPase was attached to nanopropellers on a nickel post. In 2000, supplied with a fuel of Adenine Tri Phosphate, or ATP, the micromotors rotated for several hours.

3.8. Conclusion

Modern sensor systems are, more often than not, integrated microsystems, combining mechanical components and electronic ones in the same assembly. The discussions in this chapter showed that integrated MEMS systems impose much harder electronic interface requirements than their macroscale forebears. They also restrict the electronics design and implementation options open to the designer. Moreover, some of the recently proposed sensing applications impose even more severe constraints, such as the need for extreme power economy for example.

Although different integrations strategies are available, some of which (such as multi chip integration) restrict the designer less severely, they often carry adverse cost or mechanical penalties. The art of the MEMS based sensing systems designer is to juggle all the available parameters to produce an optimum overall solution. Which solution is optimal depends often on the applications context of the system. For instance if a digital processor has been designed into a system, and it has unused processing resources, it makes sense to use digital signal processing to overcome some of the limitations of the less than optimum analogue circuit technologies which may be available, since this involves only additional software, which has a zero parts cost. However, in general, digital processing consumes more power than does the equivalent analogue processing, so the additional processing cycles resulting from the extra software might result in the consumption of more power than would additional analogue circuitry. Therefore, in a severely power limited application, the first choice may not be so sensible. The MEMS system designer is almost always forced to consider a system level, top down context to arrive at the optimum design solutions for a particular application. It is in this context that the particular solutions, surveyed in the next chapters, should be viewed.

References

1. Judy, J. W. (2001) Microelectromechanical systems (MEMS): fabrication, design and applications, *Smart Mater. Struct.* **10**, 1115–1134.
2. Dozoretz, P., Stone, C. and Wenzel, O. (2005) *Shrinking the Pirani Vacuum Guage, Sensor Technology and Design.*
3. Warneke, B., Hoffman, E. and Pister, K. (1995) Monolithic Multiple Axis Accelerometer Design in Standard CMOS, *Proc. SPIE 95.*

4. Crescini, D., Marioli, D. and Torini, A. (1996) Low-cost accelerometers: two examples in thick film technology, *Sensors and actuators* **A55**, pp. 79–85.
5. Shih, J., Xie, J. and Tai, Y.-C. (2003) Surface micromachined and integrated capacitive sensors for microfluidic applications, *The 12th International Conference on Solid-State Sensors, Actuators and Microsystems (Transducers 2003)*, Boston, USA, pp. 388–391.
6. MEMS Research Society. http://mems.interpia98.net.
7. Memsic Inc. (2001) *Accelerometer Fundamentals*, Application note AN-00MX-001.
8. Yeh, C. and Najafi, K. (1998) CMOS Interface circuitry for a low-voltage micromachined tunnelling accelerometer, *Journal of Microelectromechanical Systems* **7**.
9. Lee, C. W., Zhang, X. M., Tjin, S. C. and Liu, A. Q. (2004) Nano-scale displacement measurement of MEMS devices using fiber optic interferometry, *Journal of The Institution of Engineers*, Singapore **44**(5).
10. Brignell, J. and White, N. (1996) *Intelligent Sensor Systems* IOP Publishing Ltd, Bristol, ISBN 07503 0389 1.
11. Carro, L., Souza, A. Jr., Negreiros, M., Jahn, G. and Franco, D. (2000) *Non-linear Components for Mixed Circuits Analog Front-End*, http://sigda.org/Archives/ProceedingArchives/Date/papers/2000/date00/pdffiles/07d_3.pdf.
12. Santuria, S. D. (2001) *Microsystem Design*, Kluwer Academic Publishers, London, pp. 505–509.
13. Zhang, J., Zhang, K., Wang, Z. and Mason, A. (2002) A universal micro-sensor interface chip with network communication bus and highly programmable sensor readout, *Proceedings of 45th IEEE MWSCAS* Tulsa, OK **2**, 246–249.
14. Chavan, A. V., Mason, A., Kang, U. and Wise, K. D. (1999) Programmable mixed voltage sensor readout circuit and bus interface with built in self-test, *Proc. 1999 IEEE International Solid-State Circuits Conference.*
15. Gola, A., Chiesa, E., Lasalandra, E., Pasolini, F., Tronconi, M., Ungaretti, T. and Baschirotto, A. (2003) Interface for MEMS-based rotational accelerometer for HDD applications with 2.5 rad/s^2 resolution and digital output, *IEEE Sensors Journal* **3**(4).
16. Maloberti, F., Liberali, V. and Malcovatti, P. (1998) Signal processing for smart sensors, *Proceedings of Brazilian Symposium on Integrated Circuit Design (SBCCI '98)*, Búzios, Brazil, pp. 141–148.
17. Huddleston, C. (2003) Digital signal controllers turn thermocouples into superstars, *Sensors Online*, http://www.sensorsmag.com/articles/0103/38/main.shtml.
18. Kruglick, E. J. J., Warneke, B. A. and Pister, K. S. J. (1998) CMOS 3-axis accelerometer with integrated amplifier, *Proceedings of the Eleventh Annual International Workshop on Micro Electro Mechanical Systems (MEMS98)*, Heidelberg, Germany, January 25–29. http://www-bsac.eecs.berkeley.edu/archive/users/warneke-brett/research/cmos.htm.

19. Luo, H., Fedder, G. and Carley, R. (2000) A 1 mG lateral CMOS-MEMS accelerometer, *Proceedings of the 13th IEEE International Conference on Micro Electro Mechanical Systems (MEMS '00)*, Miyazaki, Japan, January 23–27, pp. 502–507.
20. Lee, B. N. and Park, H. D. (2000) A signal conditioning circuitry for capacitive accelerometers. *Proc AP-ASIC 2000*, Korea.
21. How sensors work, *MEMS Technology*, http://www.sensorland.com/howpage023.html.
22. Haberli, A., Paul, O., Malcovati, P., Faccio, M., Maloberti, F. and Baltes, H. (1996) CMOS integration of a thermal pressure sensor system, *Proceedings of IEEE International Symposium on Circuits and System (ISCAS '96)*, Vol. I, Atlanta, USA, pp. 377–380.
23. Maloberti, F., Liberali, V. and Malcovatti, P. (1998) Signal processing for smart sensors, *Proceedings of Brazilian Symposium on Integrated Circuit Design (SBCCI '98)*, Búzios, Brazil, pp. 141–148.
24. Kofi, A. A. and Makinwa, J. H. (2002) *A Smart CMOS Wind Sensor, ISSCC 2002*, USA, February 6.
25. Burr Brown Inc. (1994) Designing photodiode amplifier circuits with OPA128, *Application Bulletin* SBOA061.
26. Bryzek, J. (2003) MEMS-IC integration remains a challenge, *PlanetAnalog*, October 29.
27. Silicon Designs Inc, Technology Report, http://www.silicondesigns.com/tech.html.
28. Miller, M. (2004) Manufacturing-aware design helps boost IC yield, EEdesign.com, September 09.
29. Min, R., Bhardwaj, M., Cho, S., Ickes, N., Shih, E., Sinha, A., Wang, A. and Chandrakasan, A. P. (2002) Energy-centric enabling technologies for wireless sensor networks, *IEEE Wireless Communications* **9**(4), 28–39.
30. Wang, A. and Chandrakasan, A. (2002) Energy-efficient DSPs for wireless sensor networks, *IEEE Signal Processing Magazine* **19**(4), 68–78.
31. Ravindran, R. A., Senger, R. M., Marsman, E. D., Dasika, G. S., Guthaus, M. R., Mahlke, S. A. and Brown, R. B. (2003) Increasing the number of effective registers in a low-power embedded processor using a windowed register file, *Proc. International Conference on Compilers, Architectures and Synthesis for Embedded Systems, CASES03.*
32. Kham, M. N., Houlihan, R. and Kraft, M. (2004) Design & simulation of a mechanical amplifier for inertial sensing applications, *Proc. NSTI Nanotech 2004.*
33. Chandrakasan, A., Amirtharajah, R., Cho, S.-H., Goodman, J., Konduri, G., Kulik, J., Rabiner, W. and Wang A. (1999) Design considerations for distributed microsensor systems, *1999 Proceedings of the IEEE Custom Integrated Circuits Conference*, pp. 279–286.
34. Calhoun, B. H., Daly, D. C., Verma, N., Finchelstein, D. F., Wentzloff, D. D., Wang, A., Cho, S.-H. and Chandrakasan A. P. (2005) Design considerations

for ultra-low energy wireless microsensor nodes, *IEEE Trans. Computers* **54**(6), 727–740.

35. Min, R., Furrer, T. and Chandrakasan, A. (2000) Dynamic voltage scaling for distributed sensor networks, *Invited Talk at Workshop on VLSI (WVLSI'00)*, http://www-mtl.mit.edu/research/icsystems/uamps/pubs/.
36. Wang, A. and Chandrakasan, A. (2002) Energy-efficient DSPs for wireless sensor networks, *IEEE Signal Processing Magazine* **19**(4), 68–78.
37. Bhardwaj, M., Min, R. and Chandrakasan, A. (2000) Quantifying and enhancing power awarness in VLSI systems, *IEEE Trans. Very large scale Integration (VLSI) Systems* **9**(6), 757–772.
38. Williams, D. (2004) Alternative power for autonomous MEMs, *Colorado Engineer Magazine.*

CHAPTER 4

SENSOR SIGNAL ENHANCEMENT

by *Elena Gaura*

There are various obstacles to the achievement of 'perfect' sensing devices, with the imperfections tending to be more prevalent in low cost MEMS sensor designs — simply because the design features required to eliminate or lessen these imperfections tend to add to the design complexity and therefore manufacturing cost.

Whatever the sensing application, designers usually want measurement accuracy at the lowest possible cost. Whilst no measurement is ever obtained under ideal circumstances, the effects of interfering and modifying inputs, non-ideal sensing devices, process variations and time variations can be reduced either by changing the system design or by adding new elements to it. Filtering (at the input or intermediate stages of the signal conditioning chain) and negative feedback, added to compensation (in the analogue or digital domain) are the most common processing functions applied to the sensor read-out with the scope of improving sensor signal quality, following the process of calibration or measurement of the sensor's characteristics. The relative "ease" with which these add-ons are produced to accompany MEMS sensors (monolithically integrated or within multiple chip solutions) justified the use of technology, and encouraged the use of microsensors as replacements for traditionally produced products. At the same time, the improved sensor signal quality was often seen, or more precisely 'sold' as 'enhanced' sensor functionality and brought about, in the authors opinion, two of the most controversial (and confusing) terms in the sensing field, the words 'smart' and 'intelligent' as applied to sensors.

The chapter will discuss briefly sources of errors in sensor systems and put forward several techniques currently used for sensor calibration. Newer sensor system functions such as auto-calibration and self-test are also treated, together with a hopefully clarifying view on what 'smartness' and 'intelligence' for sensors actually means. For a general introduction to sensor-based measurement systems and their errors, 'Sensors and Signal' conditioning, by Ramon Pallas-Areny and John Webster, [1] is a good and thorough choice. An in-depth view on several major defects found in primary sensor mechanisms, including the time or frequency response, non-linearity, noise, parameter drift and cross sensitivity is given by Brignell and White in [2]; this is a very useful book to browse and refresh one's knowledge both on the basics of errors in micro-electronics systems and general electronics circuitry with reference to sensors.

4.1. Errors in Sensor Systems and Measurement Quality (Non-linearity, Cross-sensitivity, Offset, Parameter Drift)

Given the nature of this section, where the scope is to define what 'quality' means in the context of sensor signals and what can affect it, a few definitions are provided in Table 4.1. Several parameters, named 'quality parameters' are put forward and their values for an *ideal* sensor are given to provide the framework for signal quality enhancement discussions in this chapter.

Many factors contribute to the above ideal sensor being utopian. In reality, both the designer and the sensor user are happy if a good enough compromise can be achieved under various constraints, of which the most often ones are the cost and the sensor application requirements.

The more we know about the sensing devices however, the easier it is to design them into systems which come close to perfect, at least within certain operational ranges and for a limited range of inputs. The information as to which ranges were appropriate would come from the application designer, if he were involved with the sensor system design, or, more generally, the sensing system designer would optimise a given sensor for a certain type of inputs. For example the Motorola pressure sensors, are designed specifically, at system level, for tyre pressure monitoring and putting up their best performance for such inputs only.

Table 4.1: Quality parameters — definitions and ideal values.

Quality parameter	Definition	Parameter value for an *ideal* sensor
Full scale output (FSO)	Algebraic difference between upper and lower endpoints of output	Whatever is required by the downstream electronics, usually in the volt range, for voltage signals
Calibration	The process through which the relationship between the sensor output and the known applied physical input is established	
Error	Difference between measured physical variable and true value of the physical variable (usually expressed in percent full scale output)	0%
Offset	Sensor output for zero applied input	0 (assuming zero referenced voltage output, although some systems, for instance 4–20 mA voltage loop would require an offset of *exactly* 4 mA)
Hysteresis	Maximum difference in the sensor output when the value is approached first with increasing input and second with decreasing input, expressed in percent full scale output	0%
Linearity	Closeness of calibration curve to a specified straight line (usually measured as the maximum deviation of calibration point from straight line as percent of full scale output)	0% error

Table 4.1: (*Continued*)

Quality parameter	Definition	Parameter value for an *ideal* sensor
Sensitivity	Magnitude of change in the sensor output with respect to change in the physical variable	Whatever is required to allow measurement of the minimum input required to be detected
Accuracy	Ratio of error to FSO expressed in percent	0% error
Repeatability	Agreement between independent measurements made under identical conditions (maximum difference in output readings given as % of FSO)	0% difference
Resolution	Smallest change in the physical variable that results in a detectable change in the sensor output	Infinitesimal
Frequency response	Change with frequency of output/input magnitude ratio and phase difference for sinusoidally varying input	Flat to infinity
Cross-sensitivity	Sensitivity of sensor to another variable than the physical quantity under measurement	0%
Stability	Ability of sensor to reproduce output for identical input and conditions over time (expressed in percent full scale output)	0% error

A systematic treatment of the design process leading towards a close to ideal sensor (or a sensor "good enough" for a particular application) implies answering the following questions:

- What are the types of errors likely to affect sensor performance?
- What are the error levels in MEMS sensors?

- What do we need to know in order to aim for better measurement results?
- How can we improve the quality of the measurement/sensor performance?

In the following four sections, we consider each of these questions.

4.1.1. *What are the Types of Errors Likely to Affect Sensors Performance?*

From within a general measurement systems framework, sensor errors fall into two broad categories:

Systematic errors:
Errors such as inaccuracy of system parameters and parasitic effects, streaming from the sensor design, its fabrication processes and/or the read out electronics; in most cases these errors are measurable, sensor type specific and they apply to whole batches of sensors of a certain type produced by the same process. Systematic error batch compensation is generally applied to alleviate such undesirable effects if the errors are large enough as to take the sensor's accuracy outside the desired range. A great opportunity for implementing batch calibration (and sometimes compensation) was the development of the "smart" sensors, in the context in which "smart" equated with the ability of the sensor to connect directly to a sensor bus and had some form of memory. An automated calibration process could then be performed by placing many sensors in a controlled environment and, via a bus, measuring each sensor output and programming the integrated calibration function. The outcome was reduced costs for calibrated devices.

Random errors:
Errors arising either from random variation in the production process or from the environment the sensing device is part of (i.e. the sensor system and its application environment). Errors such as interference, noise and instability could be eliminated through chopping, dynamic amplification and division, applied to individual sensors. Other device-to-device variations could also be compensated through individual calibration.

From a systems perspective one can approach errors and their correction from the perspective of the sensor transfer characteristic (static and dynamic). Keeping in mind that all sources of measurement error cumulatively affect the accuracy and resolution of a sensing system in a

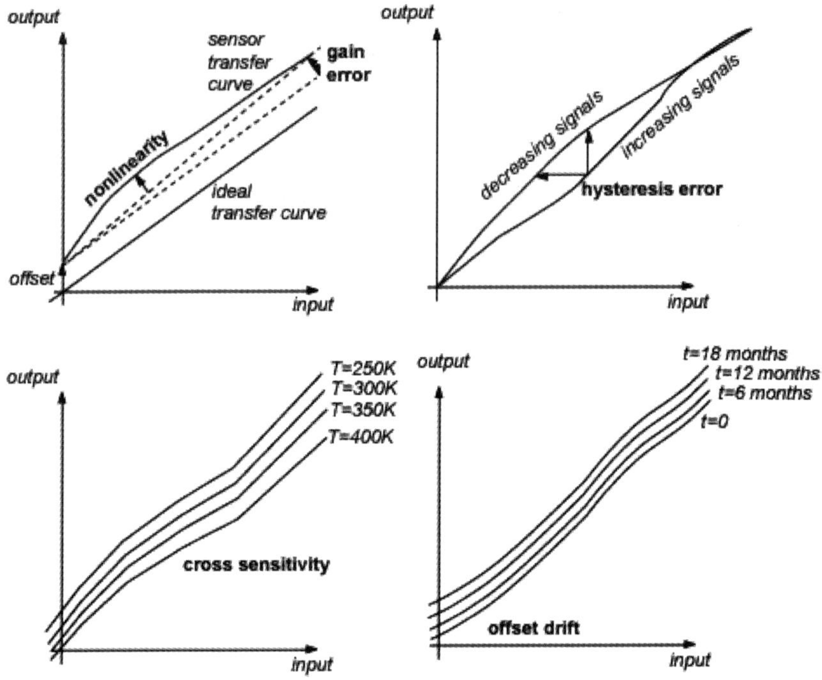

Figure 4.1: Common errors in a sensor's transfer characteristic. From [4].

negative manner, such systems obey the principle "a chain only being as strong as its weakest link" [3]. Some authors proposed approaches involving separating the building blocks within the sensor system and identifying errors block by block (i.e. errors inherent to the sensing device, electronic interface errors and errors coming from the sensing set-up environment and/or the application). Another view is, however to globally qualify, quantify and correct system level errors through means of system design. With the emergence of digital signal processing and its use with sensors, as discussed in Chapter 3, this approach is becoming the norm.* The most common sensor errors are offset, gain, range or full-scale error, non-linearity, cross-sensitivity (secondary variable sensitivity), hysteresis and drift. These errors, reflected in the sensor's transfer characteristic are depicted in Figure 4.1 [4].

*It is worth noting here that, given the relative immaturity of the MEMS field compared to microelectronics, most errors, at sensor system level are due to the sensing devices, rather than the surrounding electronics.

Whilst hysteresis and drift are common in some sensor types, all other errors are present to a higher or lesser degree in all sensor types. And again, of the above errors, hysteresis and drift are, principally most difficult to compensate. An ingenious solution for compensating hysteresis is presented in Chapter 7, applied to capacitive accelerometers but generally valid for any sensor type which exhibits hysteresis in its static or dynamic characteristic. Drift errors require "prediction" and time measurement in order to be compensated. Some sensor systems designs include an additional transducer (either sensor or actuator) with known low drift and implement self-calibration and compensation schemes based on this set-up. It is more common however, to re-calibrate sensors periodically, of course if they are accessible *in situ*. Many new MEMS sensing applications, however, of the type discussed in Chapters 9, 10 and 11 do not permit *in situ* recalibration, the sensors being permanently fitted and inaccessible after deployment.

For most traditional sensing applications, the presumption was that the sensor-associated errors are *linear* in nature and as a consequence providing the means for linear compensation was adequate. Programmable offset correction and programmable gain are features found in most sensor systems. The implementation of these functions has continuously evolved as the sensing technologies matured (in terms of fabrication technologies for the devices) and ability of integrating signal processing within the sensor systems was developed.

However, the increase in expectations from sensors on one hand, which encouraged thorough investigation of their characteristics, and the development of more accurate tools for the design and testing of sensing devices and systems on the other hand, have revealed that in reality the errors are often *non-linear*. If measurable and repeatable, non-linear errors may be compensated in a systematic manner, through linearisation followed by linear compensation (many different methods have been reported in the literature, some of which are briefly browsed later in this chapter). Another approach to compensation is by applying non-linear correction techniques, specific examples of which are treated in Chapter 7.

4.1.2. *What are the Error Levels in MEMS Sensors?*

Generally, for the "micro" technologies, the expected absolute error variations are large, with figures of up to 50% reported at times, for specific technologies and devices. The relative variations (device-to-device) are somewhat smaller.

The sources and magnitude of the errors vary with technology, sensing device type and sensor system design. If one would take for example a micromachined piezoresistive pressure sensor, the following process imperfections result in considerable errors:

- resistor mismatch and initial stress, which will generate large variations in offset;
- the position and orientation of resistors and the mechanical properties of the diaphragm, which will affect sensor sensitivity and they will also be cross sensitive to temperature by a variable degree.

Overall, for this type of sensor, one could expect an error band in the output signal range as large as 30% [4]. Another example, that of a bulk micromachined capacitive acceleration sensor is discussed in Chapter 7 and design solutions are presented motivated by the high levels of errors for this type of devices.

It is, however, rather meaningless to talk about levels of errors (or alternatively, measurement quality) except when relating this to the cost of a sensor.

In general, traditional sensing technologies have been more expensive, but more accurate, while the newer micromachining technologies have focussed on lower cost at the expense of performance. There is also a great disparity between the different types of packaging and the level of test and calibration that affect the cost differences. Continuing advances however are pushing micromachined sensors into a performance space that threatens the more mature technologies, while maintaining low cost, as shown in Figure 4.2 [5].

The microsensor performance space is a function of both historical evolution and cost. For accelerometers, for example, the charts in Figure 4.3 summarise the performance space for the last decade. Within the chart, the high performance 5 g accelerometer, XL105, was selling for $10 in OEM quantity, in 1995, could resolve 1 mg out of 5 g, had a 10 kHz bandwidth, and a flat frequency response to within 5% from DC to 5 kHz. The performance of $10 devices has essentially doubled every 18 months during the last decade and is expected to continue to do so for several more generations [5].

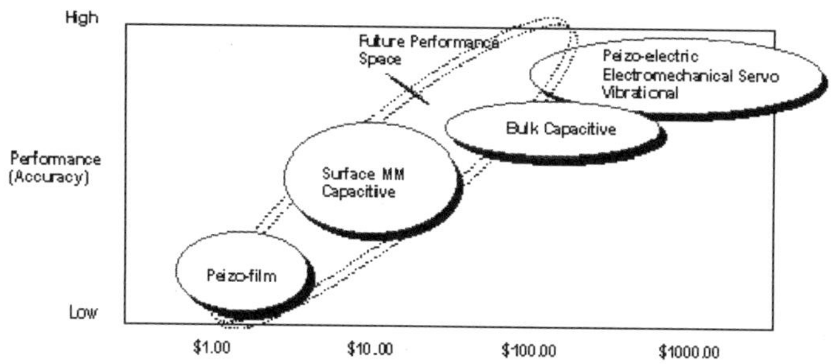

Figure 4.2: Cost/performance of some MEMS sensor technologies. From [5].

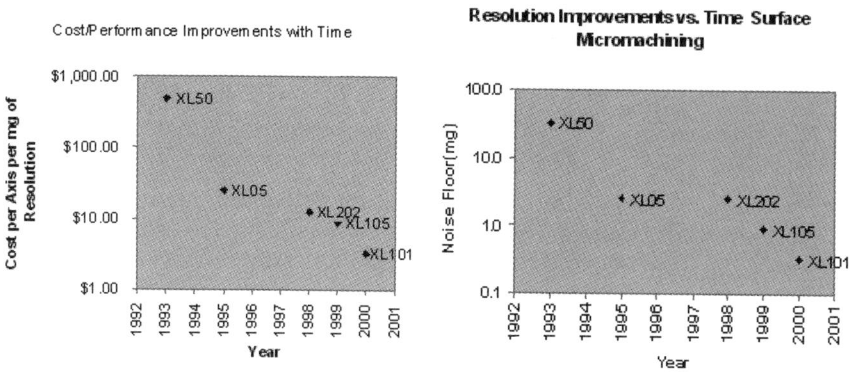

Figure 4.3: Cost and performace enhancement of Analog Devices accelerometers over time. From [5].

4.1.3. *What do we Need to Know in Order to Aim for Better Measurement Results?*

It is now widely accepted that the key to accurate sensors relies, on the one hand, on further developing the cost effective means of characterising sensors and their behaviour under various conditions, and developing standardised tests (and associated machinery) for various sensor systems at all levels of development/integration. On the other hand, accuracy can be achieved by adopting a systems based design perspective for sensors and developing the appropriate signal processing techniques for signal quality improvement.

MEMS testing is a discipline in itself and in the recent years, many of the set-ups developed by researchers have been adopted by manufacturers. Sophisticated testing equipment has been developed, but at a cost. It is a fact of life in the sensor business that packaging and test costs will often dominate the cost of the devices. Precision accelerometers for example are often referred to as 'instruments', and are often tested on a custom designed, one at a time test fixture. A customer purchasing a traditional precision sensor will receive a specification showing typical performance characteristics and then a custom calibration data sheet showing parameters for the actual device. For vibration sensors, this typically includes initial sensitivity of the device, cross axis sensitivity and a frequency plot from 0.1 to 20 kHz. From this data, the device can be calibrated or compensated.

The approach to testing varies with the fabrication technology of choice for the sensor and the level of integration for the sensor system. Analog Devices, for example, rely on an approach in tune with the mass production of integrated circuits for their iMEMS sensors. As individual hand testing of devices is expensive and keeping data with a particular device is difficult (especially for a device that may only sell for $10 and in high volume) the devices are tested on standard integrated circuit handlers that have been modified to accept shakers (Figure 4.4).

Automated testing proceeds as follows: parts are loaded in a tube to tube manner, and the machine automatically handles the tube, tests it,

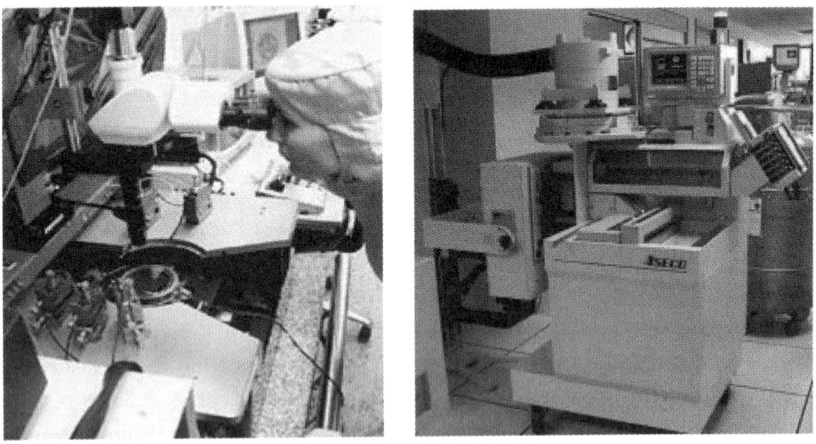

Figure 4.4: Mass production accelerometer calibration. From [5].

shakes it, and then accepts or rejects individual parts. Optical character recognition can be used to test the drift of parameters over temperature. The device, being an integrated circuit, is tested for a variety of electrical parameters at the same time. Full mechanical testing may not be done. For example, it is impractical in such systems to sweep devices from 0 to 20 kHz and impractical to keep the data with the device. Frequency response is checked at a number of key points (for example 100 Hz, 1 kHz, 5 kHz and 10 KHz) to ensure that the device is operating properly. Results are inferred in between points, and backed up by extensive process control and characterization, a hallmark of the IC business. Specification tolerances may be set at particular points so that the user knows the accuracy of the sensor for amplitude response. The other characteristic of the testing is that the devices run in much higher volume than traditional sensors, which facilitates process control and device consistency. And, as with packaging, a network of third party companies provide extra screening and characterisation for those wanting a more thoroughly tested device [5].

Although the cost of testing sensors is high, there are reasons to test and characterise sensors as early as possible in the production process. Testing early prevents 'bad devices' incurring additional production costs in finalisation and packaging. A report by MEMUNITY (the MEMS Test Community) looks at the deveopment of just such strategies with sensor manufacturers and how to integrate them within their production processes. The motivation is given below [5]:

> *Till recently companies who needed to test sensors and actuators had two possibilities: making limited electrical tests during the stages before packaging and testing fully packaged devices with the non-electrical input and/or output the sensor needed to work. An electrical test can only give very basic information about certain parameters of the die but cannot fully analyse the function of the sensors and actuators. If the MEMS device is still in the design stage time to data is of the essence and a complete analysis at wafer level and throughout the various packaging stages provides a valuable insight into the design characteristics whilst also enabling process monitoring. In production, waiting to test until the sensor is packaged means that baddies get taken right through the production process before being detected.*

Identification of Known Good Dies (KGD) before packaging will most likely form the single most profitable activity in the manufacturing process when a new microsystem is launched into the market. It has been suggested that the production costs and the high price to market are the main reasons for the slow investment into the MEMS sector. Studies show that 80% of manufacturing costs are caused by the packaging process, by testing at earlier stages of production, the packaging costs for bad dies can be saved and thus both production costs and end product price can be reduced. By testing at several production stages, possible errors can be identified and eliminated resulting in an optimised process and higher yields. MEMS manufacturers who use foundries need to build in quality control mechanisms into the incoming wafers just as the foundries need to prove the quality of their products before shipment.

The plot in Figure 4.5 shows the projected 'price of good component' decrease against yield for fabrication processes which implemented the Identification of Known Good Dies before packaging. An approximately

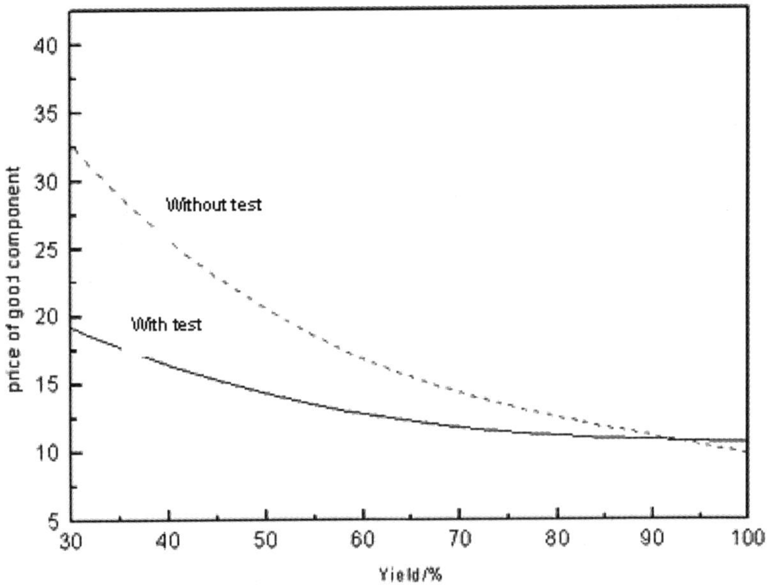

Figure 4.5: Price reduction through test is significant. From [5].

50% price reduction for very low yield processes can be observed. However vital and advantageous it seems, it is not altogether easy to test MEMS before they are packaged, but the rewards are great, as the processes can be improved and result in sensing devices which better meet their specifications and also, through the savings made, manufacturers can allow for better calibration and compensation in the final sensor system.

The conclusion to be drawn here is that 'testability', or more accurately 'the ability to calibrate', is an important design consideration if low end costs of devices are to be achieved. Thus, if additional on-chip circuitry yields a cost reduction in testing which is greater than the per die cost of that circuitry, then it is justified. This is precisely the same consideration that has led almost all current complex VLSI chips to include on the die self test and self-diagnosis capability. In making cost based design decisions, the whole production chain, from end to end, must be considered.

As previously mentioned, microsystems testing is a research domain in its own right. An outline of the challenges encountered when designing and implementing test processes for MEMS is given below.

Currently MEMS testing is mainly performed using techniques and instrumentation designed for testing of CMOS devices and packages. However, these techniques cannot test specific MEMS-related issues such as moving parts, temperature, humidity, pressure, sound, particles, gases etc. Special chambers, probes, sample holders, test structures, detection systems, sample preparation techniques and electronics are required. Equally parameter analysis is not straightforward: movable metallic parts, such as in RF-MEMS switches, may be prone to creep and fatigue.

There are many issues to be considered when testing MEMS which are substantially different to those for other electronic products, hence the expansion and imminence of the MEMS test and reliability research. Several key pointers brought up by the MEMS test community are:

- The environment at the point of sensing or actuation is often very harsh thus the test set-up needs to be able to test influences such as temperature, high-G vibrations, pressure and vacuum.
- The device often has openings through which the medium carrying the sensor/actuator signals is exposed directly to the microsystem chip inside the package. This means that the device is exposed to unwanted

environmental influences and must be tested in a completely shielded environment such as a vacuum chamber.

- The materials and packaging techniques used for microsystems devices are normally device-specific or application-specific and tend to have failure mechanisms, which differ substantially from those of other electronic components and systems. This requires a high level of equipment customisation.
- Many MEMS are used in safety-critical applications where long-term reliability is critical (e.g. medical, safety devices in cars, gas detection and aerospace applications).
- Substrate handling is often difficult because common pick and place systems and pin drives could damage the micromechanical parts.
- There is no standardisation of substrates or technology.

It is only relatively recently that the areas of test, reliability and MEMS characterisation have seen considerable investment and gained recognition. However, the leap forward is already considerable.

4.1.4. *How can the Quality of the Measurement and Sensor Performance be Improved?*

There are two distinctive schools of thought here, as follows:

- Improve the sensing devices, from design, through fabrication process, to packaging;
- Improve the sensor system by taking a systems view on the design and applying clever signal processing.

Which school of thought is followed depends very much on the sensor system designer's viewpoint. Sensor device designers spend their professional lives perfecting and honing MEMS devices, and it comes against everything they have learned to accept a 'poor' device design. However, those designing sensors at a system level often have, in any case, to correct and process out many errors in the sensing system's signal chain. Very often, in the context of a sensor system as depicted in Figure 3.1 of Chapter 3, correcting certain types of sensor error, such as non-linearity for example, take little or no additional effort. Sensor systems designers would, therefore, rather pay less for a 'less good' sensing device (as a component in their system), so long as it only suffers from the type of error amenable to correction in signal processing. The latter point of view is very much the one favoured by the authors here. The design philosophy followed throughout this book is one of integration of electronic systems with the sensing device to extract the wanted information from the unwanted. This general approach, the consequent sensor and system design requirements, and the ways in which these are addressed form the topic for the rest of the chapters.

That being said, sensor researchers and designers continue the search for better sensors: sensing devices designs are being refined to improve performance, new materials are explored and better fabrication processes are developed, as shown in Chapter 2.

Drawing from the performance of present commercial devices and linking specific sensor characteristics to sensing element design choices and the applied fabrication process is a fruitful method for optimisation and design evolution of future sensors. For a class of capacitive MEMS accelerometers, for example, a detailed performance analysis was reported in [6]. Commercial accelerometers for the same low-level input acceleration range of $\pm 10\,\mathrm{g}$, with a variety of sensing element designs, produced from different materials, through various fabrication technologies were evaluated in terms of their sensitivity, resolution, linearity, frequency response, transverse sensitivity, noise level and long term stability. The cost of the devices ranged from tens of dollars to several hundreds. The test results were compared for the Endevco 7290A-10, ADXL210A, Silicon Designs SD2012-10 and Motorola 1220D accelerometers. The results are tabulated in Table 4.2. Figure 4.6 shows the four devices and the fabrication technology involved.

Returning to the major topic of this chapter, that of signal enhancement through signal processing, one can proceed to evaluate, in the next sections, various methods and techniques currently applied to microsensors.

4.2. Sensor Calibration and Compensation — Techniques and Examples

In Section 4.1 we defined calibration as follows:

Calibration:
The process through which the relationship between the sensor output and the known applied physical input is established.

Table 4.2: Comparative performance of four commercial MEMS accelerometers.

Sensor type	Sensitivity error (%)	Non-linearity (%)	Frequency response (Hz)	Transverse sensitivity (%)	Technology
7290A-10	+1.2	0.251	1900	0.863	Out-of-plane Bulk-micromachining Two-chip Single-crystal silicon
ADXL210A	−0.3	0.441	22	1.249	In-plane Surface-micromachining Integrated electronics Polysilicon
SD2012-10	+0.5	0.531	1200	0.448	Torsional Electroforming Two-chip Nickel
M1220D	−4.8	0.694	120	2.742	Out-of-plane Surface-micromachining 'Cap' chip Polysilicon

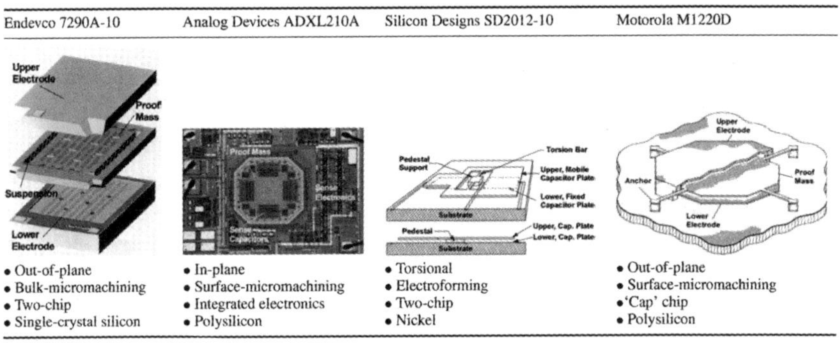

Figure 4.6: Comparative technology of the four accelerometers. From [6].

The definition aimed to bring clarity to the discussion around the linked topics of calibration and compensation, in the view of alternative definitions of *calibration* in use. Some such definitions are given below:

- As defined by ISO, calibration is "... the set of operations which establish, under special conditions, the relationship between values indicated by a measuring instrument or a measuring system, or values, represented by a material measure, and the corresponding known value of a measurand".

- It is the process of applying several physical reference signals to a sensor, and measuring the output signal, so that the input-output relationship of that sensor can be derived with a certain accuracy.
- Procedure of correcting the transfer function of a sensor, using the reference measurements in such a way that a specified input-output relationship can be guaranteed with a certain accuracy (and under certain conditions).

The definition adopted here concurs with the first two above and refers clearly to calibration as a measurement process. Note that the third definition implies not only measurement but also correction of the sensor characteristic(s). It is recommended therefore to exercise care when reading relevant literature, as different authors and commercial sensor firm do vary in their use of the term calibration.*

In practice calibration involves a reference sensor (itself ultimately calibrated against some 'official' standard) to gain the reference inputs for the sensor under calibration. The measurements of the sensor output signal for these reference inputs are used to determine the error with respect to the desired sensor transfer characteristic. This is followed by the adjustment or compensation of the sensor characteristic in such a way that an accurate transfer function is obtained. Figure 4.7 represents the process graphically.

The following observations can be made:

- a calibration may also determine other metrological properties;
- the result of a calibration may be recorded in a document, sometimes called a calibration certificate or calibration report; such "documents" take an electronic form in many "smart" sensors;
- the result of a calibration is sometimes expressed as a calibration factor, or as a series of calibration factors in the form of a calibration curve.

Most of the initial efforts in the development of "smart" sensors were in the direction of making such calibration reports available within the sensor

* *Terminology note*: Calibration, adjustment, compensation, correction and even linearisation are terms used interchangeably by some authors in the sensing field, although, as shown above they refer to very different processes relating to obtaining better measurement values from sensors.

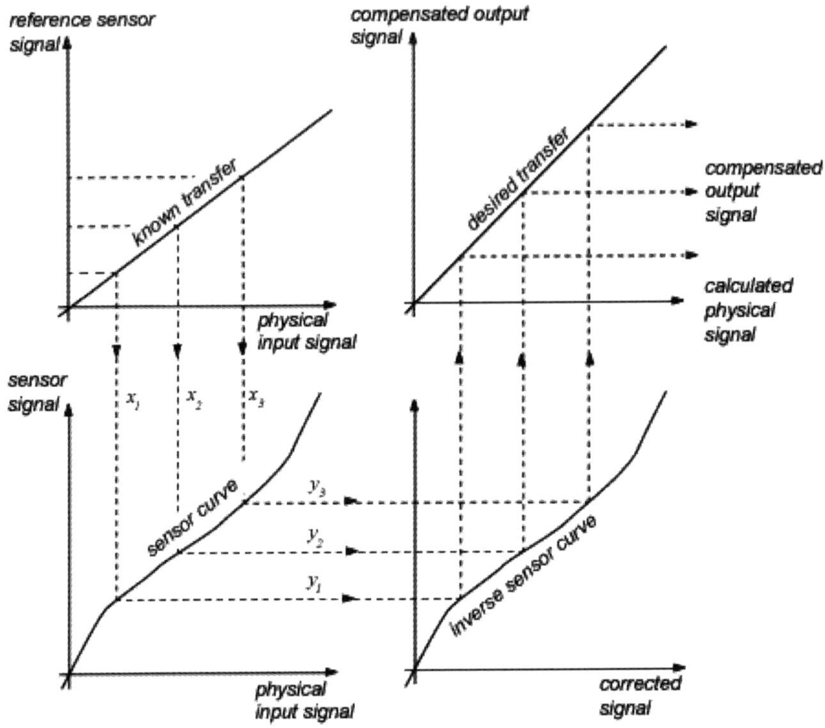

Figure 4.7: Determining the sensor transfer and calibration curve. Redrawn and modified from [4].

and accessible to the larger system that the sensor formed part of (referred to, here and in the previous chapter, as 'the application'). The interpretation of the calibration reports/curves was left to the "system controller" or decision making system of the application. Availability was what mattered and the portability of such data with the sensor.

4.2.1. *The Process of Calibration*

Within an ideal production process, the relationship between the electrical output signal of the sensor and the measured physical parameter would only need to be measured once, in order to characterise a sensor. Unfortunately this is not the case for the majority of products and as briefly shown in

Section 4.1, so much more for MEMS sensors. A certain number of measurements need to be taken in order to quantify the sensor's behaviour, or, in other words, the sensors departure from ideal behaviour. The number of measurements required mostly depends on two factors: the way in which the sensor user will be accounting for the sensor non-ideal behaviour (i.e. the correction strategy in view) and the time and cost investment in the process of calibration (which reflects in the sensor cost). Calibration is a time consuming process, particularly for sensors with a slow response time, due to the fact that measurements throughout their range have to be taken. If performed at individual device level, calibration can add considerably to the cost of MEMS sensors. For a pressure-integrated transducer, for example, calibration and test would make 20% of the cost, whilst packaging would be 45% and the silicon 35%.

To those unfamiliar with sensor systems, the time and cost involved in calibration is difficult to comprehend. An example is given here of the problems it brings, drawn from the professional experience of one of the authors. The application in question is health monitoring of helicopters, the aim being to detect mechanical faults in the rotor assembly by detecting the vibration caused by out-of-balance conditions. The sensors used to detect this vibration are accelerometers, and six are placed around the rotor bearings to detect the vibration (six sensors form two orthogonal sets of three, each of which can detect vibration along any of the three geometric axes, together the two sets of three can detect torsional vibration as well). In the original application a tape recorder was installed to record the data captured during a flight for later analysis. To be useful, the accelerometers need to operate within quite *strict error tolerances* and within certain limits *linearly*, and provide the *correct full-scale output*. If these conditions are not met, the rotor may be diagnosed as faulty incorrectly, and unnecessary repair costs incurred in stripping and checking the assembly (which is an expensive business). Thus, for this application scenario, it is required to select only accelerometers which meet the above requirements. The sensor manufacturers sell calibrated reference accelerometers, the performance of which is guaranteed within limits, which are an order of magnitude better then required by this application. However, their cost is prohibitive. The solution adopted, then, is to calibrate 'commodity' accelerometers against a reference accelerometer. The original (manual) calibration process used in this monitoring application is described below:

(i) the reference accelerometer and the accelerometer under test are mounted on a shaker, which is installed in a temperature controlled room;
(ii) the room temperature is brought to the required value for the first set of data points.
(iii) at each vibration level required for the calibration, the amplitude of the shaker is adjusted until the reference accelerometer gives the required value.
(iv) the actual value given by the accelerometer under test is recorded. Along with the value from the reference sensor, this forms an (x, y) point on the calibration curve.
(v) steps (ii), (iii) and (iv) are repeated for every temperature/frequency/amplitude point required for the calibration (maybe several hundreds in total).

All of this are undertaken by a skilled technician, and tie up expensive equipment. A calibration run as described above may last several hours (therefore, each calibration run effectively adds its cost to the cost of the sensor). The data recorder used in the application demands that sensing data falls between set limits, hence, in the calibration process, the sensors which do not calibrate within the set limits are simply rejected. Their cost, and the cost of their calibration is virtually added to the cost of the successful sensors, when assessing the overall cost of this sensing application.

A first optimisation to the process above is to calibrate several sensors in the same run, if the shaker is large enough to accommodate them. Although this requires more expensive equipment (a larger shaker and a data recording system capable of reading from several sensors) and also lengthens the run (since the technician must record the output for several sensors), the time and cost per successful sensor may be significantly reduced.

A more effective answer is an automatic calibrator, as for instance the calibration systems offered by Beran Instruments, which were originally developed for this particular vibration monitoring application. Their accelerometer calibration system, produced in 1996 was based on a "sine correlation" technique which minimised the distortion, noise (both induced by the shaker and the amplifiers within the calibration system) and dynamic range problems encountered in other calibration systems and allowed the use of lower cost shakers for automated calibration [7]. Calibration time was also reduced to approximately 3 minutes per accelerometer in order to obtain a typical 20 point frequency response and 10 point amplitude linearity calibration, also making the phase response available (this is of particular interest for rotating machinery monitoring applications). The development of accurate calibration systems is, in itself an area in which research efforts are being made and which is developing at a fast pace.

However, with regard to the application described in this section, a better solution still, is to provide the facility to also compensate for errors found in the sensor, which will allow some or all of the sensors previously rejected during calibration to be used. This topic, compensation rather than calibration, is the subject of the next section.

4.2.2. *The Process of Compensation*

Following calibration, strategies need to be developed for correcting the behaviour of the sensors and bringing them as close as possible to the requirements set by the class of applications they are intended for.

Error correction is applied to an ever-increasing amount (and types) of sensors. The reasoning for this comes from the shift in the use of sensors — many, easy to use and often inaccessible following deployment — and the increased expectations of the applications and users in terms of measurement accuracy, reliability and trust in the measurement values.

Adjustment of a sensor signal can be done in a variety of ways, depending on the sensor technology, the device structure and the level of sensor integration. Also, it can happen at different points on the sensor signal path, within the sensor system as described in Chapter 3 (Figure 3.1). A common compensation method for pressure sensors, for example, was to introduce trimmable resistors (potentiometers or laser trimmed) in the Wheatstone bridge pick off and manually adjust them. Digitally programmable transfer function correction is another commonly used way to compensate integrated sensors where signal processing capabilities exist within the sensor system. Closed loop sensor control and AI methods have also been used successfully to compensate for sensor errors.

Determining compensation requirements depends on the overall accuracy requirements and, of course, the costs entailed in achieving the target accuracies. Basic compensation typically involves correcting offset and gain (span). Higher levels of system accuracy usually require compensation for non-linearity and temperature (or other secondary/cross-sensitivity) effects.

The historic assumption and practice has been that calibration and compensation are associated with measurement instruments solely. We argue that mass produced, general use sensors, particularly the recent generations of MEMS sensors, are amenable to calibration and compensation which can allow them to perform functions and measument traditionally associated with instrumentation sensors, and additionally that means exist to make this calibration and compensation an economic possibility in a wide range of applications. This assertion depends on the existence of advanced methods of compensation which can cope with the type of error found in commodity MEMS sensors. Most departures from ideal behaviour in sensors are to a lesser or greater extent non-linear and many of the proposed compensation methods acknowledge this fact and are consequently developed to overcome/cope with it. Closed loop strategies for example limit the behaviour of the sensing element to a range where linearity is a good approximation whilst neural network based strategies tackle non-linearity in a direct manner through non-linear compensators.

Compensation or error correction strategies

Compensation strategies can be classified from several view points, as follows:

According to the way compensation is applied:

- *Systematic Error Compensation*
  Applied to all devices coming from a given process; all parameters of the "compensator block" are set at design phase. An example is the application of logarithmic compensation to a tunnelling accelerometer — the accelerometer has a known, non-linear characteristic which can be systematically compensated.
- *Batch Compensation*
  The same technique but different parameters applied to a batch of devices on the production line, all calibrated automatically. An example where this may be applicable is to compensate for wafer to wafer process variations. All devices in the same wafer will have comparable performance

(at least as regards wafer to wafer variation). The required compensation can be determined once and applied to all devices.

- *Individual Compensation*
  Designed and implemented for a sensing device at a time, responding solely to that device's behaviour. This is the default and most expensive option. It requires individual matching of the processing electronics to each sensor.

According to the signal domain in which compensation takes place:

- *Analogue Compensation*
  Performed in the analogue domain using analogue circuitry;
- *Digital Compensation*
  Performed after signal digitisation;
- *Hybrid Compensation*
  Different parts of the compensation function performed before and after digitisation.

According to the implementation of the compensation scheme:

- Hardware compensators and
- Software compensators

According to the sensor system design perspective:

- New system design steaming from the need to compensate the sensing device (such as closed loop designs, sigma delta modulators and inclusion of additional sensing elements and/or actuators);
- The linear addition of one or more processing blocks to the system — feed forward compensation;
- A combination of the above approaches.

The above provide a four-dimensional 'classification space', within which each compensation strategy can be located. A certain method of compensation, for example, might be implemented either in software or hardware. Software compensation however, presumes a digital implementation of the compensation method which, in turn, needs to suit the overall sensor system design and/or the fabrication process.* Similarly, a given method, carried

*The more complex compensation techniques are always implemented in software, and therefore in the digital domain.

out in the analogue or digital domains, could be applied systematically, in batch or individually. The reminder of this section provides some examples of compensation techniques, schemes and implementations (starting with the most common and widest used compensation techniques). Chapters 5, 6 and 7 treat separately three techniques for given case studies.

Trimming techniques — gain (span) and offset correction

The sensor gain and offset compensation were discussed in Chapter 3 (Section 3.2.4), within the functional block 'Offset/linearity compensation' in the sensor system depicted in Figure 3.1 (which is reproduced here for convenience). The concern there was only with compensation for problems of the input amplifier. Here, the discussion encompasses the gain and offset problems of the transducer as a whole.

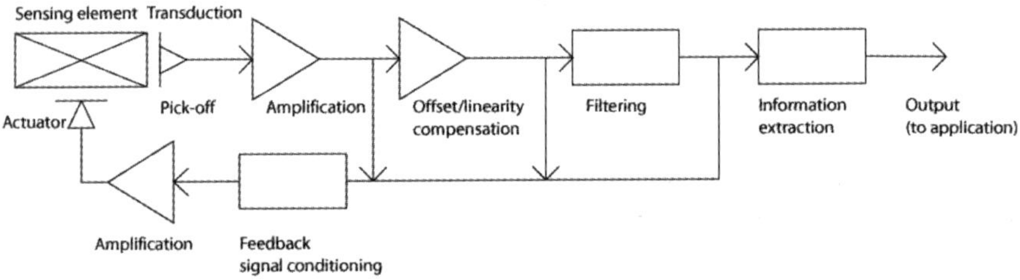

Traditional methods calibrate and compensate the sensor's gain and offset in the analogue domain using 'analogue memory' components, such as potentiometers (trimmable potentiometers on either a PCB or ceramic), capacitors and laser trimmed thin-film resistors. In the simple example in Figure 4.8, once the sensor and electronics have been assembled, the sequenced manual adjustment of the trimmable resistors will enable offset and gain (span) adjustment (R1 is adjusted to minimise the offset, R2 and R3 are then adjusted to provide the desired gain or full-scale output). In practice, when instrumentation amplifiers with two or three operational amplifiers are used for common-mode signal rejection and amplification of low-level sensor signals, adjustable gain can be achieved by adjusting one gain-setting resistor [8]. However, as for most trimming methods, compensation accuracy is restricted by non-linear sensor errors, is affected by temperature drift, and, whilst laser trimmers and other automatic equipment is very expensive, manual calibration translates to higher costs.

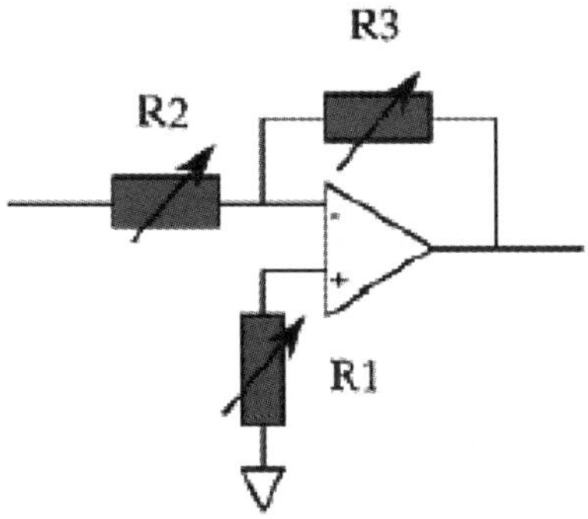

Figure 4.8: Offset and gain trimming in the analogue domain. From [8].

Figure 4.9: Printed trimming resistors on a ceramic substrate. From [8].

Another common method of compensation uses printed resistors on ceramic substrates (Figure 4.9). These hybrid circuits can allow a lower profile and smaller footprint because bare wire can be used. The main drawback of this approach is the need to have a laser-trimming system in the production line, but accurate trimming is readily achievable. This technique has also been used for many years by a few high-end analogue IC suppliers to laser trim on-chip thin film resistors in order to produce, for example, extremely low offset voltage and drift operational amplifiers [8].

For both these techniques, further adjustments cannot be made once the product is packaged. Re-calibration as a maintenance feature was however desired and made possible by the non-volatile serially programmable resistors and capacitors pioneered in 1991 by Hughes Aircraft Co. Onchip nonvolatile EEPROM was used to store the address of the desired component from a network of on-chip resistors or capacitors. Companies such as

Dallas Semiconductor (Maxim) and Xicor now have extensive product portfolios of these functions. This technique offers flexibility, but is not the most integrated or compact solution.

More recently, the digital trimming of analogue functions afforded by non-volatile memory made it possible to pack a complete analogue signal conditioning system onto a single custom IC. As with the stand-alone programmable resistors mentioned above, resistor networks can be used in this approach (see Figure 4.10). The transistor associated with the chosen resistor is turned on, allowing current to flow through the resistor. Of course, it is also possible to configure combinations of resistors in parallel to provide more flexibility. This technique can also be used to improve the accuracy of basic functions such as operational amplifiers, comparators, and voltage regulators. It can provide programmable functions such as filters (see Figure 4.11) and oscillators. AD and DA converters can also provide additional circuit flexibility and functionality [8].

The trimming function may form part of an overall system design strategy. By providing efficient methods for calibration and trimming at a system level, it may be feasible to realise precise system level performance requirements using low cost sensors. Gopel *et al.* [9] present a design for a reduced cost pressure sensor based on capacitive sensing and an efficient trimming concept. The low temperature coefficient of capacitive sensors eliminates the need for temperature compensation, as opposed to the considerable trimming efforts induced by temperature dependency in piezoresistive pressure sensors. Moreover, high stability interface circuits in CMOS technology for capacitor bridges are easy to implement. The drawback of capacitive sensing

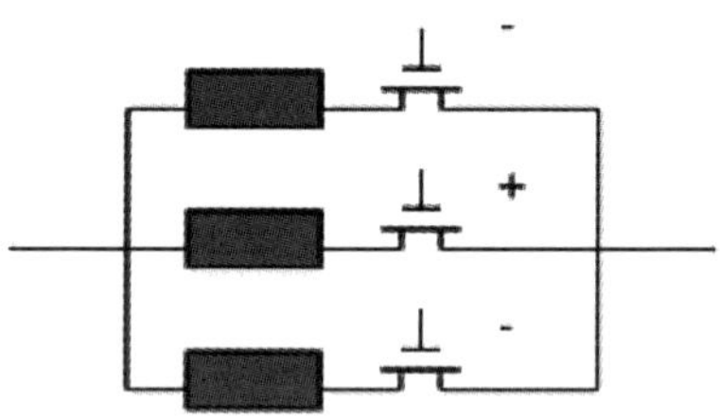

Figure 4.10: Coarse resistance setting can be achieved using the digital selection of the transistors shown. A significantly higher resolution function can be achieved by using an R-2R ladder circuit to provide 64, 128, 256, or a higher number of resistance steps. From [8].

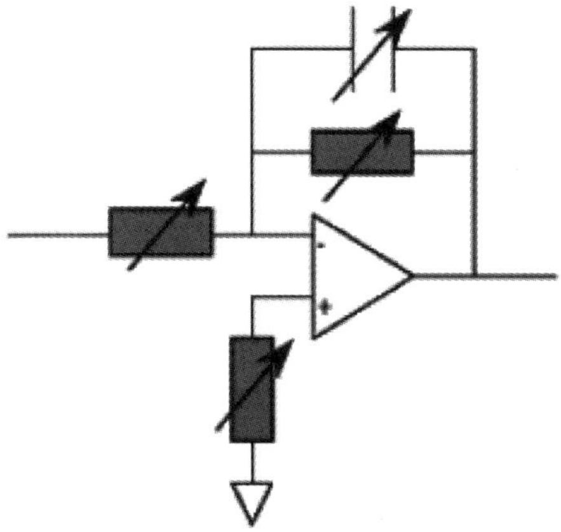

Figure 4.11: Digital selection of capacitors and resistors can provide programmable filters. Other examples of programmable functions include selectable output voltage regulators and dual-trip comparators. From [8].

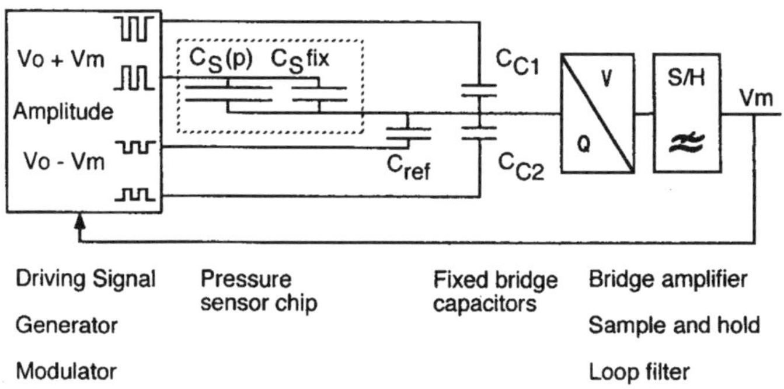

Figure 4.12: Schematic diagram of the self-balancing bridge configuration. From [9].

is however the intrinsic but predictable non-linearity, which requires an efficient linearisation scheme, implemented here based on a self-balancing capacitor bridge (Figure 4.12). A linear relationship between the pressure and the output voltage can be obtained, based on a feedback loop adjusting the driving voltages to different branches of the bridge in such a way that the net charge on the central electrode is zero. The sensor's offset and span can be adjusted by using only two parameters, as following the reasoning below [9]:

The sensor output voltage is given by:

$$V_m = V_0 \frac{C_s(p) + C_{sfix} - C_{\mathrm{Ref}} - C_{c1} + C_{c2}}{C_1(p) + C_{sfix} + C_{\mathrm{Ref}} - C_{c1} - C_{c2}},$$

where V_0 is half the supply voltage. One can now choose the capacitors C_{Ref}, C_{c1} and C_{c2} in such a way that the output voltage is proportional to the pressure. This scheme corrects the basic non-linearity of the capacitive effect and second-order effects such as the curvature of the membrane. From the above equation one notes that changing C_{Ref} and C_c also change the offset and the span. This means that the sensor can be fully adjusted using only two parameters, one constant in the numerator, $C_{\mathrm{Ref}} + C_{c1} - C_{c2}$, and one in the denominator, $C_{\mathrm{Ref}} - C_{c1} - C_{c2}$.

Optimum linearity compensation can only be obtained for one set of fixed capacitors. However, for most low-cost applications, the error budget for non-linearity is sufficiently large to trim off offset and span with the two constants. Evidently, this approach implies that the sensor design and manufacturing tolerances have to be carefully optimised and monitored. Since the circuit is implemented in CMOS technology, the capacitors C_{Ref}, C_{c1} and C_{c2} are readily integrated using arrays of capacitors and suitable switches. Trimming is done at the final testing stage. The information of the particular combination of capacitors to be used is easily stored either in non-volatile or fuse type memories.

Several further examples of compensating architectures in the analogue domain, such as DACs based programmable gain and offset control and cross-sensitivity compensation, pulse-modulated compensation using switched current DACs and implementations of polynomial calibration/compensation schemes have been described by Huijsing [4].

Compensation of temperature effects

Several transduction mechanisms are well known for their strong temperature dependency of their parameters. Piezoresistive transducers are the best examples here. Various techniques have been developed to compensate for the temperature dependency of sensor transfer characteristics. The typical temperature coefficients of the materials for example can be used to provide basic compensation. This maneuver may take the form of using resistive elements to balance temperature effects, or, in a more sophisicated scheme, could involve designing a precision bandgap reference that uses canceling techniques to provide a stable reference voltage over temperature. This becomes the stable reference for the key circuit functions.

Another approach to compensation for temperature dependence of some sensor elements is adjusting the sensor excitation voltage, current, or drive. Using EE-PROM lookup tables, D/A converters, and some logic, with the option of having on-chip or external temperature sensors, schemes for compensation can be realised.

Temperature compensation is performed either in the digital domain or the analogue one. Principally, an approach is to use a voltage reference featuring a temperature behaviour opposite to that of the sensor [10], dependence which can be digitally implemented. In the analogue domain, an example of temperature compensation of sensor offset and sensitivity is presented by Grigorie [11], an approach claimed to be suitable for low power sensing applications. The work considers capacitive sensors with charge balancing switched capacitors pick-off providing a ratiometric output (the sensor capacitor and several others are switched in order to define the full scale or to correct the offset and temperature dependency). The principle is correction of non-idealities by providing a voltage reference proportional to the relative temperature $T - T_{ref}$. The sensor calibration procedure is performed in only one step at $T \neq T_{ref}$ and precise knowledge of the sensor temperature coefficients is not required.

Non-linearity correction

Accepting that most sensors are non-linear in their characteristic (Figure 4.13), from a signal processing perspective compensation is either in the linear domain (linearization of the characteristic and linear adjustment thereafter) or the non-linear one. Generally, compensation for non-linearity can be achieved in several ways.

The first option includes determining the closest to ideal part of the sensor characteristic and designing the sensor system in such a way that this behavioural subset is the only one the sensor can exhibit. Embedding

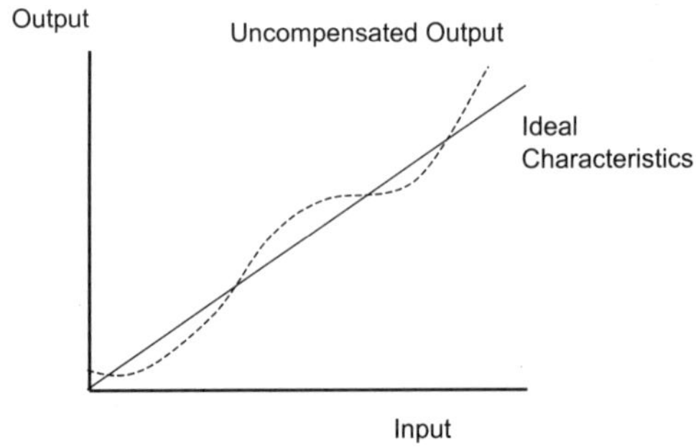

Figure 4.13: The non-linear nature of sensors, their susceptibility to environmental influences, and the effect of manufacturing tolerances make calibration and compensation through complex signal processing necessary. An alternative technique to 100% analogue signal processing is to digitise the sensor transfer function and to compare against an "ideal" transfer function. From [8].

a sensing element within a feedback system, based for example on sigma delta modulator techniques does just that. An entire chapter (Chapter 5) is dedicated to this widespread design method which found application to a variety of sensors (including commercial ones), not only due to the excellent linear behaviour obtained for the sensor system but also its good noise rejection performance and the fact that a digital sensor output is obtained.

When the non-linear sensor characteristic comes from the physical model of the sensor, a compensation module may be added to the sensor system at design stage and becomes integral part of each and every sensor of that type, produced by a given process. The correction procedure based on addition of a compensation function is illustrated in Figure 4.14. The compensation module is built as part of the interface circuitry of the sensor.

Tunnelling current sensors are a good example here [12]. The correction principle is that of inverse functions: if the non-linear characteristic of the sensor can be described mathematically for a sensor type, its inverse can be found and cascaded in the sensor signal path. Implementation wise, this can be done either in hardware or software, in the analogue or digital domain. The decision has to do more with the overall sensor system design approach and constraints than the specific sensor or specific non-linearity. If a sensor for example exhibits a logarithmic transfer function, a simple exponential circuit is cascaded in the sensor signal chain and the result is a linear transfer for the sensor system, as illustrated in Figure 4.15.

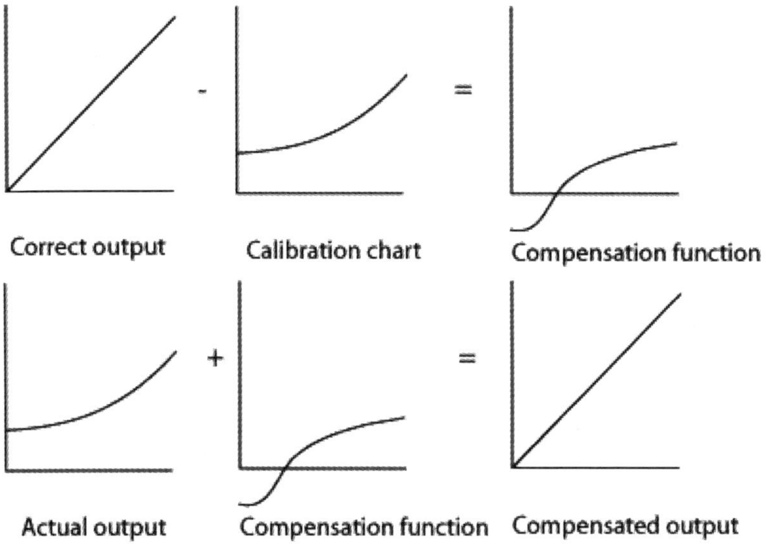

Figure 4.14: Addition of compensation function.

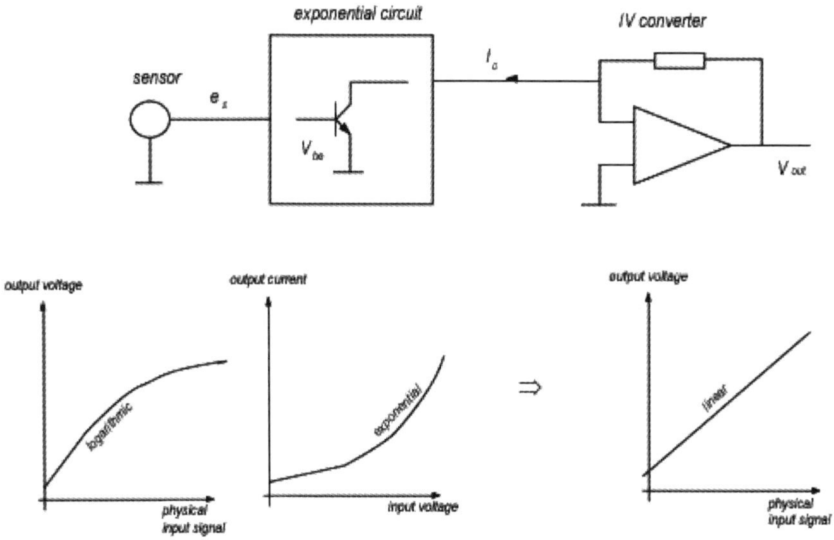

Figure 4.15: Linearisation of a typical logarithmic function. From [4].

Particularly when the non-linearity of the sensor characteristic is not known mathematically or, it is not consistent (device to device variations or random errors exist), the look-up table approach is commonly used either for calibration only or, more often calibration and compensation of the sensor errors. The method is also used in systematic/batch compensation, when the mathematical description of the non-linear transfer function of the sensor is known but not immediately and directly addressable through suitable circuitry.

For the purpose of calibration only, the look-up table contains a number of input-output points, throughout the sensor range. If calibration and compensation is desired, the inverse function of the sensor transfer characteristic built through calibration is actually stored and the sensor looks up the corrected digital value, before delivering it as its output (Figure 4.16).

The sensing element offset and gain, plus the errors of the read out electronics are automatically corrected as well as the non-linearity in this calibration based approach. Both for calibration only look-up tables and the calibration and compensation look-up table procedures, unless a very detailed, point by point table is used, between measured points, linear interpolation or extrapolation, respectively, is usually performed. In some cases, if the cost and computational restrictions are relaxed and depending on the extent at which the characteristic of the sensor is predictable, other more accurate interpolation methods might be implemented in software.

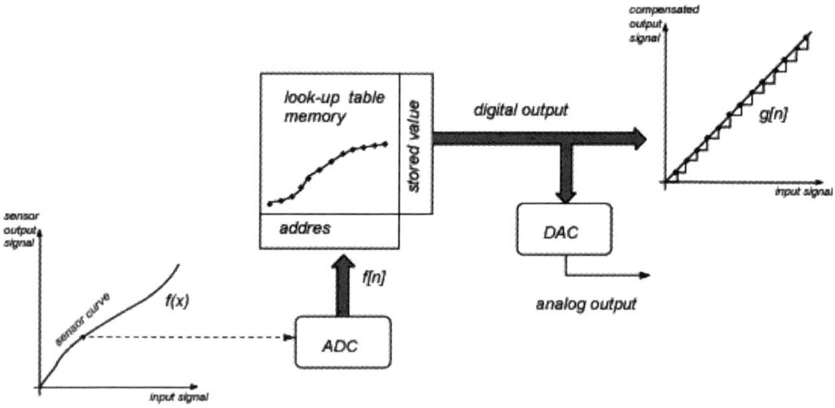

Figure 4.16: Look-up table compensation method. Redrawn and modified from [4].

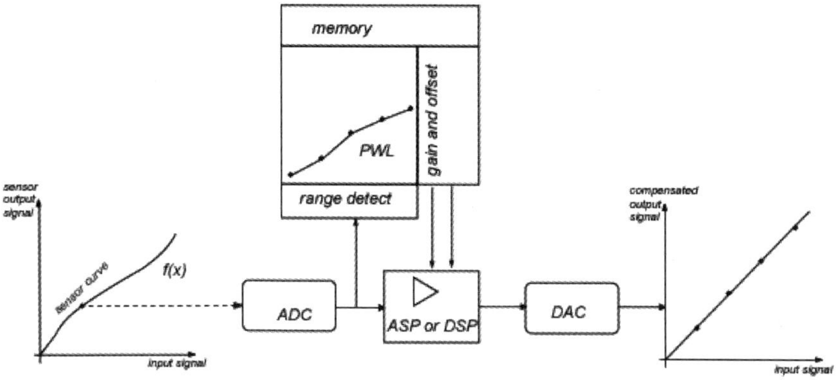

Figure 4.17: Relatively simple digital control can be used to linearise a transfer function by summing the actual output with compensated digitally generated signals. The piece-wise linear correction is illustrated here. From [4].

Figure 4.17 shows a possible arrangement for the above approach. The measurements can either be taken directly from the analogue output of the sensor or after digitisation using an on-chip AD converter. The measured digital values are stored in the lookup table. Compensation is implemented in a dedicated signal conditioning IC which, during sensor operation, compares the lookup table against an ideal set of values and provide compensation data to the DA converter, which adjusts the output of the sensor system toward the ideal values, as illustrated in Figure 4.16. From a practical viewpoint, best results with look up tables are obtained if the quantisation of the non-linear sensor output is performed using an ADC with programmable dynamic range, insuring that most of the ADC scale is used and therefore reducing quantisation errors.

The benefit of having an on-chip AD converter is that recalibration is possible in the field without connecting to the test system. The downside is that it adds to the IC complexity and therefore cost. These cost versus functionality decisions are commonplace in the definition of custom ICs [8]. However, as technology advances, the optimum choices change. Many modern IC fabrication technologies are so highly integrated, that the additional cost of adding a circuit such as an A to D to an existing design can be very small, particularly that the circuit design is available as off the shelf IPR (thus obviating expensive additional design time). Thus, if a custom IC is being used, the cost of a digital compensation scheme may be minimal, whereas additional analogue components may result in a greater cost. However, the cost of the decision (in terms of the time taken for the designer to familiarise himself with the true costs and consequence of each of the options) can be large, and very often the 'safe' option, which relies on existing, familiar technologies is taken.

A common alternative to the full (all points) look up table approach to compensation is piecewise linear interpolation of the sensor characteristic. The method is often cheaper in terms of memory requirements than the look up table approach as only the parameters of the linear sub-range approximations are stored (gain, offset, start and end points) and a simple mechanism for sub-range identification for a given measurement point is applied. When processing capabilities are at hand, the same principle can be extended to more accurate polynomial, spline interpolation or tensor spline approximation for multi-dimensional compensation (which has proven to be the most effective, according to Mozek *et al.* [13]). Once the sensor characteristic curve is built, its inverse can be found and stored, using the same interpolation method. On similar lines, a less usual but potentially powerful compensation method is the progressive polynomial type, where each calibration measurement is used directly to calculate one programmable coefficient in the correction function (the method was demonstrated at concept and implementation levels by Huijsing *et al.* [4].

All methods above are well described mathematically by Huijsing *et al.* [4] and will only be briefly explained here.

Polynomial interpolation through a set of predefined calibration points is the prevalent calibration scheme of standard (standard as opposed to 'smart') sensor designs [13]. When the compensation is to be performed in the analogue domain, the measured quantities are presented at the sensor system output independently. The most common example concerns the actual output and the additional temperature reading in a sensor system. Compensation is one-dimensional — analogue circuitry is designed to compensate the temperature effects on the measurand, in the example here. 'Smart' sensors on the other hand, could be designed to compensate various dependencies by measuring all other variables, through multidimensional calibration and compensation algorithms. For 'smart' sensors subscribing to the IEEE1451.2 standard, the correction engine is implemented digitally, providing both means for calibration and quantity calculation procedures. An example taken from Mozek *et al.* [13] is as follows: consider a pressure sensor with embedded temperature measurement ability; the correction engine takes the form of a multidimensional truncated Taylor series as given in Equation (4.1):

$$F(X_1, X_2) = \sum_{i=0}^{D(1)} \sum_{j=0}^{D(2)} C_{i,j}[X_1 - H_1]^i [X_2 - H_2]^j. \qquad (4.1)$$

The equation relates the measured quantity F (in SI units) and the raw input values to the engine from the measurement channels, X_1 and X_2 (X_1 pressure and X_2 temperature), via coefficient multidimensional matrix $C_{i.j}$, which are decisive of the type of calibration used. D is the polynomial degree, H_1 is the offset pressure value (the value of a raw pressure AD readout which corresponds to zero pressure applied) and H_2 is the lower limit of temperature measurement interval. Input quantities are normally segmented to preserve maximum accuracy of the measured quantity, hence H_1, H_2 are vectors.

Mathematically the look up table approach would involve selection of one coefficient per linear segment only. For IEEE1451.2 complying systems, the number of segments per channel is 255, yielding only 8 bits AD resolution. For other 'smart' sensors, higher accuracy can be achieved (limited only by the ADC resolution). Standard sensor systems calculate at least a linear instance of (4.1). The advantage of the method is its speed (no recalculation required) particularly when a high sensor sample rate is required. The disadvantage is the large memory consumption (a disadvantage only in older, minimal processing power systems).

Higher order calibration and compensation algorithms refer to the use of *higher order polynomials* on a single segment.

Spline approximations are used for maximum data compactisation in TEDS (Transducer Electronic Data Sheet) since it provides a straightforward description of segmented sensor characteristic. Note that the amount of calibration data must be kept to a minimum while still satisfying the criteria for desired Combined Standard Uncertainty (CSU).

Mozek *et al.* [13] present an evaluation of the three methods above (look up table, polynomial interpolation and spline approximation), for a pressure sensor with temperature dependency. The tensor-product-of-spline approach was found to be most appropriate, offering multidimensional compensation of an arbitrary number of axes and consuming the least quantity of non-volatile memory for the desired accuracy.

Minimising the time consumed on calibration and the computational time expended on correction lead to the development of various *strategies* for performing calibration and compensating smart sensors. One such strategy is the *electrical properties analyser* proposed by Mozek *et al.* [13]. The strategy involves spline approximation methods in combination with a set of pre-calibration procedures and intends to minimise the amount of data and the degree of the approximating polynomials, enabling the use of software based floating point multiplications performed by simple 8-bit microcontrollers. Given that the time-to-calibrate a single sensor is of importance, the procedure also determines sensor failures (by running a post-calibration process on a test lot of sensors and determining the tolerance limits of essential sensor properties), which are then fed back to pre-calibration procedures.

Implementation choices

Programmable electronics, combined with a capability of storing individual correction coefficients in non-volatile digital memory is now often used for trimming analogue functions in the digital domain. For sensors, electronic trimming has evolved towards the use of Digital-Sensor Signal Processors (DSSP) on the one hand and Analogue Sensor Signal Processors (ASSP) on the other hand.

DSSP techniques involve the digitisation of the sensor signals followed by calibration and compensation in the digital domain using a microcontroller with EEPROM, and the use of a DAC (if required) to convert the compensated result back to an analogue signal. The advantage of this approach is that compensation occurs after digitisation by the ADC, the signal processing occuring in the processor's zero-drift digital domain. Disadvantages include software complexity, memory requirements, and a reduced dynamic range that calls for higher resolution in the ADC. One of the several available DSSP architectures is the MAX1460 produced by Dallas Maxim and initially developed for piezoresitive pressure sensors, but available now for use with accelerometers, strain gauges and other low-level bridge type sensors [14].

By adjusting the sensor excitation and digitally adjusting the amplifier offset and gain, ASSP techniques achieve sensor calibration and temperature compensation in the analogue domain without quantising the signal. Through the use of DACs, EEPROMs, and digitally adjustable analogue electronics, this hybrid technique offers the best of the all-analogue and all-digital approaches. It allows for signal processing in the analogue domain with the "potless" ease of a digital system. To linearise the sensor, ASSP systems adjust gain and offset using feedback from the raw sensor output to the DAC reference inputs. This technique eliminates the unwieldy polynomial curve fitting required in DSSP approaches. The DAC, which multiplies a digital number by an analogue voltage (the DAC's reference input), is the key element in an ASSP electronic trimming system. High-resolution DACs are expensive, however, and a sensor requires several of them for proper ASSP compensation. This problem has been resolved by the development of a new sigma-delta technology for DACs and ADCs (MAX14xx series) that enables low-cost digital trimming. It yields 16-bit converters on very small areas of silicon, which in turn allows complex systems-on-a-chip that include multiple DACs and ADCs.

Focusing on the digital domain processing and taking a more general outlook to implementing various signal quality enhancement techniques (and for that matter potentially also building extra functionality within the sensor system), the designer has several digital signal processing implementation choices. If looking at discreet (multi-chip) solutions, the use of microcontrollers is well documented and most if not all attempts to 'serious' sensor signal processing were based on a variety of such devices. It is however claimed that microcontrollers frequently lack the horsepower required to implement sophisticated techniques and to perform the associated control and communication functions [15]. Arguably there are three basic constraints on a microcontroller's ability to support compensation algorithms: arithmetic hardware functions, memory-addressing modes, and data bus width. The ability to perform single-cycle high-precision mathematical operations is essential to implementing digital signal processing algorithms. Not only do the mathematical operations themselves need to execute rapidly, but the accumulator (register) that holds the results must be able to accumulate the results from many operations without overflow. Usually, this entails accumulating 16-bit by 16-bit multiplications (32-bit results), which requires a 40-bit-or-better register and efficient handling arithmetic overflows. Traditional, 8-bit microcontrollers can only handle high precision arithmetic using sequential single byte operations. Finally, most 8-bit microcontrollers do not support an internal bus wide enough to move 24- to 32-bit data efficiently. The overhead required to transfer high-precision data severely limits the amount of data the microcontroller can handle and unnecessarily complicates the associated code.

Digital signal processors (DSPs) are tailored for high-speed throughput and are optimised to support fast mathematical processing. The wide, single-cycle accumulators found in DSPs support the high-speed, high-precision data flow for implementing digital filters and other signal processing algorithms. Another requirement for maximum signal processing throughput is the ability to read from two separate areas of memory in a single cycle (for example, to get the filter coefficient and the associated data sample, known as a Harvard or modified Harvard architecture). With their wider internal data paths, DSPs eliminate this overhead from both a processing and coding perspective. To achieve their high performance they make extensive use of data or instruction pipelines. Once filled, these pipelines deliver all the data and the associated instruction to the processing core in a steady stream with very little overhead. Mathematical operations usually take only a single cycle, so the throughput is high *until something disrupts the pipeline.* If the instruction or data are not contained in the associated pipeline, the pipeline has to be flushed, the new information retrieved, and the pipeline loaded again. Interrupts or program branches are sources of pipeline disruption, and both are an integral part of many embedded applications.

An optimal mixture of microcontroller and DSP is claimed to come to rescue, in the form of Digital Signal Controllers (DSCs) offering a powerful, flexible platform on which to build robust measurement applications. DSCs maintain a microcontroller look and feel while adding full-featured DSP performance. This allows a designer who is proficient in microcontroller applications to quickly add DSP functionality without having to learn an entirely new architecture. Digital signal controllers tend to have much shorter pipelines or none at all, significantly reducing the disruptive effects of interrupts or program branches. Some digital signal controllers, such as the dsPIC family, Microchip Technology Inc., come with as few as 18 pins, making board layout easy and reducing manufacturing costs [16].

As a case study for the look-up table approach to compensation, a commercial implementation of a sensor signal conditioner is considered in some detail [14]. This application is of interest because it gives both an in depth view of the considerations involved in designing a comprehensive compensation strategy, but also because it is an example of a packaged digital subsystem, intended for use within an analogue system — that is, an analogue component internally realised using digital technology.

Maxim Integrated Products has introduced several ICs as interfaces to low-level bridge sensors in modern industrial systems. All of these ICs provide sensor compensation and temperature correction. The high-end device (MAX1457) shown in Figure 4.18, linearizes a sensor output by establishing 120 piecewise-linear segments, drawing on data stored in EEPROM. The resulting linearised output is accurate to within 0.1% of the sensor's

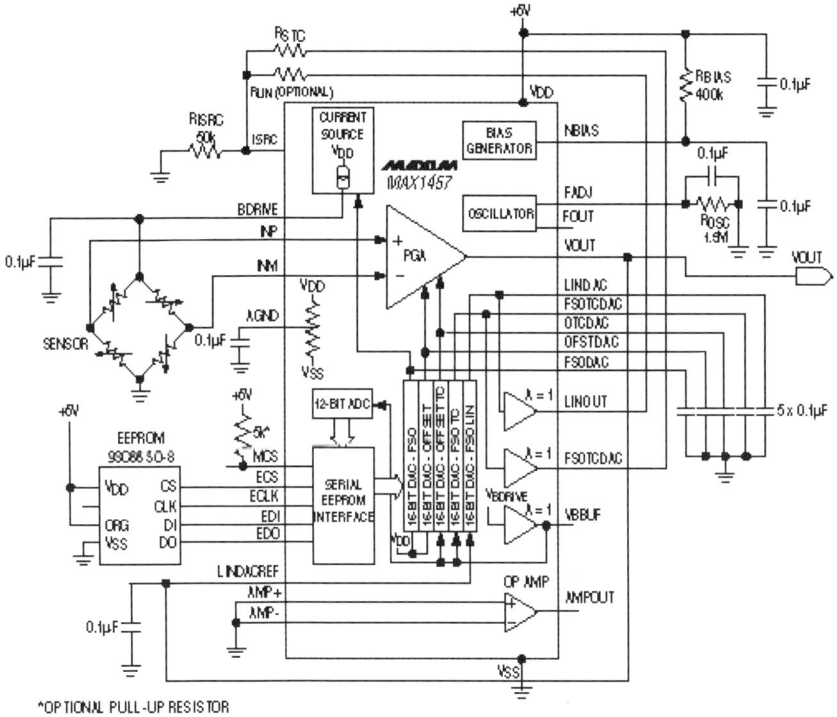

Figure 4.18: Maxim MAX1457 sensor signal conditioner. From [14].

repeatable error. These flexible signal-conditioning ICs are for use with pressure sensors, accelerometers, strain gauges, and other low-level bridge-type sensors. They can be used in an industrial sensor, in a 4–20 mA or 0 to 5 V transmitter, or in a complete instrument. Self-calibration enables these ICs to derive high accuracy from less than ideal sensors without the need for complex front-end analogue circuitry or (in the case of the MAX1457) firmware-based linearisers or multi-order polynomials. Because the IC design is based on analogue cells, the devices are easily customised for use with previous sensor types (capacitive, inductive, etc.).

All of these ICs provide a signal path that includes flexible sensor-excitation circuitry, a programmable-gain amplifier (PGA), and an analogue output. The basic device (MAX1450) includes only those functions. The midrange one (MAX1458) calibrates the gain, offset, and temperature drift of these parameters by adding four 12-bit digital-to-analogue converters (DACs), one coarse 3-bit DAC and a non-volatile, internal EEPROM for storing the DACs' calibration data. The high-end device (MAX1457) contains six 16-bit DACs and one 12-bit analogue-to-digital converter (ADC), and operates with a larger, external EEPROM.

Figure 4.19 illustrates the MAX1457's ability to compensate for temperature and linearity errors. Graph (a) shows the low-level output of an uncompensated piezoresistive sensor with its huge temperature errors of offset and gain (b). Graphs 3(c) and 3(d) show the signal after conditioning. The MAX1457 scales the sensor output in the 0.5 V to 4.5 V range (c), and limits gain and offset errors to 0.1% over a wide temperature range (d).

An important consideration in the design of the Maxim sensor signal-conditioner architecture for has been the need to support advanced manufacturing technologies. To meet that requirement, IC designers lowered manufacturing costs by integrating (along with signal-conditioning functions) the following three traditional sensor-manufacturing operations into one automated process:

Pretest:
This operation tests sensor performance over the compensated temperature and pressure ranges. The Maxim ICs' MICROWIRE interface and three-state outputs, for example, enable control by a host test computer. These capabilities enable testing of multiple transducers in a parallel connection (Figure 4.20), and allow digital communication between the test system and any specific transducer (selected through a chip-select pin).

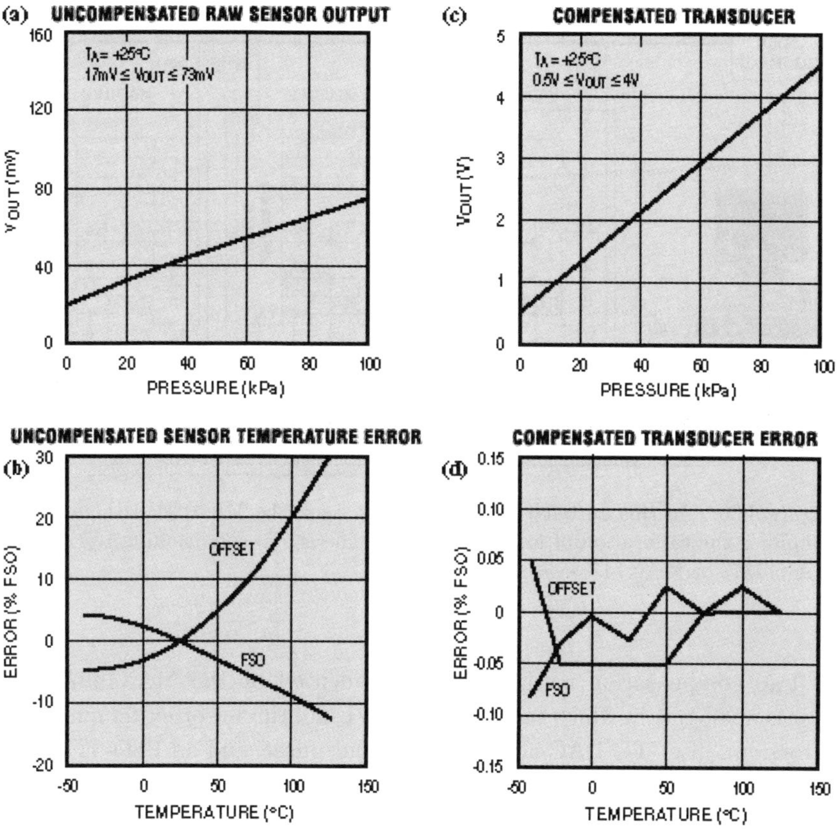

Figure 4.19: Raw output from a sensor (a) is amplified and conditioned by the MAX1457 (b), and the sensor's temperature errors (c) are compensated by the MAX1457 as well (d). From [14].

Calibration and Compensation:
This operation can be performed immediately after pretest, without removing the transducers from their test sockets. The test computer simply calculates the calibration and compensation coefficients (4kbits) and downloads them through the MICROWIRE interface to the transducer's EEPROM.

Final Test:
This operation verifies transducer performance, again without removing the device from its test socket.

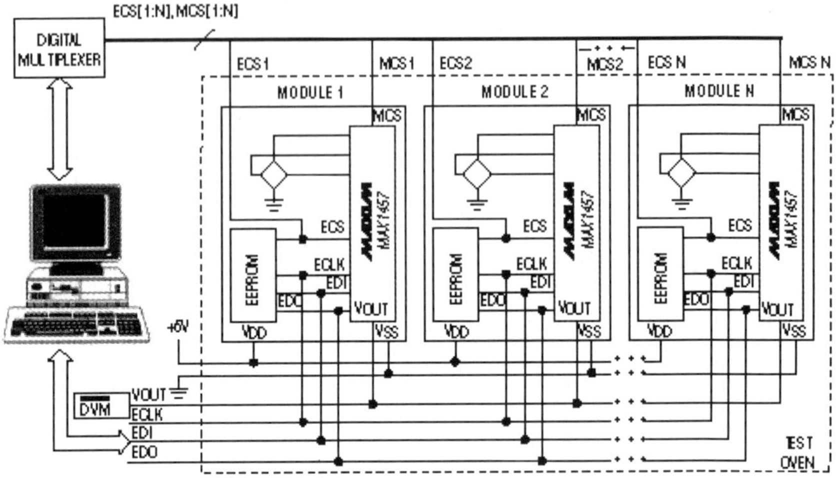

Figure 4.20: In this automated calibration system, the MICROWIRE interface simplifies the calibration of multiple sensors. The signal-conditioning ICs can be MAX1457s or MAX1458s. From [14].

Two compensation methods are implemented by the MAX1457. The first is analogue, in which two DACs compensate the 1st-order temperature errors: an offset-TC DAC adjusts the output offset, and an FSO-TC DAC adjusts the bridge-excitation voltage by adjusting its excitation current (Figure 4.21). (The less expensive MAX1458 makes these corrections and no others.)

The second method of compensation is digital. An ADC driven by the bridge-excitation voltage (which is temperature dependent) generates the EEPROM address. The EEPROM output is a multiple-segment approximation (up to 120 segments) that corrects residual higher-order errors. MAX1457-based compensation employs 16-bit DACs to provide all of the functions listed in Table 4.3. The MAX1458 employs four 12-bit DACs and a 3-bit offset DAC to provide only those functions marked with asterisks in Table 4.3.

Initial offset is corrected by feeding to the PGA's summing junction a voltage obtained by multiplying (within the offset DAC) a fraction of the supply voltage by a 16-bit word. The full-span output (FSO, or gain) is calibrated in two adjustments: coarse gain is set by feeding a 3-bit digital

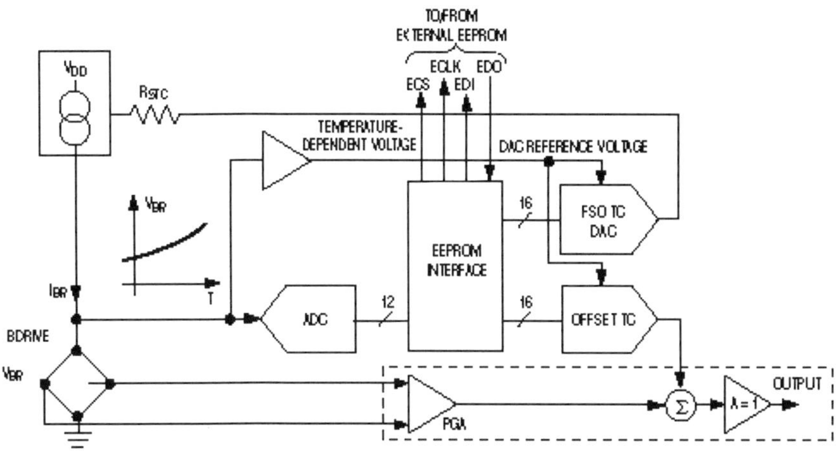

Figure 4.21: Simplified circuitry within the MAX1457 illustrates the correction of temperature errors. Analogue voltage across the sensor bridge generates the DAC reference voltages, which in turn produce the 1st-order analogue corrections. The bridge voltage is also digitised to provide fine correction through the EEPROM look-up table. From [14].

Table 4.3: Digital Compensation DAC Functions. From [14].

Function	DAC type
Initial offset calibration*	Offset
Initial FSO calibration*	FSO
Correction of TC slope for analogue offset	Offset TC
Correction of TC slope for non-linear offset	Offset TC
Correction of TC slope for analogue FSO*	FSO TC
Correction of TC non-linearity for non-linear FSO*	FSO TC
Correction of pressure non-linearity	FSO linearity

word to the PGA, and fine gain is set by adjusting the bridge current using another 16-bit word.

Two DACs connected to the bridge voltage (the offset-TC DAC and FSO TC DAC) compensate linear components of the zero and FSO TC. Bridge voltage is proportional to temperature, and a properly valued digital word (the multiplier coefficient) causes the DAC output to compensate the temperature slope by following the quasilinear change in bridge voltage.

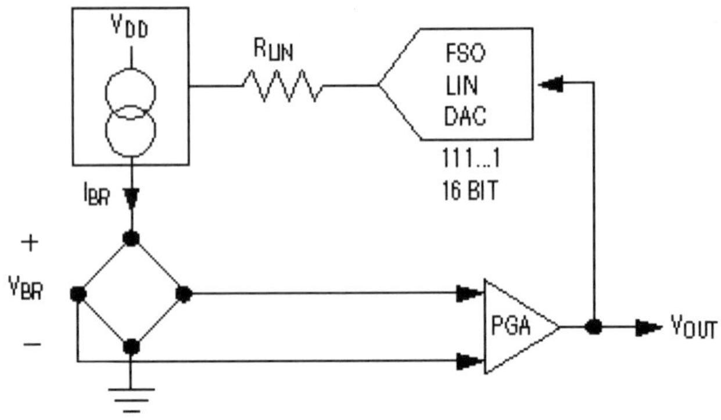

Figure 4.22: This simplified circuit, also internal to the MAX1457, demonstrates the concept of pressure-non-linearity correction. From [14].

Digital multislope temperature compensation allows compensation of arbitrary error curves, whose shape is determined only by the shape of the temperature signal and the adjustment range available in the electronics. This compensation is implemented with 120 number pairs (corrections for offset TC and FSO TC) stored in EEPROM look-up tables. The EEPROM address is the output word of a 12-bit ADC driven by the bridge voltage, which (with constant current excitation of the bridge) is temperature dependent.

When measuring pressure for example, non-linearity is corrected by feedback from the output voltage to the bridge current source. To gain control of this feedback, the output voltage is routed to the reference input of a DAC, whose output connects to the current source and is then subject to the DAC's digital input, driven by a coefficient stored in the EEPROM (Figure 4.22). Thus, coefficients delivered to the DAC can introduce a non-linearity in the bridge current that compensates (often by an order of magnitude) for non-linearity in the sensor output.*

ASIC approaches to compensation

ASIC (Application Specific Integrated Circuit) approaches are becoming an important strand of development not only for sensor compensation but also for adding extra functionality to the new sensor generations. ASIC

*The product data sheets offer further details on operation.

technologies offer part configured circuits or design libraries for the fabrication of monolithic ICs for precise satisfaction of the users specification.

A good example here is provided by Allan [17], who describes two different ASIC approaches for signal-conditioning a MEMS piezoresistive silicon pressure sensor. One is based on a digital scheme for applications requiring a low operating voltage and low power consumption. The other involves an analogue method for low-cost, high-volume applications. In both cases, the ASIC is employed to calibrate and compensate the sensor with a total error of less than $\pm 1\%$ of full scale over two different operating-temperature ranges. The total error includes effects due to offset and sensitivity, as well as the offset and sensitivity temperature coefficients. An ASIC approach was chosen because a typical output signal for a piezoresistive pressure sensor depends on temperature. The circuit was required to operate in the few-millivolt range. This level is too low for control and interface with microprocessors. Two further examples are given here. An analogue ASIC allowed integration of complex analogue compensation functions optimised for this signal level. Moreover, ASICs can offer advantages over traditional approaches. Passive laser trimming of resistor networks, for example, provides high resolution and a wide range of resistor values. But the resistors have to be serially and individually trimmed, so the approach is capital intensive. It requires complicated and expensive production fixtures, as well.

Two further examples are given here. Researchers at the Electron Device Laboratory of Fujikura Ltd. in Tohoku and Tokyo, Japan, devised a DSP-based circuit that corrects for the sensor's offset and sensitivity and their temperature coefficients. Fujikura's circuit offers a compensation range from $-30°C$ to $80°C$. Meanwhile, researchers at the Institute of Microelectronics in Singapore used a fully customised ASIC with a fusible-link array that achieves the aforementioned performance from $-40°C$ to $125°C$ [17].

Fujikura's ASIC was made on a 0.7-μm double-polysilicon, double-metal, n-well CMOS process. It consists of a sigma-delta 16-bit analogue-to-digital converter, a reference voltage with a built-in temperature sensor, the 16-bit DSP core, 101 polysilicon fuses, a step-up voltage regulator, a 10-bit digital-to-analogue converter (DAC), and a 4-MHz oscillator (see Figure 4.23).

The DSP does most of the offset and temperature coefficient corrections. Corrected coefficients are stored using the polysilicon fuses. The output code is then accessible with a serial interface or an analogue signal provided by

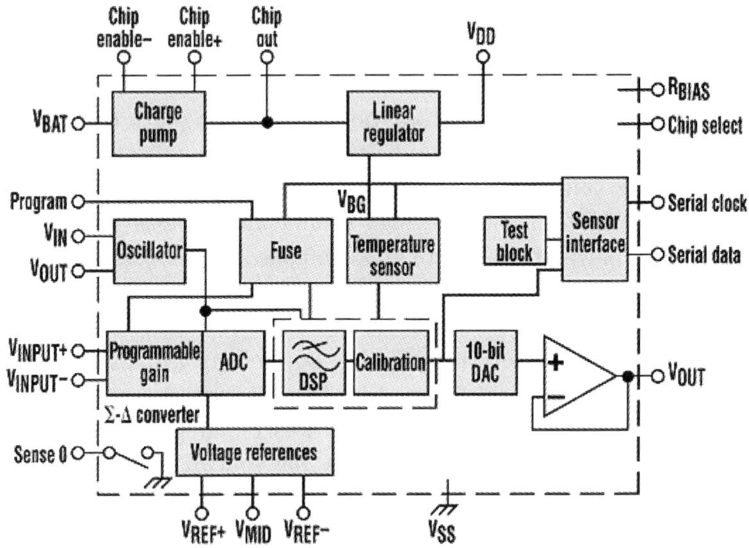

Figure 4.23: Fujikura signal conditioning ASIC. From [17].

the 10-bit DAC. This circuit also compensates for secondary temperature characteristics. It has an I2C serial interface. A built-in charge pump lets it work in circuits rated under 3 V. A "sleep" mode reduces power consumption.

The analogue-based Institute of Microelectronics' ASIC is made on a 0.8-μm double-polysilicon, double-metal CMOS process. The ASIC consists of a core analogue signal processor, a 64-bit fusible link array, and a serial fusible-link interface. The ASIC's digital portion provides the interface between the analogue signal processor and the controller (the computer). This controller writes data to the interface and reads data back from it by a serial-in and serial-out communications protocol. Data in the serial interface can be loaded into the fusible-link array to control various resistor networks in the analogue signal processor. These resistor networks are used for various programmable functions. All of these programmable elements make it possible to compensate for the non-linearity, sensitivity, temperature, and temperature coefficients effects to the first order. The ASIC features an output of 0.5 to 4.5 V using a 5-V power supply. The output is ratiometric when the power supply is varied between 4.5 and 5.5 V.

4.2.3. Sensor Integrated Compensation

The integration of compensation functions with the sensor, usually using a monolithic IC for compensation, is now an increasing trend. This can be motivated, for instance, for piezoelectric sensors, where each sensor element has individual characteristics and is sensitive to both pressure and temperature. The signal-conditioning circuitry needs to be interchangeable and compatible with most electronic control systems and the signal-conditioning functions for such sensors must include:

- calibration for offset and full-scale variations to make each unit electrically interchangeable;
- temperature compensation for offset and full scale;
- other functions, such as linearity correction, diagnostics and filtering.

Traditional fabrication technologies which marry signal-conditioning electronics with sensing elements have resulted in technical and commercial compromises in product cost and performance. Monolithic approaches have resulted in low cost and small size but in many cases have compromised device performance because of the restrictions placed on circuit and sensor design by the monolithic fabrication process, as discussed in Chapter 3. Hybrid approaches use either a dedicated ASIC for signal conditioning or a discrete circuit approach. In many cases, hybrid approaches have better performance than monolithic devices and offer flexibility for adapting new designs with simple components changes. However, they are generally not as low in cost as the monolithic approaches and are also larger and require additional assembly steps. More electrical and mechanical connections mean more concerns about reliability.

The co-integrated pressure sensor produced by Silicon Microstructures Inc. provides a good case study in the design and use of custom compensation chips for co-located integration with a single sensor (in this case a pressure sensor) and is treated below.

The product was a collaboration between MEMS sensor designers and mixed-signal IC designers. Once again the compensator is a digital component designed to fit into an analogue system. Dunbar describes the design considerations that went into the compensation chip and the integration process [18]:

> *A standard 0.65-micron mixed-signal CMOS process with E2PROM was used. All of the CMOS processing is performed at the start of the process and the MEMS processing is performed at the end. This process flow results in fewer compromises on the circuitry and better sensor performance. In addition, this "CMOS first" choice results in a better manufacturing flow because the CMOS process flow remains uninterrupted. With this approach, process controls, based on high-volume production, are maintained. The MEMS process that follows the CMOS process steps does not adversely affect the CMOS functions. These processes include the silicon etch to form the pressure-sensitive diaphragm and anodic bonding of a glass substrate for absolute pressure configurations, probing, wafer dicing and inspection. The circuit on the co-integrated pressure sensor provides for a number of adjustable parameters. Based on production test data, the following parameters can be adjusted in the ASIC by programming the device using the on-board E2PROM:*
>
> *Preamplifier gain-32 to 152 in 16 steps (16 to 76 at 2.5-volt offset);*
> *Preamplifier gain sign-1 bit (positive versus negative output);*
> *Preamplifier offset-0 to 49.5 millivolts in 16 steps;*
> *Preamplifier offset sign-1 bit (positive versus negative input);*
> *Upward of 20 coefficients in pressure and temperature correction;*
> *Output clamp-independent adjustment in 40-mV steps;*
> *Output filter-1 to 256 conversions in 1, 2, 4, 8, 16, 32, 64, 128 and 256 steps, equivalent to approximately 1 to 256 milliseconds.*
>
> *The circuitry on the co-integrated pressure sensor includes all of the functions for signal conditioning and calibration, including amplification, correction, span calibration, temperature compensation and multiorder non-linearity correction for pressure and for the temperature coefficients. The system uses an 11-bit analogue-to-digital converter with 8x oversampling to provide for an effective 14-bit data conversion resolution. Data is then processed through an on-board digital signal processor (DSP) where the A/D data is corrected, based on calibration coefficients stored in the on-board*

E2PROM. The corrected signal is then fed to a 12-bit digital-to-analogue converter, which in turn drives the output amplifier. The amplifier has been designed to be able to drive more than 2 nF of capacitance as needed for electromotive force suppression in the automotive environment.

Calibration is done by measuring the sensor at multiple temperatures and pressures. Uncorrected data from the AD converter is read out through a digital I/O for each data point. The external calibration computer then determines the minimum error and loads the order of the correction in pressure, temperature and correction coefficients into E2PROM. Verification of the calibration can then be performed to verify calibration accuracy as desired. Before linearization, the uncorrected output exhibits both positive and negative error (with respect to the ideal transfer function). In this case, a third-order pressure transfer function adequately defines the non-linearity in pressure. Using this curve fit, total error is reduced from about 1.75% to about 0.2%. The performance of the device can change over temperature, and thus calibrating at other temperatures will increase the pressure accuracy at those temperatures. By minimizing the pressure error at this second temperature, total error is reduced. The DSP algorithm used can be adapted to easily correct pressure non-linearities of various orders and further can facilitate correction that might be introduced by temperature-dependent pressure non-linearities from jelling or other media interfaces.

After completion of the CMOS processing steps, the wafers are shipped from the CMOS fab to the MEMS fab. Silicon etching is performed along with the final process steps for wafer bonding, dicing and testing. The resulting dice are then either shipped to die customers or assembled into a variety of packages.

Co-integration technologies as above can be extended beyond pressure sensors to other devices, including accelerometers, gyros and other sensor types. Although other, multichip intregration technologies are available, as was seen in Chapter 3, Section 6, there are many emerging applications which demand very high levels of miniturisation and low levels of cost. For such applications, co-integration is an attractive solution, particularly if it is possible to integrate several sensors together with the associated calibration and correction circuitry.

4.2.4. Measurement Reliability Increase Through Physical Redundancy

So far, in this chapter, the most commonly used methods of compensation have been surveyed. These ideas and methods form, however, only a tiny slice of the work one finds in this domain's research arena. Apart from highly dedicated designs and implementations, very specific to a chosen sensing element and technology, one can also identify newer strands of effort such as bringing artificial intelligence tools and techniques into the realm of sensors (initially proposed for characteristic compensation only but followed by initiatives towards fault detection and other system functions) or using system wide strategies or addition of extra devices to the sensor system to obtain better measurement results. Whilst the former will be discussed in Chapter 7, the latter is introduced below. Note that "better" here might imply more reliable and trustworthy, rather than just more accurate measurement results.

Improving measurement quality through redundancy, either by fabricating several sensors on the same silicon substrate or by mounting multiple sensors in a hybrid package is another common approach. Two examples of sensor systems which exploit redundancy are presented here, to highlight basic motivations for using this approach.

The first example is from Fraunhauer Institute of Microelectronics Circuits and Systems which designed a monolithic pressure sensor, produced by surface micromachining combined with CMOS processing. The need for multiple elements here was due to the fact that in their particular capacitive MEMS device design, a single element provided a 0.017 pF capacitance without applied pressure (a membrane of 100 μm diameter), too small to be effectively measured. 81 sensors, however, switched in parallel, offered capacitance values between 1 to 2 pF, in which case, the sensor output varied approximately 0.2 pF over a 1–6 bar pressure range [19], allowing for effective pressure measurements.

Another example is presented by Selvakumar *et al.* [20]. A complete threshold acceleration detection microsystem comprising an array of accelerometers and a low power interface circuit has been designed and implemented as a Multi Chip Module. The system contains sensors replicated thrice at each acceleration level and the output of the sensor system is weighted on the majority status of these redundant sensors. The scope of multiple sensing here is achieving fault tolerance and increased accuracy. The sensors are fabricated using the bulk-silicon dissolved-wafer process which offers a wide latitude in sensor threshold levels (1.5 g–1000 g), bandwidths of 45 Hz–40 kHz, with mass sizes between 0.015 μg to 0.7 μg. The interface circuit dissipates less than 300 μW and measures 2.2 mm × 2.2 mm. The microsystem supports communication with a standard microcontroller bus in a smart sensor network.

4.2.5. *Sensor Self-test*

By adding an actuator (or more) to the sensor system one can influence the physical input signal of the sensor during special test routines.* This usually involves a certain amount of conditioning circuitry to be present within the sensor which would command/schedule the test runs, would disable the "readable" output of the sensor (during test the sensor output is not meaningful with respect to the application the sensor is used within) and decide if the sensor is functional. Such a sensor feature is generally called 'self-test'. As with all other signal improvement functions described in this chapter, a variety of methods have been developed for realising self-test, some of which are device specific, sensor type specific or generic system level approaches.

Several newer sensing devices have been designed to implement some form of self-test within the design of their inner mechanical structure. Some sensor types are lending themselves to self-diagnostic, through the use of their internal states/signals. For such sensor types, units of diagnosis and validation of measurements (degrees of confidence) have been developed [21] and integrated within the sensor system. Other sensors have only the output signal available and therefore, the only diagnosis approaches come from the overall sensor system arrangement: sigma delta modulator sensors, for example, have an implicit self-diagnostic for 'fine/fail'. In some other cases,

*Although introducing actuators within the sensor system could improve the operational reliability of the sensor, care needs to be taken not to reduce the overall mean-time-before-failure for the product (consequently, the reliability of the actuator needs to equal or exceed that of the sensor).

self-test is a consequence of the design strategy for the sensing application a whole: within a network of collaborating sensors for example, self-test can be implemented using data mining methods for example, with information about a particular sensor being derived from the global pool of information from all sensors in the system.

Generally speaking, the ability to add self-testing functionality to a sensor system has been seen as one of the most valuable contributions an associated microprocessor could bring to a sensor. The drive to develop self-test comes partly from legislation regulations (as for example the air-bag application for accelerometers) and partly from the evolution in the sensing applications themselves, where reliable measurements are more and more critical.

4.2.6. *Sensor Self-calibration*

In principle very similar to self-test, self-calibration means further addition to the sensor system, too: accurate actuators (with precise, well specified transfer functions) need to be incorporated which can generate reference input signal for the sensor. If one takes the example of capacitive accelerometers, actuation for generating a self-test signal means applying a charge on the sensor seismic mass, which results in an electrostatic force and therefore mimics an acceleration signal.

Like much of the sensor related terminology, self-calibration as such is, generally, oversold. It does not, in reality eliminate the need for calibration (unless an actuator with a standardized reference signal can be integrated), but it is rather a method to monitor the parameters of a sensor and automatically adjust them. In this way, the time between re-calibrations can be extended or, a single calibration can be made to last for the lifetime of the sensor (in the context of battery powered sensors, inaccessible after deployment). It is more, therefore, a case of auto-zero adjust, auto-gain or auto-scale adjust rather than self-calibration. In this view, self-calibration can be realized with respect to other sensor characteristic — e.g. for sensitivity drift, cross sensitivity to a secondary variable and offset errors.

Having said all that, it must be stressed that calibration, in one form or another, is indispensable for most sensor systems. Whether it is performed individually or in batch, during the fabrication process for the device or after the sensor system has been assembled, calibration (meaning one or multiple measurement points) is part of the production process of any sensor. Whenever the sensor output signal must relate precisely to an agreed standard unit for the measured physical signal, it is necessary to correct the sensor transfer based on calibration measurements. This is unavoidable if we want to make sensors a "plug-and-play" technology [4].

4.3. System Design Choices for Compensation — Closed Loop Configurations and other Designs

The option to compensate through dedicated system design for a variety of errors in sensors and improve the quality of their signals (or indeed the reliability and repeatability of sensor measurements) has been mentioned in most of this chapter's sections. It is impossible, however, to draw generic recipes for a systems approach to compensation. This is due to the variety of existing sensing device, each with its own particular characteristics which need to be accounted for within a systems design, on the one hand, and the variety of application specific requirements that sensor system needs to meet on the other. Here we present two distinctive examples of system design for compensation, both grounded in control theory as applied to sensing systems.

Jacobson *et al.* [22] present a three-fold approach to compensate device-to-device variations in the offset voltage, sensitivity and temperature drift in sensors and optimises the resolution.† The example taken is that of a micromachined pressure sensor. Having initially experimented with hardware compensation for gain and offset, followed by fixed-hardware interface with open loop software compensation, the final system implements dynamic compensation using a closed-loop topology. The system is built around a low-cost, extremely non-ideal sensor device. The transducer consists of the sensing device, interface and conditioning circuitry, ADC and a digital processing unit (MCU) as in Figure 4.24.

Three digital-to-analogue feedback loops adjust dynamically the sensor signal (two of them maintain the zero-pressure offset level and the third provides dynamic gain control to adjust and maintain the desired sensor

†This paper is full of useful design rules of thumb and written from a good practical perspective concerning the amplification and AD conversion of the sensor signals for best results.

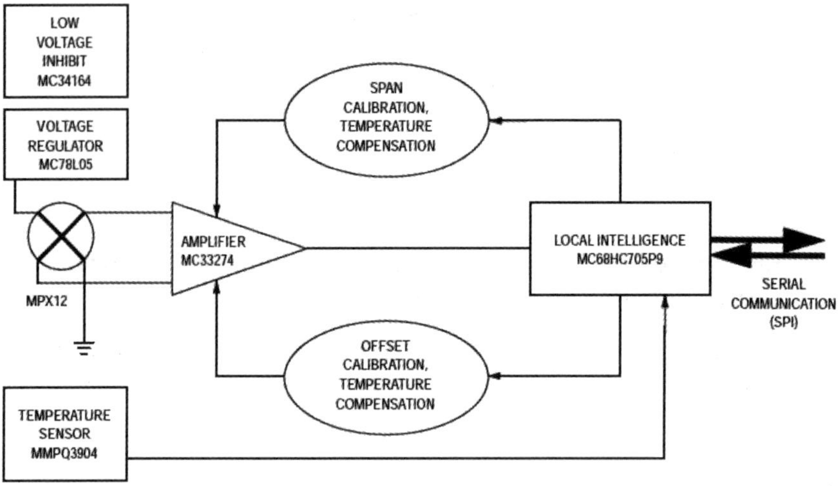

Figure 4.24: Dynamically compensated sensor. From [22].

span. The system has in-field recalibration capability, self-test and self-diagnosis features, dynamic zero, transducer electronic data sheet and serial communications interface.

More recently, an interesting view (which combines several of the signal improvement ideas in this chapter) was taken by Painter and Shkel [23] (Figure 4.25) with MEMS gyroscopes as a case study: a dual stage control strategy is adopted enabling the suppression of structural errors and enabling self-calibration of the systems. The self-calibration portion of the control circuit identifies and trims large imperfections whilst feedback control compensates for remaining small errors and in-operation perturbations. *In situ* identification of structural errors as part of a self-diagnostic calibration test is proposed based on the dynamic response of the sensing device. The work was carried out so far at simulation level with DSP implementation pending.

4.4. Summing up on Sensor Calibration and Compensation

It is clear that the error compensation for sensors is being pursued through a number of approaches, their popularity changing quite rapidly to follow advances in microelectronics and the ability to integrate more effectively

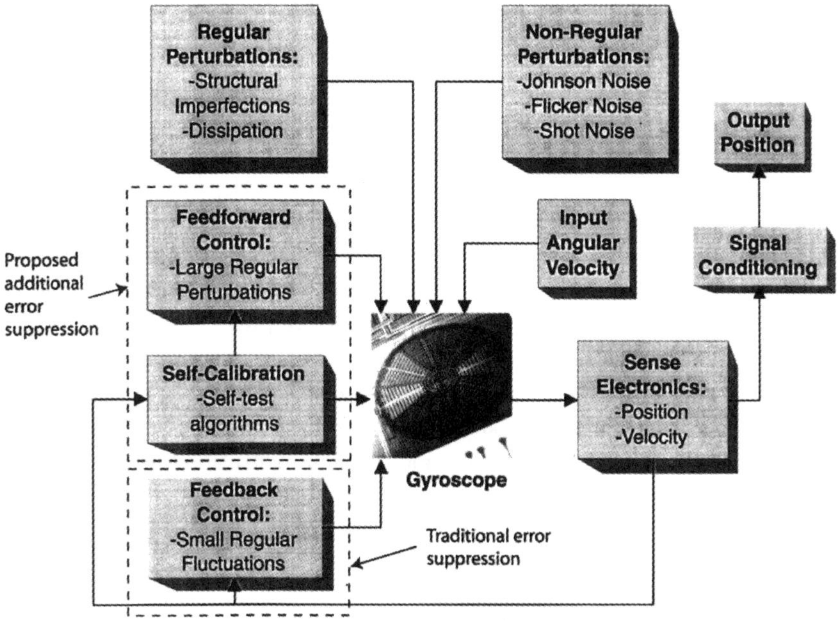

Figure 4.25: Integrated compensating electronics for vibratory gyroscope. From [23].

processing power with the sensing element or very near the sensing element. All in all, the effectiveness of these approaches to result in accurate sensors is increasing, together with the ability to apply corrections to lower and lower sensor cost bands. Whilst some still advocate that compensation is better done in hardware, the software approaches have become prevalent in the last few years, the flexibility and degree of achievable accuracy setting them as favourites. The system level approach to the sensor design is another good choice and certainly closed loop sensor system configuration have gained considerable terrain.

Given that the addition of digital signal processing (in the form of micro-controllers, DSPs or microprocessors) is likely to become common place, the chapter will end with some remarks on the digital compensation methods discussed.

Generally speaking, the use of an embedded microprocessor within the sensor system allows important changes to the design of the sensor front end circuitry. The use of digital compensation strategies implies that the designer is ultimately limited only by the stability of the analogue chain over time (hence the development of stable references and self-test techniques for the sensor front-end and analogue electronics are increasingly important). Digital compensation uses calibration data to generate a look up table or to generate a response surface that could be approximated by a mathematical function using polynomials or spline functions. Offset, slope and linearity problems are easy to deal with as the corrections are not constrained to the linear domain. There are however a couple of concerns. Firstly, if offset is not compensated at the front end, the transducer output variables might not stay within the range of the input amplifier and/or data converter or dynamic range might have to be wasted to account for it. Hence, coarse front end compensation might still be desirable. Secondly, computing the correction factors might take considerable time and hence might not be appropriate for rapidly hanging parameters. Compensation becomes a challenge at signal frequencies over 100 kHz. Speed-memory trade off between polynomial and look up table approaches can be explored [24].

Acknowledging that a good proportion of the sensors used presently by industry are still with analogue outputs, intermediate versions of digital compensation have been developed using digital inputs to generate analogue compensation voltages, leaving the output of the sensor system in the analogue world.

Sensor electronics and signal processing — introducing 'smart' sensors

Significant changes have taken place in the electronic partitioning of sensor systems and, maybe more and more of the research ventures presenting mixed analogue/digital modules employing embedded serious processing power will reach commercialisation, for a variety of sensor types. Digital compensation is effective in removing offset, slope errors, temperature sensitivities, making non-linearity less important than in the past and improving performance by orders of magnitude. Circuit stability however and the development of precision references for self-test and auto-calibration, although identified as a necessity 10 years ago, still have a long way to come. Front end interfaces do mix analogue and digital functions along with on chip E/EE/PROM or DACs, so that coarse analogue trimming can be employed to keep the signal in-range for the data converters. It is more and more the case that the entire transducer-amplifier-data converter chain is seen as a unit and compensated accordingly. Low noise, low power and small layout area are primary design goals for many sensor systems, together of course with the long running high accuracy, high reliability and low cost demands.

This chapter, aimed at describing approaches to improving signal quality from microsensors, drifted, unavoidably, into system design consideration for such sensors. This was due to that fact that little of what was discussed can be treated in isolation, from an 'added' electronics perspective. Two ideas have become clear:

- the addition of digital processing power to resolve sensors imperfections and enhance their signals is presently seen as common place, given that sensor digital output and basic communication ability is a standard (at least in the sensors R&D world);
- the efforts and achievements towards the design and manufacture of 'better' sensors in terms of their signals associated with the want for 'easier-to-use' sensors lead to new sensor applications where reliability is of major concern and consequently prompted research into exploiting further the sensor system design to allow for functions such as self-test (diagnostic) and auto/*in situ*/automatic calibration.

Both above lines of work are quite old by now, but, when they emerged they 'founded' the field of 'smart' sensors. The 'smart' sensors field has seen many developments in the past 15 years, a host of books and research literature, some standardisation and ultimately, a lack of agreement as to its proper definition. In this chapter (and maybe partly in Chapter 3 as well), we've been knowingly trespassing in the area of 'smart' sensors (as it would be seen by many), without formally introducing the concept. The choice in this arrangement was motivated by the authors' desire to clarify (if at all possible) or at least inform on some of the most prevalent views on what 'smart' sensors are (and also, what intelligent sensors — the next in the line of 'smartness' — are), at length, in a separate chapter. Following this reasoning, reinforcement of some of the ideas in this chapter are given in the case studies of Chapters 5, 6 and 7, returning to the issue of definition of the terms 'smart', 'intelligent' and 'cogent' sensors in Chapter 8.

References

1. Pallas-Areny, R. and Webster, J. (2001) *'Sensors and Signal' Conditioning*, John Wiley & Sons, ISBN 0-0471-33232-1.
2. Brignell, J. and White, N. (1996) *Intelligent Sensor Systems*, IOP Publishing Ltd, Bristol, ISBN 07503-0389-1.
3. Jacobsen, E. and Baum, J. (1997) High performance, dynamically compensated smart sensor system, *Motorola Semiconductor Application Note*, http://www.freescale.com/files/sensors/doc/app_note/AN1585.pdf.
4. Johan, H. H. and Van Der Horn, G. (1998) *Integrated Smart Sensors: Design and Calibration*, Kluwer Academic Publishers, The Netherlands.
5. Dosher (1999) Using iMEMS accelerometers in instrumentation applications, *Analog Devices*, Technical Note, http://www.analog.com/industry/iMEMS/library/imems_accl.htm (published 1999 or before).
6. Acar, C. and Shkel, A. (2003) Experimental evaluation and comparative analysis of commercial variable-capacitance MEMS accelerometers, *J. Micromech. Microeng.* **13**, 634–645.
7. Salter, T. D. (1996) *A Measurement Technique to Improve Accelerometer Calibration*, http://www.beran.co.uk/transCalAppNotes.htm.
8. McGonigal, J. (2003) *Integrated Solutions for Calibrated Sensor Signal Conditioning, Sensors Online*, http://www.sensorsmag.com/articles/0903/65/.
9. Gopel, W., Hesse, J. and Zemet, J. N. (series ed.) (1995) Micro and nanosensor technologies/trends in sensor markets, *Sensors* **8**, (eds.) Meixner, H. and Jones, R. Chapter 2: Approach to microsystem design, VCH.
10. U. Schoneberg *et al.* (1991) CMOS integated capacitive pressure transducer with on-chip electronics and digital calibration capability, *Proc. of the 6th Int. Conf. on Solid State Sensors and Actuators*, pp. 304–307.
11. Grigorie, M., de Raad, C., Krummenacher, F. and Enz, C. *Analogue Temperature Compensation for Capacitive Sensor Interfaces*, www.imec.be/esscirc/papers-96/95.pdf.
12. Yeh, C. and Najafi, K. (1998) CMOS Interface circuitry for a low-voltage micromachined tunnelling accelerometer, *Journal of Microelectromechanical Systems* **7**(1), 6–15.
13. Mozek, M., Resnik, D., Aljancic, U., Vrtacnik, D. and Amon, S. (2002) Calibration procedures for smart measurement systems, *Proceedings 9th Electronic Devices and Systems Conference 2002 and Experimental Methods in Acoustic and Electromagnetic Emission International Workshop 2002*, June 9–10, 2002, Brno, Czech Republic, Brn, Vysoké uení tecnické v Brn, 2002, str. 389–394. [COBISS.SI-ID 3188564].
14. Application Note 695 (2001) *New Ics Revolutionize the Sensor Interface*, www.maxim-ic.com/an695.
15. Huddleston, C (2003) *Digital Signal Controllers Turn Thermocouples into Superstars, Sensors* http://www.sensorsmag.com/articles/0103/38/main.shtml.

16. Application Note 695 (2001) *New Ics Revolutionize the Sensor Interface*, www.maxim-ic.com/an695.
17. Allan, R. (2000) *ASICs Used to Signal-Condition MEMS Pierzoresistive Silicon Pressure Sensor*, ED Online ID #4433, June 26, 2000, http://www.elecdesign.com/Articles/Index.cfm?ArticleID=4433.
18. Dunbar, M. L. and Allen, H. V. (2003) Performance grows with integration, *EE Times*, October 07, http://www.eetimes.com/article/showArticle.jhtml?articleId=16502355.
19. Frank, R. (1996) *Understanding Smart Sensors*, Artech House, Inc. Boston, London.
20. Selvakumar, A., Yazdy, N. and Najafi, K. (2001) A wide-range micromachined threshold accelerometer array and interface circuit, *J. Micromech. Microeng.* **11**, 118–125.
21. Henry, M. P., Archer, N., Atia, M. R. A., Bowles, J., Clarke, D. W., Fraher, P. M. A., Page, I., Randall, G. and Yang, J. C.-Y. (1996) Programmable hardware architectures for sensor validation, *Control Engineering Practice*, October 1996.
22. Jacobsen, E. and Baum, J. (1997) High performance, dynamically compensated smart sensor system, *Motorola Semiconductor Application Note*, http://www.freescale.com/files/sensors/doc/app_note/AN1585.pdf.
23. Painter, C. C. and Shkel, A. M. (2003) Active structural error suppression in MEMS vibratory rate integrating gyroscope, *IEEE Sensors Journal* **3**(5), 595–606.
24. Sansen, W. M. C., Huijsing, J. H. and van de Plassche, R. J. (1994) *Analogue Circuit Design, Mixed A/D Circuit Design, Sensor Interface Circuits and Communications*, Kluwer Academics Publisher, ISBN 0792394089.

CHAPTER 5

CASE STUDY: CONTROL SYSTEMS FOR CAPACITIVE INERTIAL SENSORS

by *Michael Kraft*

The objective of this chapter is to show how different systems level design solutions can be used to enhance the performance of inertial sensors. The technique that will be discussed in detail is the use of force feedback where the sensing element is incorporated in a closed loop control system, making the sensor a truly micro-electro-mechanical system (MEMS).

5.1. Introduction

Micromachined inertial sensors have been the subject of extensive research and development effort for 25 years. A very wide range of research proto types have been reported and many commercial sensors are available. They can be either classified by their fabrication technique or by the method of measuring the motion of the proof mass and converting it into an electrical signal (referred to in this book as the signal pick-off or interface electronics). The predominant fabrication techniques for inertial sensors are surface- or bulk-micromachining, with some sensors relying on a combination of both. The associated pick-off techniques are quite diverse but the dominant techniques are capacitive and piezoresistive; others include piezoelectric, tunnelling current, optical and thermal. For a detail survey on a variety of inertial sensor types the reader is referred to the review articles [1, 2].

Only sensors with capacitive pick-off will be considered here, as this approach lends itself particularly to force feedback, and in any case it is the pick off technique used in most integrated MEMS systems. The particular example around which the discussion will centre is a micromachined accelerometer. Various approaches for the control and interface techniques for this sensor will be described and reviewed. Accelerometers are one of the most commonly used micro-sensors, with applications ranging from inertial sensing in mass markets such as the automotive and photographic industries through to more specialised applications such as seismometry and space micro-gravity measurements.

Most commercial inertial sensors to date are open loop. However, this has some shortcomings as there are many inherent non-linear effects primarily associated with larger displacements of the proof mass. In the future, it is envisaged that higher performance sensors will need to be available to mass market applications. For example, the sensor requirements for modern automotive safety systems are becoming more and more demanding. To achieve these increased performance specifications, it is a far more promising approach to make use of circuit and systems solutions by incorporating the mechanical sensing element in intelligent control loops (as suggested in Chapter 4) than to try to improve the microfabrication process and the design of the mechanical sensing element itself. The systems solution is also beneficial from a commercial point of view, as silicon real-estate has become very cheap and circuit complexity does not contribute significantly to the cost any more.

A typical micromachined capacitive sensing element of an accelerometer is considered in this chapter, as a basis for the discussion. Table 5.1 gives the parameters of the sensing element. The values are typical for a bulk-micromachined accelerometer; if another fabrication technique was

Table 5.1: Parameters of a typical bulk-micromachined sensing element used for simulations and calculations throughout this chapter.

Parameter	Symbol	Numerical value	Comment
Proof mass	m	1.2×10^{-6} [kg]	Bulk-micromachined
Spring constant	k	5 [N/m]	
Damping coefficient	b	6×10^{-3}[N/ms]	Assumed as constant
Nominal electrode gap	d_0	3 [μm]	Distance between the proof mass at centre to either electrode
Nominal capacitance	C_0	16 pF	

used, for example surface-micromachining, leading to parameters orders-of-magnitude different, the discussed concepts still hold true, unless explicitly stated.

5.2. Open Loop Accelerometer

An open loop accelerometer is shown in Figure 5.1. The proof mass is deflected by an external inertial force and this displacement is converted into an electrical signal. For capacitive sensing the proof mass is the centre electrode of an capacitive half bridge. The displacement will cause a differential change in capacitance, which then can be measured by a charge integrator.

The electronic interface block contains an amplifier converting the imbalance in charge into an electrical signal, but may also comprise a range of other signal conditioning circuitry such as amplifiers, filters, non-linearity compensators and analogue-to-digital converters.

The mechanical sensing element of a typical capacitive accelerometer consists of a moveable proof mass that is sandwiched between two electrodes. This is shown conceptually in Figure 5.2, together with the

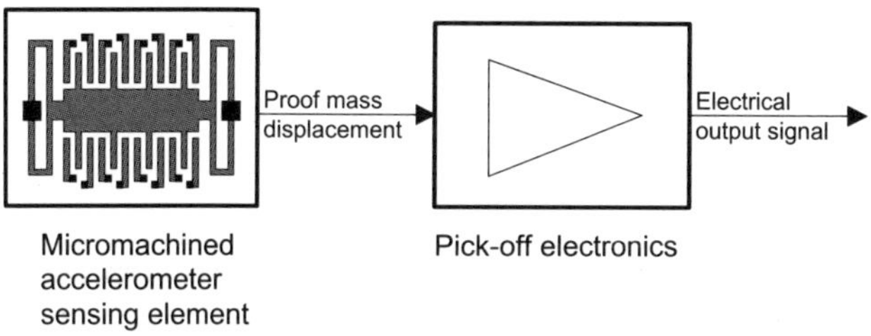

Figure 5.1: Open loop accelerometer.

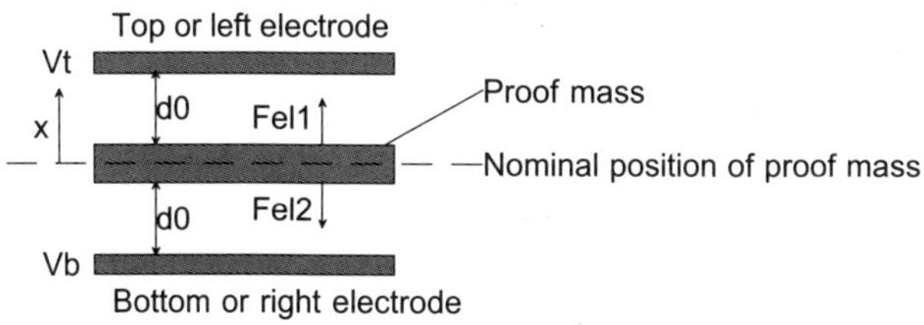

Figure 5.2: Lumped model of a capacitive sensing element of a micromachined accelerometer.

definitions of directions and coordinates used throughout this chapter. Practically, any capacitive micromachined sensing element can be represented in such a way. For out-of-plane (z-axis) accelerometers the fixed electrodes are to the top and bottom of the moveable proof mass, whereas for in-plane devices (x/y-axis) they are to the left and right, respectively. It should also be noted here that a range of other capacitive micromachined transducers such as pressure sensors, microphones and force cells, can be described by the same lumped parameter model, hence the discussions presented in this chapter are also relevant to these sensors.

The nominal gap between the proof mass at mid-position and the electrodes is labelled d_0; F_{el1} and F_{el2} are the net electrostatic force acting on the proof mass in the indicated directions. These forces are generated by voltages V_t and V_b on the electrodes, respectively. The proof mass is assumed to be electrically grounded. If the proof mass is deflected by an inertial force, the capacitances C_1 and C_2 change. Equating the difference between these capacitors yields:

$$\Delta C = C_1 - C_2 = \varepsilon A \left[\frac{1}{d_0 - x} - \frac{1}{d_0 + x} \right] = \varepsilon A \frac{2x}{d_0^2 - x^2}. \tag{5.1}$$

If one assumes that the deflection of the proof mass is small compared to the nominal gap, the differential change in capacitance becomes a linear measure for the displacement of the proof mass:

$$\Delta C = \varepsilon A \frac{2x}{d_0^2}, \quad \text{for } x \ll d_0. \tag{5.2}$$

In general, this assumption is not valid for open loop accelerometers, hence such accelerometers will exhibit non-linear behaviour. Figure 5.3 shows the differential capacitance calculated according to Equation (5.1) (solid line) and the approximation in Equation (5.2) (dashed line) as a function of the proof mass displacement. The parameters in Table 5.1 were used to build the graph.

The linear approximation only holds true for small displacements and how much displacement can exactly be tolerated depends on the linearity requirements of the open loop sensors. For precision accelerometers, this can be a major problem. For example, if linearity better than 1% is required, the maximum allowable deflection of the proof mass is only 10% of the nominal gap. This poses a severe restriction on the dynamic range of the

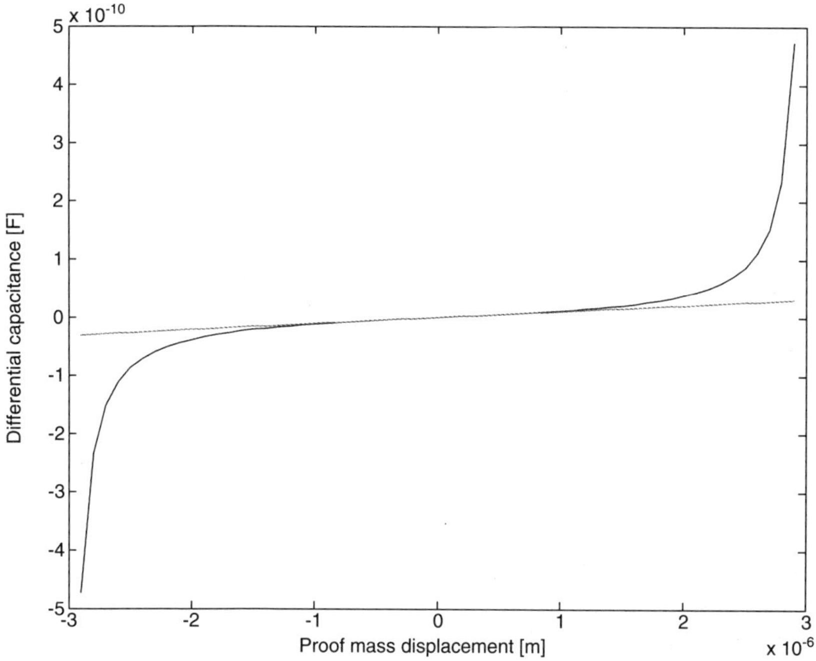

Figure 5.3: Differential capacitance as a function of the proof mass displacement. Equation (5.10) — solid line; Equation (5.2) — dashed line.

sensor. Closed loop accelerometers reduce the deflection of the proof mass, as the deflection is effectively proportional to the error signal of the control loop, hence offer an elegant solution for this problem.

Mechanically, the sensing element can be modelled to first order by a mass-spring damper system. The measurement of acceleration always relies on classical Newton's mechanics. Typically, the acceleration to which a body is subjected to, is to be measured; the accelerometer is being assumed being rigidly attached to this body of interest. This is shown conceptually in Figure 5.4.

Acceleration causes the proof mass to be deflected by a distance x, the motion of the body of interest is denoted by y. To derive the motion equation of the system, D'Alambert's principle is applied, where all real forces acting on the proof mass are equal to the inertia force on the proof mass [3]. From the stationary observer's point of view, the sum of all forces in the

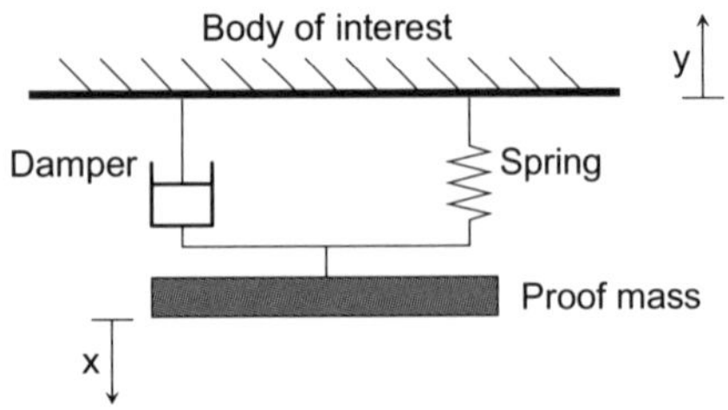

Figure 5.4: Mechanical lumped parameter model of a micromachined sensing element of an accelerometer.

y-direction is:

$$\begin{aligned} 0 &= -m\frac{d^2(y-x)}{dt} + b\frac{dx}{dt} + kx, \\ F_{inertia} &= m\frac{d^2y}{dt^2} = m\frac{d^2x}{dt^2} + b\frac{dx}{dt} + kx, \end{aligned} \tag{5.3}$$

where m is the mass of the proof mass, b the damping factor and k the spring constant. The second derivative of y with respect to time is the acceleration of the body of interest, a.

In the steady state Equation (5.3) simplifies to:

$$\begin{aligned} ma &= kx, \\ a &= \frac{kx}{m}. \end{aligned} \tag{5.4}$$

The sensitivity of an open loop accelerometer is given by:

$$S = \frac{x}{a} = \frac{m}{k}. \tag{5.5}$$

For the dynamic case it is easier to consider Equation (5.3) in the Laplace domain:

$$\frac{x(s)}{a(s)} = \frac{1}{s^2 + \frac{b}{m}s + \frac{k}{m}}, \tag{5.6}$$

which is often rewritten in the following form:

$$\frac{x(s)}{a(s)} = \frac{1}{s^2 + \frac{\omega_n}{Q}s + \omega_n^2}, \tag{5.7}$$

with $Q = \frac{\omega_n m}{b} = \frac{\sqrt{mk}}{b}$ being the quality factor and $\omega_n = \sqrt{\frac{k}{m}}$ the natural resonance frequency.

For an open-loop accelerometer the upper limit for the bandwidth is the resonance frequency. It can be seen by comparing the equations for the sensitivity and the resonance frequency that the bandwidth of the accelerometer has to be traded off with its sensitivity, as $S = 1/\omega_n^2$. This trade-off can be partly overcome by incorporating the sensing element is a force-feedback control loop.

For the dynamic performance of an accelerometer the damping factor, b is crucial. For maximum bandwidth, the sensing element should be critically damped, which is the case for $b = 2m\omega_n$. The damping originates from the movement of the proof mass in a viscous medium (air in most cases). This is a complex, flow-dynamics problem as in most inertial sensors so-called squeeze film damping effects are of importance. A thorough discussion of this phenomenon is beyond the scope of this chapter (the reader is referred to the literature [4–6]). Squeeze film damping comes from the fact that air is trapped between two plates when one moves towards the other. The air molecules cannot escaped fast enough, thus a pressure is built up in the central region. This gives rise not only to a damping coefficient but also to an air-spring coefficient. Under certain simplifying conditions, for square plates with an area A, the small signal damping coefficient can be approximated by [7]:

$$b = \frac{64\sigma P_a A^2}{\pi^6 d_0} \sum_{m,n\ odd} \frac{1}{(mn)^2 \left[(m^2+n^2)^2 + \frac{\sigma^2}{\pi^4}\right]}, \tag{5.8}$$

and the air spring coefficient by:

$$k_a = \frac{64\sigma P_a A}{\pi^6 d_0 \omega} \sum_{m,n\ odd} \frac{m^2+n^2}{(mn)^2 \left[(m^2+n^2)^2 + \frac{\sigma^2}{\pi^4}\right]}, \tag{5.9}$$

where P_a is the ambient pressure, ω the frequency of the plate movement and σ the so-called squeeze number, which is a dimensionless factor providing a measure of the pressure built-up in the central plate area. A low squeeze number means that the air molecules can escape readily, hence no pressure is built up. The squeeze number is given by:

$$\sigma = \frac{12\mu A\omega}{P_a d^2}, \tag{5.10}$$

where μ is the viscosity of air and A the area of the plates.

The above equations assume a small displacement of the proof mass, which may be a valid assumption for closed loop control, however, for open loop inertial sensors, this is not necessary the case. For larger deflections, the damping coefficient cannot be assumed to be constant anymore but increases cubically with the deflection of the proof mass [8, 9]. This introduces another non-linearity associated with larger proof mass deflections; hence it is advantageous to keep this deflection small — another argument for closed loop control. In the examples and calculations in this chapter a constant damping coefficient is assumed.

Yet another non-linear effect is introduced by the position measurement interface (pick-off) circuitry [10, 11]. Typically, for capacitive inertial sensors, an electrical excitation voltage is required, which is applied to the top and bottom electrodes. The resulting signal from the proof mass is amplified by a charge amplifier. Figure 5.5 shows a typical position measurement interface.

The excitation voltage produces an electrostatic force on the proof mass which can be calculated as:

$$F_{el} = \frac{\varepsilon_0 A}{2}\left[\frac{V_{excl}^2}{(d_0 - x)^2} - \frac{V_{exc2}^2}{(d_0 + x)^2}\right], \tag{5.11}$$

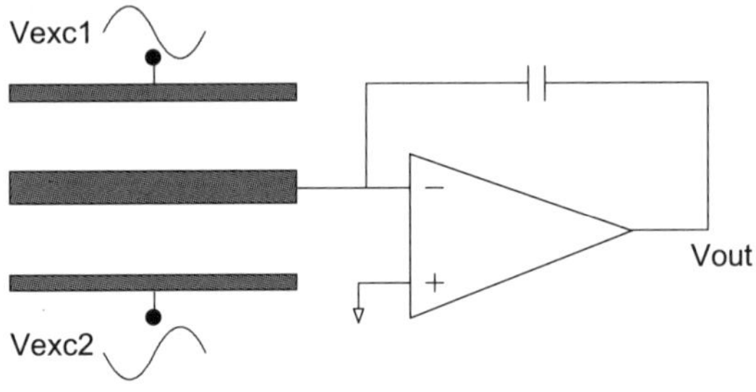

Figure 5.5: Simplified diagram of a typical pick-off circuit for a capacitive sensor. $Vexc1 = -Vexc2$ are high frequency signals to excite the capacitive half-bridge. The amplitude of the output voltage, Vout, is proportional to the imbalance in capacitance, which is a measure of the input acceleration.

where ε_0 is the dielectric constant in vacuum (assumed to be equal to that in air), V_{exc1} and V_{exc2} are the excitation voltages on the top and bottom electrodes, respectively. Usually, the excitation voltages are high-frequency (in the range of hundreds of kHz to a few MHz) signals with an amplitude of a few volts, having sinusoidal, triangular or square waveform. The signal applied to the top electrode is 180° out-of-phase with respect to the signal applied to the bottom electrode (i.e. $V_{exc1} = -V_{exc2}$). Assuming a sinusoidal excitation signal, it can be easily derived from Equation (5.11) that the electrostatic force has a constant component of half the excitation voltage amplitude and a high frequency term at twice the excitation voltage frequency. The latter can safely be neglected as the sensing element attenuates frequencies above its resonance frequency, but the constant term may be problematic. For zero proof mass displacement the net electrostatic force on the proof mass is also zero since the two terms in Equation (5.11) cancel each other. However, if the proof mass is displaced from its nominal position there will be an electrostatic force that pulls it away from its nominal position. This force increases strongly with deflection and effectively acts as positive feedback as it points always in the direction of the proof mass deflection. Eventually, it can lead to an electrostatic pull-in when the electrostatic force becomes larger than the restoring spring force. In such a case the proof mass is attracted to one of the electrodes and 'latches-up', a state that can only be overcome by powering down the system. To solve this problem most open-loop accelerometers have mechanical limit stoppers (or 'bumper') that prevent the proof mass getting too close to either of the stationary electrodes. The disadvantage is that the smallest trench width available by the fabrication process can no longer be the electrode gap (which is desirable to be as small as possible to maximise the sense capacitance) but is now determined by the gap between the proof mass and the mechanical limit stoppers.

Figure 5.6 shows the electrostatic force originating from the excitation voltage for a range of different amplitudes as a function of the proof mass displacement. Obviously, to minimise this positive feedback effect the excitation voltage should be kept as small as possible. This is in contradiction to the requirement for the electronic interface circuit measuring the differential capacitance, for which a higher excitation voltage leads to higher output signals, hence to a better signal to noise ratio. Usually, the electrostatic force originating from the position measurement circuitry is an unwanted effect,

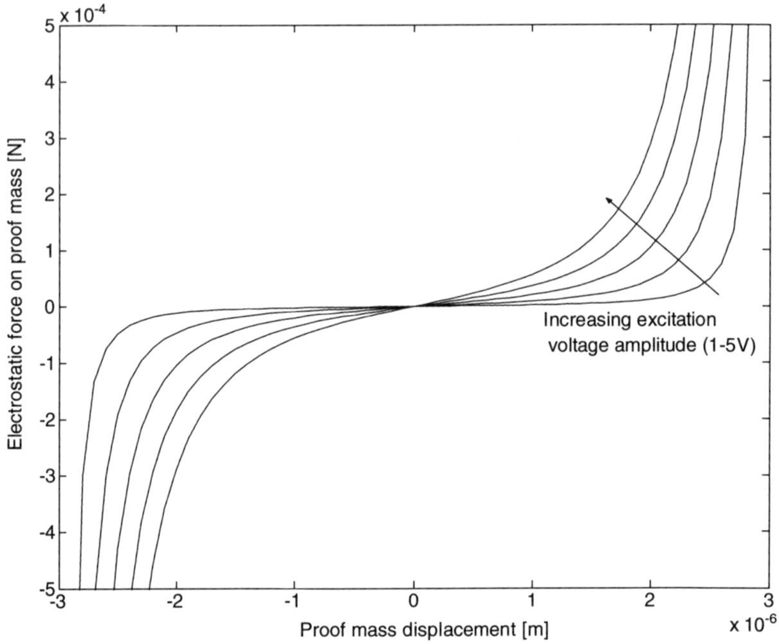

Figure 5.6: Electrostatic force generated by the excitation voltage as a function of the proof mass displacement. As can be seen from the diagram this force always points in the direction of the proof mass displacement hence deflects the proof mass further from its nominal position.

however, it may be used to electronically tune and increase the sensitivity of an accelerometer [12].

Despite the various intrinsic non-linear effects, most current commercial micromachined inertial sensors are open loop. This is due to the fact that these sensors are used in low to medium performance applications and are produced for the mass market, hence low cost is essential. Since many MEMS processes are derived from, or compatible with VLSI processes, one approach is to integrate the pick-off and some control and read-out electronics with the sensing element on the same chip. This results in a lower cost system (since the integrated, on-chip electronics add little to the cost of the sensor chip, and can lower the cost of ancillary circuits). In addition, integrated electronics can lessen induced noise and interfacing problems. A typical example is the ADXL311, which is a dual axis accelerometer. It is one of the latest Analog Devices sensors which is priced at only \$2.50 in quantities greater than 10 000 units [13]. Its noise floor is $300\,\mu\text{gHz}^{-1/2}$ and the sensor can be operated from a single 3 V power supply. Within a 10 Hz bandwidth, below 2 mG can be resolved for a dynamic range of +/−2G. For many current applications in the automotive and consumer electronics sector this performance is adequate, however, for some other applications such as inertial navigation, space micro-gravity measurement or seismometry, higher performance is required. Furthermore, and probably even more important, it is an emerging trend for mass-market applications that sensor performance is becoming more demanding. Modern cars have eight or more airbags; these systems require precise sensing about the impact of a crash to trigger airbags in the right time and sequence. Other examples from the automotive industry are dynamical stability control, active suspension and rollover sensing. Also for GPS back-up systems that take over the navigation system when a satellite view is blocked, higher performance sensors are required.

The next sections will describe different techniques to improve the performance by using a system level approach with additional electronics to boost the performance. The electronics may be implemented on the same chip, but it is more common to develop them as a separate ASIC chip and put the micromachined sensing element and the electronic pick-off chip in the same package. The issue of integration in MEMS is still open for debate, and probably will be for the time being. However, whether the electronics are integrated on the same chip or a hybrid solution is preferred does not affect the principles discussed in the next section.

5.3. Closed Loop Accelerometer

5.3.1. *Analogue Force-Feedback*

One technique to improve the performance of a sensor, which uses neural networks to linearise and otherwise improve the output signal, will be discussed in Chapter 7. Another method is to use the electronics to control the physical behaviour of the sensing element itself, and therefore make it operate close to its optimum position. This technique works by placing the sensing element inside a feedback loop, using an electrostatic force to counteract the inertial force to be measured. The result is termed a closed loop accelerometer.

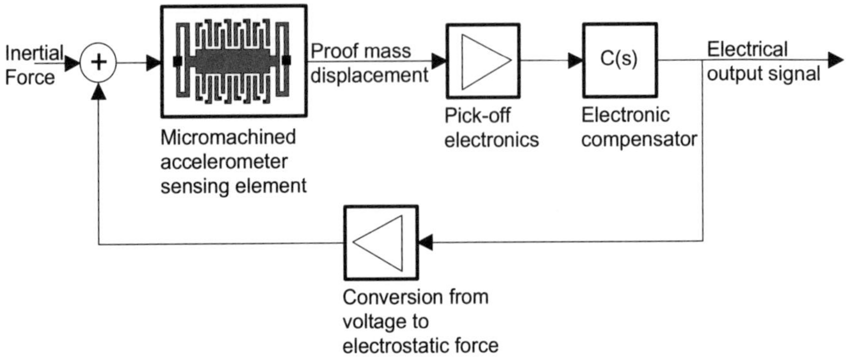

Figure 5.7: Closed loop, force-feedback accelerometer.

The voltage required to generate this feedback force is then a measure of the input inertial force. An electronic compensator may be required to stabilise the loop and improve the dynamic performance. Figure 5.7 shows a general closed loop force feedback control system.

There are several advantages associated with such an approach:

- The proof mass displacement of the sensing element is kept small, hence most inherent non-linear effects described in the previous section are greatly reduced. Consequently, the damping coefficient can be assumed constant, the relationship between proof mass displacement and differential change in capacitance becomes linear and the electrostatic force due to the excitation voltage becomes negligible.
- The sensor characteristics are now mainly determined by the feedback network which is purely electronic. The mechanical sensing element is typically subject to considerable manufacturing tolerances (e.g. the spring constant can vary of up to $+/-10\%$ from its designed value) and closed loop control achieves a large degree of independence of the absolute parameters of the sensing element, hence of these manufacturing tolerances.
- Closed loop force-feedback can increase the dynamic range and bandwidth of a sensor compared with open-loop operating mode. The dynamic range is now mainly determined by the magnitude of the feedback signal, and the bandwidth by the control method.

Obviously, there are also some drawbacks associated with closed loop control. First, the added complexity for the required interface and control electronics potentially increases the cost of the commercial devices. Secondly, from a technical point of view, it is challenging to achieve linear, negative feedback using electrostatic forces since they are always attractive and there is a non-linear relationship between voltage and the resulting force. Furthermore, they are inversely quadratically dependent on the deflection of the proof mass.

In the following, analogue force-feedback schemes are discussed, whereas the next section introduces digital force feedback based on sigma-delta modulation control systems.

The common approach to linearise the feedback relationship is to apply a fixed bias voltage, additional to the feedback voltage, to the top electrode and of equal magnitude but opposite polarity to the bottom electrode [14, 15]. The net electrostatic force on the proof mass then becomes:

$$F_{el} = \frac{\varepsilon_0 A}{2}\left[\frac{(v_F - V_B)^2}{(d_0 - x)^2} - \frac{(v_F + V_B)^2}{(d_0 + x)^2}\right], \tag{5.12}$$

where v_f is the feedback voltage and V_B the bias voltage. For small deflections, $x \to 0$ the electrostatic force simplifies to:

$$F_{el} = -2\varepsilon_0 A \frac{V_B}{d_0^2} v_F, \tag{5.13}$$

which is a linear, negative feedback relationship. The proportionality constant $k_F = -2\varepsilon_0 A \frac{V_B}{d_0^2}$ is introduced here, for later use.

If the proof mass deviates from its nominal position the feedback force becomes non-linear. In Figure 5.8 the feedback force is plotted as a function of the feedback voltage for various deflections of the mass. The bias voltage was assumed to be 10 V and the feedback voltage has a range from −5 to 5 V. It can be further shown that for a given, non-zero, deflection the feedback force has the form $F_{el} = a_1 v_f^2 - a_2 V_B v_f + a_3 V_B^2$ where a_1, a_2, a_3 are positive coefficients depending on the deflection and the geometry. The middle term is the desired linear feedback term which should be maximised, hence V_B should be made as large as possible. A too large bias voltage, on the other hand, will reduce the sensitivity, therefore a design trade off needs to be made. As a general rule the magnitude of the feedback force should be

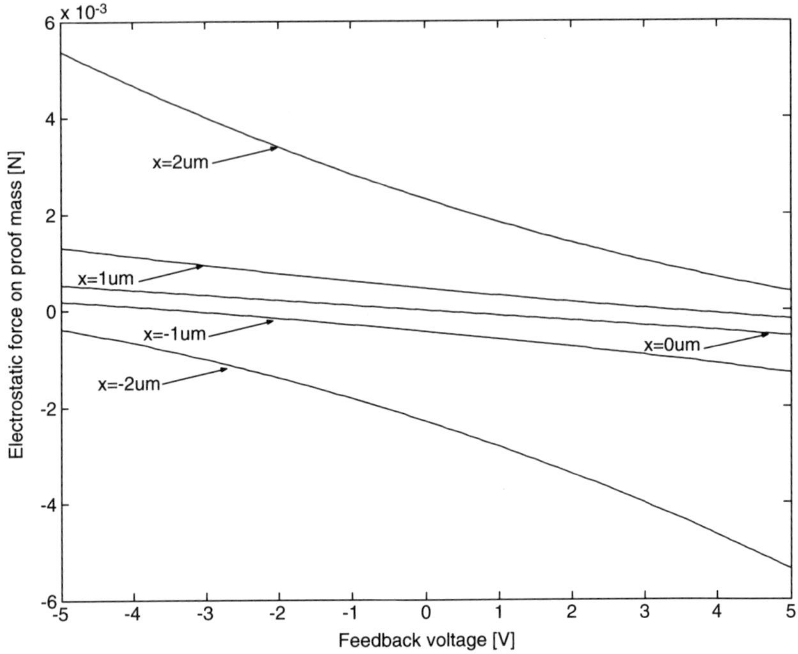

Figure 5.8: Electrostatic feedback force as a function of the feedback voltage for various proof mass deflections. Only for $x = 0$ the feedback force is negative and linear.

made larger or equal to the maximum inertial acceleration force the sensor is designed to measure.

The feedback voltage is derived from the output voltage of the position measurement interface. Assuming the simplest form of controller, a pure proportional gain, the feedback voltage can be expressed as: $v_f = k_{po}k_px$ where k_p is the proportional gain constant. k_{po} is the gain constant relating the proof mass deflection to the output voltage of the position measurement interface. This assumes the ideal case when the pick-off circuit output voltage is a linear function of the proof mass deflection. However, as it was seen from Equations (5.1) and (5.2) this is only the case for small deflections, due to the inherent non-linearity introduced by the differential capacitance to proof mass displacement relationship. For larger deflections, k_{po} becomes: $k_{po} = \varepsilon A\frac{2x}{d_0^2 - x^2}k_c$, where k_c is the pick-off circuit gain relating the differential change in capacitance to the output voltage of the position measurement interface. A typical value for k_c is approximately 10^{12} V/F.

Using the deductions above, the electrostatic feedback force can now be calculated as a function of the proof mass deflection:

$$F_{el} = \frac{\varepsilon_0 A}{2}\left[\frac{(k_{po}k_p - V_B)^2}{(d_0 - x)^2} - \frac{(k_{po}k_p + V_B)^2}{(d_0 + x)^2}\right]. \tag{5.14}$$

Figure 5.9 represents Equation (5.14) graphically for different bias voltages, including a non-linear k_{po}, hence including the non-linearity introduced by the position measurement based on the differential capacitance. Only for small deflections a linear, negative feedback relationship is maintained. For normal operating conditions under closed loop control the proof mass deflections are indeed small. However, if the sensor is subjected to a shock in acceleration (e.g. a car driving through a pothole) the proof mass may be

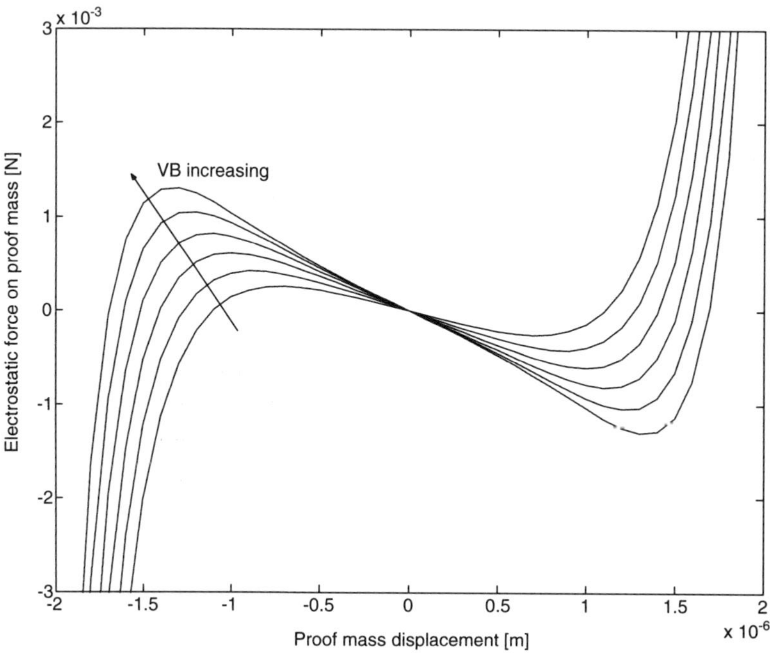

Figure 5.9: Electrostatic feedback force as a function of the proof mass deflection for various bias voltages including the non-linearity introduced from the differential capacitance measurement interface.

deflected outside the linear region. In such a case, the feedback relationship becomes non-linear in the first instance (effectively the feedback gain is reduced), for even larger deflections, the feedback force changes its polarity and drives the proof mass towards the fixed electrodes, where it will latch up. This is a non-recoverable situation requiring a sensor power shut down, which is not acceptable for most applications. This potential instability is the main disadvantage of analogue, force-feedback control. A mechanical overange limiter, as discussed previously, can again be used to prevent this situation.

For small proof mass deflections, analogue force feedback can nevertheless improve the performance of an inertial sensor considerably. In the following, a linear system transfer function will be derived, assuming small proof mass deflections, which can be regarded as a small signal model of the sensor. Furthermore, a Simulink model will be described suitable for simulating both small and large proof mass deflections.

For small proof mass deflections, the sensor system can be represented by the block diagram shown in Figure 5.10.

The forward path consists of the dynamics of the sensing element, relating any input force acting on the proof mass to its deflection, x. The gain constants k_x and k_c represent the conversion from deflection to voltage (i.e. the gain of the position measurement interface). Furthermore, there is a PID controller with gain constants k_d, k_i, and k_p for differential, integral and proportional gain. In the feedback path the gain constant k_F relates

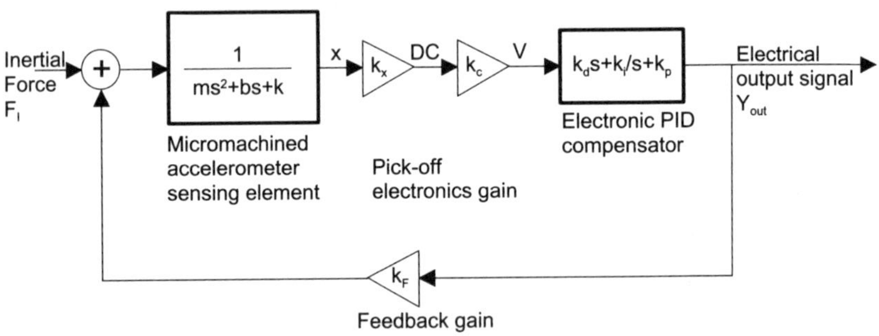

Figure 5.10: Small signal block diagram of a closed loop, analogue force-feedback accelerometer.

the feedback voltage to the electrostatic feedback force which is summed by the sensing element with the inertial input force due to acceleration.

Using standard linear control theory, the closed loop, small signal transfer function can now be calculated as:

$$\frac{Y_{out}}{F_I} = \frac{k_x k_C k_d s^2 + k_x k_C k_p s + k_x k_C k_i}{ms^3 + (b - k_x k_C k_d k_F)s^2 + (k - k_x k_C k_p k_F)s - k_x k_C k_i k_F}. \quad (5.15)$$

The PID controller constants can be used to tune the accelerometer performance to the characteristics required by the sensor application. Therefore, it is possible to use a particular micromachined sensing element for a variety of different applications. Other types of controllers are possible, for example van Kampen [15] suggests a low-pass filter as controller in the form of $K/(st+1)$ which may result in a faster response without a pronounced resonance peak. Determining the optimal controller and its coefficients depends on the specific sensing element and application in which it will be used, hence would lead away from generality as intended in this chapter. Thus, in the following, only the simplest controller, i.e. a proportional gain controller will be briefly discussed. Then, the small signal transfer function will be compared with a large-signal Simulink model that includes most non-linear effects discussed above. For pure proportional gain, Equation (5.15) simplifies to:

$$\frac{Y_{out}}{F_I} = \frac{k_x k_C k_p}{ms^2 + bs + (k - k_x k_C k_p k_F)}. \quad (5.16)$$

In Figure 5.11 the open and closed loop transfer functions are shown for $k_p = \{0.01, 0.1, 1\}$. The bandwidth is considerably increased and a progressively more pronounced resonance peak can be observed as k_p increases. The static gain of the accelerometer has decreased as well, as here, for comparison, the same values for k_{po} for the open and closed loop transfer function was assumed. In practice, k_{po} would be adjusted by an appropriate electronic gain stage.

Simulink Model

When a Simulink model is developed it should include all major non-linear effects so that the large signal behaviour (i.e. larger proof mass deflections) can be studied and the analytical, linearised model can be verified [16]. The

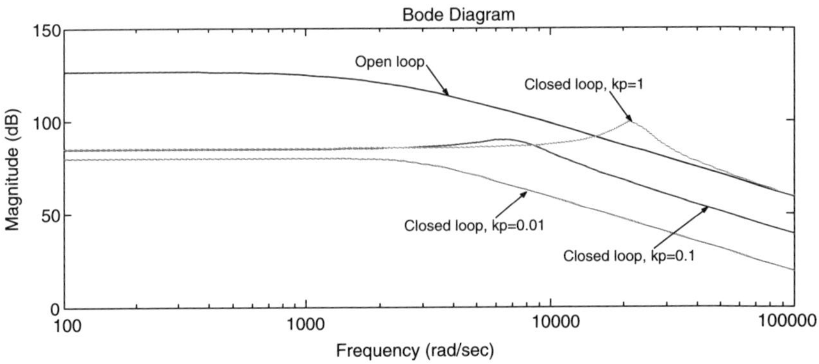

Figure 5.11: Open and closed loop responses for different proportional gain constants.

Simulink model should contain the following parts:

- Sensing element model including effects of the proof mass hitting the electrodes or overrange limiter;
- DC component of the electrostatic forces generated by the excitation voltages required for the position measurement circuit;
- Conversion from the proof mass displacement to the differential capacitance.
- Electronic compensator;
- Conversion from the feedback voltage to the electrostatic forces on the proof mass;
- Voltage limiters that model the dynamic range of the amplifiers generating the feedback voltages.

The model for the sensing element is a second order mass-spring-damper transfer function but has to include an overrange displacement limiter [17]. This could be simply the electrodes or some separate mechanical stoppers preventing the proof mass touching the electrodes. Once the proof mass hits the limiter, the integrator calculating the velocity of the proof mass is reset to zero and held there. Only when the proof mass acceleration changes polarity the integrator resumes normal operating mode. Figure 5.12 shows the model of the sensing element with its different hierarchical blocks. The other building blocks are more self-explanatory and they can be directly implemented from the Simulink library using functions, limiters and gain blocks. The full model is shown in Figure 5.13.

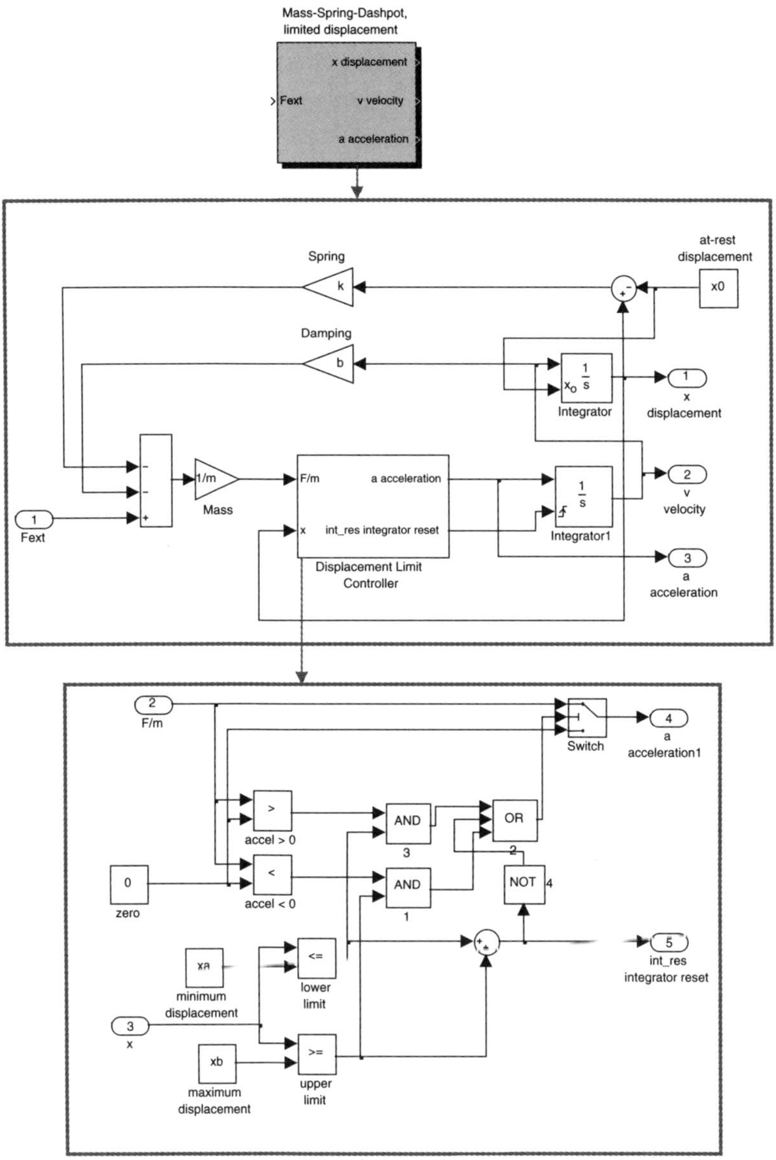

Figure 5.12: Simulink model of the mechanical sensing element including over-range displacement limiters. xa and xb are the maximum negative and positive displacement limits, respectively. $x0$ allows to set an initial displacement at the start of the simulation.

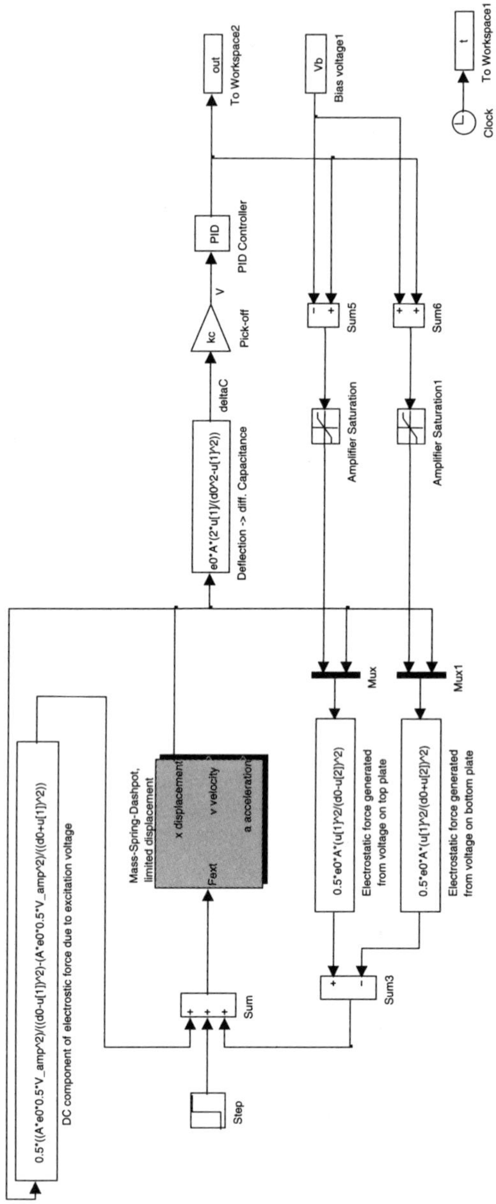

Figure 5.13: Complete Simulink model of the closed loop analogue accelerometer.

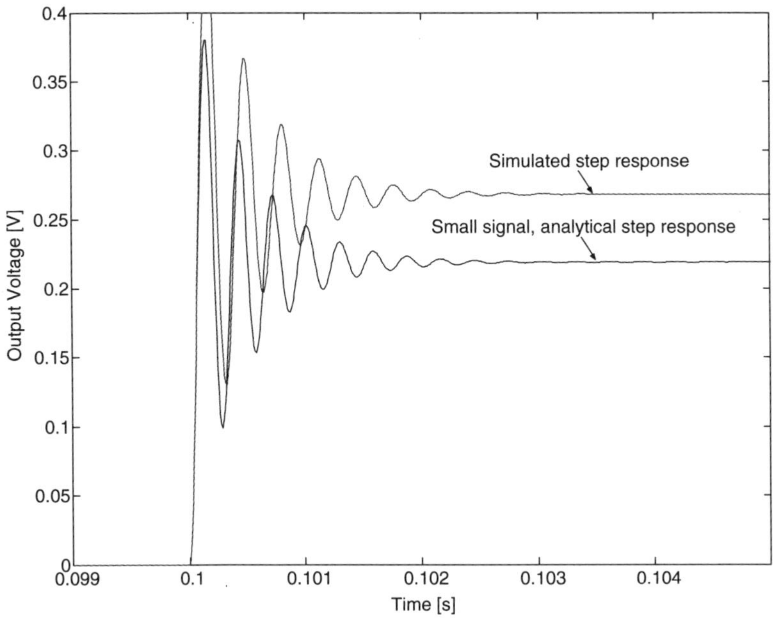

Figure 5.14: Comparison of the step response simulated with the Simulink model and the linearised, small signal model.

In the model, any compensator can be implemented, and their performance evaluated. For example, a PID controller is shown in Figure 5.13. As this model includes many non-linear effects a frequency response cannot be easily simulated. In order to compare this model with the analytical, linerarised model derived above, two step responses are presented in Figure 5.14 (for $k_p = 1$). From the graphs, it is obvious that the steady state value of the simulated response is higher compared to the analytical one, due to the electrostatic force originating from the excitation voltage required for the position measurement interface and the influence of the proof mass displacement on the electrostatic feedback force, which is neglected in the small signal model.

The Simulink model provides a useful tool for the designer to allow optimisation of the various design parameters. It also allows the investigation of the stability margins.

As a further example, an acceleration shock was assumed as the input to the model. The shock experimented with, takes the form of a sinusoidal

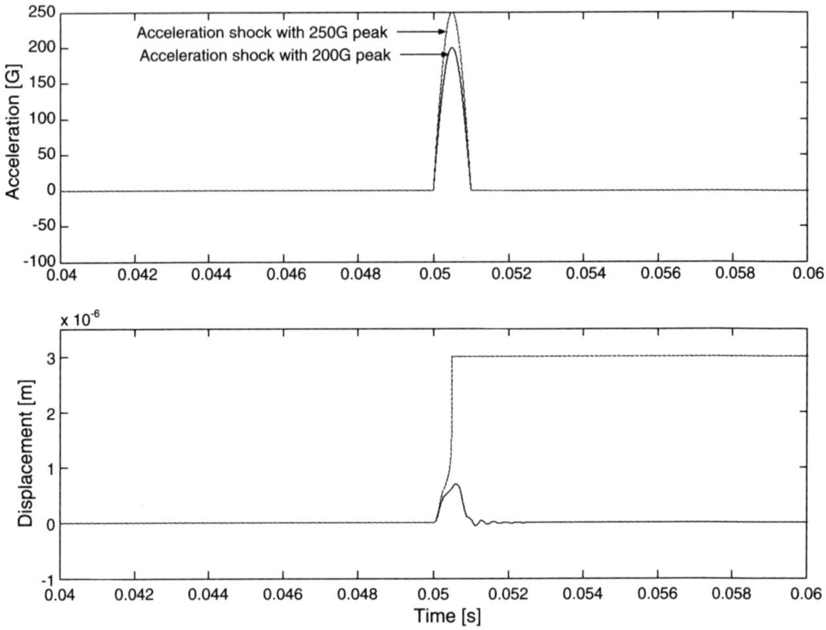

Figure 5.15: Response of the analogue, closed loop accelerometer to an acceleration shock. For a 200 G peak shock in acceleration the system recovers, but for a 250 G peak shock the proof mass latches up.

halfwave, with acceleration amplitudes of 200 G and 250 G, respectively, and a duration of 1ms. Figure 5.15 shows that the proof mass recovers for the shock of 200 G amplitude but, for a shock of 250 G, it is deflected towards the electrodes and latches-up. This is due to the feedback force changing polarity for larger deflections, as previously described.

5.3.2. *Digital Force-Feedback*

Over the last years, digital feedback systems based on Sigma-Delta Modulator (SDM) architectures for inertial sensors have gained popularity as they have a number of distinct advantages over analogue, force-feedback control:

- Electrostatic instability is impossible;
- Direct digital output signal in form of a pulse density modulated bitstream;

- Limit cycle behaviour that allows a basic self-test of the sensor.

A wide range of publications describe micromachined accelerometers designed around such a control system. In particular, researchers at the University of California at Berkeley have reported many prototypes based on surface micromachined sensing elements [18–21]. Bulk-micromachined accelerometers with SDM force-feedback were also reported, for example, in [22, 23].

The basic principle is derived from purely electronic SDM analogue to digital converters [24]. The electronic integrators are replaced by the micromachined sensing element, which acts as a double integrator, at least for frequencies above its resonance frequency. Thus, the sensing element together with the electronic building blocks form an electro-mechanical SDM.

This control approach will be explored in this section and again a Simulink model will be derived.

A basic block diagram of a closed loop accelerometer based on SDM is shown in Figure 5.16. The basic principle of operation is explained as follows: the pick-off circuit senses the displacement of the proof mass based on the differential change in capacitance, exactly in the same way as for an open loop or analogue, closed loop accelerometer. This signal is passed on to an electronic compensator that stabilises the control loop by adding some phase lead, and then to a comparator. The comparator is clocked at a frequency considerably higher than the bandwidth of the sensor. This is an intrinsic property of a SDM, as they are oversampling converters. Typical

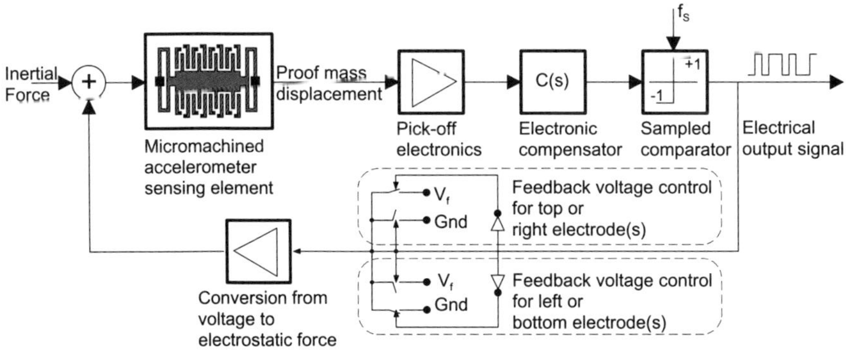

Figure 5.16: Closed loop, force feedback accelerometer based on sigma-delta modulation.

oversampling ratios are between 32 and 256. The comparator provides the output signal, which is in a digital format. This signal is a pulse density modulated serial bitstream that can be directly processed by a digital signal processor (DSP). The comparator also controls a range of switches in the feedback path, two each for the top (or left) and bottom (or right) set of electrodes of the sensing element. The comparator effectively outputs +1 or −1 depending on the sign of the differential imbalance in capacitance, i.e. whether the proof mass has moved away from the top (or left) or bottom (or right) electrodes. The electrode the proof mass is further away from, is then energised with a fixed feedback voltage, V_f, for a fixed time interval $T_s = 1/f_s$. This generates an electrostatic feedback force pulling the proof mass back towards the mid position between the electrodes. Since the electrode closer to the proof mass is always grounded, there is no possibility of an electrostatic latch-up, even for shocks in acceleration. The net electrostatic feedback force is given by:

$$F_{el} = \text{sgn}(D)\frac{1}{2}\frac{\varepsilon_0 A V_f^2}{(d_0 + \text{sgn}(D)\,x)^2}, \tag{5.17}$$

where $D = \pm 1$ is the output of the comparator.

The electrostatic force is hence always approximately constant during one sampling period, if one assumes negligible movement of the proof mass during one clock cycle. This assumption is justified by the short duration of a clock cycle compared to the dynamics of the sensing element. The feedback signal can be regarded as quantised force packets which are applied for integer multiples of the sampling time. The magnitude of the feedback force should be chosen about 10% higher than the maximum inertial force the accelerometer is specified to measure (i.e. its full scale acceleration or dynamic range). This is due to the fact that SDM control systems typically overload if the input is more than approximately 90% of the feedback signal and thus significantly degrading their performance.

Control systems based on SDM are notoriously difficult to analyse as they contain a non-linear element in form of a comparator that cannot be linearised easily. One common approach is to regard the quantiser as an arbitrary gain element plus added quantisation noise, which is white. Following a similar procedure as for the analogue, closed loop accelerometer, one can derive a linear block diagram model as shown in Figure 5.17.

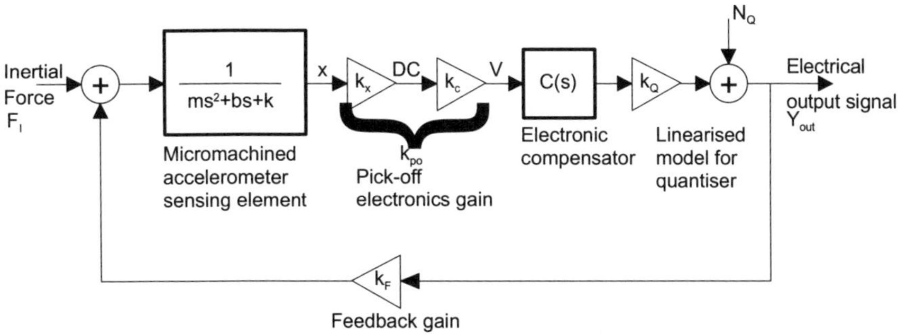

Figure 5.17: Small signal block diagram of a closed loop, digital force-feedback accelerometer.

The pick-off gain constants are defined as for the analogue, closed loop accelerometer. The feedback gain k_F is given by:

$$k_F = -\frac{1}{2}\frac{\varepsilon_0 A V_f^2}{d_0^2}. \tag{5.18}$$

The electronic compensator is usually only providing some phase lead to compensate for the phase lag introduced by the sensing element. If the sensing element is overdamped, no compensator is required at all. For the sensing element assumed for the examples in this chapter, a compensator is advantageous, having the following form in the Laplace domain:

$$C(s) = \frac{s + z_1}{s + p_1}, \tag{5.19}$$

where z_1 and p_1 are the zero and pole frequencies in radians per second, respectively. To provide phase lead in the correct frequency range, between the resonant peak of the sensing element and the sampling frequency, a pole frequency is chosen as $p_1 > 2\pi f_s > z_1$. A heuristic value for the zero frequency is to choose $z_1 = 2\pi f_s/5$ and $p_1 = 2\pi 5 f_s$. However the system performance is relatively insensitive to the absolute values of z_1 and p_1.

Using the linearised block diagram (Figure 5.17), it is now possible to derive a signal transfer function (STF), relating the output signal to the input inertial force, and a noise transfer function (NTF), relating the quantisation noise to the input inertial force:

$$\frac{Y_{out}}{F_I} = STF = \frac{k_x k_C k_Q C(s)}{ms^2 + bs + (k - k_x k_C k_Q k_F C(s))}, \tag{5.20}$$

$$\frac{N_Q}{F_I} = NTF = \frac{ms^2 + bs + k}{ms^2 + bs + (k - k_x k_C k_Q k_F C(s))}. \tag{5.21}$$

A numerical evaluation of the above transfer functions is somewhat problematic given that finding an estimation for the quantiser gain, k_Q is difficult. The easiest option is to simulate the system with the Simulink model described later in this section and calculate the root-mean-square (RMS) value of the input signal to the comparator. The quantiser gain, k_Q is then given by the output level spacing of the comparator divided by this RMS value.

Figure 5.18 shows the magnitude plots of the two transfer functions of Equations (5.20) and (5.21). The STF is flat in the signal band of interest, therefore the signal is allowed to pass through unchanged. The NTF has low gain in the signal band and higher gain for higher frequencies above the signal band. This is the typical noise-shaping characteristic of a SDM: the quantisation noise is moved towards higher frequencies, out of the signal band. However, if one compares both magnitude transfer functions with the NTF and STF of purely electronic SDM analogue to digital converters, one limitation becomes apparent. The low-frequency gain of the sensing element, equal to the inverse of the spring constant, determines the noise shaping characteristics. High low-frequency gain means high quantisation noise suppression in the signal band. For a purely electronic SDM modulator the low-frequency gain is very high, as the loop filter consists of (near-) ideal integrators resulting in much better noise shaping characteristics. This means that a SDM with a micromachined sensing element can never reach the noise shaping characteristics of a second order, purely electronic SDM.

In fact, it is possible to calculate the signal-to-quantisation-noise ratio (SQNR) of the system depicted in Figure 5.17, as shown below.

The power spectral density (PSD) of white sampled noise (i.e. the introduced quantisation noise by the quantiser) is given by:

$$E^2(f) = e_{RMS}^2 \frac{2}{f_s}, \tag{5.22}$$

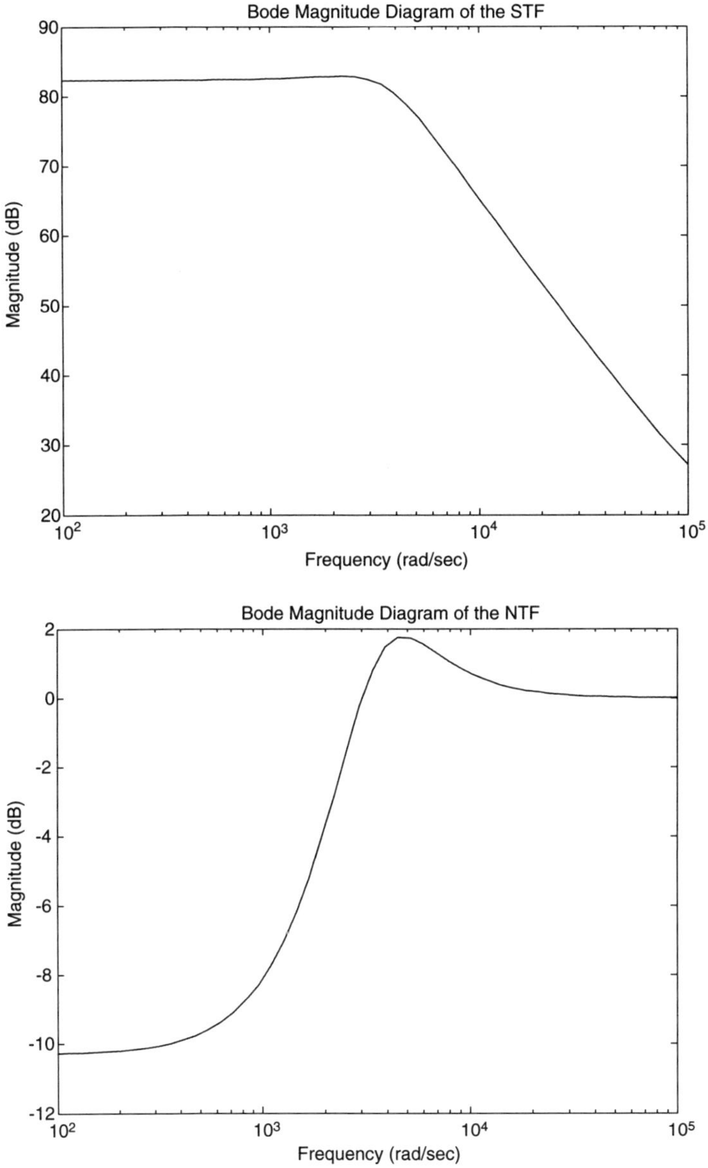

Figure 5.18: Magnitude plots of the two transfer functions of Equations (5.20) and (5.21).

with

$$e_{RMS} = \frac{q}{\sqrt{12}}, \tag{5.23}$$

where q is the quantisation level spacing of the comparator. The PSD at the output of the digital accelerometer can then be calculated by:

$$N^2_{Q,out} = E^2(f)|\mathrm{NTF}|^2, \tag{5.24}$$

and the total in-band noise is given by:

$$n_0^2 = \int_0^B |N^2_{Q,out}|df, \tag{5.25}$$

where B denotes the signal band of the accelerometer. This integral is usually difficult to solve analytically, but can be evaluated easily with a numerical integration software package.

Finally, the SQNR is then given by:

$$\mathrm{SQNR} = 20\log\left(\frac{\text{RMS of input signal}}{n_0^2}\right). \tag{5.26}$$

For a sinusoidal input acceleration signal, the RMS of the input signal is obviously $A_a/\sqrt{2}$ where A_a is the amplitude of the input acceleration.

A numerical example yields 50 dB with the parameters given in Table 5.1. The following additional parameters were used: linearised quantiser gain $k_Q = 4260$ (determined through simulation), accelerometer bandwidth $B = 1000\,\text{Hz}$, input acceleration amplitude $A_a = 1\,\text{G}$, sampling frequency $fs = 100\,\text{kHz}$ (thus an oversampling ratio of 50).

Simulink Model

The linear analysis above is only valid to a certain point, as it contains many simplifying assumptions. For a more rigorous system analysis, the designer has to rely on system level simulation. For example, the linear analysis is not suitable for predicting the stability of the SDM system. Stability, for SDM control systems, is not defined in the same way as for linear control systems, i.e. bounded input — bounded output stability. As the comparator always ensures a bounded output, stability is considered in terms of high or low time intervals. In other words, the output interval for which the comparator is in a low (or high) state has to be finite for all conditions.

In the following a Simulink model will be presented that captures the dominant effects of the sensor system. Many building blocks of the analogue, closed loop accelerometer can be re-used; in fact only the SDM building blocks comprising the sampled comparator have to be added. The full Simulink model is shown in Figure 5.19.

The sampled comparator was modelled as an ordinary comparator followed by a sample and hold controlled by a pulse generator clocked at the sampling frequency. The sampled comparator controls a switch that decides the sign of the feedback force; in other words, whether the proof mass is pulled up (or left) or down (or right). Further building blocks are required to filter the pulse-density modulated output signal. The output of a SDM is usually down-sampled and filtered; typically this is achieved by a comb-filter or moving average filter [24]. Such a filter can be simply regarded as a digital low-pass filter, which additionally, down-samples the bitstream from the sampling frequency f_s to a frequency range determined by the bandwidth of the accelerometer. The lowest sampling frequency is determined by the Nyquist frequency, $f_{NQ} = 2B$, however, in a practical implementation a five to ten times higher value would be used. In a hardware implementation, a DSP would be used which can directly interface with the output digital bitstream coming from the comparator.

Figure 5.20 shows the output bitstream and the filtered output signal for a 1G input sinusoidal acceleration of 100 Hz. The bitstream shows the typical pulse-density modulation; for the maxima and minima of the input sinewave, longer periods of high and low phases, respectively, can be observed. It should be noted that the high or low phases are always integer multiples of the sampling period. Obviously, the filtered output signal is a representation of the sinusoidal input acceleration. However, it is difficult to quantify in the time domain how well the output signal codes the input acceleration. Therefore the output bitstream is usually investigated in the frequency domain. Figure 5.21 shows the power spectral density (PSD) of the output bitstream and a magnification around the input signal frequency. The input signal peak can be clearly identified at 100 Hz and the typical noiseshaping characteristic is obvious where the quantisation noise is low in the signal band and increases at higher frequencies. The frequency spectrum can now be analysed using Matlab to calculate the SQNR. All spectral components up to the bandwidth of the sensor are quantisation noise and are summed up except for the component lying at the input

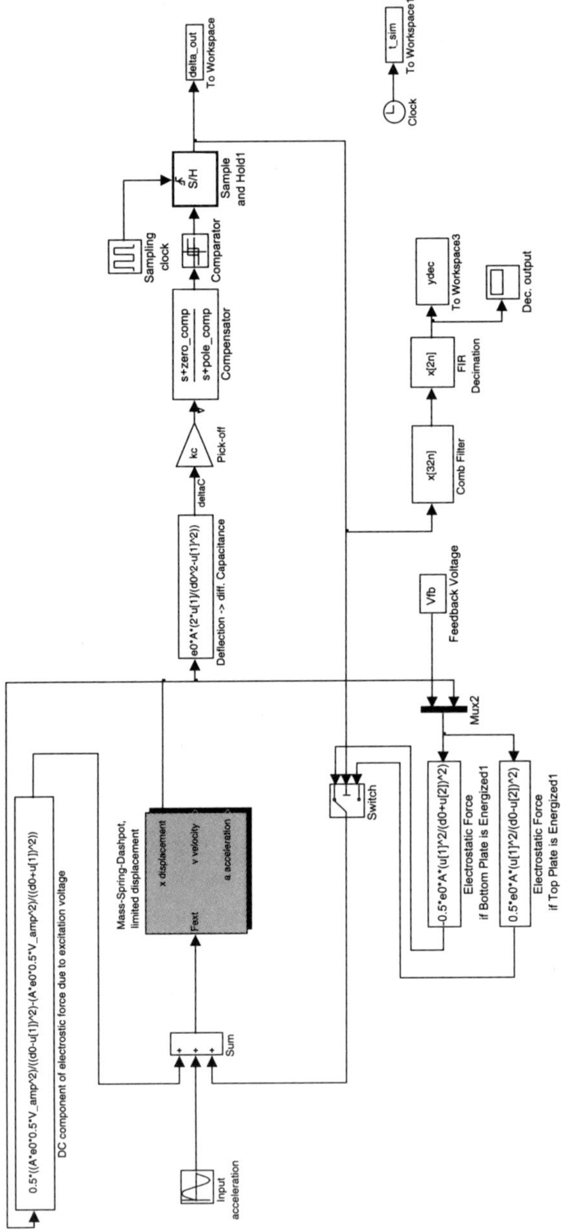

Figure 5.19: Simulink model of the digital accelerometer.

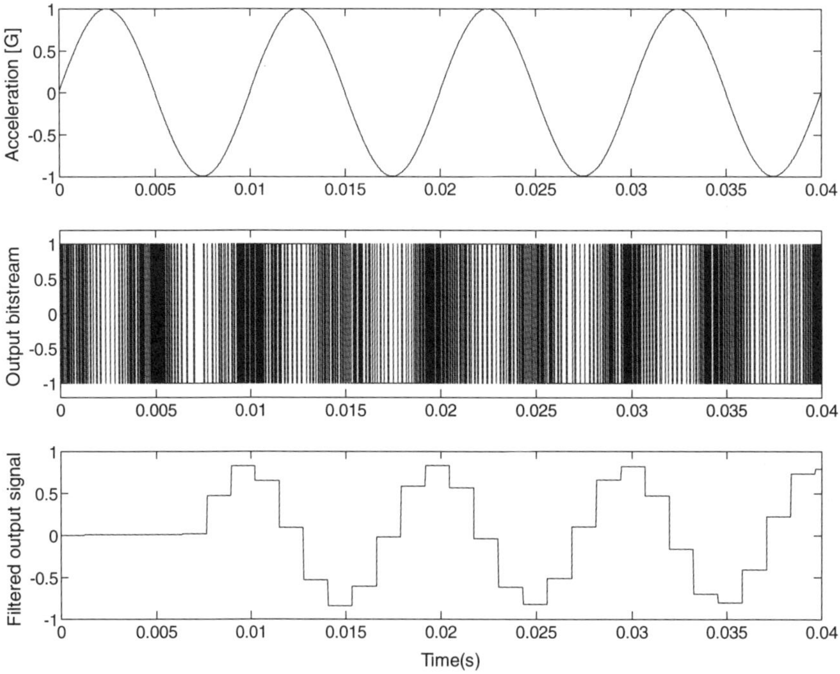

Figure 5.20: Top graph: sinusoidal input acceleration signal with 1 G amplitude and frequency 100 Hz; middle: graph — pulse density modulated output signal from the comparator; bottom graph: low-pass filtered output signal.

signal frequency. The ratio between the power spectral component at the input frequency and the sum of all noise spectral components is equal to the SQNR. For the same parameters, as for the linearised analyses presented previously, the simulation results in a SQNR of 54 dB which agrees well with the analytical result.

Another useful means of assessment of the performance of the digital accelerometer is to vary the input signal power and to calculate the SQNR accordingly. Figure 5.22 shows the results of such an assesment. Obviously the SQNR rises for increasing signal amplitude. Below an input amplitude power of about -50 dB the SQNR becomes dominated by the quantisation noise and hence approximately levels out. It should be noted here that the parameters were somewhat changed here as the Matlab script files

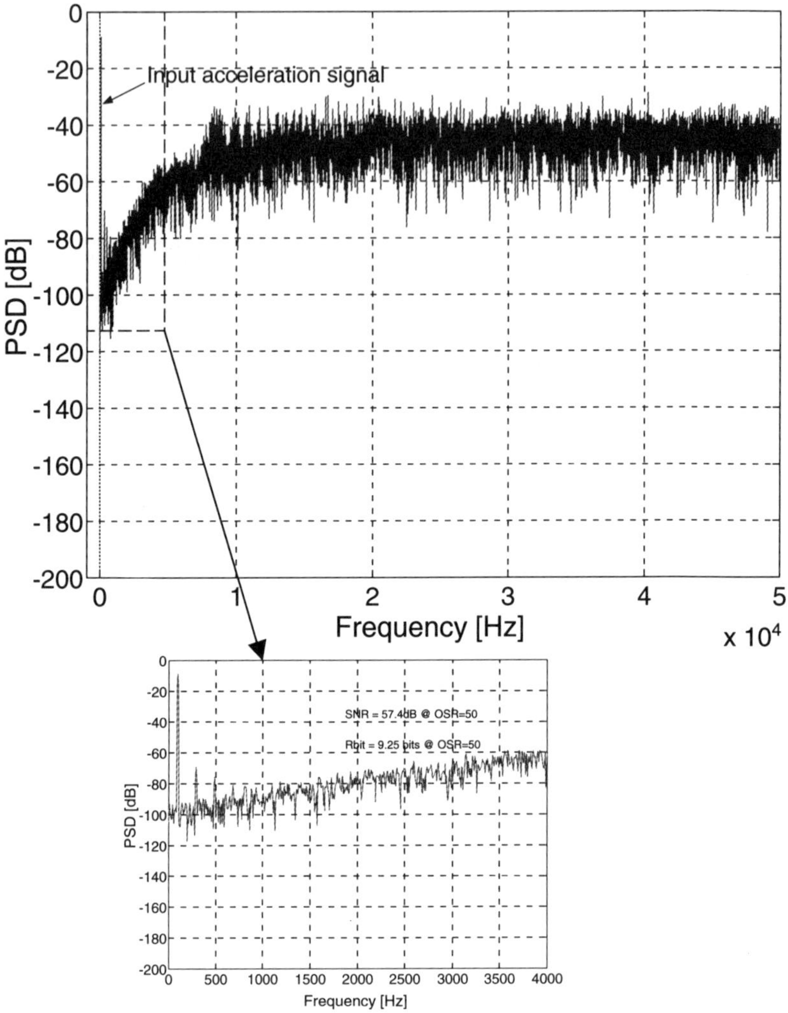

Figure 5.21: Power spectral density of the output bitstream for an input acceleration signal of 1 G and a frequency of 100 Hz.

calculating the SQNR require the sampling frequency to be a power of two since fast Fourier transformation is used in Matlab. Therefore, a sampling frequency of $2^{17} = 131\,\text{kHz}$ was used which explains the higher SQNR (around 65 dB) at the full scale input amplitude.

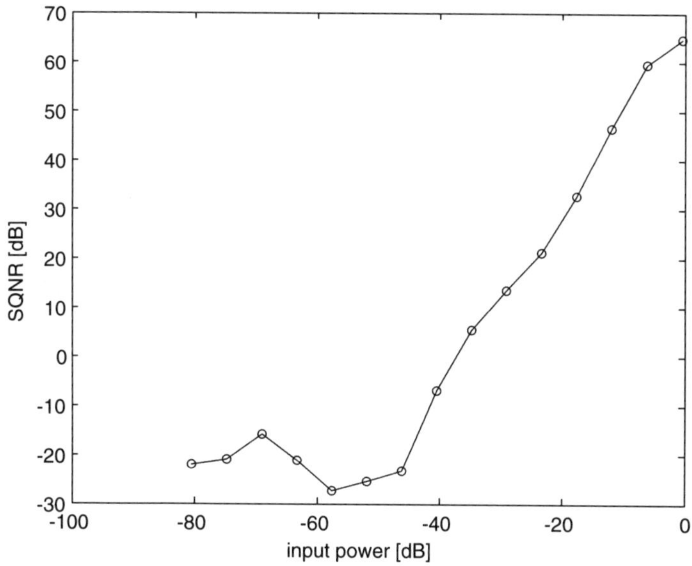

Figure 5.22: SQNR for various input signal amplitudes.

Another important issue when designing micromachined accelerometers is considering other noise sources than quantisation noise. Usually, they are Brownian noise from the micromachined sensing element and electronic Johnson and flicker noise from the electronic position measurement (pick-off) circuit. Surface micromachined sensing elements have a much higher Brownian noise floor than bulk-micromachined ones, as they are at least an order of magnitude smaller and hence the surface to volume ratio is higher, which is the determining factor. Electronic interface noise, on the other hand, can be made smaller for surface micromachined accelerometers if the electronic circuitry is integrated on the same chip. Consequently, the performance limiting factors for surface-micromachined accelerometers is usually their Brownian noise floor, whereas for bulk-micromachined accelerometers with the electronics on a separate chip, it is the thermal noise. If using a SDM force-feedback control scheme, it is important to design the system in such a way that the quantisation noise is well below the dominant noise source. This is typically achieved by choosing a high enough sampling frequency. These issues can be investigated in Simulink by adding appropriate random number generators to simulate noise sources, but this is beyond the scope of this chapter. A further interesting aspect of SDM force-feedback schemes is that they also shape the electronic noise, not only the quantisation noise. This is an area of ongoing research which may lead to further improvement in the resolution of inertial sensors [25, 26].

Another currently ongoing research topic is to incorporate the sensing element in higher order SDM structures to achieve better quantisation noise shaping at lower sampling frequencies. The architectures are borrowed from electronic analogue-to-digital SDM, where loops with an order up to seven are routinely used. Stability of higher order control loops is an important issue; pure cascading of integrators leads to unstable systems. Feed-forward and minor feedback loops topologies are used to stabilise the system. One major problem with a micromachined sensing element as part of the control loop is that the internal velocity node is not accessible, hence the standard stabilisation schemes from electronic SDM are not directly applicable.

Two approaches show promise: MASH architectures [27] and higher order, single loop architectures. The former adds an additional, purely electronic SDM loop to the first one containing the micromachined sensing element. The second loop processes the quantisation noise of the first loop, then the output of the two loops, after some appropriate digital filtering, are combined and overall noise-shaping of order three or four can be achieved. Figure 5.23 shows a block diagram of such a system. The advantage of this

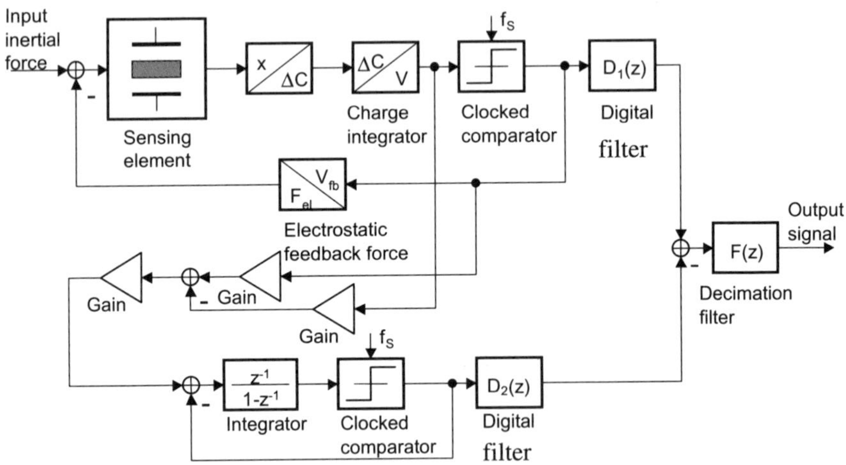

Figure 5.23: Block diagram of a micromachined sensing element incorporated in a MASH SDM force feedback architecture.

approach is that the system is always stable, the disadvantage is that the coefficients of the digital filter required to process the output of the second loop, are dependent on the parameters of the sensing element. These are often not known precisely due to considerable fabrication tolerances which are quite typical for micromachined processes. A system topology has yet to be found that is sufficiently tolerant to variations of the sensing element parameters.

The second promising approach is based on a single loop architecture, where additional electronics integrators are added to provide higher order noise shaping [28, 29]. Figure 5.24 shows a fifth order SDM control system with three electronic integrators [30]. The choice of the various gain constants is of crucial importance since they determine the system stability and performance. Finding a stable and fabrication tolerant optimum is a difficult task using analytical means therefore the usual approach is to use system level simulations. Figure 5.25 compares the noiseshaping characteristic of a fifth order loop with a second order loop with no additional integrators. It is obvious that the noise floor has been considerably reduced in the signal band for the fifth order loop. It remains to be seen whether these approaches will gain any significance in the near future.

Finally, it should be mentioned that force feedback architectures based on SDM can also be applied to the sense mode of micromachined gyroscopes. Little research has been done in this area, with the exception of a fully integrated micromachined gyroscope developed by researchers at the university of California at Berkeley [31, 32].

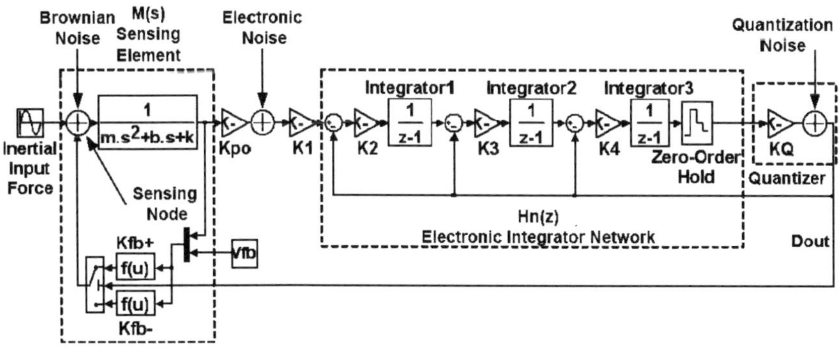

Figure 5.24: Block diagram of a micromachined sensing element incorporated in a fifth order SDM modulator loop.

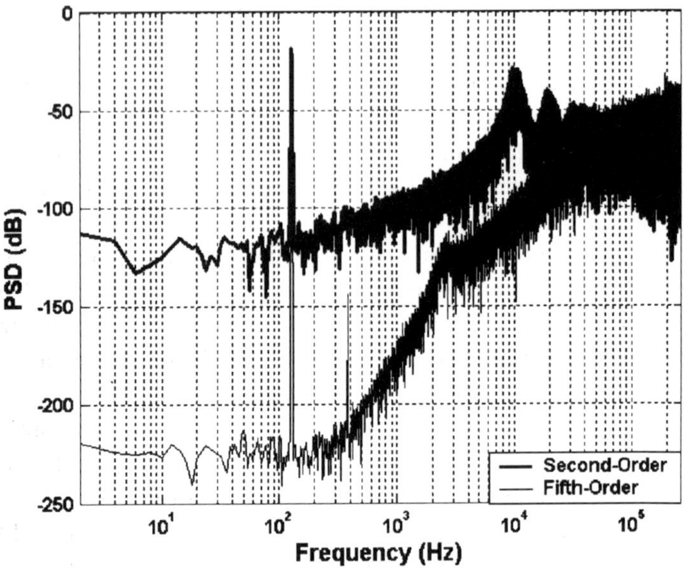

Figure 5.25: Noise shaping characteristics for a second order and fifth SDM loop.

5.4. Conclusions

This chapter introduced the most important concepts used in the design of closed loop control of micromachined, capacitive sensors. Although it mainly focussed on micromachined accelerometers, they were only used as a vehicle to demonstrate the concepts. The control systems discussed can be built around any kind of micromachined sensor with a capacitive sensing element, for example pressure sensors, gyroscopes and force sensors.

Two control system approaches were discussed: analogue force feedback and digital control relying on SDM architectures. The former suffers from a potential instability due to the pull-in phenomenon where the proof mass can be attracted to one electrode and latch up. Although this can be solved by using mechanical stoppers limiting the travel range of the proof mass, this is not an optimum solution. Digital control based on SDM system can overcome this shortcoming and additionally can result in a sensor with direct digital output. Both analogue and digital force feedback control have in common the fact that they can increase the performance of a sensor considerably compared to an open loop system, especially regarding parameters such as linearity, dynamic range and bandwidth. It is difficult to put absolute values on the achievable improvement, as this depends on the application in mind, the sensing element, available voltages to produce the feedback force and many other parameters. As a rule of thumb, it can be said that improvements of the above mentioned parameters in the order of a factor 10 should be possible. Additionally, the performance characteristics of closed loop sensors are less dependent on absolute values of the mechanical parameters of the sensing element since they are mainly determined by the feedback system which consists of electronic components. This can help relax fabrication constraints (by relaxing the requirement on small tolerance margins) and therefore increase the yield of the micromachined sensing elements. The yield increase, in turn, helps to reduce cost despite added complexity of the circuitry required for closed loop feedback control.

References

1. Yazdi, N., Ayazi, F. and Najafi, K. (1998) Micromachined inertial sensors, *Proc. IEEE* **86**(8), 1640–1659.
2. Kraft, M. (2000) Micromachined inertial sensors: the state of the art and a look into the future, *IMC Measurement and Control* **33**(6), 164–168.
3. Timoshenko, S. and Young, D. H. *Engineering Mechanics*, McGraw-Hill, London, pp. 303–304.
4. Veijola, T., Kuisma, H., Lahdenpera, J. and Ryhanen, T. (1995) Equivalent-circuit model of the squeezed gas film in a silicon accelerometer, *Sensors and Actuators* **A48**, 239–248.
5. Starr, J. B. (1990) Squeeze-film damping in solid state accelerometers, *IEEE Solid-State Sensor and Actuator Workshop*, Hilton Head Island, pp. 44–47.
6. Zhang, L., Cho, D., Shiraishi, H. and Trimmer, W. (1992) Squeeze-film damping in micromechanical systems, *ASME, Micromechanical Systems* **40**, 149–160.
7. Andrews, M., Harris, I. and Turner, G. (1993) A comparison of squeeze-film theory with measurements on a microstructure, *Sensors and Actuators* **A36**, 79–87.
8. Marco, S., Samitier, J., Ruiz, O., Herms, A. and Morante, J. (1993) Analysis of electrostatic damped piezoresistive silicon accelerometer, *Sensors and Actuators* **A37–38**, 317–322.
9. Peroulis, D., Pacheco, S. P., Sarabandi, K. and Katchi, L. P. B. (2003) Electromechanical considerations in developing low-voltage RF MEMS switches, *IEEE Trans. Microwave Theory and Techniques* **51**(1), 259–270.

10. Bao, M., Yang, H., Yin, H. and Shen, S. (2000) Effects of electrostatic forces generated by the driving signal on capacitive sensing devices, *Sensors and Actuators* **A84**, 213–219.
11. Li, B., Lu, D. and Wang, W. (2000) Open-loop operating mode of micromachined capacitive accelerometer, *Sensors and Actuators* **A79**, 219–223.
12. Cretu, E., Bartek, M. and Wolffenbuttel, R. F. (1999) Analytical modelling for accelerometers with electrically tunable sensitivity, *Proceedings of the International Conference on Modeling and Simulation of Microsystems*, USA, pp. 601–604.
13. Analog Devices, ADXL311 Datasheet, Low-cost, ultra compact +/−2g dual-axis accelerometer.
14. van Paemel, M. (1989) Interface circuit for capacitive accelerometer, *Sensors and Actuators* **17**, 629–637.
15. van Kampen, R. P., Vellekoop, M., Sarro, P. and Wolffenbuttel, R. F. (1994) Application of electrostatic feedback to critical damping of an integrated silicon accelerometer, *Sensors and Actuators* **A43**, 100–106.
16. Ping, W., Xinmin, Z., Shangshu, D. and Zhenqin, G. (1994) On modelling the dynamic non-linearity of force-balance accelerometers (FBAs), *Sensors and Actuators* **A45**, 29–33.
17. Senturia S. (2001) *Microsystem Design*, Academic Publishers.
18. Boser, B. E. and Howe, R. T. (1996) Surface micromachined accelerometers, *IEEE J. Solid-State Circuits* **31**(3), 336–375.
19. Henrion, W., Disanza, L., Ip, M., Terry, S. and Jerman, H. (1990) Wide dynamic range direct digital accelerometer, *IEEE Solid State Sensor and Actuator Workshop*, Hilton Head Island, pp. 153–157.
20. Lu, C., Lemkin, M. and Boser, B. (1995) A monolithic surface micromachined accelerometer with digital output, *IEEE J. Solid-State Circuits* **30**(12), 1367–1373.
21. Lemkin, M. A. and Boser, B. (1999) A three-axis micromachined accelerometer with a CMOS position-sense interface and digital offset-trim electronics, *IEEE J. of Solid-State Circuits* **34**(4), 456–468.
22. De Coulon, Y., Smith, T., Hermann, J., Chevroulet, M. and Rudolf, F. (1993) Design and test of a precision servoaccelerometer with digital output, *7th Int. Conf. Solid-State Sensors and Actuators (Transducer '93)*, Yokohama, pp. 832–835.
23. Smith, T., Nys, O., Chevroulet, M., de Coulon, Y. and Degrauwe, M. (1994) Electro-mechanical sigma-delta converter for acceleration measurements, *IEEE International Solid-State Circuits Conference*, San Francisco, pp. 160–161.
24. Norsworthy, S. R., Schreier, R. and Temes, G. C. (eds.) (1997) Delta-sigma data converters, *Theory, Design and Simulation*, IEEE Press, ISBN 0-7803-1045-4.

25. Gaura, E. and Kraft, M. (2002) Noise considerations for closed loop digital accelerometers, *Proc. 5th Conf. Modeling and Simulation of Microsystems*, Puerto Rico, pp. 154–157.
26. Yazdi, N. and Najafi, K. (2000) Performance limits of a closed-loop micro-G solicon accelerometer with deposited rigid electrodes, *Proc. 12th Int. Conf. Microelectronics*, Teheran, pp. 313–316.
27. Kraft, M., Redman-White, W. and Mokhtari, M. E. (2001) Closed loop micromachined sensors with higher order SD-Modulators, *Proc. 4th Conf. Modeling and Simulation of Microsystems*, Hilton Head Island, pp. 104–107.
28. Kajita, T., Moon, U. K. and Temes, G. C. (2002) A two-chip interface for a MEMS accelerometer, *IEEE Transactions on Instrumentation and Measurement* **51**(4), 853–858.
29. Petkov, V. P. and Boser, B. E. (2004) A fourth-order $\Sigma\Delta$ interface for micromachined inertial sensors, *IEEE International Solid-State Circuits Conference (ISSCC)*, pp. 320–321.
30. Dong, Y. and Kraft, M. (2004) Simulation of micromachined inertial sensors with higher-order single loop sigma-delta modulators, *Proc. 6th Conf. on Modeling and Simulation of Microsystems*, Boston **1**, 414–417.
31. Jiang, X., Wang, F., Kraft, M. and Boser, B. E. (2002) An integrated surface micromachined capacitive lateral accelerometer with 2 uG/rt-Hz resolution, *Tech. Digest of Solid State Sensor and Actuator Workshop*, Hilton Head Island, USA, pp. 202–205.
32. Xuesong, J., Seeger, J. I., Kraft, M. and Boser, B. E. (2000) A Monolithic surface micromachined Z-axis gyroscope with digital output, *Proc. Symp. VLSI Circuits*, Hawaii, USA, pp. 16–19.

CHAPTER 6

CASE STUDY: ADAPTIVE OPTICS AND SMART VLSI/MEMS SYSTEMS

by *Davies William de Lima Monteiro*

This chapter presents the desired characteristics, the operational principles, the capabilities and the limitations of a number of adaptive optical components and systems implemented in the framework of standard silicon technologies. It will be shown how to build reliable smart silicon-based systems with devices based on simple concepts and coupled by means of straightforward control algorithms.

6.1. Introduction

Applications such as turbulence-free astronomical imaging, enhanced human vision, sharp *in vivo* retinal imaging, quality inspection of automotive parts and reliable free-space optical communications can benefit from the capabilities of Adaptive Optics (AO), which is a relatively novel technology aimed at dynamically sensing and correcting the phase profile of light beams. In order to suit AO systems to the low-cost and high-speed demands common in the industrial and medical scenarios, MEMS devices and VLSI sensor systems are inescapable.

The first AO systems were conceived in order to counteract aberrations introduced by the turbulent atmosphere in astronomical observatories and military systems. The purpose was to compensate dynamic image distortions caused by fast moving atmospheric layers. Over the years, several techniques have been devised for the detection of distortions of phase profiles* [1, 2], and devices have been implemented to correct aberrations in real time [3].

AO systems are always used as a complementary part of a more general optical system and a reasonable rule of thumb states that they should at least double the system efficiency, without consuming more than one fourth of the overall system costs. The total system investments involved in astronomical and military applications are usually huge, and million-dollar figures are not uncommon. In that realm, expensive AO systems with components fabricated by means of specialised technologies are still consonant. In the past few decades, with the disclosure of classified AO achievements and results, researchers started to envision novel applications, especially in the industrial and medical fields, where budgets are orders of magnitude lower. Since then, there has been an increasing potential for inexpensive and yet high-performance AO components. An ongoing trend to meet the demands of newly devised applications is the fabrication of AO components and systems based on standard silicon microtechnologies, i.e. micromachining and VLSI processes, which are mature, widely available and constantly evolving.

6.2. Adaptive Optics and MEMS Systems

Most people do not realise how ubiquitous Adaptive Optics systems already are and how their simple concepts yield enormous benefits. Classical examples of AO systems are the human eye, capable of extraordinary tilt and defocus correction, and the lens system in a CD/DVD pick-up unit, which dynamically adjusts its focus to the data layer on the disk. The basic AO concepts are described below.

A component or medium, through which a light beam propagates, imprints changes to the phase profile of the beam (Figure 6.1). Two illustrative examples are a lens with an imperfect shape and an air path subjected to thermal variations, which mould an initially plane wavefront. The first example can be the natural lens of the human eye covered with an imperfect

*Phase profiles are more often referred to as wavefronts. In wave optics a wavefront is a hypothetical surface defined as the loci of all points of a given wave featuring the same phase. In geometrical optics, a wavefront is defined as a continuous surface having normals parallel to the light rays, where the rays represent the direction of maximum energy propagation. A mathematical representation can be found in [1, 2].

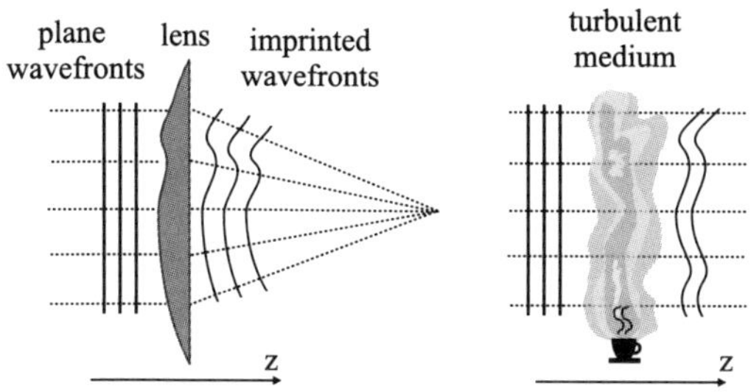

Figure 6.1: An optical element or a turbulent medium imprint a change in the phase profile of a light beam.

cornea and the latter can be an air layer close to the hood of a car on a hot day. An image observed through any of these paths will look deformed, as a consequence of wavefront distortions.

By measuring the imprinted distortions on the wavefront, one can evaluate the shape and quality of the optical component/medium and, in principle one can even use this information to restore the original wavefront/image. The most used approach to accomplish dynamic wavefront manipulation employs three key components: a wavefront sensor, a data processor and a wavefront corrector.

The successful deployment of AO in industrial and medical applications demands that the fabrication process of high-performance components meets several desirable characteristics:

Standard processes:
The sensor and corrector should be fabricated in the framework of established and widely available technologies to take advantage of well documented and tested procedures, reliable modelling, high yield, tight process tolerances, ample accessibility, affordable costs and, last but not the least, to profit from the critical mass of professionals with qualified know-how;

Portability:
A reasonably compact system concept is more likely to attend various applications with slight modifications than a room-size custom built system;

Scalability:
The chosen technology should allow for some flexibility in modifying the number and size of sensing elements or control channels without modifying process steps;

Reproducibility:
Components fabricated in subsequent process runs should match within a sensible margin;

Design flexibility:
The fabrication process should accommodate architectural changes to attend upon particular needs (for example some applications require one-dimensional wavefront correction, whereas others demand a two-dimensional scheme).

In the view of the requirements above, silicon microtechnologies qualify as an attractive option. These technologies are largely accessible and encompass various mature microelectronic and micromachining processes that guarantee reasonable manufacturing costs, design flexibility and good yield. The range of implementable features extends from digital and analogue circuitry, to photodetection and to opto-electro-mechanical structures, among others. There are, however, limitations of using silicon processes, which will be discussed when evaluating individual devices. Nevertheless, the myriad of possibilities in the mechanical and electronic domain can be used in the implementation of high-quality and yet inexpensive optical devices for adaptive optical systems.

6.3. Operational Principles

Conceptual simplicity facilitates not only the understanding of devices but also their fabrication. The physical fundamentals and architectures of both the wavefront sensor and the wavefront corrector are discussed in this section. Their concepts will prove to be intuitive and straightforward at first sight.

6.3.1. *Wavefront Sensors*

In a spectrum of techniques based on interferometry, geometrical optics and irradiance variations, the Hartmann method stands out as one of the most used wavefront sensing techniques because of its intrinsic simplicity and direct data interpretation. In this method, a light beam is sampled into several sub-beams by an opaque mask with a grid of openings. When the wavefront (beam phase profile) is flat, parallel to the mask, the resulting light spots are located perpendicularly under their respective mask openings.

This pattern of light spots can be recorded as a reference grid. However, if a distorted wavefront is sampled, each individual light spot is displaced from its reference grid point by an amount proportional to the local wavefront tilt (Figure 6.2). Therefore, the measurement of the displacements of the spots enables the evaluation of the wavefront profile.

The Hartmann mask can, in principle, be a simple randomly perforated surface made of plastic or paperboard, but more accurate masks are desirable in order to ensure more precise information about the grid, together with higher spot uniformity and symmetry. An even more attractive alternative is to substitute the mask with a microlens array, which favours better light collection efficiency and reduced overall sampling errors (aliasing).

Photographic film can be used to record the spot pattern, although an image sensor (for example based on Charge-Coupled-Device or CMOS techonology) and image-processing software evidently yield faster spot-displacement analysis. True real-time operation, however, requires customised camera architectures that circumvent both data-stream bottlenecks and lengthy image processing. Some of these latter solutions will be presented in following sections.

In general, the Hartmann method is a synonym of simplicity. It only requires a sampling element and a detector; it does not require monochromatic light and it is insensitive to vibrations.

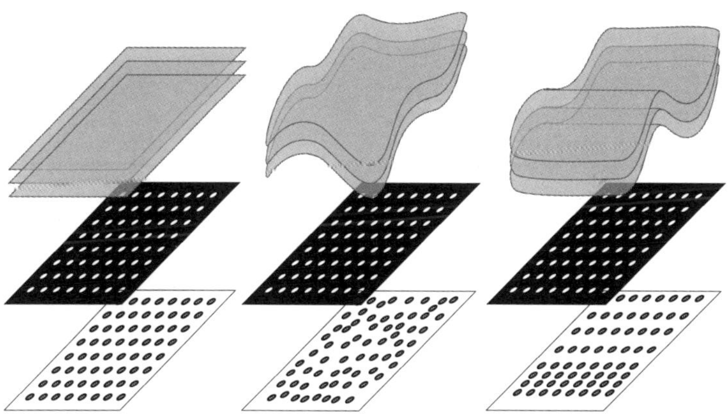

Figure 6.2: The wavefront is sampled by a perforated mask. The resulting spot pattern on a screen corresponds to local slopes of the wavefront profile.

6.3.2. *Wavefront Correctors*

Devices that manipulate wavefronts can be either static or dynamic, and either refractive or reflective. In adaptive optics, dynamic devices play a major role. In the refractive case, the optical component distorts the transmitted wavefront according to its profile and refractive index. A sheet of glass featuring a uniform refraction index n and with one non-flat surface, described by $S(x, y)$, will imprint its profile multiplied by $(n - 1)$ on a flat wavefront (Figure 6.3). In the reflective case, the profile imprinted will be $S(x, y)$ multiplied by a factor two. The same reasoning applies when a distorted wavefront is to be corrected into a plane wavefront, which actually represents the most common situation in adaptive optics.

Several pixelated and seamless correctors have been implemented over the years with different levels of performance and cost. They range from liquid-crystal and liquid correctors to stiff and flexible membranes [4–6]. Electro-optical modulation, mechanical, electromagnetic, thermal, piezoelectric and electrostatic actuation are examples of mechanisms used to put these devices into action.

A ubiquitous wavefront-corrector concept, which has been used from ophthalmology to laser-pulse compression, is the membrane mirror [7]. It consists of a thin circular membrane clamped at the edges and coated with metal. This membrane is mounted above an array of metal electrodes, with an air gap in between. The actuation principle is based on electrostatic forces between the electrodes and the membrane, and by applying DC voltages to the electrodes the membrane is pulled down. Consequently,

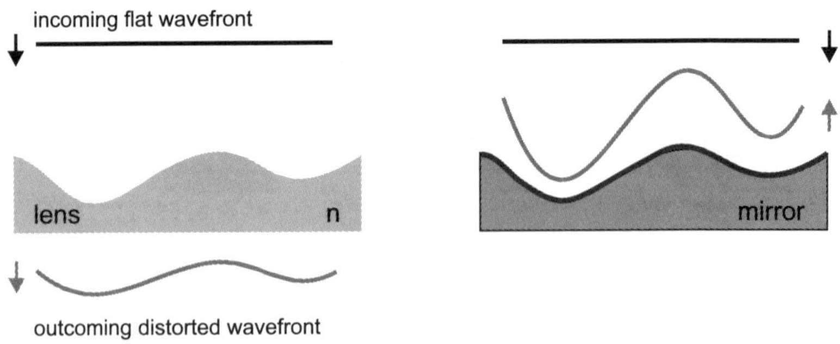

Figure 6.3: Plane wavefront imprinted by the profile of either a lens or a mirror.

the membrane can adopt different shapes depending on the set of voltages applied to the electrodes.

6.3.3. *Systems*

Adaptive optical components can be used stand-alone or coupled to each other, depending on their function in a more general optical system. As stand-alone components, they have been used either for diagnostics or for pre-programmmed feedforward control of specific parameters. When coupled, they usually operate in closed-loop feedback mode.

Wavefront sensors as a diagnostic tool

Wavefront sensors can be used to quantitatively evaluate phase profiles of laser-beams and optical components. Linear wavefront sensors, for instance, have been reported to monitor the surface quality of automotive gear-shift parts. For this application, a laser beam with a known phase profile illuminates a spinning part under test. The reflected beam is sensed by the sensor, which instantly detects deviations from the intended part shape. Two-dimensional sensors have been used for automatic characterisation of the higher-order aberrations in the human eye, which are due not only to the cornea but also to the natural lens of the eye [8–11]. A beam is directed to the retina and the reflected beam is measured by the sensor. The double pass through the aberrated path is then taken into account in the final analysis.

There are basically two ways to represent the wavefront profile based on the sensor data: zonal and modal [2, 12, 13]. In both approaches, one first calculates the x and y tilt components associated with each spot displacement. In the zonal reconstruction, the wavefront shape is approximated as if one was connecting adjoining tilted tiles, pretty much like one would sew a patchwork quilt. In principle, if one has an infinite number of sampling points, wavefronts with any spatial frequency can be reconstructed, even those with abrupt and sharp variations. The modal approach relies on the assumption that one can use a linear combination of known smooth profiles (basis functions) to reconstruct the wavefront. The tilts calculated from the measured spot displacements are decomposed as a sum of tilt components previously calculated and associated with each basis function. The advantage of this approach is that it delivers a mathematical function for the wavefront as opposed to just the geometrical representation permitted by the zonal reconstruction.

Wavefront correctors in feedforward systems

An interesting example of deformable mirrors in stand-alone operation is that of electronic spectacles, in which each lens is replaced by a pair of mirrors, one ordinary and another deformable [14]. The person wearing these spectacles sees the images reflected by the mirrors. Therefore, by actuating the mirror membrane to a pre-determined shape, one is able to compensate aberrations present in the eye. The corrected aberrations can be even more complex than simple defocus and astigmatism (these two are usually corrected for by regular spectacles). This device could still provide the person with an exceptionally accurate vision capable of distinguishing feature patterns unnoticed by a perfect eye [15]. An extremely simple demonstrator of these spectacles uses deformable mirrors with single electrodes, which can change the membrane curvature. By varying the electrode voltage, the membrane deflection allows dynamic correction of focus to read either a newspaper on a table top or a billboard at a distance (Figure 6.4). Equivalent spectacles can be produced using adaptive lenses, in which a liquid-crystal layer is confined between two transparant electrodes [16]. The voltage applied to the electrodes modulates electro-optically the orientation of the liquid-crystal molecules, therefore modulating the transmitted phase. Since the replacement of optical lens of the human eye is no longer surgically obnoxious, researchers have recently proposed the implantation of liquid-crystal lenses in the human eye and they are now evaluating the technical solutions to control the lenses [17].

Other applications involve the control of multi-actuator deformable-mirror coupled to an optmisation algorithm in order to improve a given quality parameter. This parameter can be sharpness in imaging systems, brightness in laser systems or pulse duration in ultrafast-laser tissue diagnostics. In such applications, knowledge of the wavefront can be left to oblivion, since the maximisation of the quality parameter itself is used to configure the mirror. To calibrate the system, one starts by applying random voltages to the mirror electrodes and measuring the quality parameter. The randomisation is then narrowed to those sets of voltages that guarantee the

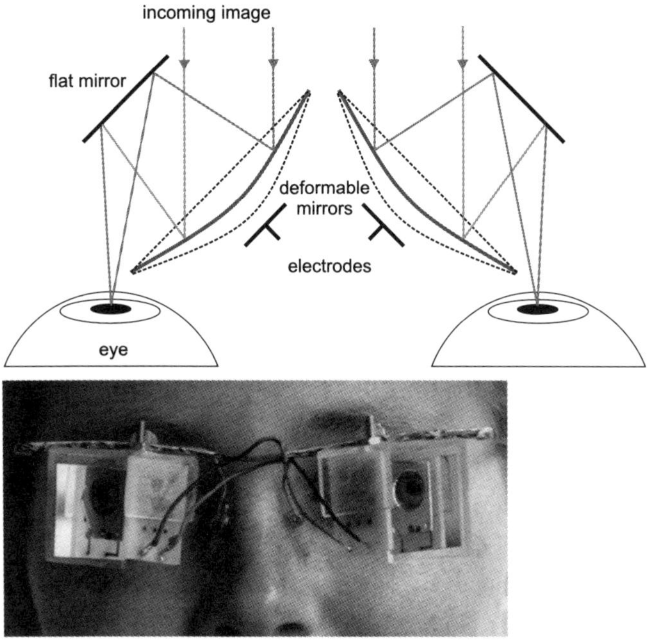

TUDelft Multimedia Services

Figure 6.4: Adaptive spectacles: a mirror pair per eye replaces a lens.

maximisation of the wanted parameter until an optimal value is reached; the corresponding set of voltages is stored in a database. The system can have a number of states and the calibration for each system state can involve thousands of iterations. This somewhat lengthy preliminary feedback process enables dynamic system operation later on by recalling optimal voltage values from look-up tables (the process is similar with the general look-up table approach described in Chapter 4).

An example of such a system is a beam-scanning optical microscope (Figure 6.5). In this setup a small focal spot (~1 μm), resulting from an objective lens or parabolic mirror, quickly scans the test sample. The size and quality of the focal spot determine the quality of the output image of the sample under test. The scanning mechanism consists in changing the angle of incidence of the collimated light beam on the lens or mirror, which consequently imparts off-axis aberrations to the focused beam, deteriorating the spot quality. A solution is offered by a deformable mirror inserted between

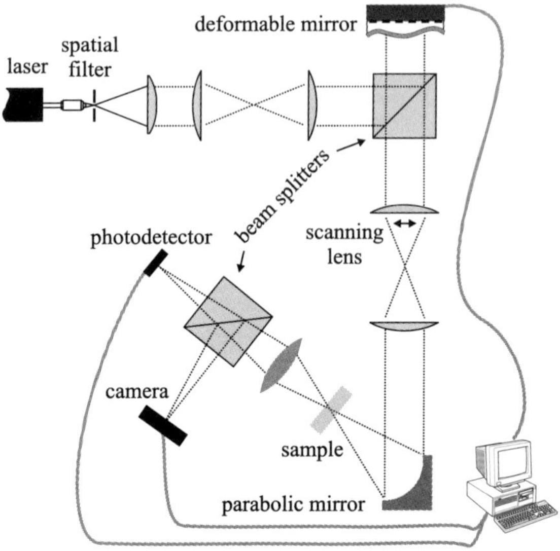

Figure 6.5: Scanning optical microscope with a deformable mirror.

the laser source and the focusing lens (or parabolic mirror) [18]. Appropriate deformations of the mirror surface counteract each off-axis aberrations introduced by the focusing element at different angles of incidence. Each scanning angle (spot position) is therefore labelled as a system state that will repeat itself in cycles, each time recalling the previously stored set of voltages associated with it.

Wavefront sensors and deformable mirrors in closed-loop systems

Several optical systems deal with rapidly changing uncorrelated aberrations. Three prompt examples are:

- highly expensive and high-precision optical systems for *astronomical observations* are left disarmed before aberrations changing as fast as 10 ms due to turbulant atmospheric layers [19];
- when the installation costs of fiber-optic cabling are prohibitively high, *Free-Space Optics* (FSO) for direct link communications represents an alternative. However, thermal fluctuations in the air cause problems coupling the source to the receiver [20];
- *in vivo imaging of the retina* is degraded by eye fluctuations [21].

Closed-loop adaptive optical systems operating in feedback mode can provide solutions to these cases. In a simple setup, a deformable mirror is controlled based on the signals from a Hartmann sensor. The reference sensor signals are measured in a situation when the mirror is in its initial state and a reference light beam, with a known profile, reaches the sensor after reflection by the mirror. For a distorted beam, the sensor signals will depart from the reference values. Therefore, the mirror surface has to be adjusted in order to force those signals to converge to the reference values, hence restoring the wavefront upon reflection. In such a system, it is not necessary to reconstruct the wavefront, i.e. to obtain either a geometrical or mathematical function for it, as will be shown later. In short, the loop forces the light spots on the Hartmann sensor to remain at the reference positions, therefore guaranteeing that the output wavefront is kept constant, regardless the profile of the input wavefront.

Take the example of an astronomical observatory. The image of a star reflected by the telescope is blurred because of atmospheric turbulence, i.e. the associated wavefronts are distorted. When this image impinges into the closed-loop system, the mirror adjusts its surface according to the sensor feedback signals in such a way that the associated output wavefronts are corrected, resulting in a sharp image of the star. To assess the distortions introduced by the atmosphere, astronomers use an artificial guided star as a reference. This is actually a high-power laser pointed at the location of the target natural star. A fraction of the beam is reflected back from the atmospheric layers, passing by the same turbulent path the natural-star image passes.

A schematic of an adaptive optical system coupled to a telescope is shown in Figure 6.6. The star image is captured by the telescope mirror and redirected to the AO system. The incoming light beam heads to a planar tip-tilt mirror which centers the beam on the surface of a deformable mirror. Then the beam is split into two branches, one that goes to a wavefront sensor and another that goes to the imaging plane (camera). The sensor measures the distortions of the wavefront and instructs the deformable mirror to adapt its surface in such a way that the reflected wavefronts are compensated.

6.4. Device Implementation

This section discusses the implementation of two devices: a Hartmann-Shack wavefront sensor compatible with standard VLSI and etching technologies and a micromachined membrane deformable mirror (MMDM).

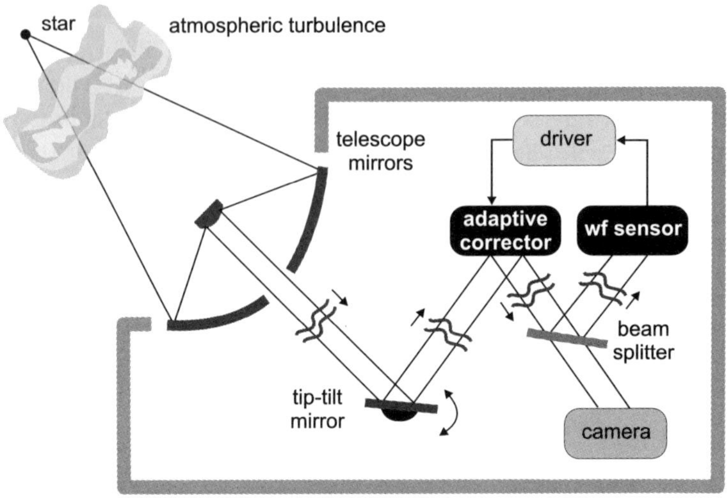

Figure 6.6: Schematic of a telescope equipped with adaptive optics.

6.4.1. *The Hartmann Wavefront Sensor*

Conceptually, a Hartmann sensor requires only two components: a sampling array and a photodetector array. The sampling array can be a Hartmann mask (opaque perfurated mask) or a microlens array. The photodetector array can be an off-the-shelf imager (CCD camera or CMOS camera) or a custom photosensitive chip. All these componets can be fabricated in the framework of silicon microtechnologies.

Hartmann mask

The optimal number of openings (sampling points) in the mask depends on the requirements of the application. The higher the wavefront spatial frequency, the more holes demanded. The sampling frequency along a given direction is given by the inverse of the pitch (separation between the centers of two neighboring openings along that particular direction). The Nyquist-Shannon theorem indicates that only wavefront spatial frequencies lower than the half the sampling frequency will be correctly sampled. The dimension and shape of the mask openings have a direct influence on the light-spot intensity profile, which changes with the distance from the mask to the detection screen (camera). The intensity profile can be calculated based on diffraction theories (Fresnel or Frauhofer depending on the distance at which the screen is placed [1]).

Using photolithography and etching one can make high precision masks, where the positions, diameters and shapes of the openings are precise. This allows the mask to be used as a reference itself, eliminating the need for a reference beam. A 1 μm thick metal layer is deposited on a 500 μm glass substrate and then openings are etched through the metal with RIE (Reactive Ion Etching).

Microlens array

A microlens array with a 100% fill factor is a more appealing sampling plane than a Hartmann mask. The individual sampling areas are larger than in the previous case, they collect more light and reduce sampling aliasing, since the wavefront tilt is averaged over each lens aperture.† Wavefront sensors using microlenses are often referred to as Hartmann-Shack sensors.

The most used fabrication methods of continuous-surface-relief‡ microlens arrays are often divided into three major categories: photoresist melting/sculpture, direct writing/etching and replication [22, 23].

Photoresist Melting/Sculpture:

The photoresist melting technique relies on the fact that surface tensions transform a photoresist pedestal into a hemispherical bump upon melting under controlled conditions. These lenses strongly absorb blue light and are only efficient for wavelengths larger than 600 nm [22]. Batch reproducibility depends on the accurate control of several parameters and the lens shapes for different focal lengths feature different surface aberrations. With this approach, 100% fill factor is not achievable, unless two arrays of cylindrical lenses are superimposed or two consecutive expositions through the same

†Aliasing occurs when there are spatial frequencies higher than the frequencies the system is able to handle. These high frequencies are sampled as spurious low-frequency components, degrading the system reliability. The smaller the sampling area, the more significant this effect is.

‡Other types of microlens arrays are based on a graded-index profile (GRIN lenses), in which the lenses are essentially plane with a variation in the refraction index in the bulk, and on diffractive optical elements, whose performance is highly dependent on the wavelength dependent.

1D mask is done, with the mask orthogonally rotated. A larger range of profiles can be achieved by photosculpture using grey-scale, or halftone, masks to control the exposure levels of parts of the photoresist pedestal, defining how much photoresist is left after development, therefore controlling the target shape. Also, laser punctual scanning, laser writing, has been used for the same purpose, and additive lithography has been recently proposed and verified [24]. Yet another option is to use an electronresist material and an electron beam, implying higher equipment costs and little advantage in the resolution due to electron scattering.

Direct writing:
Techniques as those mentioned above, comprising of laser and electron writing have been found in this category in the literature. Direct surface writing without any step for resist development involves laser ablation, focussed ion beam (FIB) laser-enhanced deposition/etching and diamond turning. Diamond turning however, is a mechanical method that only allows the fabrication of rotationally symmetric components, therefore is not suitable for monolithic microlens arrays.

Replication:
The thermoplastic or curable polymer components are replicated from a mould, which is either manufactured directly on silicon, metal or another substrate or formed by copying a master structure (e.g. a previously fabricated photoresist microlens array). The replication is often cost effective, but the master sample and/or mould can still be expensive.

A very attractive and inexpensive method to fabricate microlens-array moulds in silicon requires a single lithographic mask and an anisotropic etching solution ($KOH{:}H_2O$) [25]. A regular grid of equal openings on the mask is transferred to a thin oxide layer on silicon. Etching through the openings with KOH results in inverted pyramidal pits, whose depths equal $d_0/\sqrt{2}$, where d_0 is the initial diameter of each opening. By removing the oxide layer and etching the sample further, one achieves an area completely filled with spherical cavities with a sharp interface between them (Figure 6.7). The arrays can be used as a mould for the replication of microlens arrays.

The process is straightforward and well controlled, providing smooth hemispherical moulds for shallow lenses [26]. It is not applicable for the fabrication of lenses with a high aspect ratio (lens thickness at the centre

Figure 6.7: Hexagonal array of spherical cavities in silicon with 100% fill factor. This array serves as a mould for microlens replication.

divided by its diameter). However, it is suitable for Hartmann-Shack sensors because of the 100% fill factor, high parallelism of the optical axis (crystallographically defined), precision of the lens positions (ligthographically defined) and the sharp lens boundaries (reduction of glare) [27].

VLSI photodetector array

Off-the-shelf cameras register the spot pattern as an image, which is processed to yield the positions of the spots centroids, from which the wavefront profile can be calculated. The main limitations of this approach are the somewhat low frame rate of conventional cameras ($\leq$150 frames/s) and the relatively lengthy image processing algorithm. This brings the operational frequency of an adaptive optical system to 100 Hz or typically less. Dedicated scientific cameras are capable of faster operations, but their costs are to date exorbitant (a 0.5 Mpixel 4.8 kHz CMOS camera costs approximately USD 100 k).

There are two major silicon-based technologies complementing each other in the image-sensor arena. Charge-Coupled-Device technology (CCD) is nowadays optimised for imaging, with very low noise and extremely efficient charge transfer, and it is used in most professional imaging chips. Complementary Metal-Oxide-Semiconductor technology (CMOS), on the other hand, offers the great advantage of mixed-circuit integration on the same chip of the photodetectors, and although the noise level in standard CMOS is still high, proprietary CMOS processes have a claimed performance equivalent to some CCD lines. In any case, the architecture of a typical image sensor is appropriate for the registration of an image (detailed two-dimensional intensity map), whereas for a wavefront sensor the displacements of light-spot centroids are the only data ultimately needed.

A modified imager, with pixels clustered as optical position-sensitive detectors (PSDs), offers prompt advantages over conventional imaging chips: direct registration of spot displacements and consequently the redundancy of image processing. Each position-sensitive detector is responsible for sensing the displacement of one spot. CMOS represents a suitable choice for the implementation of a custom imager because the individual clusters (PSDs) and the array can have a much more flexible geometry, and the chip can accommodate several electronic functions, including circuitry for random access to the pixels, amplification, data processing and AD conversion. Assuming the wavefront spatial frequency is appropriately sampled (sufficient number of microlenses and correct array geometry), the root-mean-square *(rms)* wavefront detection accuracy is closely related to the noise in the spot position measurements.

Three implemented versions of a Hartmann-Shack wavefront sensor using integrated position-sensitive detectors in a CMOS chip have been reported in the literature. The arrays are orthogonal with either 16×16 or 8×8 PSDs. Each PSD yields two values, representing the x and y spot-displacement coordinates.

Two of these chips are based on the determination of the spot position by means of intensity maximum. They use winner-take-all circuitry, where the subthreshold regime of MOSFETs is used to turn on a single output bit from a number of analogue inputs [28]. Each PSD consists of a two-dimensional raster of photodiodes; columns and rows of photodiodes are connected to input circuit nodes along two independent branches (horizontal and vertical). The photocurrent levels at the input nodes compete with each other and the highest analogue level induces its respective output node to exclusively yield a digital high bit '1', among all output nodes (Figure 6.8a) [29]. When this is done independently on the vertical and horizontal branches, the coordinates of the point of maximum intensity of the spot is found.[§] Asymmetries of the spot intensity profile or noise contribute to erroneous information about the actual centroid position. To counteract this effect, each circuit branch is split into several interdigitised chains; the chains work exactly as each branch described above, but now each chain

[§]Recall that the spot is not rigorously punctual, as it might suggest, but rather a finite region with a certain intensity distribution.

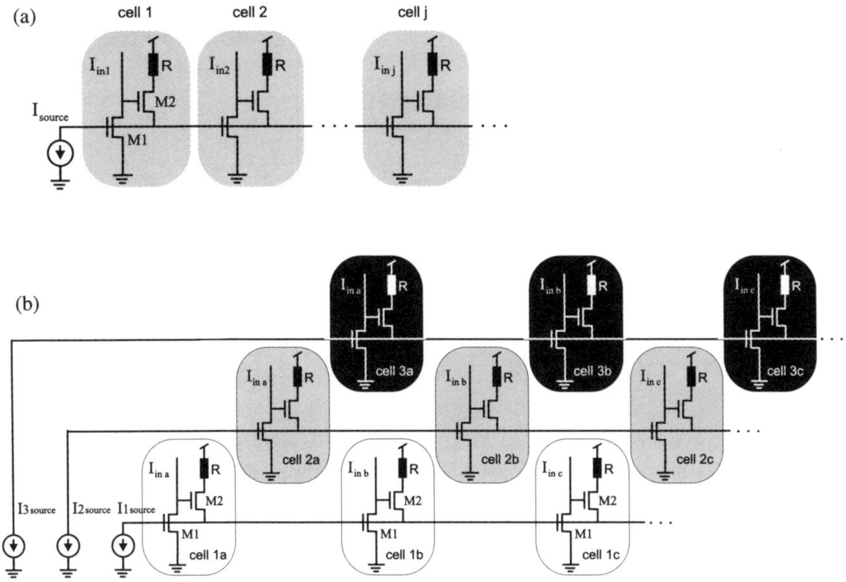

Figure 6.8: Simple schematics of (a) the classical Winner-Take-All architecture for the detection of the maximum photocurrent, (b) the alternative interdigitized WTA circuit with the possibility of pseudo-centroiding.

produces its own high bit (Figure 6.8b) [30]. Taking the most central bit of the set of high bits, promotes a more accurate determination of the spot-centroid position. This approach is labelled 'pseudo centroiding'.

The previous chips are able to operate with faint spots (spot intensity in the nW range), but the position resolution is still limited by the pixel pitch, which is determined by the technology (minimum dimensions for doped regions, wells, wires, etc.). A wavefront accuracy of 50 nm has been reported at 0.3 μW/spot for a pitch of 17 μm [31]. In this case, increasing the light level does not improve the accuracy.

A third CMOS chip consists of clusters of four pixels each (quad cells), with a multiplexed data bus [32]. There are 64 randomly accessible clusters orthogonally arranged, each aiming at the continuous detection of the centroid of a light spot (Figure 6.9). The differences in photocurrent between photodiode in a cluster indicate the x and y coordinates of the spot centroid.

A preliminary calibration of the quad cell is necessary because its response function is not linear and dependent on the intensity profile of

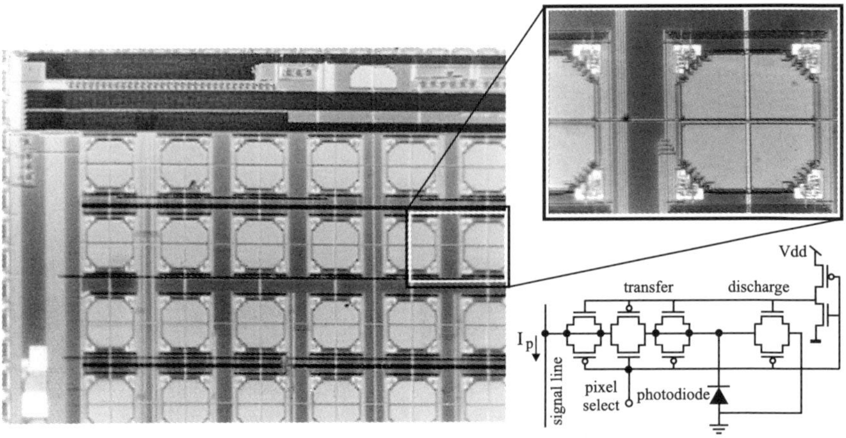

Figure 6.9: Photograph of a part of the CMOS chip, of one quad cell and the electronic schematic of the pixel.

the light spot and on the ratio between the spot size and the cluster size. The calibration identifies the slope parameter of a sigmoidal function, which is thereafter used as a characteristic function:

$$x_{QC} = A_2 + \frac{(A_2 - A_1)}{1 + \exp(x/s_x)}. \tag{6.1}$$

The pixels in this chip are passive** and because the junction capacitance of the photodiodes used is large, the capacitive noise (kTC noise) introduced in the final signal imposes a limit to the noise equivalent power, therefore limiting the minimum amount of light per spot to the microwatt range. In this chip it is not the geometry (or the minimum feature size) that limits the resolution, but the minimum noise floor. A wavefront accuracy of 12 nm has been reported at 10 μW/spot. Reduction of noise can be accomplished by using active pixels (Figure 6.10) [33, 34] and by choosing a different photodiode junction with a lower capacitance.

**Pixels consist in a photoelement and circuitry. In a passive pixel the photocurrent is transferred from the photodiode directly to the signal line through a single switch or a complementary switch set. In an active pixel the original signal is buffered to the signal line, i.e. the quantity at the signal line is a replication or a representation of the original photoelement current, voltage or charge.

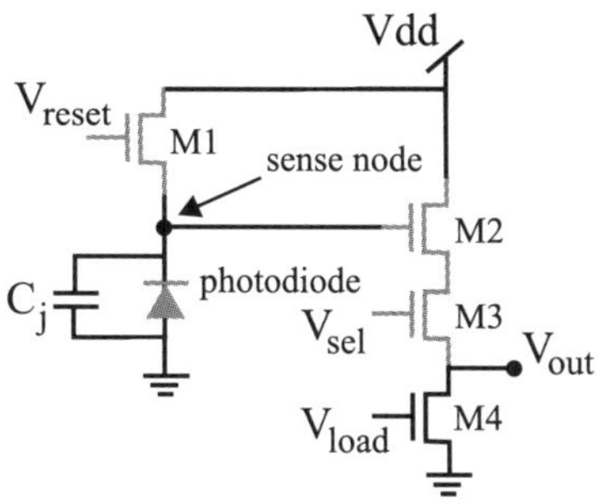

Figure 6.10: Schematic of an active photopixel. The photocurrent decreases the voltage previously set at the sense node; this voltage drop is directly observed by measuring V_{out}.

It should be noted that supra-micron CMOS processes offer more advantages to photodetection than state-of-the-art ones, especially because their junctions are deeper and the depletion layers wider, guaranteeing a better quantum efficiency over the visible spectrum. If active pixels are used, the operational voltage interval, limited at the top by the supply voltage and at the bottom by the MOSFET threshold voltage, is larger in conventional processes (in excess of 2 V), enabling a larger analogue swing [35].

Silicon is an attractive choice for applications in the visible spectrum (~400–700 nm). A number of photodetector structures can be implemented, mostly either taking advantage of junctions between doped regions or of depletion layers upon the application of an external voltage. Each type of structure has a particular footprint and yields a different performance. The ultimately acceptable wavelength range is limited either by the silicon bandgap energy (1.12 eV) or by surface recombination. Photons with very short wavelengths (<300 nm) are absorbed so close to the surface, that most of the generated carriers diffuse to the surface and end up recombining there. On the other hand, photons with a wavelength larger than 1100 nm go throughout silicon because they do not have enough energy to be absorbed.[††] Enhancement of detection at the sidebands of the visible spectrum (near ultra-violet and near infra-red) are possible by using special techniques, like phosphor coatings, metal/semiconductor interfaces

[††]The photon energy is not enough to induce an electron transition from the valence band to the conduction band followed by thermalisation (emission of a phonon).

and back-side illumination, for instance. Alternatives to different wavelengths are heterostructures involving combinations of different semiconductor materials.

When chosing a particular technology, one should take into account the level of functionality that can be embedded on chip (e.g. digital functions, analogue functions, wavelength range of photodetection), the existence of reliable modelling, the required circuit density, the feasible noise and dynamic range performance, and the compatibility with system costs.

6.4.2. *The Membrane Deformable Mirror*

The membrane of a bulk micromachined deformable mirror consists, in its simplest case, of a silicon nitride layer coated with aluminum. This membrane is continuous and fixed at the borders. It is mounted on a set of electrodes which can, in principle, be patterned on a silicon substrate, on a glass substrate or on a PCB (Printed-Circuit Board). The typical thickness of the air gap between the membrane and the mirror ranges from 20 to 100 μm and the DC electrode voltages necessary for full deflection are proportional to the square of this thickness, and can typically amount to 300 V.

A thin silicon nitride layer, 0.3 to 0.8 μm, is deposited by LPCVD (Low Pressure Chemical Vapor Deposition) on a silicon substrate. Because of the direct contact, this layer replicates the excellent finish of the silicon-wafer surface. The silicon is etched from the other side through a window till the membrane is released at the centre. Tensile stressed membranes are obtained, since the control of stress in the nitride deposition is very precise. Silicon nitride is an advantageous choice because it is a mechanically strong material, not brittle, and compatible with IC (Integrated Circuit) fabrication processes.

If anisotropic etching with a high selectivity is used (e.g. KOH aqueous solution to etch a $\langle 100 \rangle$ wafer), the nitride membrane is kept unaffected by the etching of the silicon substrate. However with anisotropic etching, regardless the shape of the etch window, the final etch contour is rectangular (i.e. the membrane boundaries are rectangular), which results in a cushion-like surface response when the central electrode is actuated. It is optically attractive to have a parabolic response, which is accomplished by having a circular boundary of the membrane, whereas rectangular membranes can only be used for this purpose at their very centres or when a large number of actuators is present.

It is possible to achieve a step-wise approximation to a circular contour by designing an etch window as a gear-wheel. The remaining sharp edges can be rounded by an isotropic etching step. Isotropic etching could be used for the whole process, resulting directly in a circular membrane edge. However, this type of etching attacks silicon nitride, spoiling its high surface quality. Pure nitride membranes are strong enough to fabricate membranes with diameters up to 25 mm. Composite membranes, using a 10 μm polysilicon layer covered with nitride on both sides, are appropriate for diameters up to 50 mm.

Once the membrane is ready, a 0.2 μm aluminum or gold layer is evaporated on it to ensure reflectivity and conductivity (the voltage is applied between the electrode and the membrane. These coatings are sufficient for a large spectral range in the visible (Al) and the infrared (gold). Coating the membrane with a Cr/Ag layer plus a dielectric film with up to 12 layers can result in close to 100% reflectivity in a narrow spectral band. This improvement is mandatory for relatively high laser loads (e.g. 0.5 kW in a 5 mm circular beam, $\lambda = 1.06\,\mu\text{m}$) [36].

The electrode structure is easily implemented on a PCB. Although the feature resolution of metal structures in typical PCB processes is usually not better than 20 μm, and more often in the range of 100 μm, this is a reliable and inexpensive solution suitable for membrane mirrors with 1 to 100 actuators. A hole in the centre of each electrode helps in minimising the air damping of the membrane, therefore extending the linear frequency operational range to a couple of kHz.

One obvious fundamental limitation of the electrostatic actuation is that the membrane can only be pulled towards the actuators and not pushed away from them. However, optical aberrations often have both concave and convex components, requiring bidirectional mirror compensation. This drawback is compensated by biasing the membrane. Bidirectional correction is achieved by applying a set of voltages to the electrodes, so that the mirror initial state corresponds to a parabolic profile with half the maximum deflection magnitude. In such a situation, decrementing or incrementing the electrode voltages emulates the push or pull action, respectively.

Compactness, ease of manufacture, low cost and high optical quality, added to quick response, hysteresis-free operation and smooth modal response, make micromachined membrane deformable mirrors a suitable device for a number of optical applications in the laser optics, imaging, optical testing and astronomy [37–40].

6.5. Closed-loop Adaptive Optical System

The combination of a VLSI device (wavefront sensor), a MEMS device (deformable mirror) and a very simple control algorithm constitute a smart system that corrects optical wavefronts dynamically. This system can be visualised as a black-box device in which the input is a distorted wavefront and the output is a corrected wavefront. The coupling of these components and the control algorithm for a closed loop AO system are presented in this section.

6.5.1. *Setup*

The input wavefront is reflected by the mirror surface towards the wavefront sensor, whose signals indicate how the mirror membrane has to be reshaped in order to yield a corrected output wavefront. However, because an aberrated wavefront changes as it propagates from the mirror to the sensor, optical plane conjugation is necessary. This is done by positioning lens pairs, called relay systems, which reproduce the image of the input pupil (plane where the aberration to be corrected enters the system) on the mirror surface and on the sampling plane of the sensor. This guarantees that the wavefront at these three planes correspond to one another. A schematic of the setup is shown in Figure 6.11.

6.5.2. *Control Algorithm*

A reference wavefront reflected by the mirror in its initial state results in a spot pattern on the CMOS wavefront sensor. This pattern defines a reference grid. When a distorted wavefront[‡‡] is reflected by the mirror, in the same state, the sensor registers displacements of the light spots from the reference grid. In order to correct the distortions, the mirror membrane should adapt its shape in such a way that the the displacements on the sensor are minimised. Therefore, the sensor signals are used as feedback data to the mirror control.

[‡‡]A distorted wavefront is any wavefront whose profile deviates from the reference one.

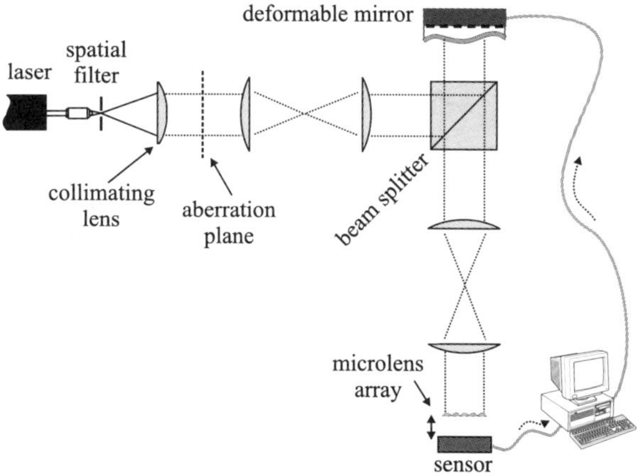

Figure 6.11: Adaptive optical setup with a deformable mirror and a wavefront sensor.

Prior to tackling the problem of how the membrane is linked to the wavefront-sensor signals, the system must be calibrated by establishing a reference wavefront (beam profile). This can be done by optimisation of an input beam upon reflection by the mirror. The idea is to have a reference beam with a phase profile as plane as possible, to which any introduced aberration will be compared. If the reflected beam is redirected to a micro-objective, the output image is the far-field intensity distribution (Fraunhofer diffraction) corresponding to the beam at the mirror plane (Figure 6.12). If the far-field image of a diffraction limited beam is projected on a screen, the intensity pattern is seen as a central circular maximum and fading concentric rings around it, separated by dark rings. Aberrations in the beam profile reduce the intensity in the central ring and spoils the symmetrical ring pattern. Therefore, one way to generate a reference beam is to change the mirror surface in order to maximise the intensity of the central peak in the corresponding far-field image.

To achieve this in practice, an opening is drilled in the projection screen and a photodiode is positioned behind it. The photodiode signal indicates the optical intensity through the hole and serves as feedback to establish the optimal mirror shape. A downhill simplex control algorithm [41] is used to find the best possible combination of voltages to apply to the mirror. The mirror is set initially to its initial state (biased) and the optimisation process starts by attributing random sets of voltages to the mirror

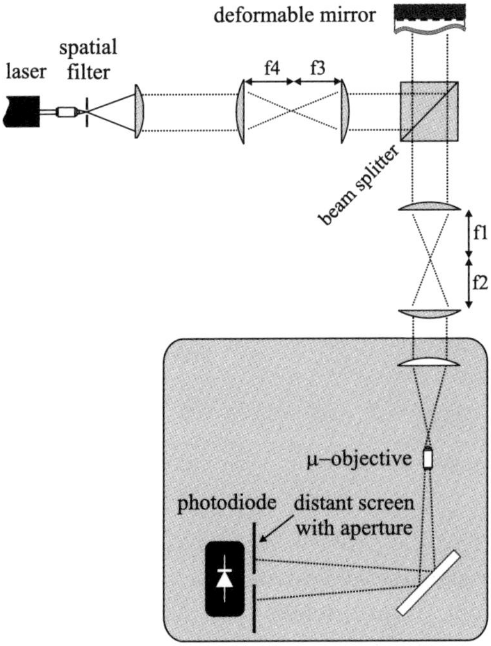

Figure 6.12: Optical setup for the optimisation of the reference beam.

electrodes and storing those which result in the highest photodiode signals. Fine tuning assesses each of these winning sets and creates subsets with different combinations of voltage increments, from which the best are chosen. The process is often lengthy (several minutes) and requires many iterations to reach a stable result. Since the solution is not unique, we always need to crosscheck the best solution with the intensity pattern observed on the projection screen.

The next step involves understanding the feedback control, which assesses the voltages to be applied to the mirror according to the wavefront-sensor signals. Actuation of a single electrode modulates the whole mirror surface and the actuation of all electrodes imparts a profile to the mirror surface, whose deflection is a linear superposition of each individual deflection associated with individual electrodes. The vector of spot displacements $\boldsymbol{d}$ is related to the vector of mirror control signals $\boldsymbol{c}$ by means of a feedback matrix $\boldsymbol{M}$. This matrix consists of sensor responses to different characteristic shapes of the mirror, also referred to as influence functions. When an incremental unity voltage is applied to a single electrode, the corresponding x and y displacements of all spots are registered on a column; this is done for all electrodes. If the sensor has N sampling points and the mirror has L electrodes the $\boldsymbol{M}$ matrix has $2N \times L$ elements. The relationship between the parameters can be expressed as a matrix equation:

$$\mathbf{d} = \mathbf{M}\mathbf{c}.$$

$$\underbrace{\begin{bmatrix} \Delta x_0 \\ \Delta y_0 \\ \Delta x_1 \\ \Delta y_1 \\ \vdots \\ \Delta x_i \\ \Delta y_i \\ \vdots \\ \Delta x_{N-1} \\ \Delta y_{N-1} \end{bmatrix}}_{\substack{\mathbf{d} \\ \mathbf{2N \times 1}}} = \underbrace{\begin{bmatrix} \Delta x_0|_0 & \Delta x_1|_0 & \Delta x_2|_0 & \Delta x_3|_0 & \cdots & \Delta x_{L-2}|_0 & \Delta x_{L-1}|_0 \\ \Delta y_0|_0 & \Delta y_1|_0 & \Delta y_2|_0 & \Delta y_3|_0 & \cdots & \Delta y_{L-2}|_0 & \Delta y_{L-1}|_0 \\ \Delta x_0|_1 & \Delta x_1|_1 & \Delta x_2|_1 & \Delta x_3|_1 & \cdots & \Delta x_{L-2}|_1 & \Delta x_{L-1}|_1 \\ \Delta y_0|_1 & \Delta y_1|_1 & \Delta y_2|_1 & \Delta y_3|_1 & \cdots & \Delta y_{L-2}|_1 & \Delta y_{L-1}|_1 \\ & & \vdots & & \ddots & & \vdots \\ \Delta x_0|_i & \Delta x_1|_i & \Delta x_2|_i & \Delta x_3|_i & \cdots & \Delta x_{L-2}|_i & \Delta x_{L-1}|_i \\ \Delta y_0|_i & \Delta y_1|_i & \Delta y_2|_i & \Delta y_3|_i & \cdots & \Delta y_{L-2}|_i & \Delta y_{L-1}|_i \\ & & \vdots & & \ddots & & \vdots \\ \Delta x_0|_{N-1} & \Delta x_1|_{N-1} & \Delta x_2|_{N-1} & \Delta x_3|_{N-1} & \cdots & \Delta x_{L-2}|_{N-1} & \Delta x_{L-1}|_{N-1} \\ \Delta y_0|_{N-1} & \Delta y_1|_{N-1} & \Delta y_2|_{N-1} & \Delta y_3|_{N-1} & \cdots & \Delta y_{L-2}|_{N-1} & \Delta y_{L-1}|_{N-1} \end{bmatrix}}_{\substack{\mathbf{M} \\ \mathbf{2N \times L}}} \underbrace{\begin{bmatrix} C_0 \\ C_1 \\ \vdots \\ C_k \\ \vdots \\ C_{M-2} \\ C_{M-1} \end{bmatrix}}_{\substack{\mathbf{c} \\ \mathbf{L \times 1}}} \tag{6.2}$$

This allows the control coefficients to be calculated from the spot displacements. The deflection of the mirror surface $\delta(x, y)$ is proportional to $V^2(x, y)$ over the electrode array. Therefore, the linear system we introduced above implies that the control coefficients should be the square of the incremental voltages to be applied to the mirror electrodes. In the initial state, each electrode has already a non-zero bias voltage applied to it, in order to guarantee bidirectional membrane operation and to establish the reference beam. The incremental voltages are either added or subtracted from these reference voltages.

In the system of linear equations given in (6.2), $\mathbf{d}$ is often dimensionally higher than $\mathbf{c}$ and the system has more equations than unknowns. This is an overdetermined system, which does not have an exact solution [42]. However, by using the least-square approximation method, it is possible to find the control coefficients in $\mathbf{c}$ that more closely satisfy the set of linear equations. This method determines which vector $\mathbf{c}$ best fits $\mathbf{d}$ through $\mathbf{M}$, and is equivalent to solving the following equation:

$$[\mathbf{M}^T\mathbf{M}]^{-1}\mathbf{M}^T\mathbf{d} = \mathbf{c}, \tag{6.3}$$

where $\mathbf{M}^T$ is the transpose and $[\mathbf{M}^T\mathbf{M}]^{-1}\,\mathbf{M}^T$ is the pseudo-inverse of the feedback matrix $\mathbf{M}$. Equation (6.3) is valid on the condition that $\mathbf{M}^T\mathbf{M}$

is invertible. Then, direct inversion methods such as Gaussian ellimination can be promptly used. If, however, $\mathbf{M}^T \mathbf{M}$ is singular or nearly singular, i.e. if the equations are (close to being) linear dependent, iterative methods or Singular-Value-Decomposition (SVD) must be called upon.

The pseudo-inverse matrix is calculated at the system start-up. During system operation, the computational load is reduced to the decomposition of an arbitrary sensor response over the responses generated by the individual mirror modes (influence functions), which yields the incremental voltages to be applied to the mirror in order to minimise the spot displacements from their reference positions. The spot displacements are evaluated again after each iteration and, if necessary, a new set of incremental voltages is applied. A feedback coefficient (<1) attenuates the magnitude of the control voltages and improves the chances of convergence of the algorithm. If the spatial frequencies of the input distortion lies within the correction range of the mirror and complies to the sampling capability of the sensor, the optimal correction becomes only limited by noise in the wavefront sensor and fluctuations of the mirror membrane. Figure 6.13 illustrates the system operation, including the calibration (generation of a reference beam), basis generation (registration of sensor responses to each characteristic mirror surface) and loop operation. Note that the wavefront does not need to be reconstructed at any point.

A photograph of a table-top adaptive optical system is shown in Figure 6.14. The source beam passes through a plane where aberrations

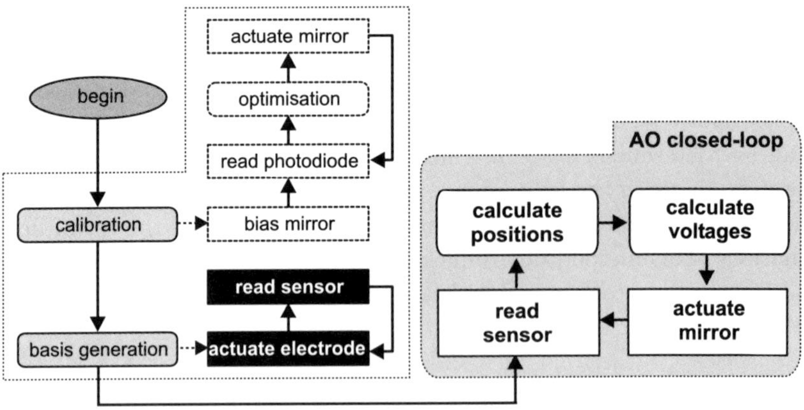

Figure 6.13: Block diagram of the adaptive optical system operation.

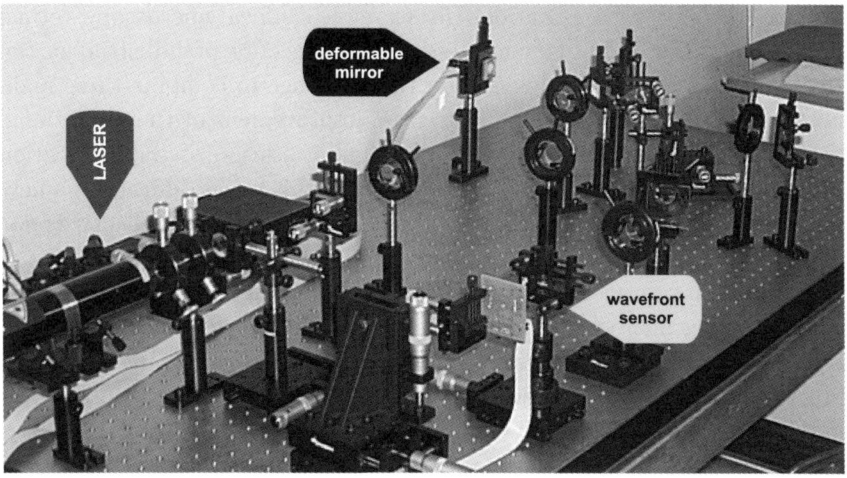

Figure 6.14: Photograph of a closed-loop adaptive optical system.

can be inserted (turbulence, irregular glass plate, etc.) and heads towards the deformable mirror. The beam is reflected by the mirror and is split in two. Part of it enters a micro-objective that will image the far-field pattern corresponding to the mirror surface at a distant screen, and part of it goes to the wavefront sensor.

6.6. Conclusions and Future Trends

Smart systems are undoubtedly pertinent to Adaptive Optics (AO), which is a technology used for dynamic optical compensation and optimisation. The applications of AO vary from the correction of turbulence-distorted images to laser optimisation. Simple device concepts realised in standard VLSI and micromachining processes constitute a corner stone for the further expansion of AO technology into the industrial and medical scenarios, where there is a potential for low-cost, compact and easy-to-handle systems with a reliable performance and a demand for a consistent component supply.

Despite the large number of components and systems built from scratch to meet the demands of specific applications, wavefront sensors, adaptive correctors and complete AO systems are already available commercially and they have encouraged the identification of adaptive optics as a solution in a number of novel applications (for example confocal microscopy, retinal imaging, femtosecond-pulse compression and inspection of industrial parts). Recent trends are mainly focused on the challenge to combine three main issues: reduction of the component costs, improvement of the operational bandwidth (temporal/spatial) and the standardisation of the fabrication process. Two other parameters can be added to this list as desirable, namely improved light sensitivity, for wavefront sensors, and high reflectivity for deformable mirrors.

It has been proven that good quality optical components can be manufactured using microelectronics fabrication technologies, which are widely available, constantly improving and offer reliable modelling. Besides, these processes offer a great number of structures and functions, which are permutable in multiple ways. These are features one can take advantage of in order to promote smart optics as an attractive solution to the optimization of optical systems. Moreover, smart optics provide an opportunity to create some very sophisticated sensing devices, which can sense remotely a number of phenomena, by measurement of the optical signatures caused by these phenomena. Some of these discussed in this chapter, include the sensing of an object's shape, its surface characteristics, its distance and the optical wavefront produced by reflected or transmitted light.

The operation of smart optics requires a high level of control and data analysis, and therefore entails a considerable amount of processing ancilliary to the sensing assembly. The advent of integrated intelligence which can be fabricated along with an active optical assembly, and the use of some of the processing techniques discussed in Chapters 7 and 8 will allow active optical sensing devices to be included within the sensing networks discussed in Chapters 9 and 10.

References

1. Hecht, E. (1998) *Optics*, 3rd edition, Addison Wesley Longman Inc., Reading.
2. de Lima Monteiro, D. W. (2002) *CMOS-based Integrated Wavefront Sensor*, Delft University Press, Delft, http://www.library.tudelft.nl/dissertations/PDF-files_2002/its_lima_20021104.pdf.
3. Tyson, R. (1998) *Principles of Adaptive Optics*, 2nd edition, Academic Press, San Diego.
4. Tyson, R. (2000) *Adaptive Optics — Engineering Handbook*, Dekker, New York.

5. Kotova, S. P., Clark, P., Guralnik, I. R., Klimov, N. A., Kvashnin, M. Y., Loktev, M. Y., Love, G. D., Naumov, A. F., Rakhmatulin, M. A., Saunter, C. D. and Vdovin, G. V. (2003) Technology and electro-optical properties of modal liquid crystal wavefront correctors, *Journal of Optics A — Pure and Applied Optics* **5**(5), S231–S238.
6. Borra, E. F. (1995) Liquid mirrors, *Can. J. Phys.* **73**, 109.
7. Vdovin, G., Middelhoek, S. and Sarro, P. (1997) Technology and applications of micromachined silicon adaptive mirrors, *Opt. Eng.* **36**(5), 1382–1390.
8. Fernández, E. J., Iglesias, I. and Artal, P. (2001) Closed-loop adaptive optics in the human eye, *Optics Lett.* **26**(10), 746–748.
9. Liang, J. Z., Williams, D. R. and Miller, D. T. (1997) High resolution imaging of the living human retina with adaptive optics, *Investigative Ophthalmology & Visual Science* Part 1 **38**(4), 55.
10. Zhang, Y., Ning, L., Rao, X., Li, X., Wang, C., Ma, X. and Jiang, W. (2002) A small adaptive optical system on table for human retinal imaging, *Proc. 3rd Int. Workshop on Adaptive Optics for Industry and Medicine*, Albuquerque, USA, pp. 97–103.
11. Iglesias, P. A. (2000) High-resolution retinal images obtained by deconvolution from wavefront sensing, *Opt. Lett.* **25**, 1804–1806.
12. Tyson, R. (2000) *Introduction to Adaptive Optics*, SPIE Press, Bellingham.
13. Southwell, W. H. (1980) Wavefront estimation from wavefront slope measurements, *J. Opt. Soc. Am* **70**(8), 998–1006.
14. Vdoving, G. (1997) Quick focusing of imaging optics using micromachined adaptive mirrors, *Optics Communications* **140**(4–6), 187–190.
15. Liang, J. Z., Williams, D. R. and Miller, D. T. (1997) Supernormal vision and high-resolution retinal imaging through adaptive optics, *J. Opt. Soc. Am. A* **14**(11), 2884–2892.
16. Loktev, M. Y., Belopukhov, V. N., Vladimirov, F. L., Vdovin, G., Love, G. and Naumov, A. (2000) Wave front control systems based on modal liquid crystal lenses, *Rev. Sci. Instrum.* **71**(9), 3290–3297.
17. Vdovin, G., Loktev, M. and Naumov, A. (2003) On the possibility of intraocular adaptive optics, *Opt. Express* **11**(7), 810–817.
18. Albert, O., Sherman, L., Mourou, G. and Norris, T. B. (2000) Smart microscope: an adaptive optics learning system for aberration correction in multiphoton confocal microscopy, *Opt. Lett.* **25**(1), 52–54,
19. Roddier, F. (1999) *Adaptive Optics in Astronomy*, Cambridge University Press.
20. Levine, B. M., Martinsen, E. A., Wirth, A., Jankevics, A., Toledo-Quinones, M., Landers, F. and Bruno, T. L. (1998) Horizontal line-of-sight turbulence over near-ground paths and implications for adaptive optics corrections in laser communications, *Applied Opt.* **37**(21), 4553–4560.
21. Hofer, H., Artal, P., Singer, B., Aragon, J. L. and Williams, D. R. (2001) Dynamics of the eyes wave aberration, *J. Opt. Soc. Am. A* **18**(3), 497–506.

22. Herzig, H. (1997) *Micro-optics: Elements, Systems and Applications*, Taylor and Francis, London.
23. Daly, D. (2001) *Microlens Arrays*, Taylor and Francis, London.
24. Pitchumani, M., Mohammed, W., Hockel, H. and Johnson, E. G. (2004) Presculpting of photoresist using additive lithography, *Proc. SPIE* **5347**, 85–94.
25. Kendall, D. L., Eaton, W., Manginell, R. and Digges, Jr. T. (1994) Micromirror arrays using KOH:H20 micromachining of silicon for lens templates, geodesic lenses, and other applications, *Opt. Eng.* **33**(11), 3578–3588.
26. de Lima Monteiro, D. W., Akhzar-Mehr, O., Sarro, P. and Vdovin, G. (2003) Single-mask microfabrication of aspherical optics using KOH anisotropic etching of Si, *Optics Express* (www.opticsexpress.org) **11**(18), 2244–2252.
27. de Lima Monteiro, D. W. and Nirmaier, T. (2003) CMOS technology in Hartmann-Shack wavefront sensing, to be published in the *Proc. 4th Adaptive Optics Conference for Industry and Medicine*, Münster-Germany.
28. Lazzaro, J. *et al.* (1989) Winner-take-all networks of O(N) complexity, *Proc. Neural Information Processing Systems (NIPS)*, Denver-USA, p. 703.
29. Droste, D. and Bille, J. (2002) An ASIC for Hartmann-Shack wavefront detection, *IEEE J. Solid-State Circ.* **37**(2).
30. Nirmaier, T. Pudasaini, G. and Bille, J. (2003) Very fast wavefront measurements at the human eye with a custom CMOS-based Hartmann-Shack sensor, *Opt. Express* **11**(21), 2704–2716.
31. Nirmaier, T. (2003) *A CMOS-based Hartmann-Shack Sensor for Real-time Adaptive Optical Applications*, PhD Thesis, University of Heidelberg.
32. de Lima Monteiro, D. W., Vdovin, G. and Sarro, P. M. (2004) High-speed wavefront sensor compatible with standard CMOS technology, *Sensors and Actuators A* **109**(3), 220–230.
33. Fossum, E. R. (1993) Active pixels sensors — are CCDs dinosaurs? *Proc. SPIE* **1900**, 2–14.
34. Mendis, S. K., Kemeny, S., Gee, R., Pain, B., Kim, Q. and Fossum, E. (1994) Progress in CMOS active pixel image sensors, *Proc. SPIE*, **2172**, 19–29.
35. Wong, H. (1996) Technology and device scaling considerations for CMOS imagers, *IEEE Trans. Electr. Dev.* **43**(12), 2131–2142.
36. Vdovin, G. and Sarro, P. M. (2002) Micromachined membrane deformable mirrors for laser applications, *Proc. 3rd Int. Workshop on Adaptive Optics for Industry and Medicine*, Albuquerque, USA, pp. 35–48.
37. Dayton, D., Restaino, S., Gonglewski, J., Gallegos, J., McDermott, S., Browne, S., Rogers, S., Vaidyanathan, M. and Shilko, M. (2000) *Optics Communications* **176**, 339–345.
38. Gonte, F., Courteville, A. and Dandliker, R. (2002) *Optical Engineering* **41**(5), 1073–1076.
39. Druon, F., Cheraux, G., Faure, J., Nees, J., Nantel, M., Maksimchuk, A., Mourou, G., Chanteloup, J. C. and Vdovin, G. (1998) *Optics Letters* **23**, 1043.

40. Vdovin, G. and Kiyko, V. (2001) *Optics Letters* **26**(11), 798–800.
41. Press, W., Teukolsky, S. Vetterling, W. and Flannery, B. (1992) *Numerical Recipes in C — The Art of Scientific Computing*, 2nd edition Cambridge University Press, Cambridge, pp. 408–412.
42. Williams, G. (1984) *Linear Algebra with Applications*, Allyn and Bacon Inc., Newton.

CHAPTER 7

ARTIFICIAL INTELLIGENCE TECHNIQUES FOR MICROSENSORS IDENTIFICATION AND COMPENSATION

by *Elena Gaura and Robert Newman*

The chapter aims to discuss and justify the feasibility and usefulness of integrating MEMS sensors and Artificial Intelligence (AI) to enhance the quality of sensor signals and improve, as a consequence on other performance aspects of sensing systems. The focus is on a particular AI tool, i.e. Artificial Neural Networks (ANNs), whose non-linear mapping abilities are exploited. A brief introduction to ANN techniques used generally in non-linear systems identification and control is provided with the purpose of illustrating the properties which make Neural Networks (NNs) suitable for integration with MEMS sensors, from a systemic viewpoint. Further, a capacitive accelerometer is considered as a case study and a methodology for identifying and controlling the sensor is presented. The material in this chapter forms a reference point for the review of some of the successful NN based sensor systems designs presented in Chapter 8. The chapter does not treat the implementation implications of an AI based sensor system design approach, as the view taken is that the system will contain the means to digitise the sensed signals and has the computation abilities to process them.

7.1. Artificial Neural Networks: What They are and How They are Used for Microsensor Control and Identification

Biology has continually given models from which engineers have evolved some design approaches [1]. Neural Networks and knowledge-based systems have already made a contribution in a variety of engineering areas (particularly automation and control). It is only natural therefore, that, amongst the "borrowed" macrosystems design techniques and tools, AI in general and ANNs in particular are presently receiving increasing attention from researchers in the area of microsensor systems.

A number of successful applications of AI to sensors and microsensors have been reported in the last few years, including:

- sensor metrological performance enhancement — calibration, non-linearity correction, offset and identification,
- actuation control,
- sensor fault detection and classification,
- sensor data validation — analysis of sensor data, data restoration and validation, signal failure detection and reconstruction,
- sensor data mining — used for with example electronic noses.

The above, together with projected future applications (for example within very large scale networks of sensors), make ANNs a tool which cannot be neglected in the context of this book. Many of the AI applications above will be exemplified in a separate chapter (Chapter 8) in the context of intelligent and autonomous sensors. Here the discussion is limited to ANNs' ability to enhance sensor signal quality.

Thorough treatments of ANNs, from a mathematical or an engineering perspective are largely available. A good choice for the uninitiated reader is Simon Hykin's book, "Neural Networks. A comprehensive foundation" [2]. The book is written from an engineering perspective and provides a detailed treatment of most common NN types and training algorithms, supported by examples and applications. Here, neural network techniques will be introduced in the context of identification and control of non-linear dynamic systems. This is the approach of choice, as ANNs entered the area of microsensors primarily with the aid of control engineers. Moreover, this perspective should be sufficient for the non-AI specialists to gather a system level understanding of the applications presented. The authors hope that the interested reader will gain the ability to assess, in the context of their work, the feasibility of using NNs and maybe to devise (at the same systems level) their own NN and sensor based integrated applications. The properties of neural networks, which make them suitable for use in conjunction with microsensors (particularly for control and identification tasks) will be highlighted. Both static and dynamic neural network architectures are considered here and ways of integrating dynamics into the network and the neuron itself are described.

7.1.1. *Control of Non-linear Systems Using Neural Networks*

Intelligent control became a common tool in many engineering and industrial applications due to its ability to comprehend and learn about systems, disturbances, environment and operating conditions [3]. It is widely believed that in the context of engineering, intelligent control should have the following features [4]:

- a learning ability and adaptability;
- robustness;
- a simple control algorithm for relatively user-friendly man-machine interface;
- the ability to be integrated within existing hardware/software structures to enhance the control system performance.

Neuro-control is defined as the use of fully specified neural networks to issue actual control signals [4]. The success of natural neural networks in controlling complex biological systems, suggests that artificial neural networks might be suited for control of complex manmade systems, thesis which has been proven by their application in not only difficult non-linear applications, but also in systems for which no other solutions are known [4–6].

There are several reasons behind the extensive research interest in the application of neural networks for control purposes, as alternatives to traditional control methods. The following are the main points [4]:

- Neural networks can be trained to learn any function, provided that enough information is given during the training process, coupled with judiciously selected neural models. This self-learning ability of the neural networks eliminates the use of complex and difficult mathematical analysis, which is dominant in many traditional adaptive and optimal control methods.
- The inclusion of a semilinear sigmoidal activation function (or certain general non-linear functions) in the hidden neurons of multilayered neural networks offers a non-linear mapping ability for solving highly non-linear

control problems where, to date, traditional control have produced no practical solutions. This, perhaps, is the most significant advantage of using neural networks from the theoretical control viewpoint.

- Before they can be implemented, traditional adaptive and optimal control techniques require a large *a priori* information pool regarding the system to be controlled; due to the self-learning capability of neural networks such extensive information is not required for neuro-controllers. Thus, provided that they are adequately trained, neuro-controllers are able to perform the control task under a wider range of system uncertainty.
- The massive parallelism of neural networks offers a very fast multiprocessing technique when implemented using neural chips or other types of parallel hardware [7].
- Damage to some parts of the neural network hardware may not affect the overall performance badly due to its massive parallel processing architecture.

It has been shown that, the neural network approach for systems control contains no restrictions about system linearity, performs well in noise and can be implemented very efficiently in real-time [8].

These very characteristics made them attractive for use as a tool in microsensor system design where the sensing devices behaviour is not always fully known and is, certainly in most cases, non-linear in nature.

7.1.2. *Identification of Non-linear Systems Using Neural Networks*

Since the early 1970's numerous model-based control algorithms have been proposed [9], with the prime development objective being to increase performance and robustness of system regulation. In the development of such a control strategy, the choice of the system (sensor in this research) model is of paramount importance [10, 11]. A common practice is to describe the system by a linear representation which might reflect the pertinent dynamics around the expected operating region. Many sensing systems designs are based on this approach for dimensioning their closed loop components parameters. An example here is the PID contolled accelerometer presented in [12]. If the system dynamics are relatively linear, then the use of a linear model based control algorithm may lead to acceptable performance. However, in situations where the system is highly non-linear, the local linearity assumption could be detrimental to the overall control system robustness [13]. In this situation, if a representative non-linear system (sensor) model could be developed, then the need to make the compromise between robustness and loss of system performance would be reduced. However, the development of accurate non-linear mechanistic models for sensors can be an extremely time consuming task. A generic modelling technique which provides for rapid non-linear model formulation, whilst also allowing the capture of essential sensor characteristics, would therefore be an exceedingly valuable tool. If the claims concerning the powerful non-linear modelling capabilities of artificial neural networks are to be believed, [4, 10, 14, 5] then, this technology could provide such a tool and through the enhanced models generated, provide a means by which to achieve sensor system improvements.

7.1.3. *Network Types*

There are two general structures of neural networks used for identification and control tasks:

Static (feedforward)networks:
The input and intermediate signals are always propagated forwards. The flow of information is directed towards the output and no returning paths are allowed. Static networks are widely used for pattern recognition/classification and also for image processing. Another area is approximation theory, where the problem may be stated as follows: given two signals $x(t)$ and $y(t)$ find a 'suitable' approximation of the functional relation between the two. This aspect is of interest in the context of the present chapter. Although the static approximation capabilities of feedforward networks are of limited use for dynamic system identification, they can be successfully used for control. A static network can be used as a non-linear gain state feedback controller for example, within the design of closed-loop sensor systems. An accelerometer design based on this idea is presented in the Section 7.3.

Dynamic networks:
Either states or output are fed back. The signals are re-used, thus their current values are influenced by the past ones [14]. These networks are qualitatively different from static ones, as their structure incorporates feedback.

They are universal, parametric, non-linear dynamic systems of possible use as identifiers or dynamic controllers [10, 5, 15]. One way of introducing dynamics in the network is the time series approach which makes use of time histories of the data as network inputs by using Tap-Delayed-Lines (TDL). When applied to multilayer perceptron (MLP) type networks, this procedure is called "backpropagation through time" [4, 5]. An alternative philosophy is to extend the standard neuron processing function and incorporate dynamic effects, i.e. in addition to the sigmoidal processing, the neurons (or the transmission between neurons) can be given dynamic characteristics [16, 17]. Both approaches are effective, and the choice of one over the other depends upon the characteristics of the system under consideration. The time series approach results in significantly larger networks than the dynamic neuron method, especially where there is uncertainty over the system time delay. As a result, a greater number of parameters must be determined. This has consequences on training time and may result in problems due to over-parametrisation. The dynamic neuron approach results in a network with less parameters, however, problems arise when attempting to model systems with varying 'time constants' [18].

7.1.4. *Neuro-identification, Control Strategies and Structures*

In this context, neural networks are viewed as a system modelling formalism or even a knowledge representation framework; the knowledge about the system dynamics and mapping characteristics is implicitly stored within the network. In the same way that transfer functions provide a generic representation for linear black box models, artificial neural networks potentially provide a generic representation for non-linear black box models [17, 19]. The ability of networks to approximate non-linear mappings is thus of central importance in this task.

System identification

The process consists of constructing a suitably parameterised identification model and adjusting the parameters of the model to optimise a performance function based on the error between the real system and the identification model output. The structure of the identification model is chosen to be identical to that of the system, as far as the system order is concerned

[5, 14]. Both problems of forward and inverse modelling are presented below, as they are at the basis of most neuro-control algorithms.

Forward modelling is the procedure of training a neural network to represent the forward dynamics of a system. The problem of identification can be formulated as follows: the input and output of a time-invariant, casual discrete-time dynamical system are $a_s(.)$ and $x_s(.)$, respectively, where $a_s(.)$ is a uniformly bounded function of time. The system is assumed to be stable with a known parametrisation but unknown values of the parameters. The objective is to construct a suitable identification model, which when subjected to the same input $a_s(k)$ as the system, produces an output $\hat{x}_s(k)$ which approximates $x_s(k)$ in the sense described by Narendra and Parthasarathy [5]:

$$\|\hat{x}_s - x_s\| = \|\hat{S}_{sys}(a_s) - S_{sys}(a_s)\| \leq \varepsilon, \quad a_s \in A_s, \tag{7.1}$$

where $\|\|$ denotes a suitably defined norm, $\varepsilon > 0$ is the identification error, $S_{sys}(a_s)$ and $\hat{S}_{sys}(a_s)$ represent the actual system and the identification model, respectively and A_s is the input space.

A structure for forward modelling is shown schematically in Figure 7.1, where a_s is the input of the non-linear system, d' represents the disturbances acting on the system output, e_i is the identification error, x'_s is output of the non-linear system including the effect of disturbances, x_s is output of the non-linear system, and x_m is the identification model output.

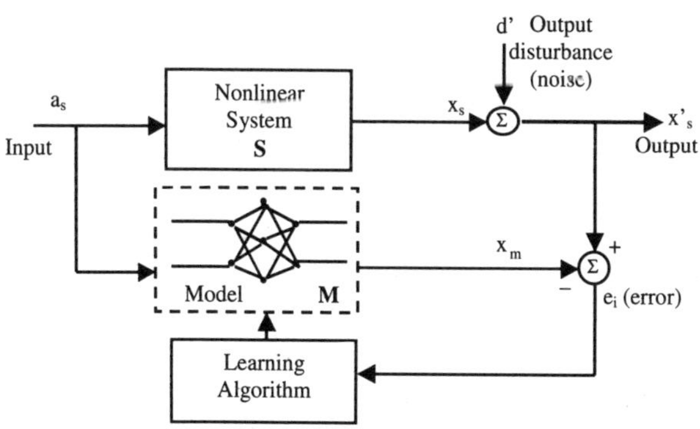

Figure 7.1: System identification structure.

The neural network model is placed in parallel with the system and the error (e_i) between the system and network outputs is used as the network training signal. This learning structure is a classical supervised learning problem where the teacher (i.e. the system/sensor under investigation) provides target values (i.e. its outputs) directly in the output co-ordinate system of the learner (i.e. the network model).

Narendra and Parthasarathy proposed four non-linear system models, for the purpose of identification (for the representation of discrete-time systems), of which, the most general (and least tractable) is given by:

$$\begin{aligned} x_s(k+1) = f[&a_s(k) + a_s(k-1), \ldots, a_s(k-m+1); \\ &x_s(k), x_s(k-1), \ldots, x_s(k-n+1)], \end{aligned} \tag{7.2}$$

where $[a_s(k), x_s(k)]$ represents the input-output pair of the single-input-single-output (SISO) system at time k, and $m \leq n$.

Under fairly weak conditions on the function f, multilayer neural networks can be constructed to approximate mappings such as those in Equation (7.2) over compact sets. Weight matrices of the neural networks in the model are assumed to exist so that, for the same initial conditions, both system and model have the same output for any specified input. Hence, the identification procedure consists of adjusting the parameters of the neural network in the model. The training algorithm is based on the error between the system and model outputs. However, care must be taken to ensure that the procedure results in convergence of the identification model parameters to the desired values.

Determining an identification model corresponding to a system represented by (7.2) could be done in several ways, of which the series-parallel architecture is presented in Figure 7.2. 'TDL' in Figure 7.2 denotes a tap-delayed line whose output vector has as its elements the delayed values of the input signal. Hence the past values of the input and output of the system form the input vector to a neural network whose output $\hat{x}_s(k)$ corresponds to the estimate of the system output at any instant of time k. Since there are no feedback loops in the model, static back error propagation (BKP) can be used to adjust the network parameters, substantially reducing the computational overhead of training dynamic networks. A common choice in building the network training set with the purpose of identification is perturbing the system with white noise, Gaussian centred [4, 5]. The noise

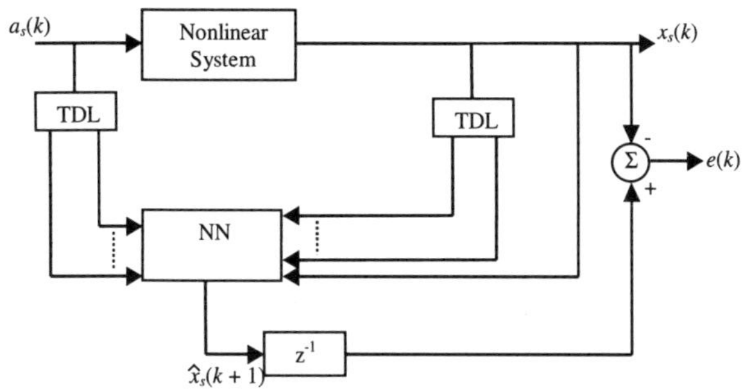

Figure 7.2: Identification of non-linear systems using tapped delayed lines. From [5].

should cover the whole dynamic range of the system and should be subsequently scaled to a unit standard deviation, to form the identification network input.

Inverse modelling

Conceptually, the simplest approach to inverse modelling is presented in Figure 7.3 (this structure is sometimes referred to as generalised inverse learning). Here, a synthetic training signal is presented to the system. The system output is then used as input to the network ($a_r = x_s$). The network output is compared with the training signal (the system input) and this error is used to train the network. This structure will clearly tend to force

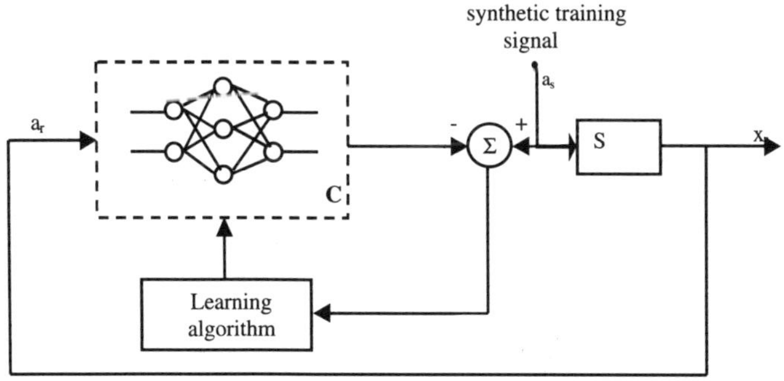

Figure 7.3: Direct inverse modelling (generalised inverse learning). (S = the actual system; C = the controller).

the network to represent the inverse of the system. However, there are drawbacks to this approach [14, 4] which can be overcome by specialised inverse learning.

Direct inverse control

Direct inverse control utilises an inverse system model [20, 21]. Specialised inverse learning architectures are commonly used for building the inverse system model (Figure 7.4). In this approach, the network inverse model precedes the system and receives as input a training signal, which spans the desired operational output space of the controlled system. This learning structure also contains a trained forward model of the system placed in parallel with the actual system. The error signal for the training algorithm is the difference between the training signal (a_r) and the system output. In the case of noisy systems, the error signal may also be the difference between the training signal and the forward model output. This obviates the need for the real system in the training procedure which is important in situations where using the real system is not viable.

Only the inverse network weights are adjusted during the training procedure. Thus, the procedure is effectively directed at learning an identity mapping across the inverse model and the forward model; the inverse model is learned as a side effect.

Although this type of control is common in robotics applications, for general-purpose use, serious questions arise regarding its robustness. The

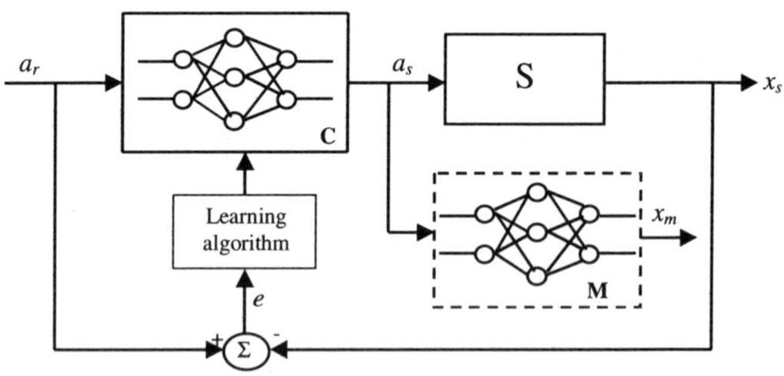

Figure 7.4: Specialised learning method for inverse identification (C = the controller, M = the system model, S = the actual system).

lack of robustness can be attributed primarily to the absence of feedback. The problem can be overcome to some extent by using on-line learning, where the parameters of the inverse model are continuously adjusted on-line [4].

A number of assumptions have been made throughout the section regarding the system to be controlled in order for the methods presented to prove successful. These include stability properties [22], observability, controllability and identifiability of the models suggested as well as the existence of non-linear controllers to match the response of the reference model. At the present stage of development of the non-linear control theory, few methods exist for checking the validity of these assumptions in the context of general non-linear systems [5]. However, the large number of publications considered by the authors proves that a variety of structures have already been successfully used to identify and control various industrial processes and/or engineering systems. All configuration in this section have been used in conjunction with micromachined accelerometers and are detailed in the following sections.

7.2. Open Loop, Neural Transducer Prototype for Static/Low Frequency Applications

This section describes an open-loop, compensated acceleration transducer prototype, developed by one of the authors. Direct inverse control was chosen as a method of compensation and implemented using a novel type of dynamic neural network. The design procedure was based on measured data and validated initially in simulation, followed by the implementation of a hardware prototype, aimed at static and low frequency applications. Ways of further improving the performance of the smart transducer through additional functions, by means of adaptive control are also speculatively considered here.

7.2.1. *Static Measurements for Acceleration Sensors*

The static behaviour of the micromachined accelerometers considered here was characterised by mounting them on a dividing head (angular resolution of 1°) (Figure 7.5) and rotating them in the gravitational field. Several sensors were measured. It was found that all sensors exhibited a non-linear

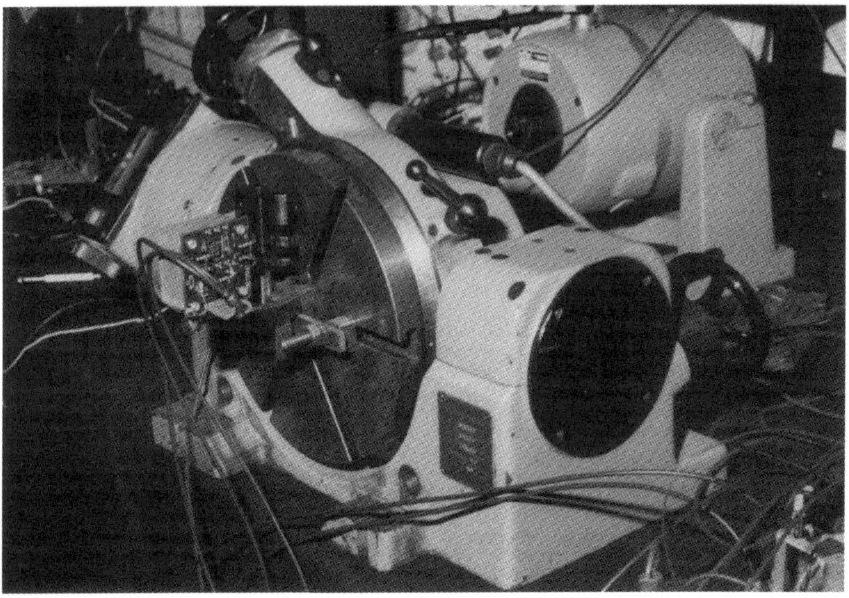

Figure 7.5: Experimental set-up for static measurements with a dividing head.

behaviour and their sensitivity varied quite drastically from one to another. Two 'articulated' sensing elements, with a solid seismic mass and identical manufacturer specifications were randomly selected from the available set of devices and 73 measurements were taken for each device, for a complete 360° rotation. The static characteristics of the two sensors are depicted in Figures 7.6 (Sensor 1) and 7.7 (Sensor 2), respectively.

Over the acceleration range tested, both sensor characteristics exhibited mainly offset and hysteresis. For Sensor 1, for example the average offset was 270 mV (offset error of 61% for the 440.5 mV/G device sensitivity) and the hysteresis was 60 mV (13.6% error over the ±1 G range). The measured offset due to the pick-off electronics was only 6 mV, which is negligible compared to that of the sensing element itself. For Sensor 2, the offset was 230% and the hysteresis error was 15.7%. The sensitivity of the device was calculated as 1.4 V/G.

The magnitude of the above errors were calculated here as:

$$\frac{(V_{out\_actual} - V_{out\_ideal}\,[V])/S\,[V/G]}{|a_{mes\_range}[g]|},$$

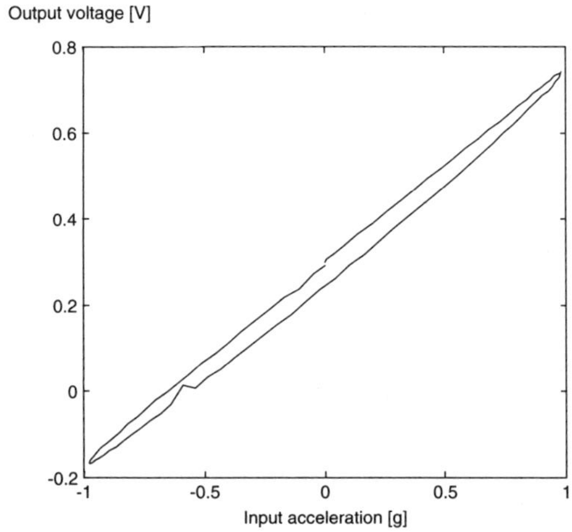

Figure 7.6: Sensor 1. Measured static characteristic.

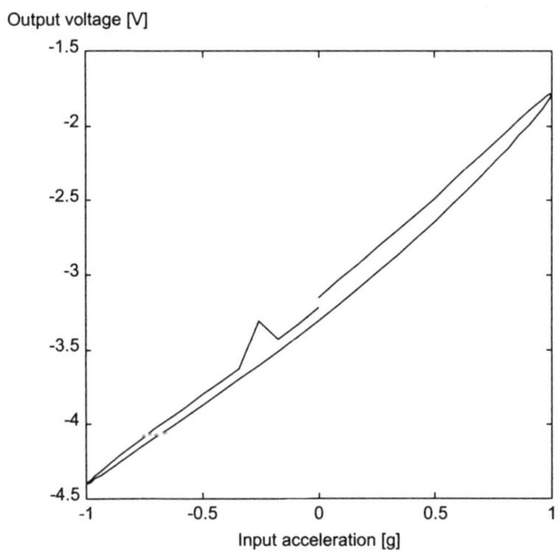

Figure 7.7: Sensor 2. Measured static characteristic.

where V_{out_actual} is the actual, measured value of the transducer output at the point where the error is estimated, V_{out_ideal} is the ideal output of the transducer at the same point, S is the sensitivity of the tranducer (defined as the swing of the actual output voltage over the acceleration measurement range, divided by the acceleration measurement range) and a_{mes_range} is the acceleration measurement range.

Although manufactured to have identical parameters and performance, the two sensors differed greatly, in terms of offset and hysteresis. Moreover, the most important parameter of a sensor, its sensitivity, varied by more than 300% from one sensor to the other. Hence, these types of devices are in great need for calibration and subsequent compensation.

7.2.2. Open-loop Compensation of Micromachined Accelerometers for Static-low Frequency Applications

A direct inverse control strategy was adopted in order to correct the behaviour of the sensing elements above. Although the chosen control strategy required only the production of the sensor inverse model, forward modelling of the sensor was also necessary, in order to be able to validate by simulation the compensation procedure. Due to the presence of hysteresis, a dynamic type of neural network was needed for modelling both the inverse and forward characteristics of the sensor. As only the static behaviour of the sensor is considered here, the well-established method of using tap-delayed-lines (TDL) dynamic networks [17] was not suitable. The TDL approach embeds the sampling period/delay time within the network structure and noise training is therefore necessary in order to build suitable sensor models. In fact, this means modelling the dynamic behaviour of both the sensor and its inverse. Although this approach extends the use of the compensated system over the whole dynamic range of the 'off-the-shelf' sensor, the effort involved in gathering the data sets through dynamic measurements, training the forward and inverse networks and implementing the sampled data system in hardware is not justified for an open-loop transducer design. The aim here is to develop a straightforward calibration and compensation procedure leading to a simple implementation, specifically for static and low frequency applications.

Following on the reasoning above, a novel approach to the design of neural networks able to identify and compensate history dependent non-linearities such as those exhibited by the sensor was needed. Consequently, a new dynamic network type was specifically developed and used for both the direct and inverse models. The networks are of a MLP type, with two inputs, a single output and two layers of hidden neurons. The novelty consists in using a 'flag' in order to account for the one-step-back history of the

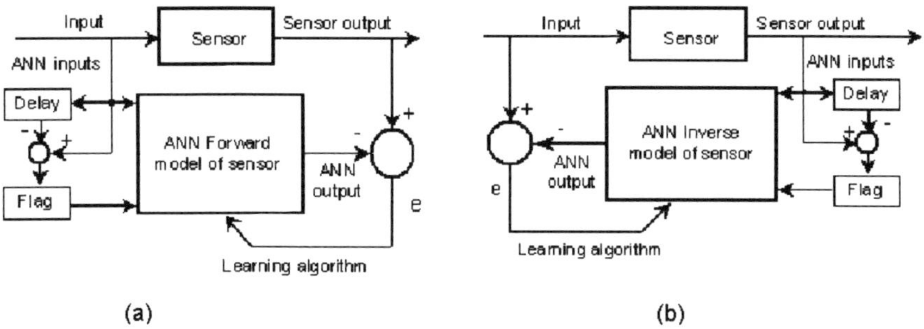

Figure 7.8: (a) ANN training for forward modelling. (b) ANN training for inverse modelling.

signal to be processed by the network, as opposed to the TDL approach. Hence, one network input is the current value of the input signal, whilst the other is the 'flag' whose value depends on the evolution of the input signal. The 'flag' takes arbitrarily chosen values of 0.99 if the current input is greater than or equal to its previous value and -0.99 if it is less. Figure 7.8 shows the block diagrams of the training schemes used for the forward and inverse neural models. The 'delay' blocks in these diagrams signify that the last measurement value taken from the sensor is to be compared with the current measurement value.

The initial approach to constructing the training sets was to record the readings directly from the sensor and then manually assign the appropriate 'flag' value. Networks were subsequently trained, using the Matlab environment, to approximate the direct and inverse sensor characteristics. A dynamic error-back-propagation training algorithm was used, which included both a variable learning rate and momentum term [23, 24]. The performance of the inverse networks on the training sets for Sensor 1 and Sensor 2 are shown in Figures 7.9 and 7.10, respectively.

The inverse network for Sensor 1 consisted of $2 \times 9 \times 5 \times 1$ neurons (layer by layer, starting with the input- *Input* $\times$ *Hidden Layer 1* $\times$ *Hidden layer 2* $\times$ *Output*) and was trained to a sum-squared-error (SSE) of 0.05 over 73 samples in approximately 20 000 epochs. Similarly, for Sensor 2, an inverse network of $2 \times 11 \times 7 \times 1$ neurons and was trained to a SSE of 0.08 in approximately 14 000 epochs.

The direct networks performance on the training sets for the two sensors are shown in Figures 7.11 and 7.12. SSEs of 0.003 and 0.0019, respectively were obtained over the 73 samples training set.

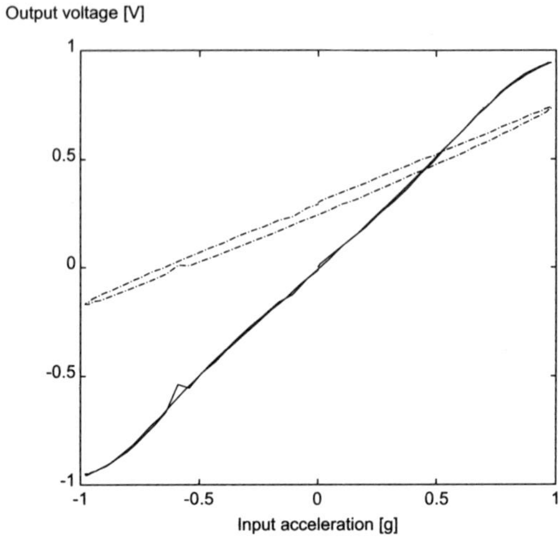

Figure 7.9: Sensor 1. Static characteristic of sensor (dotted line) and inverse network performance on the training set (solid line).

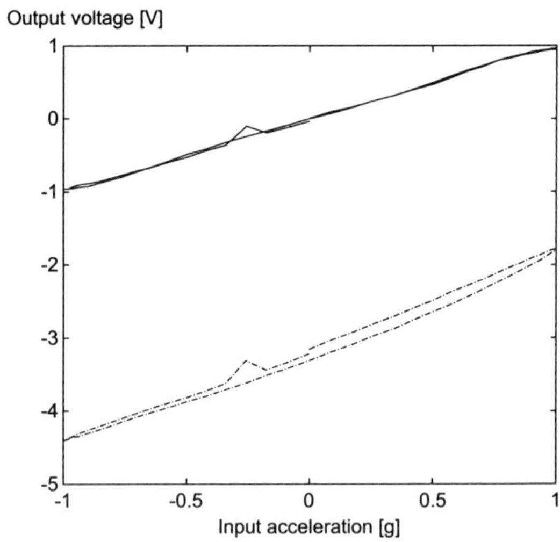

Figure 7.10: Sensor 2. Static characteristic of sensor (dotted line) and inverse network performance on the training set (solid line).

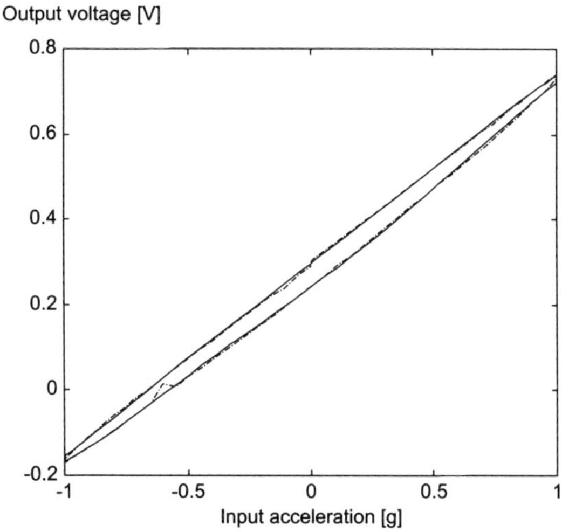

Figure 7.11: Sensor 1. Direct network performance on the training set (dotted line); Measured sensor characteristic (solid line).

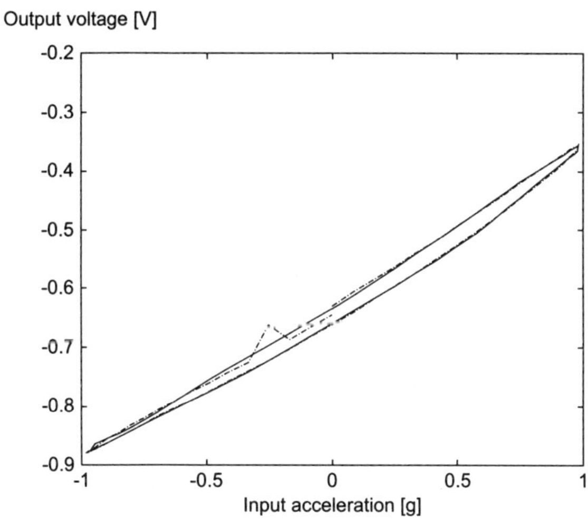

Figure 7.12: Sensor 2. Direct network performance on the training set (dotted line); Measured sensor characteristic (solid line).

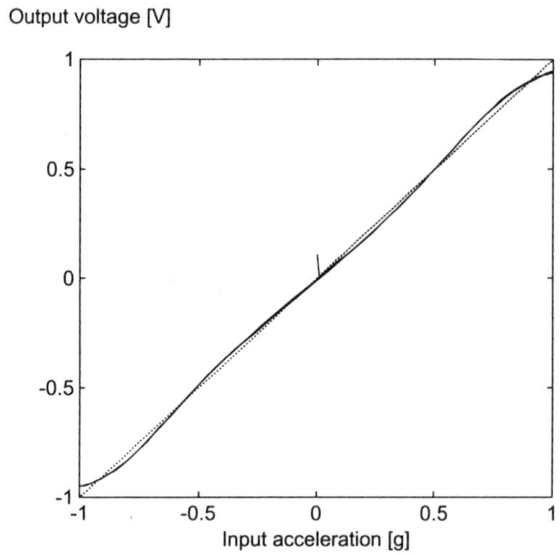

Figure 7.13: Sensor 1. Simulated transducer performance on the test set (solid line); Ideal (desired) transfer characteristic (dotted line).

Assessing the performance of the overall compensated system in simulation implied cascading the direct and inverse neural network models and applying excitation accelerations corresponding to a full rotation of the sensor in the gravitational field, at $\theta = 3°$ steps. Hence, the test set contained 121 points, with only 24 of them having been part of the training set. The transfer characteristics obtained for the two simulated compensated transducers (based on Sensor 1 and Sensor 2) are presented in Figures 7.13 and 7.14, respectively. The dotted lines in these figures represent the ideal behaviour of the transducers (i.e. identical mapping between the input acceleration and the transducer's output voltage). Both offset and the hysteresis have been almost entirely compensated. The functionality of the two 'off-the-shelf' sensors has therefore been significantly improved, although a slight departure from linearity remains. The linearity errors will be quantified in the next section, for the prototype transducer implemented in hardware.

7.2.3. *The Open-loop Transducer Prototype*

The success of the procedure described above encouraged the implementation of the compensated transducer as an embedded system with the

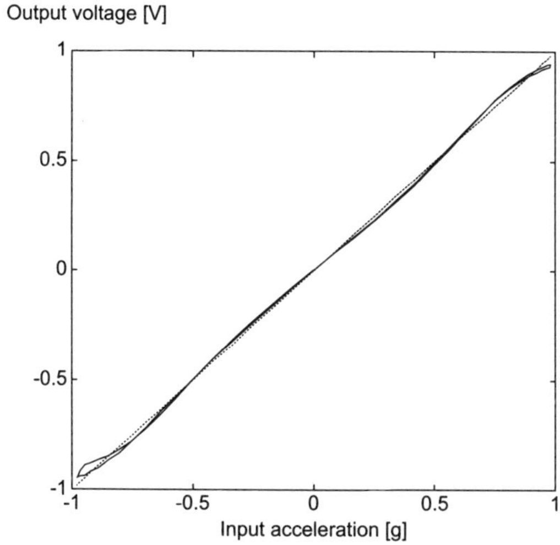

Figure 7.14: Sensor 2. Simulated transducer performance on the test set (full line); Ideal (desired) transfer characteristic (dotted line).

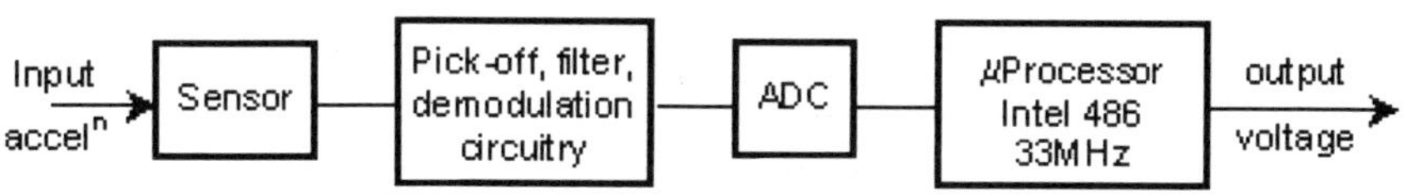

Figure 7.15: Block diagram of the smart transducer hardware implementation.

neural processing being supported by a remote Intel 486 microprocessor as in Figure 7.15. The analogue to digital conversion was performed by a ZN427 8-bit, successive approximation converter, with a clock frequency of 1 MHz and 10 μs conversion time, 39 mV/bit resolution and ±1 bit accuracy. An input/output (I/O) board was designed to interface the ADC with the microprocessor.

A program was developed to read the input from the ADC, filter and 'flag' the data and perform the neural processing. The filtering was performed for each individual input acceleration by storing 100 readings, calculating the mean and standard deviation (σ), rejecting those readings which fell outside of the range $\pm 1\sigma$ and then recalculating the average. This value, along with the appropriate 'flag' value was then processed through the

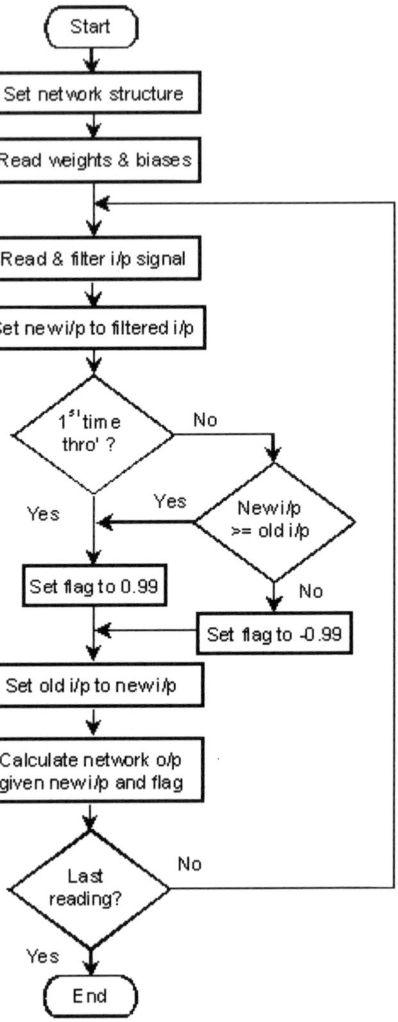

Figure 7.16: Flow chart: software filtering and neural processing.

inverse network. The inputs and output were then saved to file. The flow chart representing the operation of the program is shown in Figure 7.16.

Testing the hardware implementation of the smart transducer requires the parameters of the trained network representing the inverse model of the sensor to be loaded into the program. Although the performance of the

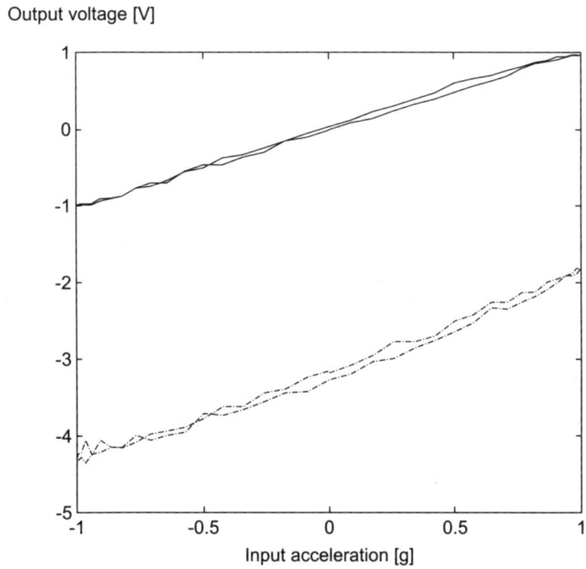

Figure 7.17: Sensor 2 characteristic after analogue-to-digital conversion (dotted line); Measured smart transducer characteristic (full line).

compensating networks presented in Section 7.2.2 were very good in simulation, the hardware test results proved rather unsatisfactory, due to the noise introduced by the ADC. In order to reduce its effect on the overall system performance, a new approach to gathering the network training data was adopted, and the results are presented here for the transducer based on Sensor 2. The training data were collected after the analogue to digital conversion had taken place and the inverse of this latter characteristic (represented by the dotted line in Figure 7.17) was reproduced by the network. The measured transfer characteristic of the smart transducer obtained is shown in Figure 7.17 with the full line.

The offset was completely compensated while the hysteresis has been reduced to 70 mV (equivalent to 70 mG). This residual hysteresis is due to a combination of effects, the training error of the network and the limited accuracy of the ADC being the main ones. Moreover, it was noticed that Sensor 2 did not have good repeatability properties (measurements taken during the same day had variations of up to 2%, due to changes in temperature and humidity levels). Repeatability deficiencies cannot be compensated for using the procedure described above; instead, an adaptive control strategy (involving end-user recalibrating ability) is necessary. Such an adaptive control procedure is proposed in Section 7.2.4.

With regard to the errors arising from the ADC, improved performance for the compensated transducer can be obtained by modifying the converter design. Redesigning the converter such that its input range is [−4 V; −1 V] rather than ±5 V, would lead to an increase in the converter accuracy by a factor of 3. Such an ADC was built and a conversion accuracy of approximately ±17 mV was obtained. The measured characteristic of Sensor 2, after analogue to digital conversion, using the new converter (and without any network re-training) is presented in Figure 7.18(a) and the compensated smart transducer characteristic is shown in Figure 7.18(b) (solid line). The dotted line in Figure 7.18(b) represents the ideal sensor behaviour.

The error between the 'off-the-shelf' sensor output and the ideal output is presented in Figure 7.19(a) and that between the compensated sensor output and the ideal output in Figure 7.19(b).

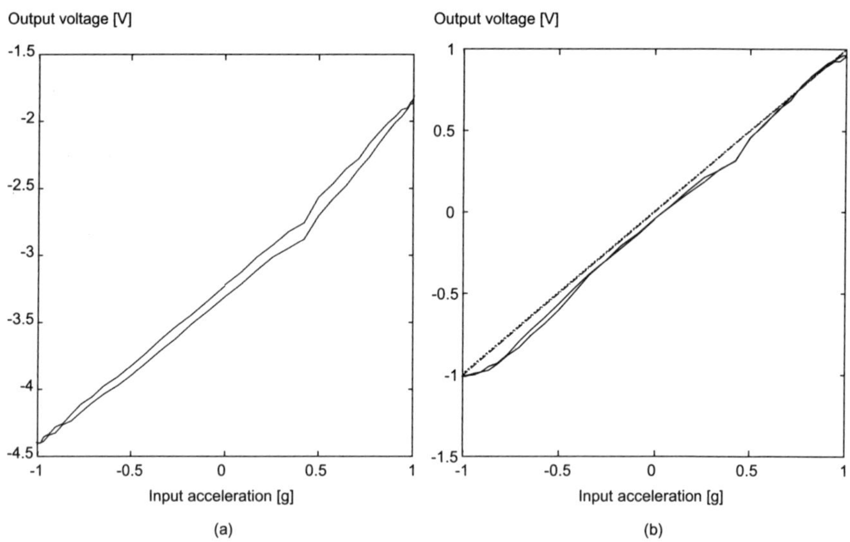

Figure 7.18: (a) Sensor 2. Static characteristic measured using the modified ADC. (b) Sensor 2. Measured static characteristic of the final smart transducer (solid line); Ideal transducer characteristic (dotted line).

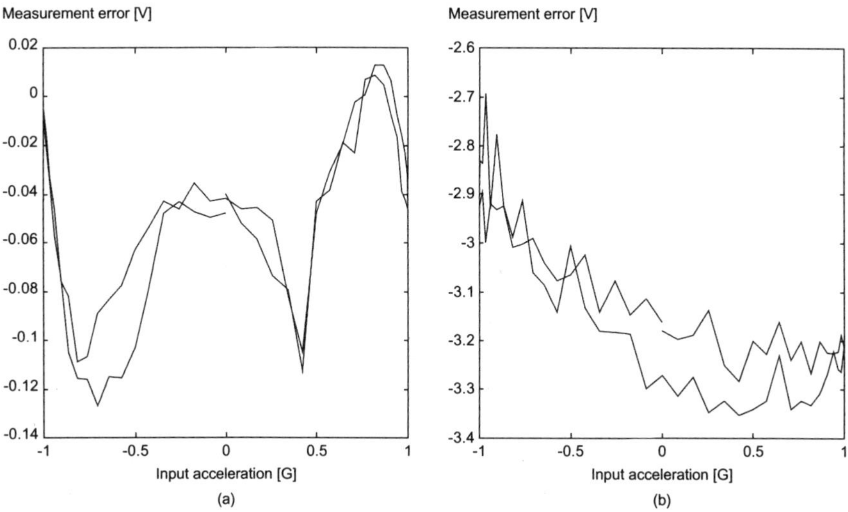

Figure 7.19: (a) Sensor 2. Error between the measured compensated sensor output and the ideal, desired response (exhibiting different values of error for the same input acceleration, due to the effects of hysteresis). (b) Sensor 2. Error between the measured "off-the-shelf" sensor output and the ideal, desired response (exhibiting different values of error for the same input acceleration, due to the effects of hysteresis).

The smart transducer prototype, in its final version, exhibited a maximum positive error of 1.2%, a maximum negative error of 12.5%, an offset error of 4.5% and hysteresis error of 4% over the ± 1 G range. The 'off-the-shelf' sensor used had an offset error of 230% and a hysteresis error of 14.3%, a maximum positive error of over 100% and a maximum negative error of over 200%. The compensated transducer errors are still large, due to the unusually large initial figures for both offset and hysteresis. The error reduction factors were of 3.5 for hysteresis and 45 for the offset. Finer compensation could have been achieved if the initial range of errors was smaller or, by creating larger compensation networks.

It could be argued that the offset could have been compensated using simple linear techniques (with a summing amplifier following the pick-off circuit). Assuming this was the case, the hysteresis would still remain at 14.3% and the maximum positive and negative error would still be up to 33%.

Hence, the prototype has been successful in its task of compensating the static characteristic of the sensor. Only the time-invariant non-linearities have been compensated for. Correction of the time-variant non-linearities, arising from the lack of repeatability in the sensor behaviour and drift of parameters over time involves an adaptive control procedure which is discussed next.

7.2.4. On-line Adaptive Open-loop Control of Accelerometers

In a context in which the repeatability and drift problems can be identified and the means for end-user re-calibration are available, adaptive compensation procedures could be applied. Addressing random errors such as those discussed below through self-test and auto-calibration is an area of current research, already mentioned in Chapter 4. No generic solutions are available yet, so the procedures described here are, as many other, mere feasibility studies.

The process of compensating for in-use random errors proposed here is presented as a flow chart in Figure 7.20, within the scenario of offset and hysteresis compensation. Note that the procedure cannot be applied *in situ*. The sensor has to be removed from its environment and involves the user having access to a measurement set-up (a dividing head for example). The adaptive procedure actually involves going through the "Read i/p signal-Filter-Associate flag" loop of the flow chart in Figure 7.16, initially with the purpose of forming the training set for the neural network. Once the training set is complete, the training of the network could start (the Matlab based training used in the previous section being replaced by its embedded software counterpart). The training ends when the SSE reaches the preset value imposed by the user and the network weights and biases are saved. At this point, the adaptively-compensated sensor is back in fully functional mode. The sensor output "reading" procedure, after the network training ended is identical to the procedure described in Figure 7.16.

Although not fully worked out here, theoretically, due to the structure of the sensing element it is possible to design an *in situ* re-calibration followed by adaptive compensation procedure for this particular type of accelerometers, as the sensor communicates with the outside world through three electrodes.

Voltages could be applied to the top and bottom electrodes respectively, in order to emulate different acceleration magnitudes (voltages applied to

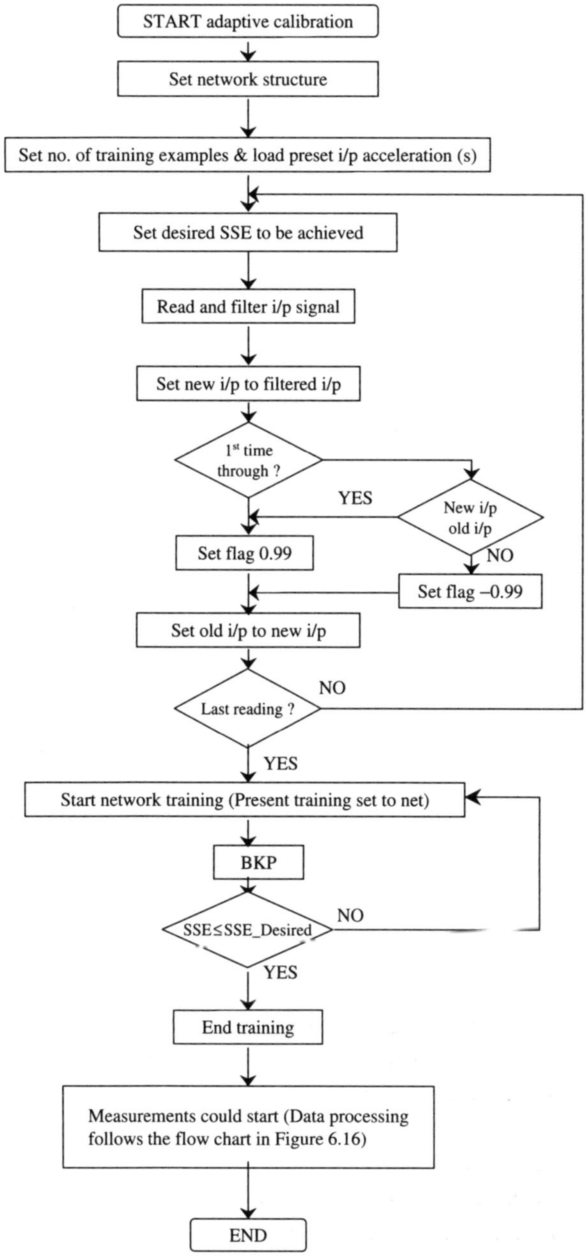

Figure 7.20: Flow chart: Adaptive control procedure.

the top electrode would emulate positive accelerations and those applied to the bottom electrode would emulate negative accelerations). The relationship between voltage and acceleration can be deduced quite simply from the electrostatic force expression, as both an acceleration and a corresponding voltage have to generate the same electrostatic force on the seismic mass:

$$F_{el} = \frac{1}{2}\varepsilon_0\varepsilon_r A\frac{V^2}{d_0^2} = ma, \tag{7.3}$$

where A is the area of the plates, d_0 is the distance from the seismic mass to each plate, m is the mass of the seismic mass, a is the acceleration to be emulated and V is the voltage to be applied to the corresponding electrode.

The relationship between the acceleration and the self-calibrating voltages is:

$$V = d_0\sqrt{\frac{2ma}{\varepsilon_0\varepsilon_r A}}. \tag{7.4}$$

An input acceleration to top/bottom electrodes DC voltage look-up table can be established, with the reservation that the manufacturer specification for the distance between the electrodes was used to establish the acceleration-voltage relationship. In practice, the real value for d_0 is not precisely known, this being the origin of the offset. Therefore, look-up tables could be useful only if the sensor does not exhibit offset and the values for area of the plates and the mass of the seismic mass are accurate. Provided that such a look-up table is delivered together with the sensor, the recalibration procedure at the user end would involve successively applying around 50 DC voltage levels to the outer electrodes, as opposed to using a dividing head for the formation of the training set. The output of the sensor after analogue to digital conversion and filtering is automatically recorded by the adaptive program described above, the training set for the network established, and the training/compensation procedure starts.

To conclude, the compensation of both time invariant and time variant effects was based on a direct inverse control strategy and implemented with 'flag' networks. The approach taken has been successful and uncomplicated for this particular problem. For applications where accuracy and linearity is needed over a larger dynamic range and at higher frequencies, the formation of training set, gathering of measurement data and training of the networks may not, however, be straightforward. Also, it may be noted that the system is not robust to the incidence of extraneous disturbances, due to the open-loop nature of the control system. Moreover, exposure to accelerations greater than a threshold value causes irreversible latch-up of the seismic mass to one or other of the outer electrodes, as detailed in Chapter 5. One way of increasing the system robustness and stability is to apply some form of feedback. This approach has been considered at simulation level in the next section.

7.3. Closed-loop Neural Network Controlled Accelerometer

As already demonstrated in Chapter 5, negative feedback can be used to increase sensors' linearity, bandwidth and dynamic range. The starting point in this section is the mathematical model of a conventional closed-loop accelerometer which uses a PID based control strategy. Figure 7.21 shows the block diagram of the transducer, comprising of the mechanical sensing element (the model of which includes the effect of non-linear damping), conversion from the mechanical to the electrical domain, the PI

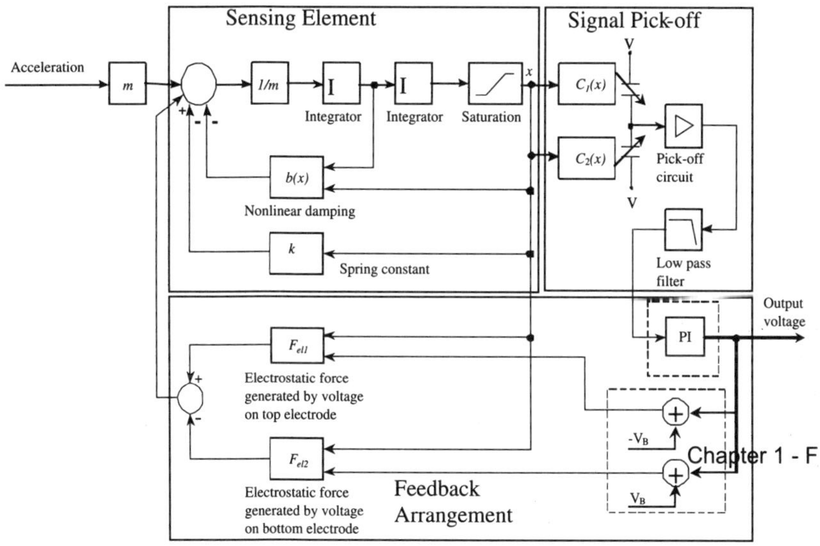

Figure 7.21: Block diagram illustrating the mathematical model of closed-loop accelerometer (FNN is the Feedback Neural Network and CNN is the Controller Neural Network).

controller and the conversion of the controller output voltage to an electrostatic force [25, 26]. The method of reset (the feedback effort necessary to bring the seismic mass back to its rest position in the central position between the top and bottom electrodes) involves separating the sensing and feedback signals in the frequency domain. Consequently, the feedback voltage is applied continuously to the top and bottom electrodes. This feedback arrangement, taken in conjunction with the non-linear nature of electrostatic forces, requires an undesirable high bias voltage V_B to be applied to the outer plates, in order to linearise the feedback signal [12]. The resultant electrostatic force on the seismic mass is the superposition of two forces generated by potentials applied to the top and bottom electrodes respectively. Mathematical analysis reveals that this approach results in a desired linear relationship between the controller output voltage and the net force on the mass under the condition that the mass motion is restricted to maximum 20% of the distance to either electrode. Whilst this is the case under normal operation, certain conditions (such as large input accelerations, shocks in acceleration and unknown/offset mass position at power up) may arise in which this constraint is violated. This leads to an irreversible electrostatic latch-up (lock-up) of the mass to one electrode. Normal operation can be only recovered by switching-off the power supply to the sensor, option which is unacceptable for many applications. Despite this, the above approach was used in many devices described in the literature [27, 28] since it nevertheless improves the sensor performance compared to its open-loop, 'off-the-shelf' operation.

For very high integrity systems however, a more robust design is needed. The method proposed here aims to improve the transducer performance through the use of Artificial Neural Networks both for replacing the PI controller and generating the feedback electrostatic forces (Figure 7.21). Two modular Neural Networks have been designed and trained using the error-back-propagation algorithm [29].

The compensating neural network (CNN) performs a static mapping, replicating only partially the behaviour of the PI controller.

The feedback neural network (FNN) has two functions:

- Firstly, it calculates the square root of the output voltage, in order to obtain a truly linear feedback relationship between the system output and the electrostatic forces acting on the electrodes;

- Secondly, the network demodulates the output signal in order to apply the feedback to only one electrode at a time: the bottom electrode will be activated if the proof mass has moved towards the top electrode and vice-versa (this new reset concept theoretically eliminates the possibility of the feedback becoming positive).

The two neural networks are subsequently included in a novel closed-loop transducer design.

7.3.1. *The Feedback Linearisation Procedure*

The network used for this purpose is of MLP type, has one input (the output voltage of the transducer), one hidden layer and two outputs, connected to the outer electrodes of the sensing element. Both the hidden and the output neurons are governed by a sigmoid-type transfer function, with bias. The network was trained to approximate the following input-output function:

$$
\begin{aligned}
output_1 &= \begin{cases} \sqrt{input} & \text{if } input > 0 \\ 0 & \text{if } input \leq 0 \end{cases} \\
output_2 &= \begin{cases} \sqrt{-input} & \text{if } input < 0 \\ 0 & \text{if } input \geq 0 \end{cases},
\end{aligned}
\tag{7.5}
$$

with *input* being scaled to the range ± 1. The desired mapping and the approximation performed by a trained network are shown in Figure 7.22(a) and (b), respectively.

Some scaling was necessary in order to integrate the trained network into the closed-loop transducer structure: the output of the transducer (which is the input to the feedback network) was divided by 100 and the outputs of the network were multiplied by 10 (in view of the required square root law). The testing and validation of the network performance was done using the behavioural description of the whole closed-loop transducer in the SPICE environment. Figure 7.23 shows the SPICE behavioural description of the neural network, with the "Function" blocks obeying Equation (7.6) and the weights being represented by "Gain" blocks.

$$
Function\_out = \left(\frac{2}{1 + \exp(-2(Function\_in) + bias)} \right) - 1. \tag{7.6}
$$

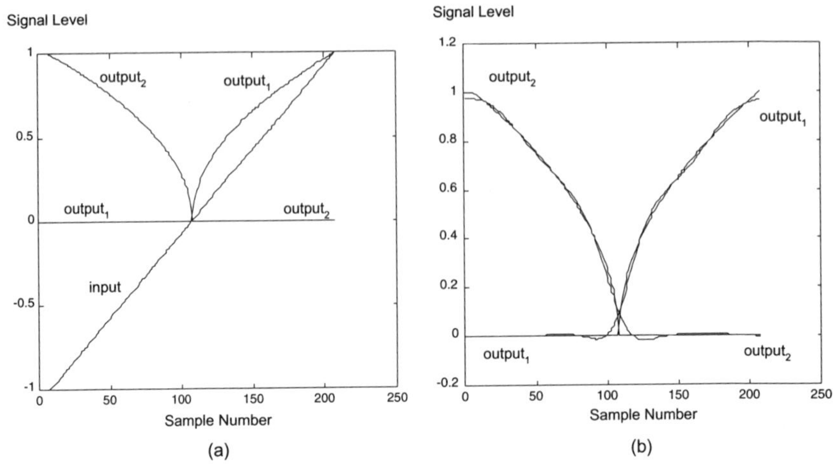

Figure 7.22: (a) The FNN inputs and desired outputs. (b) The approximation performed by the trained network.

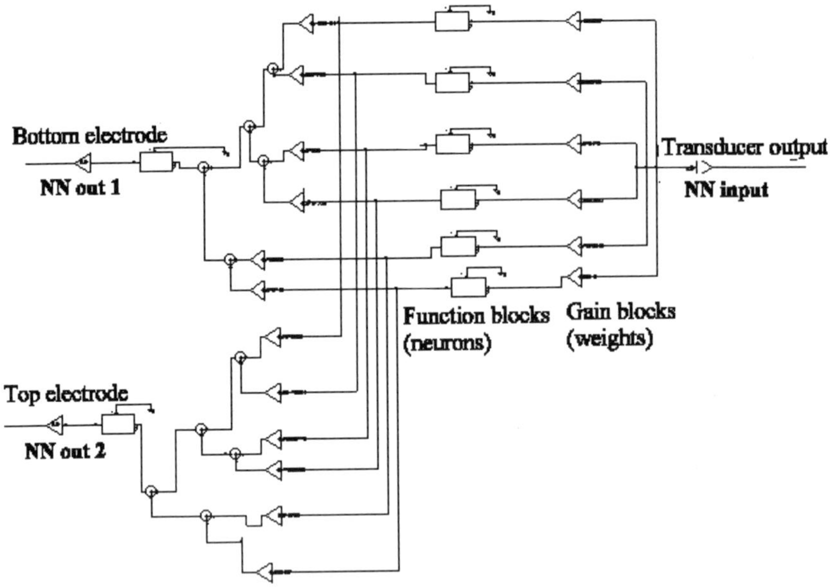

Figure 7.23: Neural network behavioural description in SPICE.

The behaviour of the conventional and FNN transducer was studied by subjecting both systems to different types of acceleration inputs. Comparative simulation results for the two transducers, for a sine wave input acceleration with a frequency of 1 Hz and an amplitude of 7 G are shown in Figures 7.24 and 7.25 respectively. The scale ratio for the input acceleration and the mass displacement in the SPICE simulations is 1 G ≡ 10 V and 1 uV ≡ 1 μm, respectively. Figure 7.24 shows that, for large accelerations, the conventional closed-loop transducer is driven into a latch-up state, where the seismic mass becomes attached to one of the electrodes. Conversely, the results in Figure 7.25 indicate an approximately linear performance of the neural transducer for this type of input acceleration.

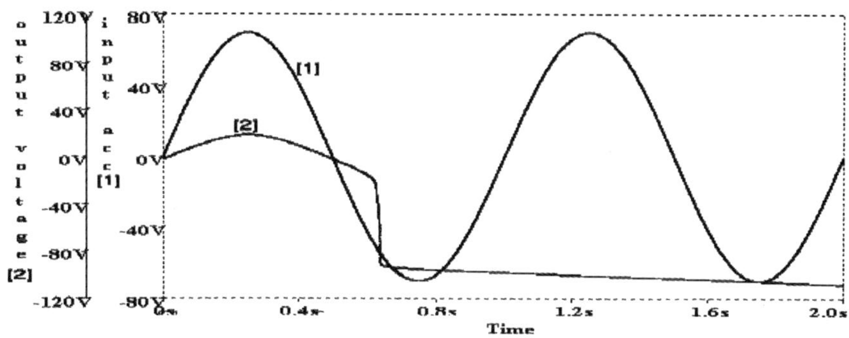

Figure 7.24: Input acceleration and output voltage for the conventional transducer (1 G = 10 V).

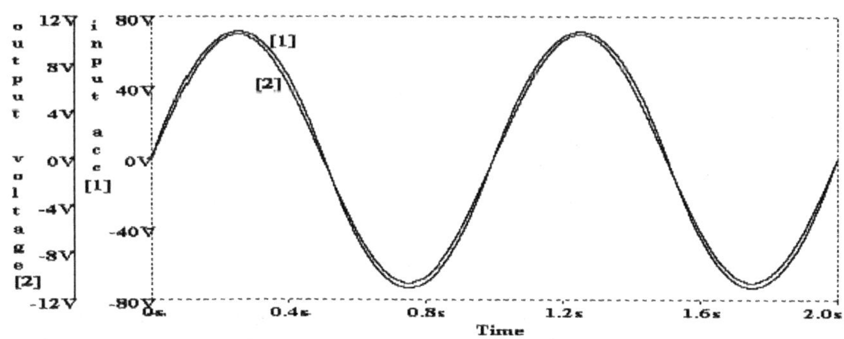

Figure 7.25: Input acceleration and output voltage for the neural transducer (1 G = 10 V).

More insight into the behaviour of the two systems can be obtained by exciting them with a ramp type of input, ranging for example from 0 G to 10 G (Figures 7.26(a) and (b)). Three ranges of deflection can be distinguished in Figure 7.26(a):

- A small signal range (mass displacement smaller than $1\,\mu$m), where the system behaves approximately linearly;

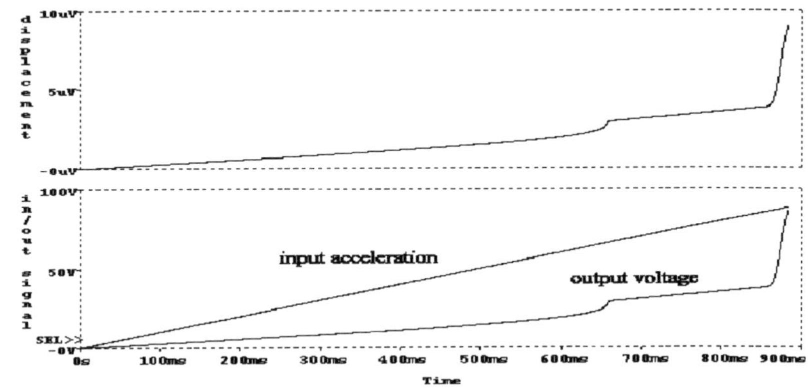

(a)

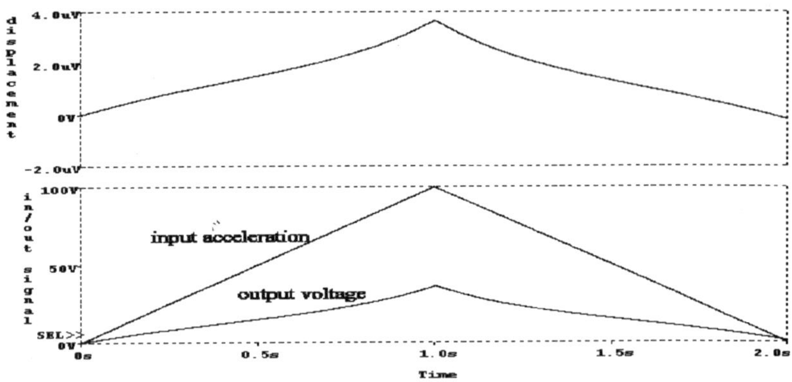

(b)

Figure 7.26: (a) Conventional transducer: seismic mass displacement ($1\,\mu$V $=$ $1\,\mu$m) and output voltage for a ramp type of input acceleration ranging from 0 to 10 G (10 V $=$ 1 G). (b) Neural transducer: seismic mass displacement ($1\,\mu$V $=$ $1\,\mu$m) and output voltage for a ramp type of input acceleration ranging from 0 to 10 g (1 G $=$ 10 V).

- A medium deflection range (displacement between 1 μm and 3 μm), where the system exhibits non-linear behaviour with the net electrostatic force on the seismic mass still providing negative feedback but the feedback gain is reduced;
- A large deflection range (displacement larger than 3 μm), where the system becomes unstable, due to the net electrostatic force changing polarity and deflecting the seismic mass even further until it touches one of the electrodes.

As opposed to this conditionally stable behaviour, in the FNN transducer case (Figure 7.26(b)), only the first two ranges of deflection appear: a linear behaviour up to approximately 3 μm displacement (corresponding in this case to a 8.5 G acceleration) and a stable but non-linear behaviour for accelerations between 8.5 G and 10 g. The maximum mass displacement is approximately 3.8 μm for 10 G input acceleration, with the mass returning to the rest position once the acceleration starts decreasing. Theoretically, positive feedback can never occur in this system, due to the time domain separation of the feedback signals: only one electrode is activated at one given instant, namely the one the mass is moving away from. Practically, however, due to the training error (albeit small) of the FNN, the system could latch-up for some extreme input stimuli. An example of such stimuli are shocks in acceleration larger than 20 G with a duration of more than 30 ms.

7.4. The Neural Network Non-linear Gain Controller

The Compensating Neural Network (CNN) — acting as a non-linear gain feedback neural controller — performs a static mapping, replicating only partially the behaviour of the PI controller: the integral and derivative actions are ignored and the linear proportional action is replaced by a non-linear gain. The training set was formed by subjecting the conventional transducer to ramp type of input accelerations rising from −6 G to +6 G (which is the full working range of the conventional accelerometer, before the latching-up of the seismic mass takes place). The resultant mapping (after suitable scaling) desired to be performed by the network is presented in Figure 7.27. The CNN transfer characteristic contains a linear region, (equivalent to a purely proportional gain) corresponding to small input signals, followed by a soft-limiting transition to a saturating region,

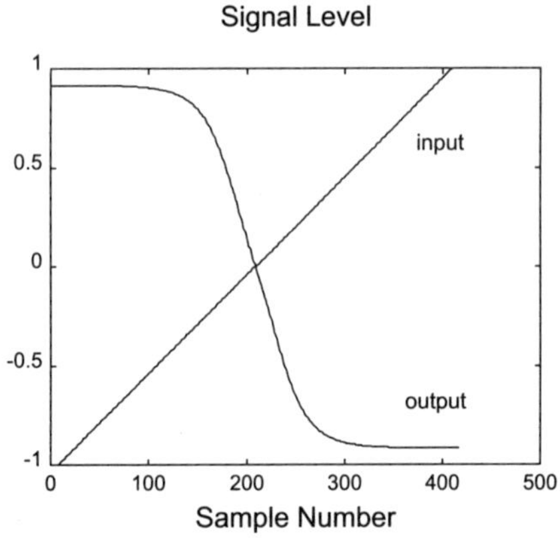

Figure 7.27: The network input and desired output.

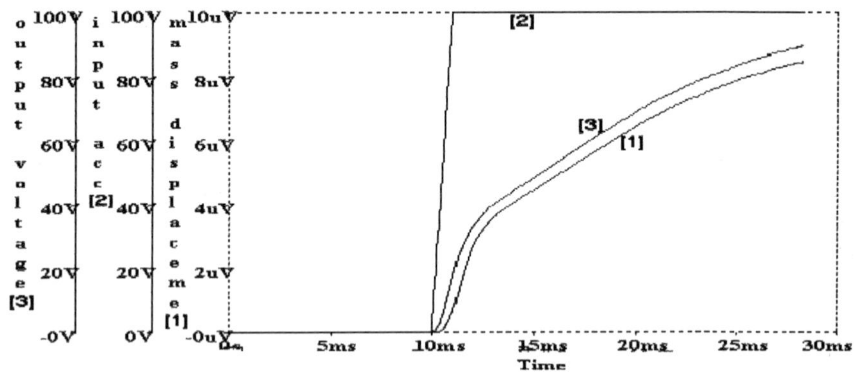

Figure 7.28: Input acceleration (10 G) (1 G ≡ 10 V), output voltage and seismic mass displacement (1 μV ≡ 1 μm) for the conventional transducer.

corresponding to larger input signals. A 1 × 6 × 1 MLP was successfully trained to perform this mapping.

Once trained, the network was integrated into the transducer structure (using adequate scaling factors) and SPICE simulations were performed. As the behaviour of the new transducer for large input accelerations is of primary interest, comparative results for the case of shocks in acceleration of 10 G are presented in Figures 7.28 and 7.29 for the conventional and the CNN transducer, respectively.

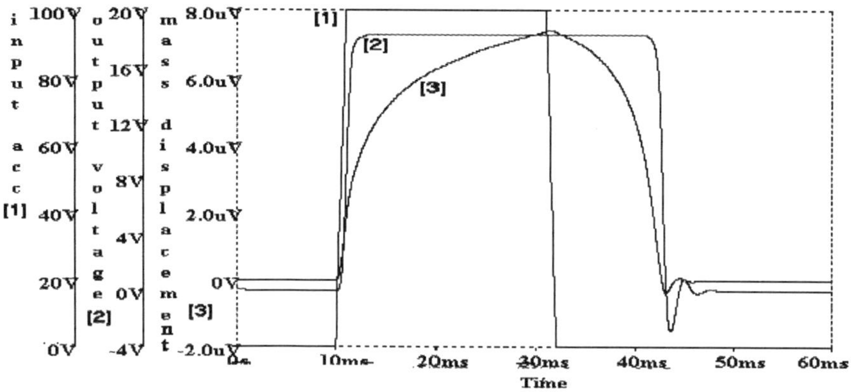

Figure 7.29: Input acceleration (10 G) (1 G ≡ 10 V), output voltage and seismic mass displacement (1 μV ≡ 1 μm) for the variable feedback gain transducer.

For the conventional transducer, the shock in acceleration leads to the seismic mass being deflected towards the top electrode and irreversibly latching up (this is illustrated in Figure 7.28 by the seismic mass displacement reaching values as high as 9 μm). In contrast, the latch-up situation was eliminated for the neural network controlled transducer. The CNN allows the mass to deflect reversibly by as much as 6 μm without the feedback becoming positive, based only on the variable feedback gain this controller provides. The seismic mass, in this case, after being deflected by the shock in acceleration, returns to its central position between the outer electrodes (0 μm displacement in Figure 7.29) upon the disappearance of the shock.

Further simulations with the neural system showed that the transducer can withstand shocks up to 20 G, for short periods of time. However, for shocks longer than 30 ms, the saturation effect provided by the neural controller is not appropriate any longer and the feedback eventually becomes positive, irreversibly attracting the mass to the top or bottom electrode.

For normal operating conditions (input accelerations up to ±5 G), the conventional and the CNN systems have similar behaviour. Some difference in the frequency behaviour of the two transducers was expected due to the removal of the integral gain in the new design; it has been however shown that, apart from a gain reduction, the two transducers have the same behaviour in terms of the amplitude frequency response and phase shift.

Design and performance of the full closed-loop neural transducer

By incorporating both the FNN and the CNN previously designed into a new system, the advantages of linear feedback electrostatic forces, time domain separation of feedback signals and soft-limiting non-linear gain control will be successfully combined. In the following, the functionality of the system is studied by subjecting the transducer to a variety of stimuli and establishing both its advantages and limitations. The non-linear control action imposed by the CNN is evident from the simulation results presented in Figure 7.30, where the neural transducer was excited with a sine wave of 6 G amplitude and 1 Hz frequency. Although the mass displacement is highly non-linear, the transducer output closely resembles the sine type of input. On the other hand, the saturating properties of the CNN and the demodulating action of the FNN are revealed when a 12 G input acceleration is imposed on the system (Figure 7.31). In this case, the seismic mass is deflected to a maximum of $\pm 8\,\mu$m, followed by a sharp return to the rest position, in the middle of the distance between the top and bottom electrodes. The output voltage is therefore reaching its limit values of ± 20 V.

Shocks in acceleration of up to 25 G can be withstood by the transducer, without irreversible displacements of the mass, provided that the duration of such a shock is less than 30 ms (amplitudes of this level and duration

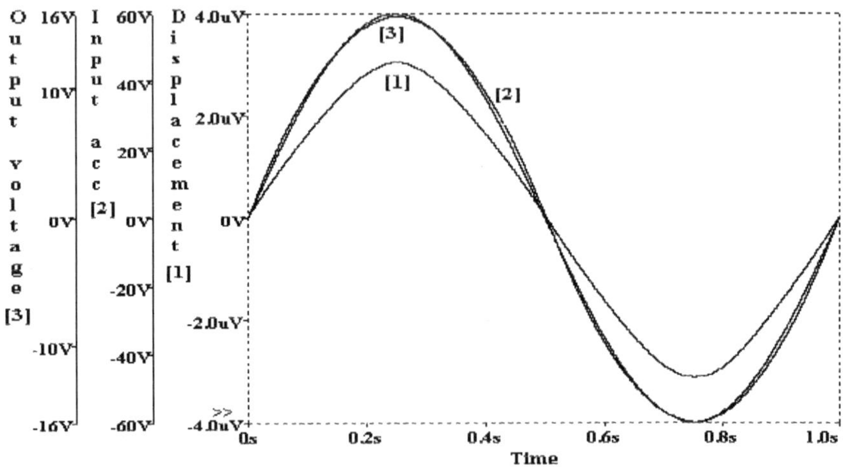

Figure 7.30: Mass displacement ($1\,\mu$V $\equiv 1\,\mu$m) and output voltage for the fully neural transducer for a 6 G input acceleration (1 G $\equiv$ 10 V).

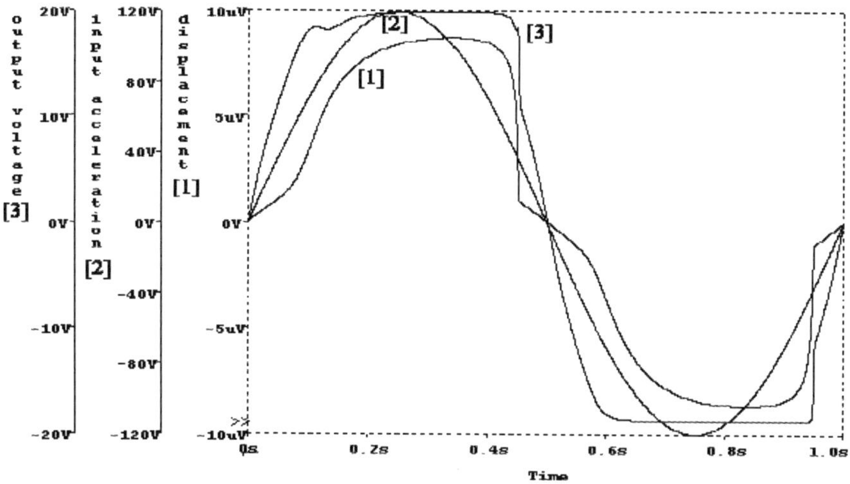

Figure 7.31: Mass displacement ($1\,\mu\text{V} \equiv 1\,\mu\text{m}$) and output voltage for the fully neural transducer for a 12 G input acceleration ($1\,\text{G} \equiv 10\,\text{V}$).

greater than 30 ms cannot be classed as shocks and therefore fall outside of the dynamic range specification). The stable region of the transducer has therefore been increased from ± 7 G in its conventional design to ± 25 G in the neural design.

The transfer characteristic for the new system is presented in Figure 7.32. The system behaves linearly up to approximately ± 6 G, exhibits a slight hysteresis between ± 6 G and ± 8 G, followed by saturation. The system exhibits a maximum departure from linearity of 3.8% over the range ± 6 G, maximum hysteresis of 5% between ± 6 G and ± 8 G, followed by saturation for acceleration magnitudes in excess of $8|\text{G}|$. This performance compares with a departure from linearity of 8% for the conventional PI transducer, over its entire dynamic range of ± 4 G. Therefore, the dynamic range of the transducer was increased from ± 4 G in the conventional design to ± 6 G in the neural design.

According to the application requirements for the acceleration sensor, the design can be easily altered by modifying the CNN scaling factors: improved linearity can be obtained for a restricted range (precision applications), or the whole dynamic range extended, by reducing accuracy in linearity [30].

The frequency behaviour of the system was assessed through a parametric analysis for a given magnitude of the input signal, at several frequencies. Bode diagrams were then drawn for the fundamental component of

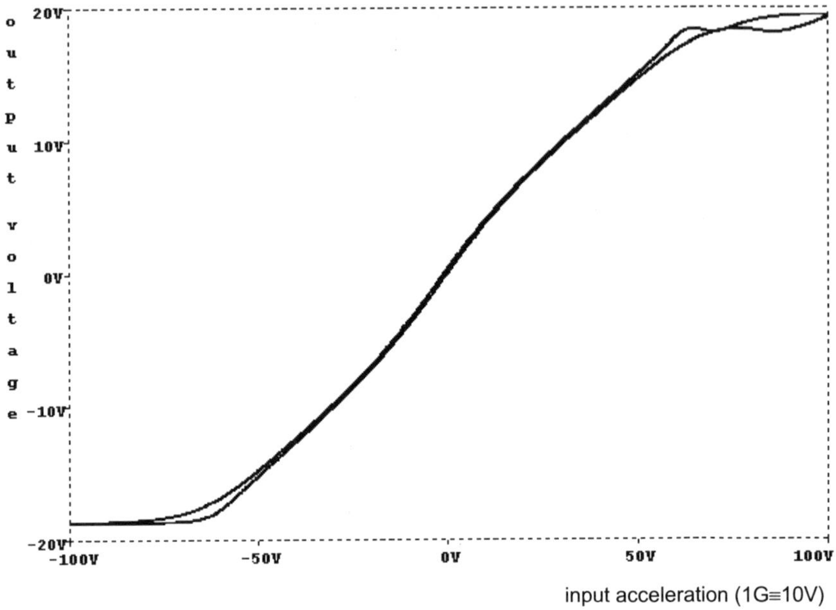

Figure 7.32: Transfer characteristics of the fully neural transducer.

Table 7.1: Transducer gain and bandwidth for different input acceleration magnitudes.

Acceleration magnitude	Gain [dB]	Bandwidth [Hz]
1 g	11.74	350
2 g	10.88	350
4 g	9.76	300
6 g	8.99	60

the transducer output. For an input acceleration of 4|G| magnitude and variable frequency in the range [1 Hz; 100 Hz], the transducer has a flat frequency response up to approximately 300 Hz, but the phase shift reaches quite large values at frequencies above about 30 Hz. Phase compensation is therefore necessary if the transducer is to be used outside the [0; 30 Hz] range, for phase sensitive applications. Results of several other similar studies are given in Table 7.1.

A major decrease in the bandwidth takes place for accelerations around 6 G for this particular design. However, the general performance of the transducer has considerably improved compared to the conventional transducer where the gain variation for the [1 G; 4 G] magnitude range was 6 dB [12] (compared to less than 3 dB for the novel design).

7.4.1. *The Effect of Unknown Initial Conditions and Manufacturing Tolerances on the Closed-loop Neural Transducer Behaviour*

Given the large manufacturing tolerances many other sensing devices suffer from, an assessment of the effect of such tolerances on the behaviour of both the conventional and the neural transducer is called for. Three particular types of tolerances were investigated, namely:

- initial displacement of the seismic mass at switch-on;
- offset of the seismic mass;
- spring constant variations.

Initial displacement and offset of the seismic mass

At the instant when the accelerometer is powered up, it cannot be guaranteed that zero acceleration acts on the transducer, thus the seismic mass is not necessarily at the central position between the electrodes and its initial velocity is not necessarily zero either. These two situations can be simulated by setting non-zero initial conditions on the two integrators in the mathematical model of the sensing element in Figure 7.21. $x(0)$ represents an initial deflection of the seismic mass and is set as an initial condition on the second integrator whilst $dx/dt(0)$ represents an initial velocity of the seismic mass and is set as an initial condition on the first integrator. Figure 7.33 presents the simulation results.

In the case of the conventional accelerometer, it has been shown that initial mass deflections up to 7 μm result in a stable behaviour of the system, the seismic mass being attracted to the central position between the electrodes, which is a stable attractor of the system [12]. For larger deflections (7 μm and above), the positive feedback term of the electrostatic force attracts the seismic mass to the top electrode, which can be regarded as an unstable attractor of the system. The neural transducer exhibited similar

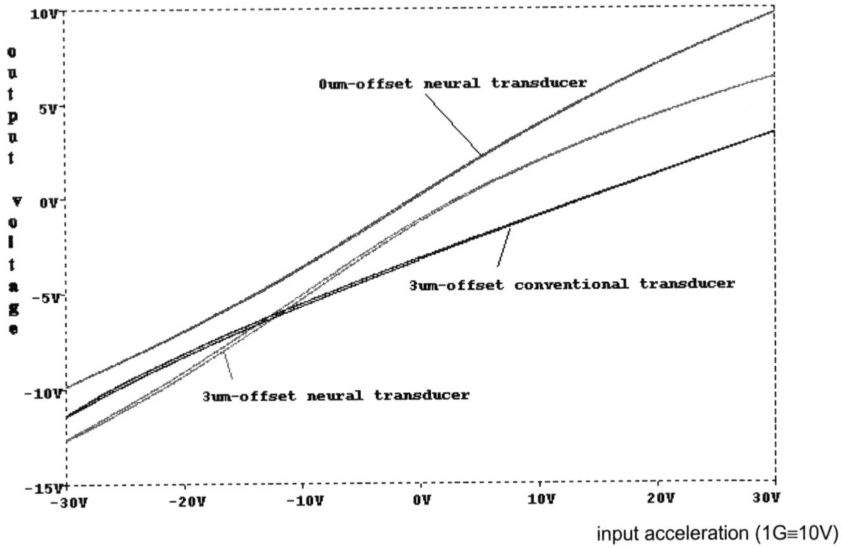

Figure 7.33: Transfer characteristics of zero-offset neural transducer, 3 μm-offset conventional transducer and 3 μm-offset neural transducer for the ±3 G range.

behaviour, although much larger input accelerations needed to be applied in order to deflect the seismic mass to such an extent in the first place. A previously applied acceleration with a magnitude of 8|G| deflects the seismic mass of the conventional transducer by 7 μm, whilst a shock in acceleration of around 20 G has the same effect on the neural transducer. Therefore, the neural transducer is more robust to the effect of unknown power up circumstances.

The offset of the seismic mass from the middle position between the outer electrodes is discussed here in more detail as the conventional and neural transducers exhibited very different behaviour. In order to simulate such a manufacturing fault, several blocks of the mathematical model of the sensor in Figure 7.21 were modified (the variable capacitors block, the non-linear damping block and the electrostatic forces block) [30] and two offset values (3 μm and 6 μm) were simulated. For the 3 μm offset, both transducers are stable for input accelerations up to ±3 G. Their transfer characteristics are presented in Figure 7.33 (for comparison, the characteristic exhibited by the zero-offset neural transducer is also shown).

The neural transducer output has an offset error of −2 V, while the conventional one has an error of −4 V. However, the conventional transducer maintains its linearity better than the neural one.

For the same offset of the seismic mass but higher input accelerations (above 3 G magnitude), the conventional transducer latches up, whilst the neural one remains stable, but its transfer characteristic becomes highly non-linear.

For the second case, of 6 μm offset, the conventional transducer was stable for the very limited ±2 G acceleration range; outside this range, it became unstable. In contrast, the neural transducer remains stable over the whole of its dynamic range. The transfer characteristics of the two transducers are presented in Figure 7.34, for the ±2 G acceleration range.

Both the offset and the non-linearities in the neural transducer characteristic which are due to device offset can be compensated by cascading the CNN with appropriate networks of the type described in Section 7.2. Alternatively, the CNN could be allowed to adapt to the new model of the sensor. The first option implies that the offset and the severity of the non-linearity are determined first, the networks trained, and a fixed control structure delivered with a particular sensor. The second option offers more

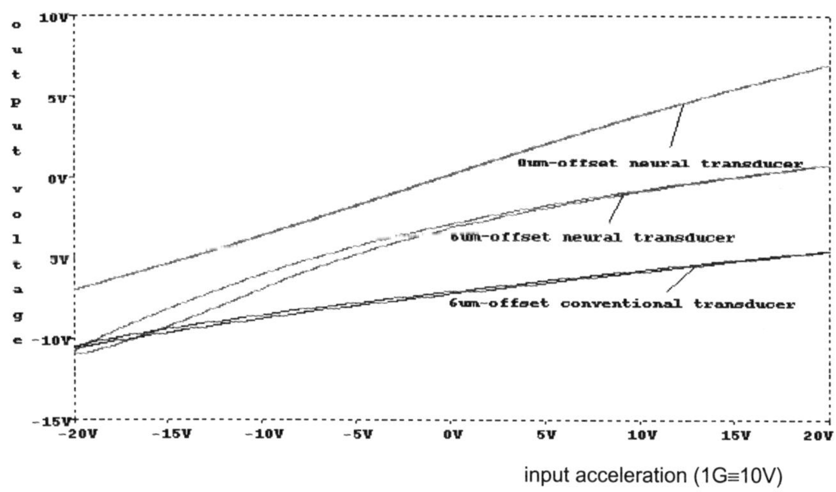

Figure 7.34: Transfer characteristics of zero-offset neural transducer, 6 μm-offset conventional transducer and 6 μm-offset neural transducer for the ±2 G range.

flexibility in that a unique controller could be determined (rather than the original CNN being cascaded with compensation networks) by means of on-line training of the CNN. Hence, the controller would need to have a hardware trainable configuration [31].

Spring constant variations

The spring constant departure from its nominal value is probably the second most important source of errors in micromachined devices [12]. The neural transducer behaviour was analysed for 15% tolerances in the spring constant parameter. The mathematical model of the sensor was modified to include a $k = 63$ and a $k = 100$ spring constant (the nominal value was $k = 83.3$). The transfer characteristics of the corresponding neural transducers are presented in Figure 7.35. No major changes in the transducer behaviour take place and therefore, it can be concluded that the neural transducer designed is robust as far as variation in the spring constant as above are concerned.

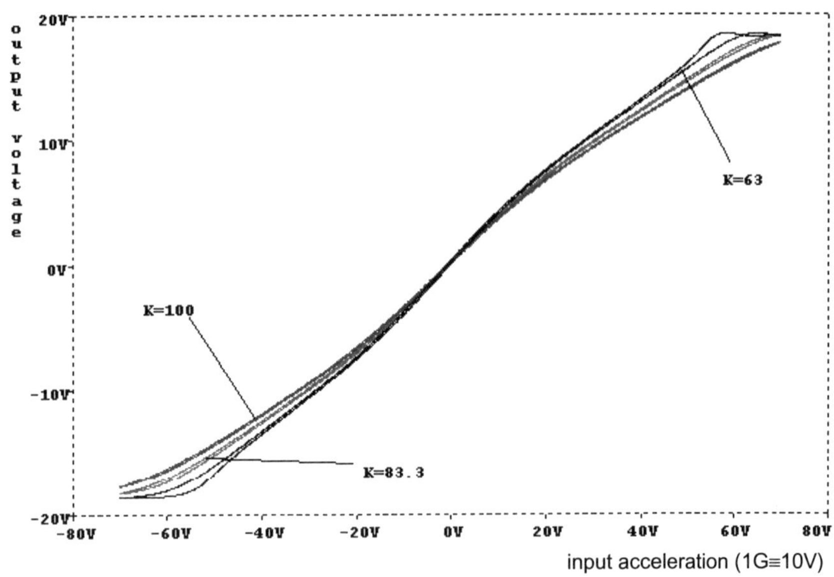

Figure 7.35: Transfer characteristics the neural transducer for $k = 63$, $k = 100$ and $k = 83.3$.

To summarise, a fully neural, closed-loop accelerometer was developed based on a mathematical model of an acceleration sensing element. The aim of the design work was to improve the functionality of a conventional, PI controlled accelerometer previously developed. A non-linear control strategy was proposed, based on MLP type neural networks, as opposed to the linear approach previously adopted. Electrostatic forces were employed to provide the feedback action for the closed-loop system. A truly negative, linear feedback relationship was ensured in the new design by inserting a neural network in the feedback path of the system. A new concept for the feedback action was used: the feedback electrostatic force is applied in a discontinuous manner, only one electrode being activated at a time. Therefore it is possible to maintain negative feedback for a wider range of accelerations. It should be mentioned here that, although other ways of implementing this feedback arrangement exist, — for example by using dedicated integrated circuits to perform the square root function and using logic gates for the feedback signal demodulation — the neural network implementation was preferred, due to its simplicity both in the design phase and from integration point of view.

The performance of the transducer in terms of bandwidth and linearity was further improved by replacing the PI controller by a neural network non-linear feedback gain controller. It was shown that the novel transducer proposed has a stable region extended by 150% compared to the conventional one, a dynamic range increased by 50% and a bandwidth of around 300 Hz. Possible application oriented modifications to the novel design were suggested in order to increase the dynamic range or to increase accuracy, as required.

As far as the hardware implementation of the transducer is concerned, the new methods of feedback and control could significantly reduce the complexity of the electronics associated with the sensing element, by eliminating the need for bias voltage on the top and bottom plates and PI setting adjustments. Consequently, a step forward has been made towards simpler integration of the transducer, provided that an appropriate choice of hardware is made for the implementation of the two modular neural networks. Hardware devices for implementing neural networks such as dedicated VLSI circuits [7, 32–35], DSPs or even small microprocessors [36] usually support a large number of neurons, therefore, both the FNN and the CNN could be physically supported within a single chip.

7.5. Micromachined Sensor Identification Using Neural Networks

Whilst enhancing the measurement performance of non-linear sensors implies applying a form of control, if a model-based control approach is chosen, the development of an accurate sensor model is of paramount importance [11, 37]. The most natural strategy would be to use a detailed mechanistic model of the sensor as the basis of the controller. Accurate mechanistic models which would include manufacturing tolerances and faults (in particular offset of the seismic mass and hysteresis) are difficult to generate for micromachined devices [12]. Alternatively, the use of a generic non-linear process modelling technique could be considered, for example, artificial neural networks [9], as proposed here. The starting point of the modelling process is an available mathematical model of the sensor [12]. Several local neural network models of the sensor are presented in this section and a generic model proposed.

7.5.1. *The Identification Problem and the Structure of the Identification Model*

The problem of identification consists of setting up a suitably parameterized identification model and adjusting the parameters of the model to optimise a performance function based on the error between the real system and the identified model output [11, 38]. As stated before, the neural network approach represents a potential generic modelling technique, which provides for rapid non-linear model formulation, whilst also allowing the capture of essential process/system characteristics [15, 39].

A good approximation of the sensor behaviour in the analogue domain is given by:

$$ma = m\frac{d^2x}{dt^2} + \frac{\mu A}{2}\left(\frac{1}{(d_0 - x)^3} + \frac{1}{(d_0 + x)^3}\right)\frac{dx}{dt} + kx, \qquad (7.7)$$

where a is the input acceleration; m is the mass of the proof mass; x is the movement of the proof mass relative to casing; A is the area of the proof mass; d_0 is the distance between the seismic mass and either of the outer plates at rest; k is the spring constant and μ is the viscosity of air.

As the process of forward identification is of interest here, the output of the system must be expressed as a function of input and previous outputs.

Equation (7.7) can be re-written, in the discrete time domain as:

$$x_{k+1} = \frac{\begin{array}{c}(d_0 - x_k)^3(d_0 + x_k)^3\left[2mT^2a_k - \frac{m}{2}x_{k-1} + (m + 2T^2kx_k)\right] \\ -\mu ATx_k[(d_0 - x_k)^3 + (d_0 + x_k)^3]\end{array}}{\mu AT[(d_0 - x_k)^3 + (d_0 + x_k)^3] + 2m(d_0 - x_k)^3(d_0 + x_k)^3}, \quad (7.8)$$

where T is the sampling interval. Equation (7.8) is non-linear in the input as well as the output signal history.

The structure of the identification model was chosen to be identical to that of the system, as far as the system order was concerned [5, 17]. Thus, the identification neural network (INN) has three inputs (the sensor input at time $(k-1)$ and the sensor outputs at time $(k-1)$ and $(k-2)$) and one output (the sensor output at time (k)). In other words, the INN can be seen as a one-step-ahead predictor.

A series-parallel identification procedure was adopted: the output of the sensor is fed back into the identification model as shown in Figure 7.36. Tap-delayed-lines (TDL) were used to incorporate the dynamic behaviour of the sensor into the model. Two delay units are necessary, to generate the one-step and two-steps back output signal, respectively. Hence, the values of the input and the past values of the system output form the input vector to the neural network, whose output $\hat{x}_s(k)$, corresponds to the estimate of the system output at any instant of time (k). Since no feedback loop exists in the model, static error backpropagation (BKP) can be used to adjust the TDL-MLP network parameters [5, 40].

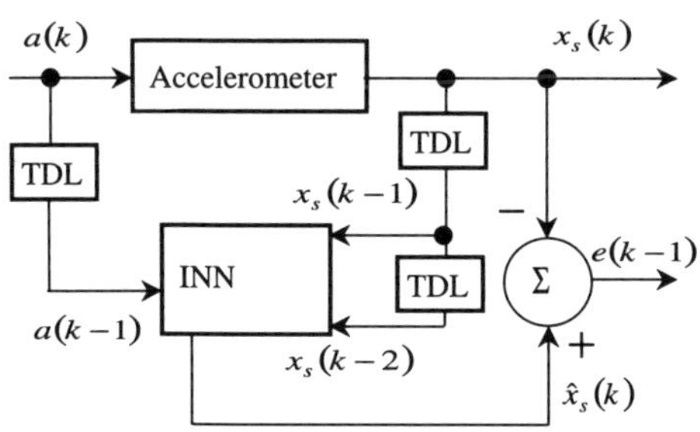

Figure 7.36: Series-parallel identification model of the sensor.

7.5.2. The INN Training, Testing and Validating Procedure

A common technique in building the network training set with the purpose of identification is to perturb the system/sensor with white noise, covering the whole dynamic range of the system [5]. Although the identification procedure developed here is based on a mathematical model of the sensor, the ultimate aim of this section is to provide a step-by-step strategy readily applicable to the identification of a physical sensor. Therefore, the subject of identification is the mechatronic sensor system composed of the sensing element, the pick-off circuit and all the electronics associated with transducing the seismic mass displacement into a readable voltage signal. Based on these considerations, the behavioural SPICE implementation of the mathematical model of the sensor was chosen for gathering the input-output training data. The aim was to identify the sensor over the $\pm 5\,$G range and $\pm 7\,$G range, respectively and [0.5 Hz 80 Hz]. Different noise signals were generated, which resulted, after suitable INN training, in four different local models and a global sensor model. The test sets were obtained in a manner similar to that followed for obtaining the noise training signals. The trained INN was tested in Matlab. The individual SSE for each test signal was compared with the SSE obtained on the training signal. It should be noted here that this process does not assess fully the performance of the INN, as the output signal history is still provided by the sensor rather than the trained network.

A recurrent network was consequently designed according to the diagram in Figure 7.37 and loaded with the parameters (weights and biases) obtained for the INN in the step above. At this point, the INN constitutes a stand-alone system, able to represent the sensor off-line. New tests were run and

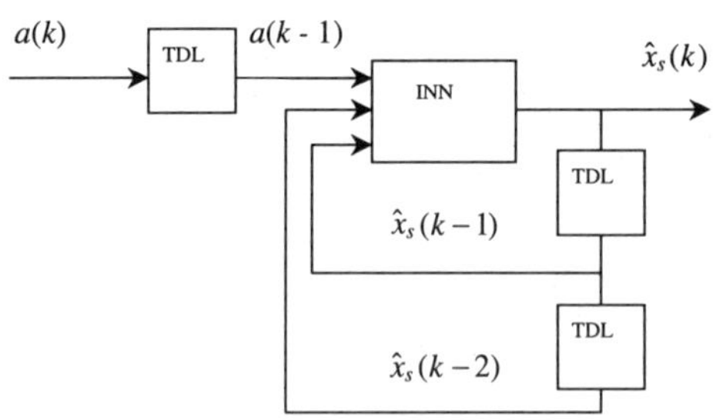

Figure 7.37: Recurrent stand-alone INN.

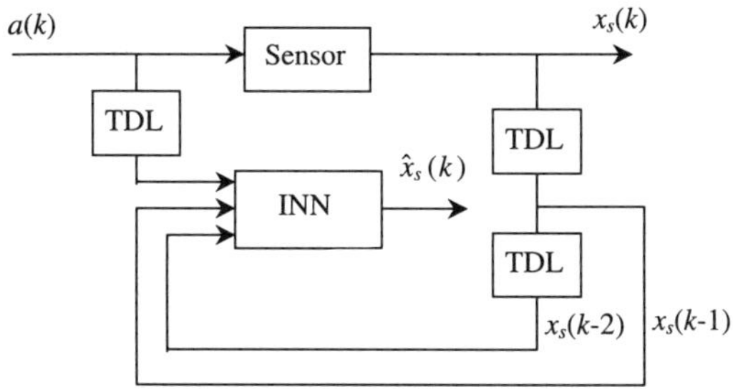

Figure 7.38: Series-parallel sensor-INN configuration.

the error between the network output and the sensor output was measured to assess the suitability and modelling accuracy of the network.

Having obtained a satisfactory network, its validation was performed in SPICE. The block diagram of the composed series-parallel sensor-INN system is shown in Figure 7.38. Various stimuli were applied and the network performance monitored.

For the identification procedure to be complete, the stand-alone INN had to be implemented in SPICE and its modelling accuracy, over the sensor dynamic and frequency ranges, assessed. The modelling errors were expected to reach higher values here, due to an "avalanche" process of error propagation.

Following the procedure described above, several neural models were produced. One of these models is described here and simulation results, at different stages during the design process, are presented.

The block diagram of the sample-data sensor system, used for the generation of the training sets and later for the validation of the series parallel INN configuration, is presented in Figure 7.39, where *ck1,2,3* represent the system clocks and S/H are sample-and-hold elements. The input and output of the sensor are buffered in order to prevent changes in the sensor output level due to the changes in the sensor load, during the validation process.

The circuit is a sampled data system, with a sampling frequency of 1 kHz for all clocks. The timing diagram for the three clocks used in the system (given in Figure 7.40) was designed so that, following the SPICE simulation, all necessary information for INN training was recorded (the sampled input, the sampled output and delayed versions of the latter) and included

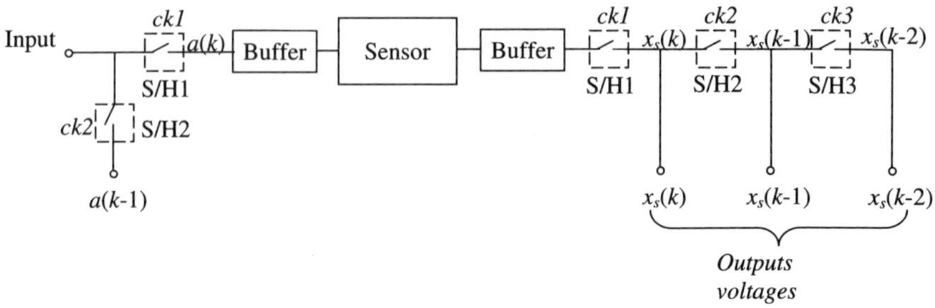

Figure 7.39: Block diagram of the sample-data sensor system.

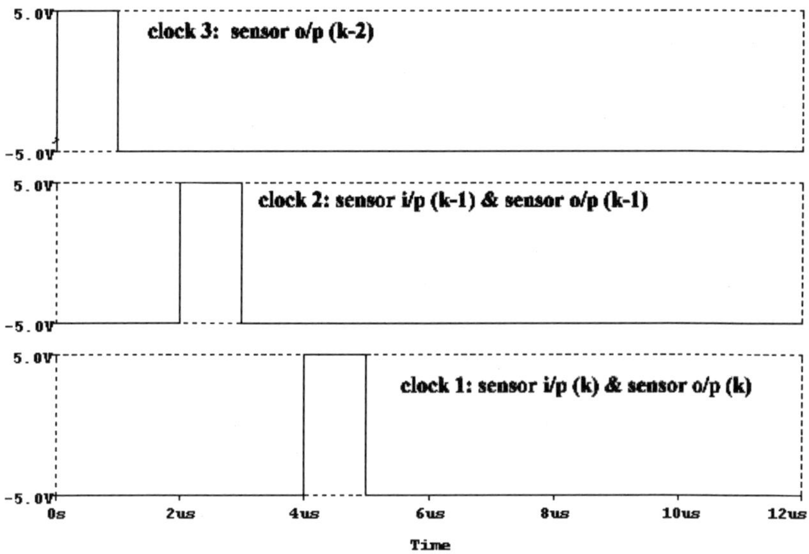

Figure 7.40: Clock timing diagram (clock 1,2,3 are ck 1,2,3 in Figure 7.39).

the effects of the electronic sample-and-hold devices. The SPICE implementation of a sample-and-hold circuit is presented in Figure 7.41.

In the circuit of Figure 7.41, *Sbreak* is an 'on-off' switch with an 'on' resistance of 1 Ω and an 'off' resistance of 10 MΩ. The capacitor C_1 has a capacitance of 10 nF, leading to a charging time constant of 10 ns and a discharging time constant of 10 ms. *VS/H* is the clock, with an 'on' duration

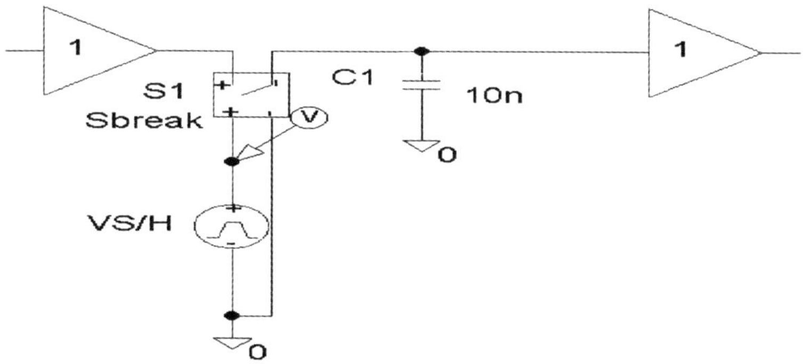

Figure 7.41: Sample-and-hold circuit (S/H in Figure 7.39).

of 2 μs and a period of 1 ms. These values are in agreement with the chosen sampling frequency of 1 kHz.

7.5.3. *Local and Global INN Models*

Low frequency sensor model

The first training set, built with the purpose of sensor identification, was based on 100 random points, generated over a time interval of 1 s. The resulting signal, after SPICE processing, filtering and sampling with a frequency of 1 kHz is presented in Figure 7.42. The frequency content of the signal in the range from DC up to and including the Nyquist frequency was estimated by calculating the power spectral density and plotting it against frequency. For the system considered in this research, the Nyquist frequency is 500 Hz. Figure 7.43 shows the power spectral density of the noise type input acceleration signal, displayed only up to 100 Hz, as no components existed above this frequency.

Low frequencies, up to about 15 Hz are well represented in the above noise signal. It is therefore expected that the INN model produced by training on such noise would perform well for low frequencies. The signal in Figure 7.42 was scaled to span over a sensor dynamic range of ±6 G (the signal was multiplied by a factor of 60, where 1 G ≡ 10 V). Following simulation runs in SPICE, the output of the sensor and its delayed versions also need to be scaled to fall within the ±1 range. The scaled input/output sensor characteristic to be modelled by the INN is presented in Figure 7.44.

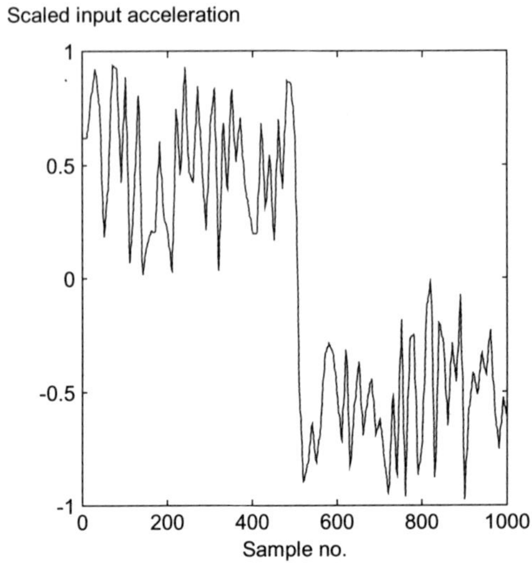

Figure 7.42: Sensor scaled input acceleration.

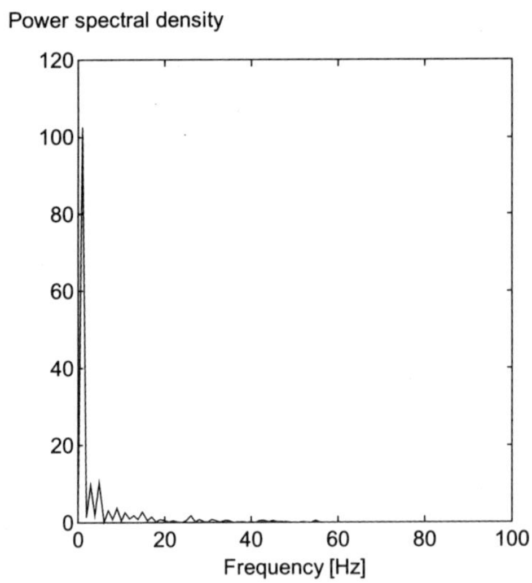

Figure 7.43: Input acceleration signal — power spectral density.

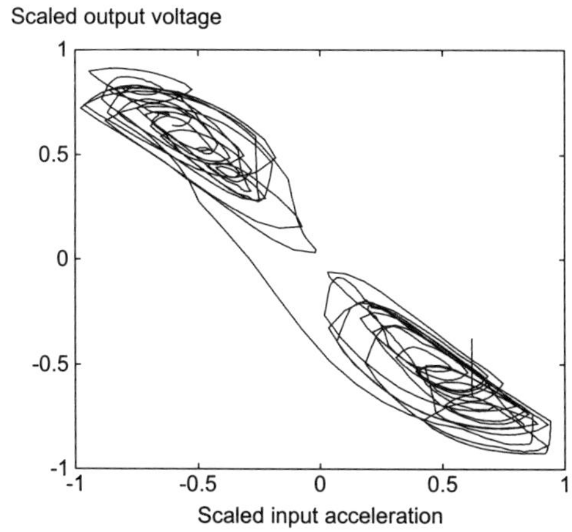

Figure 7.44: Input/output, scaled, sensor characteristic.

A $3 \times 9 \times 5 \times 1$ network was successfully trained to approximate this characteristic. Good modelling accuracy was obtained over the [0.5 G; 5.6 G] and [−5.6 G; −0.5 G] dynamic ranges and for frequencies up to 20 Hz. As expected, outside of these specified ranges, the network behaviour was inadequate. The network inability to replicate the sensor behaviour in the ±0.5 G range is due to a lack of training data for that range. The best (optimal) modelling capabilities of the INN are exemplified here for input accelerations of 5.6 G amplitude and 1 Hz, and 10 Hz frequencies, respectively (Figures 7.45 and 7.46, Sensor output (dotted line); INN output (full line)).

A low frequency model of the sensor has therefore been obtained, adequate for characterising the sensor for acceleration magnitudes between 0.5 G and 5.6 G and frequencies up to 20 Hz.

High frequency sensor model

A second training set was built with the purpose of identifying the sensor at higher frequencies. The set was based on 1000 random points, generated over a time interval of 1 s. The resulting signal, after SPICE processing and filtering is presented in Figure 7.47. The input/output sensor characteristic is presented in Figure 7.48.

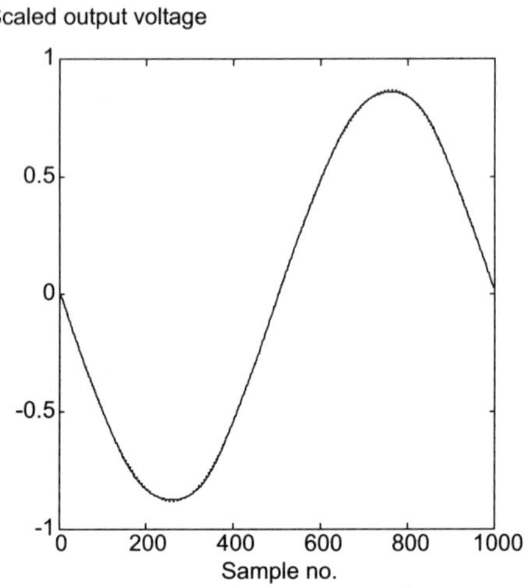

Figure 7.45: Input acceleration: 5.6 G, 1 Hz.

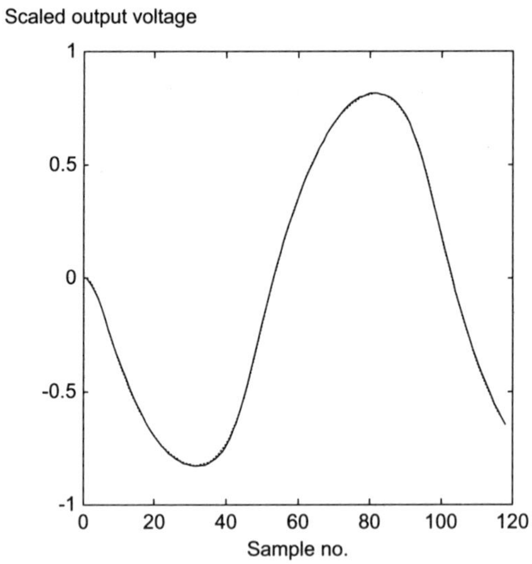

Figure 7.46: Input acceleration: 5.6 G, 10 Hz.

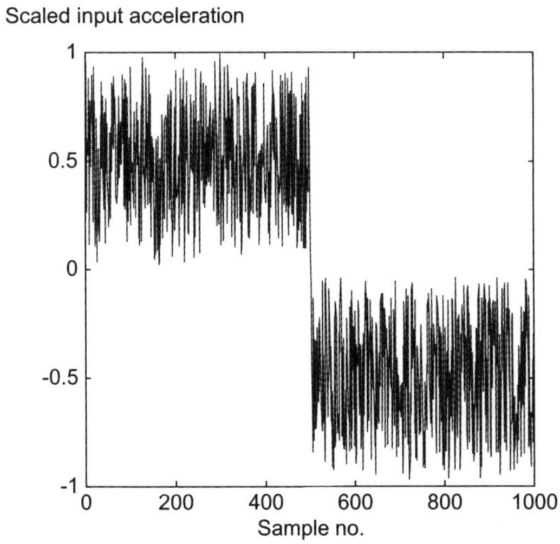

Figure 7.47: Scaled input acceleration.

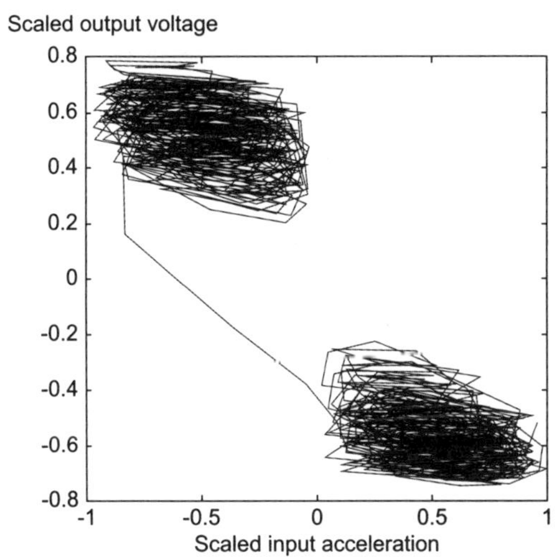

Figure 7.48: Scaled input/output characteristic.

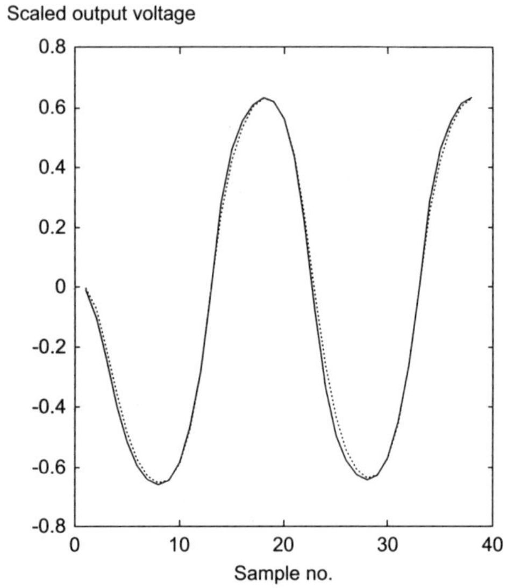

Figure 7.49: Input acceleration: 5.6 G, 50 Hz; SSE = 0.027 over 38 samples.

A $3 \times 11 \times 7 \times 1$ network was trained to a SSE of 0.22 over 1000 samples. The best performance from the INN was obtained for input accelerations in the frequency range of [1 Hz; 80 Hz] and amplitudes up to ± 4.9 G, excluding the ± 1 G range. Examples of excellent INN modelling capabilities are presented in Figures 7.49 and 7.50 (the dotted lines in these figures represent the sensor output and the full lines the INN output).

Compared with the low frequency model, this model has the ability of representing the sensor over a wider frequency range but with a lower overall accuracy.

Global sensor model

Although the local models for low and high frequencies have adequate sensor modelling abilities, their use is limited to the specified dynamic and frequency ranges. Their limitations are imposed by the noise signals used for network training. Provided that training signals with more homogenous frequency and amplitude distribution can be generated, global sensor modelling can be attempted. Based on the above considerations, a new sequence of 200 random points between ± 1 was generated over a time interval of

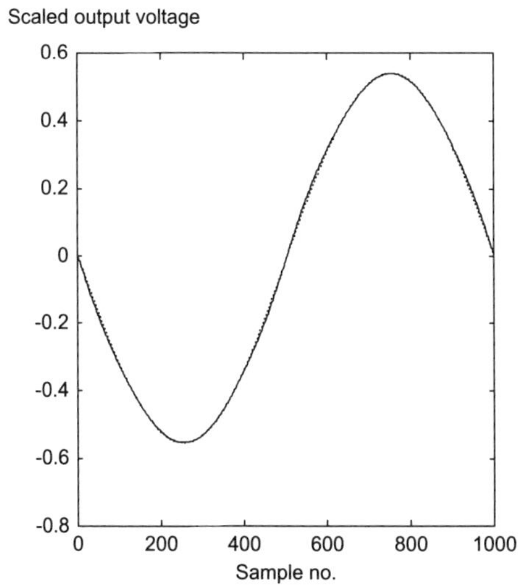

Figure 7.50: Input acceleration: 3.5 G, 1 Hz; SSE = 0.04 over 1000 samples.

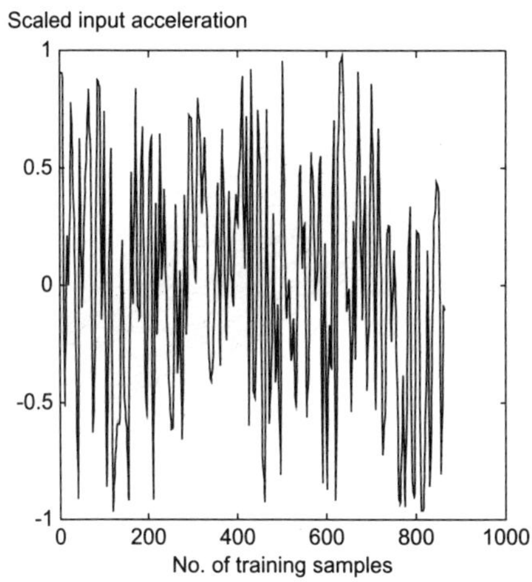

Figure 7.51: Scaled input acceleration.

0.85 s. The noise signal obtained (consisting on 850 points after sampling with a frequency of 1 kHz) (Figure 7.51) differs from the previous training signals in that the positive and negative input accelerations are not applied to the sensor in a separate manner, but in a random sequence, increasing

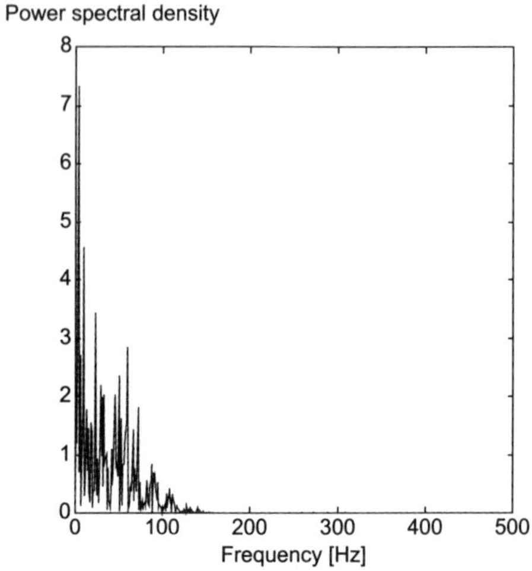

Figure 7.52: Power spectral density of the input acceleration signal.

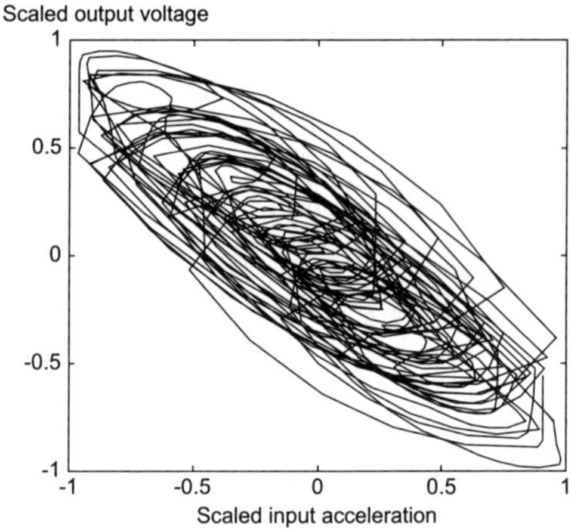

Figure 7.53: Scaled input/output sensor characteristic.

therefore the chances for the network to achieve better approximations. Moreover, this new noise signal has a more homogeneous frequency content in the range 0–80 Hz (Figure 7.52). The scaled input/output transfer characteristic of the sensor is shown in Figure 7.53.

The desired identification range was set to be $\pm 5\,\mathrm{G}$. A $3 \times 9 \times 5 \times 1$ was trained, reaching a training error of 0.057 over 850 samples in less than 50 000 epochs. Matlab tests with the INN in a series-parallel configuration shown that the behaviour of the INN was satisfactory over the entire range.

As a next step in the identification process, the stand-alone INN (parallel model of the sensor for off-line use) was simulated in Matlab. As expected, the identification error increased due to an 'avalanche' process of error propagation. However, the results were still acceptable for the frequency range [15–80 Hz] and accelerations magnitudes of up to 4 G. An example of the series-parallel INN performance versus stand-alone INN performance is shown in Figure 7.54, for a sine wave input acceleration of 2 G amplitude and frequency of 50 Hz. The SSE has increased by about 60%, for the stand-alone configuration.

Once satisfactory results had been obtained in Matlab, both sensor-network system (series-parallel) and the stand alone INN were implemented in SPICE, and new tests were run. The maximum error for the working range of the accelerometer was 2%, thus the INN had an accuracy of 0.1 G for the whole dynamic and frequency range. An accuracy of 0.05 G was achieved for the range [15–80 Hz].

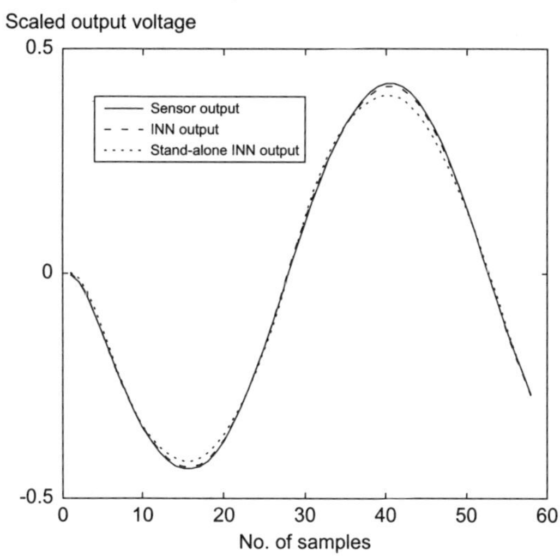

Figure 7.54: Sensor output, series-parallel INN output and stand-alone NN output (Matlab simulation results).

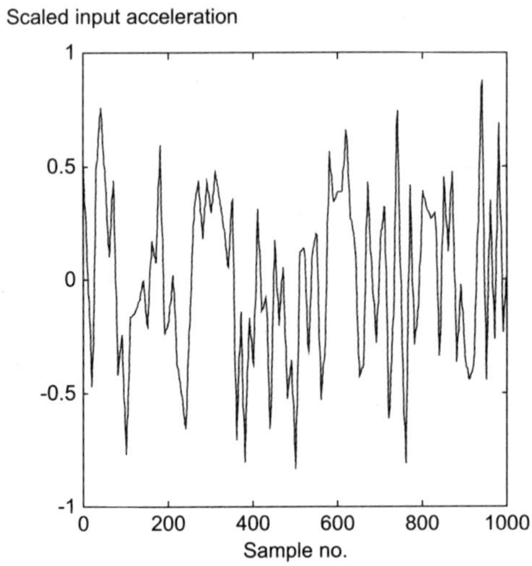

Figure 7.55: Scaled input acceleration.

A further attempt to improve global modelling was made by developing a training set generated by superimposing a low frequency noise signal and a high frequency one. The low frequency signal was produced from 100 random points and the high frequency one from 1000 points, both over a time interval of 1 s. The resulting scaled input acceleration signal is shown in Figure 7.55 and its power spectral density is presented in Figure 7.56. The INN was trained to model the sensor over the $\pm 7\,$G range. A $3 \times 15 \times 8 \times 1$ was successfully trained to reach a SSE of 0.05 over 1000 samples. The error between the stand-alone INN and the sensor output, on the training set is shown in Figure 7.57.

On the training set, the average error is within the 5% limit. Of the four models, it was found that the global model offered the best modelling performance over the widest dynamic and frequency sensor ranges. The approximating capabilities of this last INN could potentially be further improved by additional training on signals of high frequency and high magnitudes (over 5.6 G).

To conclude, in the context of sensor identification, the aim of this section was to assess the feasibility and the appropriateness of using TDL-MLP networks for modelling purposes. Four neural network identification models

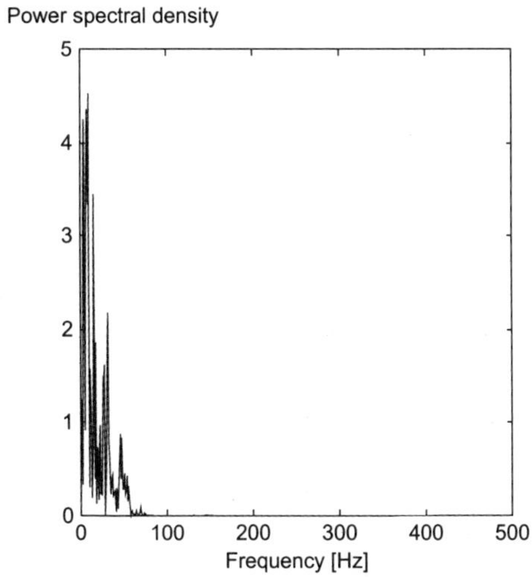

Figure 7.56: Power spectral density of the input signal.

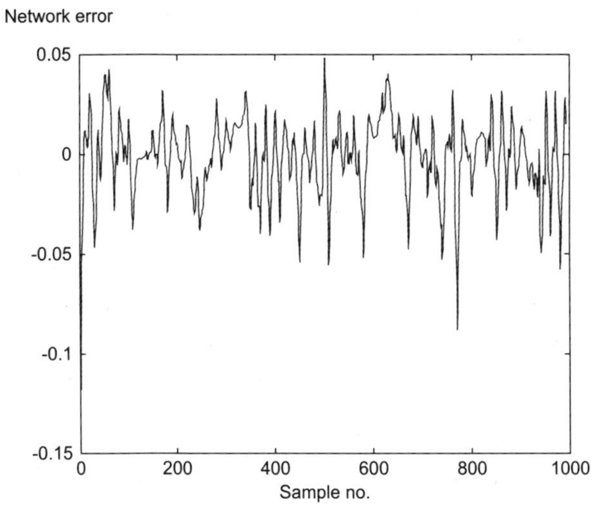

Figure 7.57: Stand-alone INN performance on the training set; The error between the stand-alone INN output and the desired output.

were built for a micromachined acceleration sensor, based on its mathematical model. The sensor identification procedure was set and several noise type signals were generated and used for developing local/global models of the sensor. The neural models obtained were successful in identifying the sensor over the prescribed ranges with an accuracy between 2% (for a restricted range) and 10% (for a wide range). If the identification of the physical sensor was to be performed rather than that of the mathematical model, only slight modifications to the step-by-step procedure given here are required. Such a process would require the use of a high resolution vibration table and noise generators, together with accurate voltage recording equipment. Drawing from the expertise acquired in modelling the mathematical model of the sensor, it is recommended that local, rather than global, models be developed, as local network requirements in terms of training effort, the formation of the training sets and the network size are minimum. In the work presented here, a heuristic approach was taken for generating different training signals. However, for high accuracy applications, the training set for the INN must be selected with particular care. Also, as the sensor is highly non-linear, on-line identification is likely to lead to better results (i.e. the INN always used in a series-parallel configuration).

7.6. Concluding Remarks

Having arrived at the end of several sections describing three particularly useful opportunities for integrating microsensors and AI, some general remarks are in order.

Firstly, the chapter is rather different in style from most of its companions in the book. The amount of detail, the personal perspective the developer took towards the developmental strategies and methodology, the particular rather than general discussion (due to the precise choice of the sensing element) and lastly its length, set this chapter apart. There are two main reasons for this. Firstly, a more general view on the opportunities to use AI in the context of sensors and sensing (and some of such practice and achievements) are treated separately in Chapter 8, from a higher system level perspective. It was felt that a good understanding of AI's potential in the context and a boost of confidence on the validity of these methods could only be insured by a thorough presentation of one particular such application. Secondly, although the chapter refers in its entirety to micomachined, capacitive pick-off accelerometers, the methodology for integrating NNs with microsensors for both identification and sensor signal quality enhancement is transferable to many other sensor types with only (if any) slight modifications. The detailed descriptions here could well inform where and how such modifications to the designs should be made in order to adapt the work to any other specific sensor. Imprecise recipes are characteristic as far as the development of NN applications are concerned and NN applications to sensors are no different. We have tried to take away as much as possible from the "trial and error" frame of mind the developer must be in, when working with NNs, and leave the lessons and guides as clear as possible.

AI is definitely an area worth getting acquainted with in the context of smart and intelligent sensing, particularly following the current trends for the development and use of large scale distributed sensor networks where AI comes in by many doors, as we will attempt to show in the last part of the book.

References

1. Culshaw, B. (2001) Complex adaptive structures: design considerations, complex adaptive structures, W. B. Spillman, Jr., (ed), *Proc. SPIE* **4512**, 13–24.
2. Hykin, S. (1994) *Neural Networks, A Comprehensive Foundation*, Macmillian College Publishing Company, Inc. New York, USA, ISBN 0-02-352761-7.
3. Yabuta, T. and Manabe, T. (1994) Learning control aspects in terms of neurocontrol. In: Zurada, M. (ed.) *Computational Intelligence Imitating Life*, IEEE Press, New York, pp. 328–339.
4. Omatu, M. K. and Yusof, R. (1996) *Neuro-Control and its Applications. Advances in Industrial Control*, Springer-Verlag, London.
5. Narendra, K. S. and Parthasarathy, K. (1990) Identification and control of dynamical systems using neural networks, *IEEE Trans. Neural Networks* **1**(1), 4–27.
6. Mitchell, R. J. and Bishop, J. M. (1995) Speech, vision and colour applications. In: Irwin, G. W., Warwick, K. and Hunt, K. J. (eds.) *Neural Networks Applications in Control*, IEE Control Engineering Series 53, Short Run Press Ltd., UK.
7. Faggin, F. and Mead, C. (1990) VLSI implementation of neural networks. In: Zorneter, S. F. (ed.) *An Introduction to Neural and Electronic Networks*, Academic Press Inc., Arlington, Virginia, USA, pp. 275–300.
8. Kraft, L. G. and Campagna, D. P. (1989) A comparison between CMAC neural networks control and two traditional adaptive control systems.

In: *Proceedings of American Control Conference*, Pittsburg, Pennsylvania, Paper No. 3.13, pp. 483–489.
9. Warwick, K. (1995). Neural networks: an introduction. In: Irwin, G. W., Warwick, K. and Hunt, K. J. (eds.) *Neural Network Applications in Control*, IEE, London, UK, pp. 1–16.
10. Zbikowski, P. J. (1992) A survey of neural networks for control. In: Warwick, K. Irwin, G. W. and Hunt, K. J. (eds.) *Neural Networks for Control and Systems*, Peregrinus on behalf of IEE, London.
11. Billings, S. A. and Chen, S. (1992) Neural networks and system identification. In: Irwin, G. W., Warwick, K. and Hunt, K. J. (eds.) *Neural Networks Applications in Control*, IEE, London, UK.
12. Kraft, M. (1997) *Closed-loop Accelerometer Employing Oversampling Conversion*, PhD Thesis, Coventry University, UK.
13. Gaura, E., Kraft, M., Steele, N. and Rider, R. J. (1999) A comparison of approaches for the design of closed-loop micromachined accelerometers, *Journal of Systems Science*, Poland **25**(4).
14. Hunt, K. J. and Sbarbaro, D. (1995). Studies in artificial neural networks based control. In: Irwin, G. W., Warwick, K. and Hunt, K. J. (eds.) *Neural Networks Applications in Control*, Chapter 6, IEE Control Engineering Series, London.
15. Kuschewski, J., Hui, S. and Zak, S. H. (1993) Application of feedforward neural networks to dynamic systems identification and control, *IEEE Trans. Control System Technology* **1**(1), 37–49.
16. Montague, G. A., Willis, M. J., Tham, M. T. and Morris, A. J. (1991) Artificial neural networks based multivariable predictive control. In: *Proceedings of the Second International Conference on Artificial Neural Networks,* UK, pp. 119–123.
17. Turner, P., Morris, J. and Montague, G. (1995) Applications of dynamic artificial neural networks in state estimation and non-linear process control. In: Irwin, G. W., Warwick, K. and Hunt, K. J. (eds.) *Neural Networks Applications in Control*, IEE Control Engineering Series 53, Short Run Press Ltd., UK.
18. Willis, M. J., Montague, G. A., DiMassimo, C., Tham, M. T. and Morris, A. J. (1992) Artificial neural networks in process estimation and control, *Automatica* **28**(6), 1181–1187.
19. Adam, O., Zarader, J. L. and Milgram, M. (1993) Identification and prediction of non-linear models with recurrent neural networks. In: *Proceedings of New Trends in Neural Computation International Workshop on Artificial Neural Networks. IWANN '93*, Springer-Verlag, Berlin, Germany, pp. 530–535.
20. Omatu, S. (1994) Learning of neural controllers in intelligent control systems. In: Zurada, M. (ed.) *Computational Intelligence Imitating Life*, IEEE Press, New York, pp. 285–292.

21. Tzirkel-Habcock, E. and Fallside, F. (1991) A direct control method for a class of non-linear systems using neural networks. In: *Proceedings of the Second International Conference on Artificial Neural Networks*, UK, pp. 134–138.
22. Darlington, P. (1991) Estimation and neurocontrol in the presence of feedback. In: *Proceedings of the Second International Conference on Artificial Neural Networks*, UK, pp. 300–303.
23. Gaura, E. and Burian, A. (1994) Dedicated environment for BKP neural networks synthesis. In: *Proceedings of Medical Informatics Conference*, Iasi, Romania, pp. 79–82.
24. Gaura, E. and Burian, A. (1995) The backpropagation algorithm and its derivations: comparative performances. In: *Proceedings of the Second International Symposium of Economic Informatics*, Bucharest, Romania, pp. 95–101.
25. Kraft, M., Lewis, C. P. and Hesketh, T. G. (1998) Closed loop silicon accelerometers, *IEE Proceedings, Circuits, Devices and Systems* **145**(5), 325–331.
26. Lewis, C. P., Hesketh, T. G., Kraft, M. and Florescu, M. (1999) A digital pressure transducer, *Trans. Inst. Meas. Control* **20**(2), 98–102.
27. Zimmermann, L., Ebersohl, J., Le Hung, F., Berry, J. P., Baillieu, F., Rey, P., Diem, B., Renard, S. and Caillat, P. (1995) Airbag application: a microsystem including a silicon capacitive accelerometer, CMOS switched capacitor electronics and true self-test capability, *Sensors and Actuators* **A46–47**, 190–195.
28. Chau, K., Lewis, S. R., Zhao, Y., Howe, R. T., Bart, S. F. and Marcheselli, R. G. (1996) An integrated force balanced capacitive accelerometer for low-g applications, *Sensors and Actuators* **A54**, 472–476.
29. Gaura, E. and Burian, A. (1995) A dedicated medium for the synthesis of BKP networks, *Romanian J. Biophysics*, Bucharest, Romania **5**(15), 26–32.
30. Gaura, E. (2000) *Neural Network Techniques for the Control and Identification of Acceleration Sensors*, PhD Thesis, Coventry University, UK.
31. Graf, H. P. and Henderson, D. (1990) A reconfigurable CMOS neural network. In: *Proceedings of IEEE International Solid-State Circuits*, pp. 145–146.
32. Akers, L. A., Ferry, D. and Grondin, O. (1990) Synthetic neural systems in VLSI. In: Zorneter, S. F. (ed.) *An Introduction to Neural and Electronic Networks*, Academic Press Inc., Arlington, Virginia, USA, pp. 317–337.
33. Gaura, E., Festila, L. and Lupea, D. (1996) Studies on a VLSI neural network for solving linear equations systems. In: *Proceedings of Mixed Design of Integrated Circuits and System, Education of CAD of Modern IC's and Devices*, Lodz, Poland, pp. 431–434.
34. Graf, H. P. and Jackel, L. D. (1989) Analog electronic neural network circuits, *IEEE Circuits and Devices Mag.* pp. 44–55.

35. Tombs, J. and Tarassenko, L. (1991) A fast, novel, cascadable design for multilayer networks. In: *Proceedings of the Second International Conference on Artificial Neural Networks*, UK, pp. 64–68.
36. Atlas, L. E. and Suzuki, Y. (1989) Digital systems for artificial neural networks, *IEEE Circuits and Devices Mag.* pp. 20–24.
37. Kosmatopoulos, E. B. and Christodoulou, M. A. (1992) *Dynamical Distributed Neural Networks for Non-linear System Identification, Neural Networks World* **3–4**, 241–268.
38. Irwin, G. W., O'Reilly, P., Lightbody, G. and Brown, M. (1995) Electric power and chemical process applications. In: Irwin, G. W., Warwick, K. and Hunt, K. J. (eds.) *Neural Networks Applications in Control*, IEE Control Engineering Series 53, Short Run Press Ltd., UK.
39. Gaura, E., Steele, N. and Rider, R. J. (1999b) A neural network approach for the identification of micromachined accelerometers. In: *Proceedings of the Second International Conference on Modelling and Simulation of Microsystems, MSM'99*, San Juan, Puerto Rico, pp. 245–248.
40. Poopalasingam, S. (1995) *Neural Network Based Digital Compensation Schemes for Industrial Pressure Sensors*, PhD Thesis, Coventry University.

CHAPTER 8

SMART, INTELLIGENT AND COGENT MEMS BASED SENSORS

by *Elena Gaura and Robert Newman*

8.1. Introduction

This chapter draws on the first part of the book which looked at stand alone sensors and provides the link to the reminder of the chapters which are concerned with multi-sensor systems. It can be argued that, in some ways, the development of sensor networks and arrays has been prompted by the designers' ability (and success) in enhancing the functionality of stand alone sensors beyond insuring mere measurement accuracy. Newly designed sensor systems have entered (or are about to enter) the realm of 'intelligent products' which could be used to build 'intelligent', large scale sensing applications.

To start off with, the chapter develops new terminology to describe sensors which have been functionally enhanced in some way through the integration/addition of supplementary processing circuitry. Several terms, current in the literature, including 'smart sensors' and 'intelligent sensors' are discussed and the 'cogent' sensor is introduced. Added sensor functionality is discussed in the context of growing expectations from sensors and the blurred boundaries between them and measurement instruments. The chapter reconsiders some of the issues in Chapters 3–7 under the umbrella of 'smartness' and attempts to set the role of 'intelligence' within a sensor system at different levels. The scene is set for further chapters by opening a discussion on 'cogent' type of sensors in the context of large scale systems of such sensors.

A particular route towards implementing smartness, intelligence and cogency is through AI technologies. The enabling value of this set of tools is argued through a review of existing and *potential* AI applications to microsensors, in terms of technology integration, device level performance enhancement and added functionality at various level within sensor(s) systems. Examples of AI applications to the design of smart, intelligent and cogent microsystems will include sensor data validation, correction and missing data restoration, sensor fault detection, intelligent actuation and information inference from sensor data. Hardware implementations of ANNs to support the functions above are also brought forward and the future of AI for sensors is discussed.

8.2. Smart, Intelligent and Cogent Sensors — What do the Terms Mean

8.2.1. *Preamble*

Traditionally, the main sensor requirements were in terms of metrological performance, i.e. the (most often) electrical signal produced by the sensor needed to match relatively accurately the measurand (the relative accuracy was strictly a function of cost, up to the domain crossing line between sensors and measurement instruments). If such basic sensor functionality was adequate several years back, this is no more the case. A well written appraisal of this trend is given by White in [1]:

> *The boundaries between sensors and instruments, which once seemed so firm, are now quite blurred. Processes that were once confined to physically large electronic instruments are now available within the sensor housing. Thus, a sensor is now regarded as a system that inputs information and serves a host system. A complete intelligent sensor may therefore comprise: a primary sensing element, amplification, excitation control, active feedback control, analogue filtering, data conversion, local digital information processing and external information procession such as data fusion, neural networks or self-evaluation techniques.*

The use of sensors by industry has seen a gradual shift, away from large systems incorporating relatively few and expensive transducers towards the utilisation of more and more sensors as components, or in subsystems. A new set of requirements for sensors and more generally for measurement systems was therefore generated. Such requirements include long mission duration, reliability and availability, operation in non-benign, unstructured environments, real time operation and flexibility of use. These requirements lead the research community to pursue the development of sensing components with increasingly autonomous functioning capabilities, based on decentralised-distributed systems architectures principles. Such systems would need to offer a wide range of quality attributes such as ease of systems integration, interoperability, scaling, portability and modularity, inherent robustness and survivability. It follows that the quality of information exchanged by these distributed components has to be high, which in turn involves designing accurate, robust, reliable and resource saving sensing/measurement modules within these components.

It is clear to the reader by now that MEMS technology in its present form, supported by the wealth of related research in neighbouring disciplines seems to provide an answer for fulfilling many of the above (sometimes contradictory) requirements. The large number of existing, commercial MEMS products shows that MEMS devices could be designed and produced to achieve: low cost, low mass and low power consumption, plug and play, digital output, enhanced reliability, sensitivity and selectivity and high accuracy. Traditionally fulfilling these requirements has been subscribed to the effort area of sensor metrological requirements improvement and was aimed at by:

- advances in the manufacturing techniques of the sensing element itself (to produce linear, accurate, highly sensitive and reliable sensing devices),
- the addition of some compensating electronics or control features (most times through bottom up developments, as illustrated in previous chapters) and
- the provision of (sometimes) standardised sensor system interfaces.

The newer pool of potential 'big' sensors applications, however, need more than this — the inherent, natural MEMS properties of size and potentially low cost was quickly linked to liberal usage of these devices in applications (for example smart skins with thousands of embedded devices, deployable sensor webs, etc.) which in turn lead to the need of relying on, or adding *efficient* and *clever* processing of data generated by the sensing device, before such data reaches the outer world. Technology perfection and linear translation of control principles from the macrosystem domain might not, therefore be, in the new light, the primary aim in developing successful MEMS sensors and particularly multi sensor systems.

It is here, in the area of *efficient* and *clever* data processing and extraction of information, that the authors propose to clarify some commonly used terminology and introduce new terms. The definitions are supported by examples. Some of the arguments presented in support of the new terminology are taken up again in Chapters 9 and 10 from a higher level system perspective, that of the sensing application itself and the nature of the delivered outputs.

8.2.2. *Smart, Intelligent and Cogent Sensors — Definitions*

One must recognise that the MEMS/micro sensors field has become a highly interdisciplinary one, pulling together researchers from the materials domain, microelectronics, mechanics, physics, computer sciences and newly, from a host of application domains (military, environment, transport and space are first to come to mind). The widening of the share holders pool in MEMS meant that 'borrowed' technologies, methods and terminology from the macrosystems domain (the 'home' for most of the domain specialists above) began to be used in conjunction with the development of microsystems. Several terms are current in the literature, including 'smart sensors' and 'intelligent sensors'. 'Adaptive', 'distributed', 'autonomous' and other adjectives are also routinely applied to pick out a particular sensor from the common herd of 'dumb' sensors. (Such terminology, a few years ago, was solely dedicated to macrosystems, with a much stronger meaning.) These are the terms the discussion in this section starts with and centres on.

By comparison with the usage of these terms in other fields, it would appear that the sensor community is over-selling the 'intelligence' of their products. The phrase 'intelligent sensor' often merely indicates that the sensor is integrated with a digital processor; it may say nothing about the intuitive abilities of the functionality programmed into the sensor. It can be argued that the 'misuse' of the 'smart' and 'intelligent' terms has to do with the somewhat parallel developments in *designing* sensor systems and *producing* microsensors, the achievers in each of the two areas using the terms to position the work in the leading edge (which was the 'smart', followed by 'intelligent' product world during the last decade).

Given that the terms 'smart' and 'intelligent' have become somewhat confusing and meaningless with respect to sensors, these terms are re-defined in the context of this book and a new term, the 'cogent sensor' is introduced.

The meaning of the three terms relates here to what a sensor *does*, rather than what came before it or how it is constructed.

Smart sensors

In the most general definition, smart sensing devices and systems are those produced by integrating sensors and actuators with electronic circuits [2, 3].

According to the IEEE 1451 smart transducer interface standards (which describe a set of open, common, and network-independent communication interfaces for smart transducers, detailed in Chapter 9), 'smartness' means on-board data storage/processing capability, interfaced/integrated with the digital sensor [4]. Variations of the term refer to the hardware implementation of the sensor, as follows: when a microsensor is integrated with signal processing circuits in a single package, it is referred to as an integrated sensor. A monolithic integrated sensor has the signal processing circuitry fabricated on the same chip as the sensor, while a hybrid integrated sensor has the signal processing circuit on the same hybrid substrate as the sensor chip. A hybrid implementation (where the microsensor fabrication processes are incompatible with the electronics fabrication, resulting in a hybrid chip) is seen as being less smart.

Given the fuzziness of the above definitions and the continuous evolution of integrating capabilities in the MEMS domain, 'smart' has become a term which, although used in conjunction with nearly every newly designed or produced sensor, means different things to different people. To some, it just means sensors which can communicate digitally, while to others it means sensors that have serious computing power integrated within them, to include one or more of the following functions: self-calibration, non-linearity correction, offset elimination, failure detection, communication, and even decision making ability. (It is important to mention that, designs to cater for individual functions listed above have been developed strictly for a sensor type and manufacturing technique and are, mostly, highly application oriented, so much so that, it is often difficult to assess whether it is the sensor or the application that is 'smart'.)

Looking at 'smart' sensors since the term was coined in the 1980's, Gardner [5] separates them in two classes:

- sensor + preprocessor = smart sensor I
- sensor + preprocessor + processor = smart sensor II (higher degree of integration, i.e. to include a microcontroller or microprocessor)

Further on, he differentiates 'smart' and 'intelligent' sensors as follows [5]:

> *Sensing devices which have a part or all of the processing functions integrated onto the same silicon substrate are called smart sensors. The label 'intelligent' is reserved for devices that have, in addition, some biomimetic function such as self-diagnostic, self-repair, self-growth and fuzzy logic.*

The perspective here is a mixed technological and functional one.

Other views equate 'smartness' with 'ease of use'. This is the common 'commercial' line, enabled mostly by the introduction of the related IEEE smart sensor standards family. In this case, the reasoning focuses on the architectures and protocols which were introduced on a higher systems level (filedbus, pci, etc.) and have proved advantageous. Features like reduced wiring, low level bus set up in data acquisition, exchangeability of devices, multiplexing options, easy maintenance and expansion, simplifying the division of complex systems with high level bus communication into subsystems using low level bus communication, make sensors *easily* amenable to a variety of new applications, therefore, they deserved the branding of 'smart'. The additional benefits 'sold' with smart sensors as defined above (the 'easy to use' and/or 'plug and play' components) are justified from a process instrumentation perspective. For example, the following merits of a *field bus* are presented as 'smart sensor characteristics':

- Increased amount and improvement of the quality of information — more information can be obtained by a communication link and the quality of the information will be improved by its check function;
- Improvement of maintainance ability — it becomes easier to access the field instruments from a control room; the wiring check and start-up check will become easier;

- Reduction of installation costs — plural instruments can be wired using a single cable.

The discrepancy of meaning of the term 'smart' is obvious between macro and micro scales, between older and newer generation products. In support of this, Gardner takes as an example 'smart electronics' which is defined as 'electronic systems that have some form of embedded intelligence (i.e. a neuronal chip for example made of analogue VLSI)' [5].

Thinking now from a clear 'systems' perspective, the authors' proposed definition of a 'smart sensor' is as follows:

> *...a sensor including signal processing capability integrated into it to make good physical deficiencies in the intrinsic sensor hardware or to ease its interface with other parts of the system. The signal processing capability may be either analogue or digital. 'Integrated' here means that the smart sensor has been designed as a system, top down, to fulfill a set of system requirements which referred mostly to metrological performance and sensor usability. The implementation aspects are not dealt with by this definition.*

With this in mind, one could revisit the sensor systems presented in Chapters 5–7 and clearly class them as 'smart sensors', together with many of the 'compensated' sensor examples referenced in Chapter 4.

Intelligent sensors

As the capacity of VLSI techniques increased, it became possible to integrate substantial digital or analogue processing capability onto a sensor.

The term 'intelligent sensors' was often used incrementally from 'smart sensor'. Initially 'intelligence' meant and served the same purposes as the electronics in a smart sensor: enhancement of the measurement function of the sensor itself. For macrosensors, the term intelligent sensor was re-coined in 1992 [6], with an early definition of intelligent sensors having been produced in 1978 by Breckenbridge and Husson, as follows [7]:

> *The sensor itself has a data processing function and automatic compensation function, in which the sensor detects and eliminates abnormal values or exceptional values. It incorporates an algorithm which is capable of being altered, and has a certain degree of memory function. Further desirable characteristics are that the sensor*

is coupled to other sensors, adapts to changes in environmental conditions and has a discrimination function.

This definition does not distance itself from the 'smartness' defined by the authors here. In his 'Microsensors, MEMS and Smart Devices' book, Gardner agrees with authors' view in that intelligence must be associated with functionality rather than form, to allow differentiation from smart [5]. He drew the following classes of intelligence, starting with the lowest:

Signal compensation:
The device automatically compensates for changes in an external parameter (for example a temperature compensated silicon accelerometer);

Structural compensation:
The physical layout is designed to reduce signal-to-noise ratio (and therefore enhances signal quality);

Self-testing:
The device tests itself out and therefore has self-diagnostic capability;

Multisensing:
The device combines together many identical or different sensors to improve performance (for example the electronic noses);

Neuromorphic:
The device shares characteristics with a biological structure, such as parallel architectures or NN processors (e.g. Neuro VLSI chips).

Hence, according to Gardner's classification, a two-chip microaccelerometer with self-test would be called a smart sensor with intelligent features (the self test is seen as providing a certain level of intelligence and it is important in applications where sensor failure is safety critical).

Several observations are required here:

- The first two functions 'make good' sensor deficiencies which we classed as 'smartness';
- Self-diagnostic in its full meaning has not been resolved for sensors at a generic level; it is, as such, very different from 'self-test' which provides a binary type of information (the sensor is functioning or not) and which, as a feature, is usually provided as a 'side effect' of the sensor design (see the sigma-delta modulator configurations). Diagnostic would certainly

mean much more than that; it would mean an ability of the sensor to scrutinise its functionality and pin-point the precise 'illness' (if any), also assessing how badly its measurements are affected. Whilst 'self-test' could be seen as just 'smart', self-diagnosis as described above would mean a high degree of intelligence and decision making ability, located within the sensor itself;
- Both multisensing and neuromorphic sensing devices and systems could mean less than intelligent: redundancy in its simplest form is downright 'dumb'. Or, it could mean, much more: multisensing of various parameters to infer new information is a highly intelligent feature, for example. This idea will be explored further in the following sections.

With a view to the above, it can be argued that the use of the word 'intelligent' is subjective and evolving, with devices reported as intelligent a few years back being now no more than dumb, compared with newly designed devices. Trade journal and popular news articles use smart and intelligent interchangeably, for example, and there is no secret that they are both highly fashionable words.

Having said all that, there are several examples of work, where by using the term intelligent the authors implied the use of macro-scale intelligent techniques (i.e. artificial intelligence and most often neural networks). Such works are discussed in Section 8.4. It should be noted however that the nonlinear signal processing abilities of NNs are merely exploited in these works rather than more evolved 'thinking/decision making in new situations' aspects of the technique, this setting most designs and prototypes in the category of 'smart' sensors. In these cases, the perspective in naming the sensor intelligent is taken from the tools used in design and again has little to do with the precise way the results of applying the tools reflect in the final sensor product. This is similar with naming a sensor 'smart' when a standard digital interface is attached to it.

In the authors' view, a low level of the following characteristics of human like intelligence may need to be embedded in sensors to make them intelligent:

- fault-tolerance,
- adaptive learning and
- some basic decision making ability (possibly hard rule-based).

Hence,

> *...an intelligent sensor is a sensor integrated with signal processing which includes the capability to present the data from the sensor hardware in the form required by the application or system. The semantic content of the data remains essentially that from the sensor hardware. The intelligent sensor therefore would for example energy-save if its measurements are event-less for a given time span, without any intervention from the application which it is part of (low level decision making ability); it might furnish the application with a 'confidence' index for its measured data, provided that the sensor system is complex enough to allow assessment based on accessible internal states or has multiple outputs (fault tolerance); together with peers, for example, within an application, it can acquire some new forms of behaviour adaptively through 'learning'.*

Some examples to support the definition above will be presented in Section 8.5.

Cogent sensors

What is common to the previous classes of sensors is that they provide raw data. Essentially, the readings of the original sensor are passed on, albeit linearised, temperature corrected, hysteresis corrected, packetised, network routed or re-packaged in one of many ways specific to the sensor in question. What these sensors do not do is reduce the data to information.

We term a sensor that performs the data to information transformation a 'cogent sensor'. Removing unnecessary data for example and converting the remaining data to a format amenable to the application the sensor sits in is an example of 'cogency'. One situation in which this might occur is where the application requires information in the frequency, rather than time domain. A cogent sensor may implement decimation-in-time (removing unnecessary data) and a fast Fourier transform (format conversion) to provide the required information.

Formally, a cogent sensor would be

> *...a sensor including processing and decision making capability in order to convert the raw data from the sensor hardware to the particular information that the application requires (implying most often that the application has the ability to query the sensor and*

request the information). Performing semantic transformations if necessary by removing unneeded data, producing inferred, calculated or derived information rather than provide the raw sensor data are amongst most common 'cogency' characteristics.

Some examples of existing and proposed cogent sensors and a discussion on various ways in which artificial intelligence techniques may be used to implement them are given in this chapter. However, 'cogency' is best highlighted and indeed most needed when the aim is to design multiple sensor systems, usually organised in the form of sensor networks, which are treated in detail in Chapters 9 and 10.

8.3. What and Where is the Added Value Brought by Intelligence?

Given the proliferation of 'intelligent' sensors publications, and for the benefit of those who might find the proposed definitions in Section 8.2.2 constraining, another view is offered here on what has been seen as 'intelligence' and 'intelligent' features in sensors throughout their development.

Starting with the level closest to the sensing device and working up (from a systems perspective) towards the sensing application as a whole, the roles of 'intelligence' and the weaving of 'machine intelligence' with sensors have evolved on the lines presented below.

8.3.1. *The Bottom Layer*

In the bottom layer (close to the sensing element), intelligence would have the following roles:

Reinforcement of inherent characteristics of the sensor device:
The most popular operations here are the compensation of characteristic and suppression of the influence of undesirable variables on the measurand [8]. Whilst many solutions have been found for eliminating measurable effects of undesirable variables (as shown in Chapter 4 for example), the problems become more complicated if such effects are difficult or impossible to quantify. In this case, the use of machine intelligence techniques could be beneficial, as described in Chapter 7. This is the most fundamental role for machine intelligence in relation to sensors. The range of techniques

in this category are considered to be a static approach to the selectivity improvement of the signal. Some examples relating to this role are given in Section 8.5.

Signal enhancement for the extraction of useful features:
This is slightly an overstatement as the goal of the type of signal processing referred to here is to eliminate noise and make the signal clear. Such signal processing techniques would mostly utilise the differences in dynamic responses to signal and noise. They are broadly divided into frequency domain processing and time domain processing, as shown in Chapter 3. This approach, in general, can be seen as a dynamic approach for selectivity improvement of the signal.

8.3.2. *The Middle Layer*

In the middle layer, the role of intelligence is threefold:

To organise multiple output from the layer and generate intermediate output:
Here, the output signals of multiple sensors are combined or integrated and the extracted features are then used by the upper layer intelligence to recognise given situations/scenarios;

To perform the function of sensor signal fusion and integration:
Signals from the sensors for different measurands are combined in the middle layer and the results provide new, useful information; ambiguity or imperfections in the signal of a measurand can be compensated for by another measurand; this processing creates a new phase of information;

To perform optimisation or parameter tuning of the sensor to optimise the total system performance:
This is done based on the extracted features and knowledge of the target signal; the knowledge comes from the higher level (generally the application level) as a form of optimisation algorithm.

8.3.3. *The Top Layer*

In the context of systems composed of networked, autonomous sensors, which are discussed in Chapters 9 and 10, the role of the 'intelligence' within the sensor is largely to do with collaboration between sensors. Sensor networks are typically too large and complex to be feasibly configured and maintained manually. Thus, it is natural, when contemplating the problems of designing self organising systems to look for inspiration from the self organising systems in nature, namely organisms. The functions to be performed by nodes in sensor networks are often compared with those performed by 'collaborating' neurons in the brain, and architecture for such systems based on these models are currently being developed [8]. In this scenario, the 'brain' model is applied with respect to network organisation, rather than aiming to achieve inherent 'intelligent' behaviour. Other models for self organisation are discussed in Chapter 9.

Moreover, transmitting data around the network, or extracting data from it, is not always straightforward, in terms of the communication facilities and bandwidth available. Thus, in these sensor networks there is a strong motivation to use 'sensor intelligence' to reduce data at the sensor. Yamasaki summarises this application of 'intelligence' [8]:

> *. . . The roles of the dedicated signal processing functions are to enhance design flexibility of the sensing devices and realise new sensing functions. Additional roles are to reduce loads on central processing units and signal transmission lines by distributing information processing in the sensing system.*

The next stage from 'data reduction' is 'information extraction', that is, one is no longer interested merely in minimising the data in the network, but the concern is now with 'querying' the network to extract only the information required by the application context. Yamasaki summarises this idea as follows [8]:

> *Technological change brought with it an increasing capability to manipulate, store, display and communicate large amounts of information. This is causing a shift in the computing paradigm from 'computational speed' to 'information-oriented' computing.*
>
> *The trend towards improved human-machine interfaces and intelligent machines will increase the need for sensors that will do more than just acquire and condition the data. These sensors will perform feature extraction and preprocessing for the incoming signals.*

The system level architectures required to realise systems with this type of capability are discussed in Chapter 10.

At the top of the 'intelligence tree', so far as sensors are concerned, are those proposed designs where some or all of the complex application or user level processing is performed by the processors integrated within the sensor network, either individually or with the tasks distributed so as to be shared by a number of sensors. One established context in which this type of capability is proposed is for 'data fusion' (or, in the realm of sensor technology 'sensor fusion') whereby the data derived from a number of sensors is processed to calculate some derived value, or even the value of the measurand at positions between sensors — to produce 'virtual sensors'.

As a cross mapping with the intelligence levels above, the reasons for making sensors intelligent are summarised in Table 8.1.

To conclude, this section looked at the issue of 'sensor intelligence' from a different viewpoint to that usually taken. Rather than defining 'intelligence' as an attribute linked with a particular form of hardware, the approach here was to consider the 'added value' that 'intelligence' could bring to sensor nodes. This 'added value' could be measurable in nature — improvements in sensing accuracy, operational capability, reliability (at a system level) and ease of use of a sensor system, or, most importantly of an 'enabling' nature — allowing essentially for the design of conceptually new sensing applications and catering for evolved user-application interactions.

The rest of this chapter looks at ways in which these forms of 'intelligence' may be provided.

8.4. ANNs and MEMS

8.4.1. *The Motivation for Integrating ANNs and MEMS*

AI techniques are amongst the 'borrowed' macrosystems design tools that are presently receiving increasing attention from researchers in the area of microsensor systems. A variety of successful applications of AI, and in particular ANNs, to sensors and microsensors have been reported, including sensor metrological performance enhancement (calibration, non-linearity correction, offset, identification, as shown in Chapter 7), actuation control, sensor fault detection and classification, sensor data validation (analysis of sensor data, data restoration and validation, signal failure detection and reconstruction), sensor data mining and sensor fusion.

Table 8.1: What intelligence can bring to sensors.

Improvements in measurement accuracy	Improvements in operational capability and maintenance ability	Improvements in communication function and reduction of system failure	Improvements in ease of use or capability of large sensor networks
Linearisation of the relationship between the input and output signals	Remote maintenance operation using digital communication functions (zero and span adjustment, change of measurement range, etc.)	Monitoring of important data (input, output and self-check signals, etc.)	Self configuration and maintenance
Automatic zero-point calibration	Integration of different range sensors by widening sensor's rangeability (improving the flexibility to user specification changes)	Communication with upper system and surrounding systems	Data reduction
Automatic compensation of errors caused by environmental disturbances and changes (ex. temperature)	Storage and read-out of sensor data and process control data	Fault detection and fault prediction of control system	Information extraction
Automatic compensation of errors caused by changes in the process conditions (sensor working within a process)	Self check and self learning functions, emergency detection and alarm operation (out of range of measurement versus unusual environmental conditions, etc.)		Sensor fusion, application level processing

The first (and largest number of) successes come from the area of sensor fusion, where ANNs enable the extraction of a linear measurement of a hidden quantity (indirect measuring) from an array of sensors, each of which may exhibit non-linear and noisy behaviour [9].

More generally, from AI viewpoint the multisensor applications of NNs could be categorised as follows:

Classification:
The need is to relate the multidimensional input data to a predefined class that represents the state of the input space. A typical macrosystems example would be reaching a diagnosis decision on a process/plant, based on measurements from many sensors. Microsensors wise, sensor health decisions would fall in this AI application category.

Quantification:
Processing the information describing the input space in order to extract the values of primary variables within that space. An example here is the measurement of vehicle exhaust emission using cross-sensitive sensors, or any other odour quantification/inference from multiple measurements.

Description:
The need is to extract and present meaningful features or concepts that are representative of the input space, in other words, assigning qualitative descriptors to quantitative data. Calculating the overall risk of an event at application level from independent and/or individual sensory information is a possible application. Enhancing a sensor's functionality with the means of producing measures of 'trust' leading to decisions about the quality of its own data and its worthiness is another challenging application in this category.

The literature is rich in examples of all types above, particularly as applied to electronic noses. In this section, however, the interest lays with looking at how AI was used to realise smart, intelligent and cogent sensors, in terms of end-system (sensing system that is) functionality enhancement, quality of data and information extraction/extrapolation, rather than considering how the application the sensor is part of, was enhanced by AI.

ANNs, as a particular technique within the AI domain have been with us, actively researched and used successfully in a variety of applications, for more than 20 years. In some circles, they grew an often undeserved reputation of 'all problems solvers', maybe due to the 'brain like' paradigms which govern their functionality, or due to the little understanding of their mathematical functionality by people outside the AI community. Or maybe, it was just the hope that a new, 'all purposes' powerful tool for complex data analysis has been established and is ready for generic use. (It is true that, in general, if applied correctly, NNs outperform all other data analysis methods [10].) However, even after a fruitful period in the 90's when AI was applied to every domain with some success, their credibility for critical applications has not reached the desired level to enable many NNs based end-products. In the authors' opinion, one of the reasons could have been the fact that not many generic hardware implementations of NNs have been produced to exploit one of their main advantages, that of parallel, embedded processing ability. The most successful instances of neural computing, so far, have operated in association with a supporting computer system, restricting therefore their use to applications where the external computer was already part of the set-up, necessary for other functions. Another reason could be the 'fashion' accessory view on this technology.

Presently, however, VLSI technology is ready for NN implementations (several examples will be briefly reviewed here), hence, NNs have a great potential for becoming part of the intelligent and cogent sensors of the future. In the context of the proposed discussion, some of their main advantages are:

- the majority of intensive computation takes place during the training process;
- due to the self-learning capability of neural networks, no detailed *a priori* information regarding the system under analysis or design is required;
- the massive parallelism of neural networks offers a very fast and robust multiprocessing technique when implemented using neural chips or other types of parallel hardware.

Once the ANN is trained for a particular task, operation consists of propagating the data through the mapping produced by the ANN, thereby making possible the addition of real-time functions to systems, including microsystems.

MEMS technology is able to produce the micro sensors and actuators, and modern analogue VLSI NN implementations are able to match the time scales of such devices. The need for fast, powerful, complex processing associated with microdevices can therefore be fulfilled by integrating NNs and MEMS.

8.4.2. VLSI NN Implementations

In the sensing context, at present, there appears to be a strong tendency to develop high performance NN chips with embedded neural learning at chip level, rather than implement already trained, fixed structure NNs. Two such recent examples, presented below, come from NASA and are still in the realm of research. Other implementations have already made their way from private companies to the consumers, as described further in this section.*

As a first example, one can consider the chip design proposed by Dong [11], which is suitable for implementation in analogue VLSI. Although the aim was to produce a general purpose NN chip, its specification fits well with possible uses within the sensing domain. The work focuses on MLPs, which are the simplest and easiest to hardware implement NNs, but whose training time has always been seen as a drawback. Dong produced a conceptual design for actually accelerating learning. The circuitry supporting the learning processes would be fabricated in the same chip with the NN, eliminating the need of training the network via software and further implement the dedicated hardware. Fast learning is achieved by combining the standard back-propagation (BKP) algorithm with the cascade correlation functions. The network is able to evolve in hardware (hidden neurons are added when the rate of learning falls below a specified level, as one would do in software) providing the application designer with the same options as a software based NN implementation.

As opposed to the generic need for evolving NN hardware which motivated the chip design proposed above, in the second example here, the chip design steamed from the necessity of performing fast, a variety of computation-intensive tasks for maximising the return of useful scientific information from future multiple-sensor systems. The chip is dedicated to sensing applications where synergy among multiple sensors, limits on the power and bandwidth available for transmission of data, the need to recognise significant data in the absence of prior explicit instructions are

*Several general purpose ANNs implementations are also mentioned here to complete the depiction of the topic.

stringent requirements, as described in [12]. The novel type of compact analogue/digital electronic data processor devised is based on the concept of optimisation neural networks (OCNN) and designed around cellular NNs which are suitable for VLSI implementations in real-time, high-speed applications, due to their local connectivity, simple synaptic operators and massive parallelism. The OCNN brings together the ability to perform programmable functions for fine-grained processing with annealing control to enhance the quality of the output (hardware annealing processes are included in the chip). An application based on the chip has already been developed by Fang [13]. The system could be characterised as an eye/brain machine, as the conceptual design is intended to mimic basic functions of biological vision systems. The system performs computations 100 times faster with 10 times less power (only about 10 W) and 10 times size reduction than achievable with state-of-the art microcomputers, DSP chips and off-the-shelf components. In the application (Figure 8.1), an active-pixel-sensor camera collects the raw data, the neural computer generates synthetic images, fuses all images and analyses the fused images at high speed (achieved through parallelism and programmability). Further, a

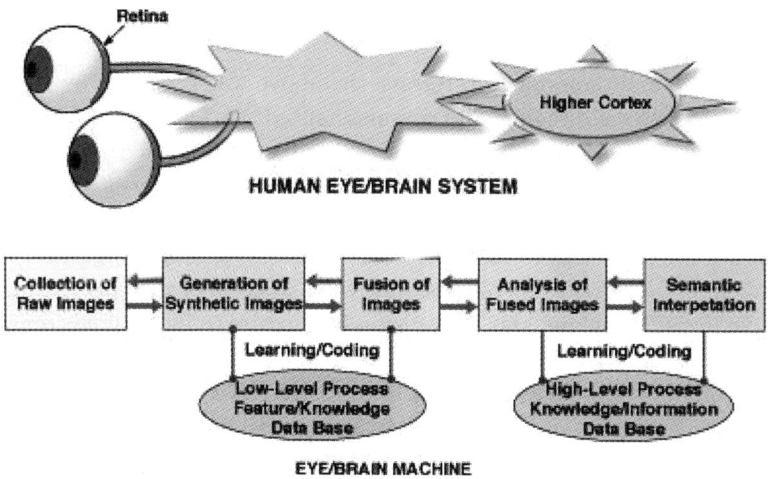

Figure 8.1: The computational aspect of the proposed system is based on a simplified model of the human vision system, with five stages of processing of image data. From [13].

microcomputer controls the overall operation of the system and performs the scene interpretation functions.

Overall, this is a good example of a 'cogent' sensor system, as defined in the previous section, as it has all the named 'cogency' characteristics: it performs semantic transformations on the sensor raw data and produces inferred information.

Moving on to the wider scope for NN hardware implementations, it is argued, for example that the development of smart sensing applications often involves processing of complex signal data into 'metric' or 'signature' form that can be compared against templates or archetypes representative of conditions to be recognised. These applications usually comprise of many sensor types and are a mixture of micro and macro scale systems.

Whilst traditionally, Fourier based methods have been used for such tasks, they are presently proving inadequate for the following (mainly implementation related) reasons:

- Producing a good metric requires a considerable amount of digital signal processing — therefore power;
- Exploiting the power of tools which allow nonlinear mappings, curve fitting, etc., in the frequency domain, subject to complexities such as dynamic time wrapping and impulsive noise artefacts that spread across the whole spectrum.

A time domain approach to resolve the above was proposed by Domain Dynamics [14] who designed and commercialised the TESPAR/FANN technology. Several new sensing applications have already used this technology, particularly in the area of biometrics. TESPAR is most suitable for applications where automatic signal recognition is needed. It involves the integration of novel time encoded signal processing and recognition with orthogonal, fast, ANNs in purpose designed structures that permit highly flexible decision making and data fusion hierarchies to be tailored to match the needs of the recognition or classification task. TESPAR coding was made available both as a software algorithm and as a low power ASIC. The ANNs implementation and training is foreseen to be supported by pRAM technology.

A variety of chips such as the three described so far have been reported, associated with very specific sensors or sensing applications. A detailed presentation of such works falls outside the scope of this section, where the intention was merely to bring the issues of hardware ANNs forward and possibly demonstrate their usefulness and worth in the context of sensing. One can't however end the section without reminding the reader of the most generic NN chip of all, the Neural instruction Set Processor (NiSP) which was the world's first dedicated neural computer on a single chip [15]. NiSP provided a plug-in (standard microprocessor interface) integrated, compact and cost effective (<£100) solution to NN implementations, supporting up to 8000 neurons and 65 000 synapses. Ideas from the conceptual NiSP design could be adapted for microsensor integration.

Whilst the main purpose of implementing ANNs on hardware is to realise their full potential as parallel distributed processors, hardware implementations introduce various processing errors, for instance due to synaptic weights quantisation and neuron characteristic variations. Particularly for applications where the networks needed involvement of large numbers of neurons, the use of simple synapses and neurons with low precision weights can have a noticeable detrimental effect. Various solutions have been proposed, resulting in hybrid networks composed for example of analogue sigmoid neurons and digital weights [16], although it must be said that, it is not due to such errors and variations that we see so few microsensor/NNs integrated designs. Rather, it is more likely that, due to a lack of basic research establishing the general applicability of NN solutions to these problems, integrated system designers have not been motivated to include specific NN processing capability into the design of their systems. Once the role of ANNs is established in relation to microsystems, efforts will certainly follow from the VLSI-NN research community to produce designs in technologies compatible with the ones chosen for MEMS and electronics integration.

Considering the simplicity of Mead's transistor level NN VLSI designs [17], there are comparatively few applications involving already-trained integrated sensor and NN implementations. In the authors' opinion, using simple, pre-trained NNs in conjunction with microsensors would be a feasible first step and would bring, for now, sufficient functionality enhancements.

A more detailed, albeit application independent, view on various options for hardware implementations of NNs can be found in Lindsey's lecture notes [18], and from Liao's survey [19] and a more specialist treatment of such implementations in conjunction with sensors and sensing systems is given in [5].

8.5. AI for MEMS Intelligence

Given the variety of AI and MEMS references in the literature, a possible route to a survey on 'AI for MEMS intelligence' seemed to be that of providing examples of applications for each sensor category defined in Section 8.2, and showing how 'smartness', 'intelligence' and 'cogency' have been achieved by means of NNs. While building the subsections here, however, it became apparent that, despite the promise of application of 'intelligence', in many of the cases surveyed, the AI techniques contribute only to a 'smart' sensor (in the authors' definition), let alone a 'cogent' one. There are several exceptions to this (as for example the case of virtual and software sensors, where AI techniques act on a system or array of microsensors and infer new information from multiple sensor data), highlighted in Section 8.5.3. The task was made even harder by the blurriness of the borders between 'smartness' and 'intelligence', when looking at the proposed sensors from an AI perspective.

8.5.1. *AI in Smart Sensors*

We classify the applications below as 'smart' as no additional functionality has been added to the sensor system, rather the quality of the data was enhanced by applying AI techniques.

Sensor metrological performance enhancement
With auto-calibration and self-test of intelligent transducers becoming a topic of major interest, developing analogue and digital methods for a transducer's characteristic interpolation and/or linearisation is an area of sustained research. Given the inherent non-linear behaviour exhibited by most sensors and the desire (or more often, in newer applications — necessity) to perform fast calibration, which effectively means reducing the number of calibrating points, curve fitting is an appealing strategy. A number of calibration methods exist for macrosensors which could be extrapolated to microsensors, for example based on Newton, Lagrange, spline functions and LMS regression — polynomials and ANNs. Of all, studies revealed that ANN interpolation was the most accurate, especially when multivariable extrapolation or non-linearity characteristics were under analysis, considering the number of calibration points for a given accuracy requirement, extrapolation capabilities and processing load [20]. Upon successful

training, the extrapolation errors with ANNs are lower both inside and outside the calibration range. The computational load with ANN interpolation is of the same order as polynomial interpolation. Although ANN training requires more computational resources than polynomial methods, this is not important in NN applications where training can be performed in a central host or implemented locally based on previous training weights and biases.

Following calibration, the adjacent issues are those of linearizing and compensating the sensors characteristic. ANNs were successfully integrated with micromachined capacitive acceleration sensors for such purposes, resulting in full hysteresis compensation and a considerable increase in the measurement accuracy, as shown in Chapter 7. The same method, based on simple multilayer perceptrons (MLPs) was also applied by one of the authors to tunnelling accelerometers [21]. Patra applied similar techniques to nonlinear pressure microsensors [22, 23]. Both MLPs and Functional Link ANNs have been considered and the resulting smart sensors implemented as 3-chip solutions — microsensor, switched capacitor interface and microcontroller unit to support the NN. The NNs were able to fully compensate for the temperature effects and provide reliable linear measurements.

As opposed to getting 'better' data from sensors, Almodarresi and White used AI to obtain sensor data 'faster' [24]. NNs coupled with pre-processing of sensor data was used to deduce and predict the sensor measurement values whilst still under transient behaviour.

Sensor Data Validation

Here, intelligent techniques are applied to validate or restore missing or corrupted data, by exploiting either physical sensor redundancy or using other observable states within the sensor or the application itself. The output is still data, as opposed to information.

Narayanan and Marks [25] exploited partial physical redundancy to recover failed sensor data in a closed environment. The assumption is that the sensor array generates interdependent readings among the sensors. If the dependence is sufficiently strong, the readings from one or more lost sensors can be estimated from the remaining ones. An autoassociative regression machine was trained from historical data. A solid theory was developed for the case where the sensor data are linearly related. The data from the sensors is processed globally, no local decisions are taken.

A software based, plant wide sensor monitoring system was developed by Hines *et al.* [26] and NN calibration methods were developed in [27]. The monitoring system is composed of an autoassociative NN, a statistical decision logic module, a faulty sensor correction module and a network tuning module. In respect with its overall functionality, the system lays at the border with 'cogency', however, it is not a generic, portable system but rather a strongly application specific one.

Sensing array calibration

Although individual sensing devices might have good metrological performance, when building measurement systems composed of a number of such devices, the measurement performance of the system does not necessarily equal that of the devices themselves. When building an integrated three-dimensional tactile sensor for use in space robotic applications and designed to meet specific constrains of robustness, reliability, compactness and low mass, Mei *et al.* encountered this problem [24, 28]. Their tactile sensor includes 4×8 sensing cells, each exhibiting a linear response to the three components of forces. Post-bulk micromachining was performed on foundry fabricated CMOS chips to produce the sensor cells. The sensor is constructed with five structural layers: rubber surface, force concentrating columns, sensing array, protection base and circuit base. Each cell has an E-shape square membrane structure fabricated by bulk micromachining. Strain induced by the applied forces on the membrane is sensed by three groups of integrated piezoresistors in the silicon membrane. The circuit base is made of a PCB for off-chip wiring. A metal case encloses the tactile sensor (except the rubber layer). The total sensor volume is $20 \times 50 \times 7\,\mathrm{mm}^3$ while the sensing area is $16 \times 32\,\mathrm{mm}^2$. The sensing cells were calibrated to determine their sensitivity before the rubber was used to cover them. However, once the rubber layer was mounted, the sensitivity of the cells was affected. As mechanical analysis of the multilayer structure has proved difficult, a MLP was used to compensate differences in individual cells sensitivity, compensate for temperature effects on the overall array sensitivity, and as a mechanism for avoiding both time and computation extensive mechanical analysis and the need for individually designed compensators later on. The NN had $3 \times 32 + 1$ inputs (32 three dimensional forces and the temperature parameter), one hidden layer and three outputs providing the correct force reading. The sensor was connected to the NN, and controlled, known forces were applied for training. Although not mentioned, it is presumed that, in use, the NN is implemented in software and hosted by a computer.

8.5.2. *AI in Intelligent Sensor Systems*

In this category the system as a whole uses sensors and AI but the sensor itself is not endowed with the data to information transformation.*

The adventurous Silicon Active Skin project [29, 30] lead by the Center for Neuromorphic Systems Engineering at Caltech, provides one of the best examples of intelligent Microsystems design. The work aimed to integrate MEMS sensors and actuators, neural network sensory processing, and control circuits all on the same silicon substrate to form a 'smart skin', capable of reducing drag on an aircraft wing. The system senses the shear stress, while a neural controller with feedback mechanism efficiently actuates robust micro actuators for surface stress reduction (Figure 8.2).

A sensor array containing a large number of sensors was initially developed for measuring the near-wall stream wise streaks. The sensors, produced by using surface micromachining have a sensitivity of 15 mV/Pa and, within the array, they are arranged such as to provide adequate spatial resolution in measurement. Flap type actuators were produced, driven by electromagnetic force. The neural network controller is of a MLP type (shown in Figure 8.3), with hyperbolic tangent non-linear hidden units and linear outputs and is trained off-line to predict actuation using data from near-wall controlled experiments. Furthermore, the controller is allowed to

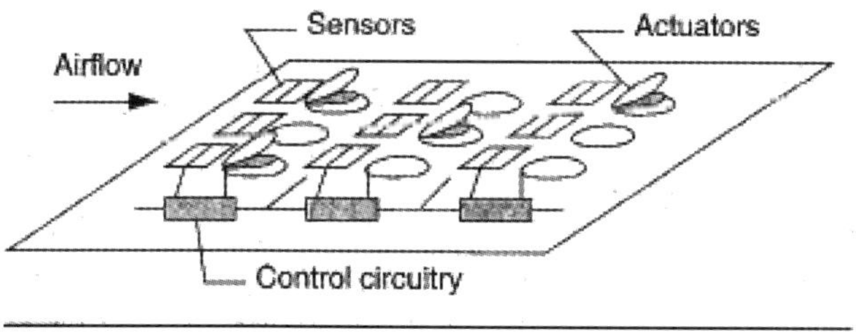

Figure 8.2: Simplified diagram of the hardware system. From [30].

*A single example is provided here, as, as mentioned before, the boundaries between 'smart' and 'intelligent' sensors and 'intelligent' applications are fuzzy and clear 'intelligent' sensor examples are hard to identify.

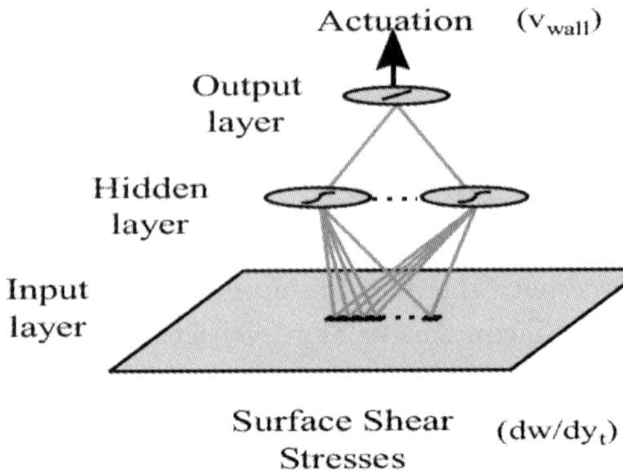

Figure 8.3: Active drag reduction control network architecture. From [30].

adapt on-line, as it is included in an on-line adaptive inverse model scheme (of the type discussed in Chapter 7). In simulation, a 20% shear stress reduction was obtained. The M3 system (1 cm × 1 cm die) monolithically integrates 18 micro shear sensors, 3 micro flap actuators, circuits for logic, sensor driver and actuator drivers. Most of the VLSI neural computation is performed in current mode. This example was classified as 'intelligent' as, following NN processing, actuation takes place within the sensor system, therefore, at some level, decisions are taken by the sensor system itself.

8.5.3. *AI in Cogent Sensors*

Our category of 'cogent sensors' includes sensors integrated with the means to reduce raw data to 'information' of the type required by a specific application. Up to date, most examples in the literature are of a multi-sensor nature, so they are, strictly speaking cogent sensor systems. Two categories here are: mono-type sensor systems (all sensors measure the same physical quantity) and multi-type ones. Examples follow for each category. At individual sensor level, cogent functions have been reported as work in progress and one example is presented here.

Multi-type sensor systems

In some applications, necessary system level information has to be inferred from available measurements of observable quantities using a statistical model. Such a model is referred to as a 'software' or 'virtual' sensor [31]. The actual measurement devices within such virtual sensors are often MEMS components. Software sensors were the first and major ANNs application to the instrumentation field. Today, they continue the lead in terms of number of research contributions, extended to microsensors and arrays, by continuous new developments in electronic noses, gas sensor arrays and image sensor arrays. The innovative aspects of such works are primarily application specific and reside in the integration of various techniques in a global system allowing for prediction of the quantity of interest. Some contributions extend the role of ANNs (or add on specific NN modules) to enhance the designed system with functions such as data validation, reconstruction, and analysis of uncertainties. The dominant architecture is a two-level one, with the first level consisting on the array of physical sensors and the second level consisting on a pattern recognition unit, implemented most often in software but occasionally in hardware. It is not within the scope here to survey the multitude of such developments (a good survey on electronic noses and particularly CMOS gas sensors incorporated in smart devices is given by Gardner *et al.* [32]), but rather to mention efforts oriented towards real-time, MEMS sensor-ANN embedded systems. Examples were sought where the role of the NN is not only or not mainly that of creating a software/virtual sensor but the NNs contribute or enable the designs or implementations of systems which would fulfil most integrated microsensor systems requirements.

In this context, Roppel *et al.* successfully designed a low-power, portable sensor system using mixed analogue and digital VLSI circuitry for on-board data pre-processing together with pulse coupled neural networks for feature extraction and also for pattern recognition [33]. The design is aimed at minimising cost, size, weight, power and post-sensing computational burden. The sensor test bed is a 30 nodes, MEMS sensor array consisting of tin-oxide gas sensors and the target is to discriminate among 7 odours (acetone, ammonia, beer, etc.). Spatio-temporal encoding is used for pre-processing of the dynamic sensor outputs. A 50-element integer valued signature vector for each sensor, for each odour is obtained. A further MLP acting as a classifier provides correct identification rates of 96% for the odour data sets considered.

Another good example is the gas sensing array reported in [34]. The array has 10 different SnO_2 microsensors fabricated on a substrate for the purpose of recognising various kinds and quantities of combustible gas leakages. A two-step recognition system is employed, where both the gas classifier and the gas concentration determination are implemented using NN techniques, supported by a DSP.

A thorough survey on electronic noses, with or without a NN component is given by Gardner [5].

Mono-type sensor arrays, sensor webs

The military (particularly DARPA) has historically funded much work in the NN field and continues to do so with several programmes running at present looking at the potential of NN in different aspects of design and implementation of large intelligent sensor arrays. With MEMS technology providing the means for deploying thousands of wireless sensors, inspiration is sought even more now from biological intelligent systems, as the amount of data streaming in from sensor networks would be impossible to digest in real-time by a central computer. It has been suggested that, if the sensor networks are wired in a way similar to that in which the spinal cord sends information up to the human brain from the eyes, ears, nose, etc., by placing next to each sensor a neural network memory element that 'learns' what is normal, the problem could be solved. Since the network manager (brain for humans) also has a copy of what is normal, it can refer to its copy rather than congest the network with redundant reports. The sensors would only transmit updates to the manager's copy when something abnormal happens. Neural learning could exponentially decrease network traffic and management overhead, resulting (hopefully) in only linear overhead increases for each added node [35]. Another project [36], aimed at exploiting the proven organisational abilities of natural NN, led to the proposal that functions be distributed across a network of sensors. The network itself then, collaboratively extracts information from the data field provided by the sensor network. The sensors are seen as ANN nodes. The ANNs are designed here to have architectures approximating those of biological NNs that performs specific functions. Three such architectures have been proposed:

- Centre Surround Architecture — suitable for the detection of movable edges across a receptive field (for example spread of seismic activity, radiation and toxic chemicals);
- Autoassociative NN — used for recalling old memories from partial or noisy stimulus (for example search for known gaseous, biological or geological signatures);

- Hypernetworks — groups of NN that can cooperate on vaguely defined tasks.

An example of Center-Surround Architecture is featured in Figure 8.4. The view is to use swarms of neural networks to accomplish tasks that would be impossible for a single large neural network. Here, a swarm could spread out to cover a large area or move in single file to go through a small opening. Swarms of flying sensor pods, organised with simple hyperneural rules similar to those of flocks of birds, could perform a wide variety of exploratory and data-collection tasks, based on a retinal neuron and its nearest neighbours. The balance between the stimulatory effect of light and the inhibitory inputs from the nearest neighbours is such that the central neuron is active only when the edge of the shadow crosses the central neuron sufficiently close to the centre. In a field of center-surround neurons, only those along the edge of the shadow are active.

Such systems are part of a 'third wave of computing' that could use NN to build sensors and other machines capable of the unsupervised learning exhibited by the human brain [37].

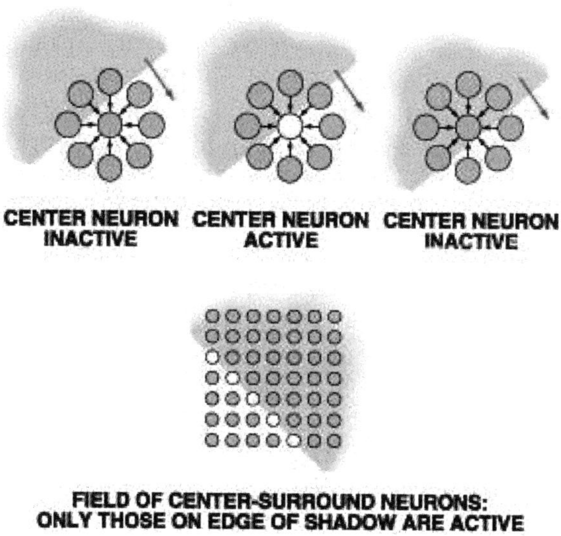

Figure 8.4: An example of Center-Surround Architecture. From [37].

Fault detection and classification

Over the past 10 years, neural networks have found wide application in systems that are designed to recognise fault conditions from sensor data, the derived information being whether the data is trustworthy or not [28, 38, 34]. If such information on the trustworthiness of data is derived by a microsensor about its own data, and independent of the application of which the sensor is part of, the microsensor would be subscribing to our 'cogent' property.

Although the literature abounds in ANNs aided sensor fault diagnosis, the approach normally taken is mostly to validate measurements obtained from sensors and to diagnose sensor faults by using a mixture of mathematical and knowledge-based modeling of the system under measurement (see for example [39]) on the one hand and the healthy sensor model information on the other hand. Hence, inconsistencies in measurement data are identified by comparing the data with the NN predictions. The preferred architecture is that of autoassociative networks, where a mapping is performed of actual sensor signatures with the learned and predicted healthy signatures for the given system or application inputs and state. The schemes are not portable and cannot be made generic as the underlying design principles involve the application itself. Another line of work exploits sensor physical or analytical redundancy to predict sensor measurements and compare them with the sensor being monitored. The work described in [40] is taken as an example here.

The objective is to apply NN techniques to sensor fault detection, isolation and accommodation on an aeronautic system in order to increase the system reliability and safety, extend its useful life, minimise maintenance and maximise performance. The system structure is presented in Figure 8.5.

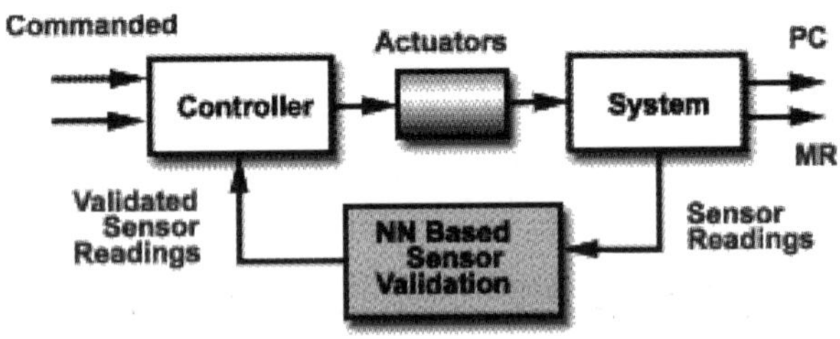

Figure 8.5: System structure for NN based sensor validation. From [40].

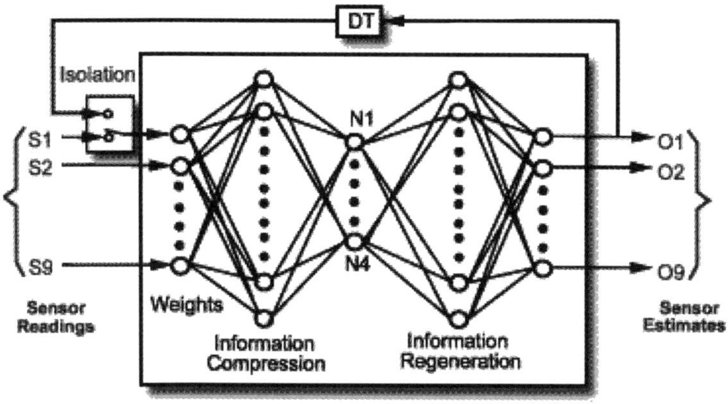

Figure 8.6: An auto-associative network is trained to produce an output vector that is equal to its original input vectors. From [40].

The approach uses analytical redundant sensor information and is based on learning from experimental data or simulated data.

The method is as follows: an auto-associative network (as the one in Figure 8.6) is trained to produce an output vector that is equal to its original input vectors. The operation of the auto-associative network is based on the principle of dimension reduction. The input information is compressed by a process of dimension reduction, before it is regenerated to recover the original information. The redundant sensor information is compressed, mixed and reorganised in the first part of the network. In the compression process, the sensor information is encoded into a significantly smaller representation. The compressed information is then used to regenerate the original redundant data at the output. Because of the information mixture, if a sensor fails, other redundant sensor data can still provide enough information to regenerate a good estimate for the faulty measurement. Because of its parallel-processing capability, the neural network can process real-time data for time-critical applications. Also, because it learns by example, the neural network does not require a detailed system model for sensor validation as is often required. After the training, the neural network can be implemented in a closed-loop configuration to validate the control sensors for system performance purposes. During operation, if a sensor signal is significantly different from the corresponding estimated value, the sensor signal is considered incorrect and the failed sensor is identified. The failed sensor reading is isolated by feeding the neural network its previous estimated value. The isolation of a failed sensor enables the neural network to detect subsequent sensor failures. Successful demonstration on various simulations including implementation in a real-time demonstration system of the Space Shuttle Main Engine (SSME) has been accomplished [40].

As opposed to the above approaches, the authors' microsensors work includes the development of a truly generic self-diagnosing sensor [41]. The fault detection function is intrinsic to the sensor functionality and provides the cogent sensor not only with the means of detecting its healthy/faulty state but also with the power of decision about its appropriateness of contributing its data to the specific application. The approach can be applied when two or more sensors of the same type are linked in an intelligent array. The method is based on the use of NNs both for detecting and classifying faults and is presented as a case study in Section 8.6, as the work has considerable bearing on the design issues discussed in Chapters 9 and 10.

Although not directly within the scope of this section, the review of AI applications to MEMS sensors and systems would not be complete without at least a brief mentioning of the suitability of AI techniques to design and modelling of MEMS devices. A few successes on these lines which have been reported in the literature, include the work of Ahmed and Moussa [42] in the area of MEMS modelling and synthesis, the NN based model reduction techniques developed Liang *et al.* [43] and the efforts to develop a CAD package for robust and efficient MEMS design based on evolutionary algorithms at Berkeley [44].

8.5.4. *Further Thoughts on how Biology Could Inform the Design of Large MEMS Based Systems*

Without the intention of trespassing the domain of 'Futurology', it is worth mentioning some further 'grander' applications where biologically inspired models could inform the design of advanced MEMS based systems. Probably the most serious study in this direction was produced by Price *et al.* [45], where the technological needs for various levels of built-in intelligence in future machines was analysed. The discussion here focused on the design principles for an intelligent integrated health monitoring system for potentially ageless aerospace vehicles (pictured in Figure 8.7) although the conceptual thinking behind is generally applicable.

Figure 8.7: Artist impression of the ageless vehicle — the morphcraft (image from NASA). From [45].

The perspective taken is that of a vehicle which may contain thousands to millions of sensors (enabled by MEMS and nanotechnology), each providing continuous streams of data. Inevitably, some sensors will malfunction, some will be out of calibration and some will loose data due to faulty communication paths. At a systems level, a health monitoring system has to deal continuously with dynamical environmental and overall system changes. It is foreseen that, in order for such a monitoring system to be able to cope with processing vast amounts of data to produce reliable information, several attributes linked to a formal definition of intelligence have to be integrated within the system:

- flexibility in responding to situations,
- opportunism (take advantage of fortuitous circumstances),
- extracting meaningful information out of ambiguous data,
- find similarities between situations and extrapolate to new situations and maybe highest of all,
- adaptively respond to produce original solutions to the situation (although the ability of AI in the near future to do anything close to this is doubtful).

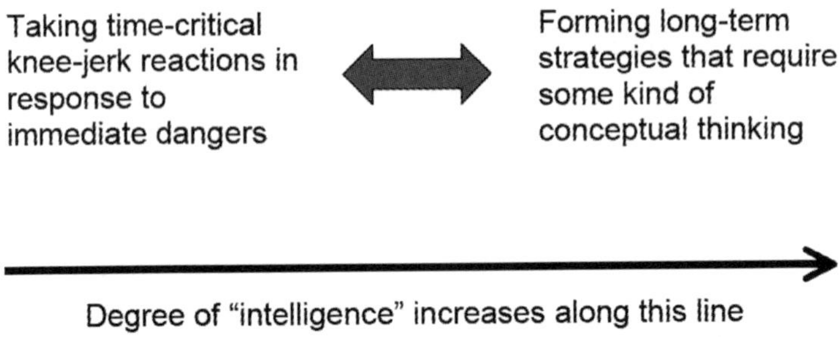

Figure 8.8: Various degrees of intelligence need to be built in various subsystems of an 'ageless' vehicle. From [45].

When viewing the overall system proposed here as a collection of subsystems, it is clear that different levels of intelligence are required in various parts, from fast automatic response to sudden stimuli, to long term, conceptual thinking based strategies, as expressed in Figure 8.8.

At macro scale, several approaches of achieving intelligence have been suggested, which include rule-based systems, systems of agents and control of emergent behaviour. One such approach is exemplified graphically in Figure 8.9.

At micro scale, one could implement adaptability (if not learning) by pursuing well known strategies as feedback (already extensively used to increase sensor performance and integrate sensors and actuators). However, to achieve conceptual thinking at micro level, learning is essential.

In most contexts, achieving intelligence through learning is attractive as it removes responsibility from the system designer. In some designs, a deterministic solution is perhaps not practicable due to the limits of our own conceptual abilities and our ability to write robust software. Instead, a learning system requires setting up a robust framework in which learning can take place. One major advantage of such a design approach is that it provides a way of producing useful and generic system components at the single device level, analogous to the way that the basic structure of a biological cell remains invariable over all the many different organs and organisms that are built from them. The adaptive behaviour of biological systems allows the configuration of the basic component without committing the overall design of the higher level system.

For future advanced sensor systems, it is clearly impossible to produce enough detailed information on their every likely function of to allow a

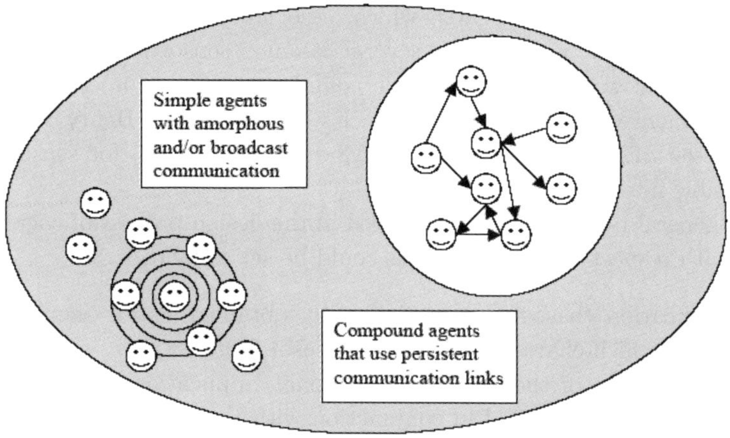

A range of intelligent agents may coexist to build the intelligence required for an ageless vehicle. The complexity of their interactions is governed by the media available for communication. Here, a group of agents using broadcast communication, with a connectivity that depends on proximity, is contrasted with a group of agents tied into a network of defined links, which itself may be considered an agent. On the one hand, having persistent links reduces flexibility; on the other hand, the information content of the link structure adds to the compound agent's intelligence.

Figure 8.9: Systems of agents providing control of emergent behaviour. Figure and caption from [45].

classical top down approach, which could result in a detailed requirements specification for the basic components. This problem can be addressed by viewing the whole system as a collaboration of autonomous, adaptive devices, namely cogent sensors.

8.6. 'Cogent' Sensors — Fault Detection Case Study

One of the design requirements of sensors and sensor systems for critical applications (and recently not only, but for many other general applications) is to enable unique sensor failure diagnosis and measurement validation with minimal sensor requirements, i.e. no hardware redundancy. In the view of the authors' definition of a cogent sensor, this should be achieved by exploiting the information content of readily available signals: the sensor output signal and contextual information gathered from the sensor's working environment. These capabilities could be incorporated into a validation and diagnosis module (VDM), which, associated with the sensor, should be able to detect, in real-time, several common sensor faults and failures, issue specific warnings and provide confidence indices for each validated measurement value. A similar approach was proposed by Henry, who introduced the SEVA-Sensor Validation Module concept [46] for sensors with accessible internal states.

In general terms, the steps involved in the design process of cogent sensors with respect to fault diagnosis could be set as follows:

- Identification (based on experimentally obtained sensor signatures) of features which characterise common sensor faults;
- Determination of the nature of additional, application related information, which can be used in conjunction with the sensor output for fault diagnosis and measurement validation purposes;
- Identifying the appropriate techniques for qualitatively and quantitatively representing the information gathered above;
- Identifying the optimal implementation strategy for the VDM;
- Producing and implementing a fault management strategy for the system the sensor is part of (this mostly refers to multi-sensor systems, i.e. sensor networks where the failure of sensors needs to be managed as well as identified).

A design example for the strategy described above is presented here, based on acceleration sensors, which were chosen due to their lack of available internal signals (hence, the fault identification and sensor health diagnosis task are so much more challenging).

8.6.1. *Sensor Faults*

In the development of a reliable signal-based diagnosis and validation strategy for individual sensors, one needs to consider not only sensor failures (soft and hard) but also situations which can give rise to faulty (false) measurement data, in the specific application where the sensor is used. For the chosen case study, a set of fault signatures for acceleration sensors could be obtained under laboratory conditions for several different scenarios.

For the initial proof of concept work presented here, different types of sensor fault conditions were modelled, using electrical equivalent circuits, to gain an initial idea of the nature of the fault signatures.

The test structure is represented by a beam, which will have some characteristic vibration frequency. The loosely mounted sensor is fitted on this beam by a compliance or spring. This is in itself a resonant system, with some characteristic resonance proportional to the mass of the sensor and the stiffness of the spring. An electrical analogue of this system can be produced, as shown in Figure 8.10.

Case 1:
The sensor being not rigidly attached to the structure under test. One can imagine a simplified version of this case as illustrated in Figure 8.11.

So long as the resonant frequency of the poorly mounted sensor is widely separated from that of the beam, the symptoms of the mounting fault can be clearly differentiated. Since a micromachined sensor is likely to be very

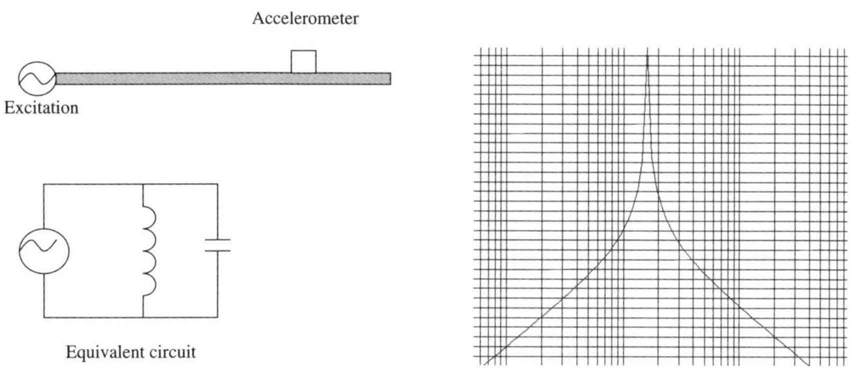

Figure 8.10: Beam model of structure under test, equivalent circuit and resonance.

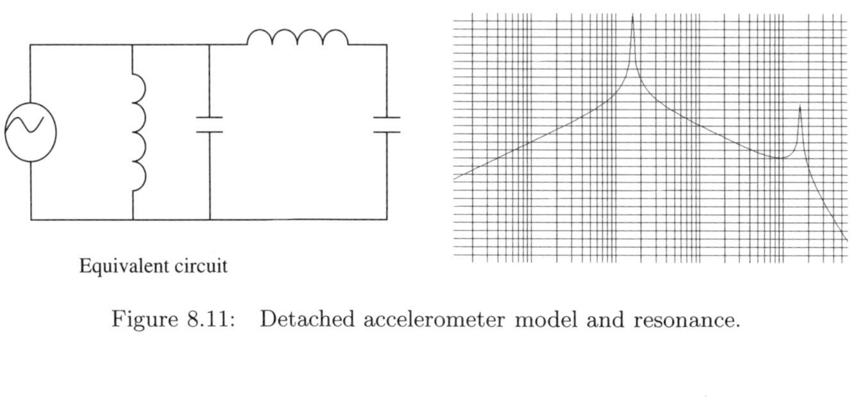

Figure 8.11: Detached accelerometer model and resonance.

light, and even a poor mounting quite stiff, it can be expected that the symptomatic resonance of the sensor mounting will be very high. Thus, at first sight it appears that a properly designed sensor should be capable of self-diagnosis of such a fault, given the opportunity to compare its frequency signature to that of a properly mounted counterpart.

However, one should also consider the way in which faults in the structure are likely to manifest themselves. Imagine that the beam shown above suffers an incomplete fracture. This results in two beams, connected by a compliance, giving an electrical analogue shown in Figure 8.12. This circuit is similar to that devised for the loosely mounted sensor, with the exception of the coupling compliance. Now, there are two beam resonances of concern, but in most cases it is likely that they will be at a much lower frequency than the symptomatic resonance of the sensor. The only case in which there might be some scope for misidentification is if the sensor is located close to the end of the beam and the fracture is located close to the sensor, resulting in a very high resonant frequency for the piece of the beam close to the sensor. If this frequency is in the range of that expected from a loose sensor, then the difference between the two signatures is the compliance linking the two resonant systems which occurs if the fault is in the beam, rather than the sensor mounting. The effect of this compliance is shown in the frequency plot in Figure 8.12. It is possible that a suitably designed diagnostic function could detect such a characteristic, but the presence of information from adjacent sensors as to the magnitude of such a resonance would certainly make diagnosis easier.

Case 2:
An extreme case of the above is if the sensor detaches completely from the structure under test, in which case no resonances will be detected.

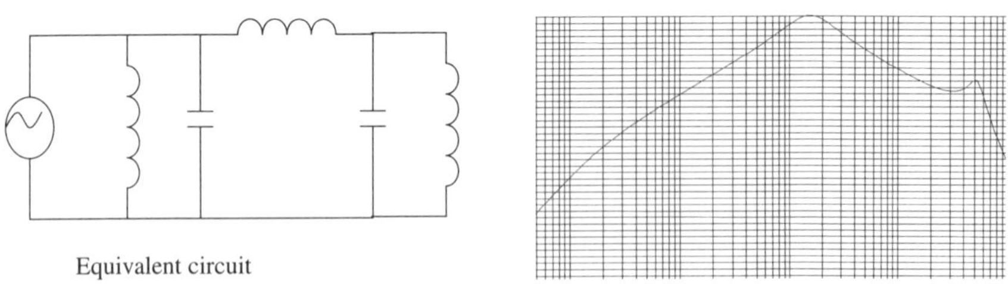

Equivalent circuit

Figure 8.12: Model of fractured beam.

Such a condition may occur simply because there is no stimulus to cause resonance. Information from other sensors in the locality can be used to identify whether this is the case, the fault condition 'detached sensor' being indicated by a single sensor detecting no resonances surrounded by ones which do. Complete loss of function of a sensor, for instance due to loss of power supply, would cause similar symptoms, except that there would not even be residual noise detected.

Case 3:
Another related case is if the sensor housing suffers some structural damage. Such a problem may include a number of situations, the simplest of which is a small part of the housing becoming partially detached. Once again, this takes the form of a small mass attached to the sensor by a compliance. The difference from *Case 1* is that the mass is still smaller and so the resonance will be higher, thus the discussion above applies, although the two resonances should be more easily separated.

Case 4:
Changes in ambient conditions can affect the sensor and could be detected and, if the sensor is properly equipped, compensated for appropriately.

Case 5:
Parameter drift of individual sensors can generate faulty measurements. Detecting this fault type, as well as faults induced by internal damage to the sensor structure would depend on detecting the variation of the output of that sensor from its neighbours, once again in a way which could not be confused with symptoms of the structure under test. The nature of these variations would need to be precisely characterised to be detected.

Faulty and healthy sensor signatures obtained for the cases above can be analysed with a view to extracting their characteristic features. An example of a partial analysis is proposed below.

8.6.2. *Neural Network Based Sensor Health Diagnosis*

Following on from the discussion in Section 8.6.1, it becomes clear that sensor self-validation on the basis of the sensor output only is impossible. However, within an application, a set of rules based on physical principles can be deduced for the data expected from neighbouring sensors. For example, acceleration measurements from sensors situated at adjacent locations along a cantilever will only be permitted to be different within certain limits imposed by the expected accelerations to which the object is subjected. Whilst any multi-sensor application would offer opportunities for discovery of relationships between sensor signals, it should be noted that the application-related reasoning should be kept to a minimum, in order to maintain the overall generality of the fault detection procedure and insure portability of the method and its implementation. Moreover, as will be discussed in Chapters 9 and 10, the associated fault detection management protocols are influenced by the specific detection method which in turn should not be to tightly linked to the application in hand.

For the study presented here, it is assumed that two neighbouring sensors, S1 and S2, are measuring, for simplicity, the same acceleration.* The diagnosis system under design is associated with sensor S1 (DIAGNOSTIC NETWORK 1) and the contextual information is provided by sensor S2 (Figure 8.13). For design purposes, the input acceleration for the sensors ($a_1(k) = a_2(k)$) is a filtered white noise signal, with a frequency range of 0–100 Hz, sampled with a rate of 4 kHz. Only one sensor fault was considered at this stage, corresponding to *Case 1* in the previous section.

The scaled input acceleration, healthy sensor output and faulty sensor output are shown in Figures 8.14(a) and (b). The diagnosis module to be designed will consist of an ANN, whose task is to identify the healthy/faulty response of sensor S1 in the following situations:

- S1, S2 *Healthy*;
- S1 *Healthy*, S2 *Faulty*;
- S1 *Faulty*, S2 *Healthy*.

The ANN proposed is of a tap-delayed feed-forward type, with two hidden layers, trained by dynamic backpropagation (with a momentum term and a variable learning rate). Use is made of tap-delayed-lines (TD) in order to incorporate the dynamic behaviour of the sensor into the model. Two delay units are necessary, to generate the one-step and two-steps back sensor output signals ($S1(k-1), S1(k-2), S2(k-1), S2(k-2)$), respectively. Hence, the present and the past values of the sensors S1 and S2

*Note that the fact that both sensors sense the same acceleration does not bring any limitations to the methodology presented here; the important aspect is that the output of two sensors (at least, as it is going to be discussed later) are needed to accomplish the task.

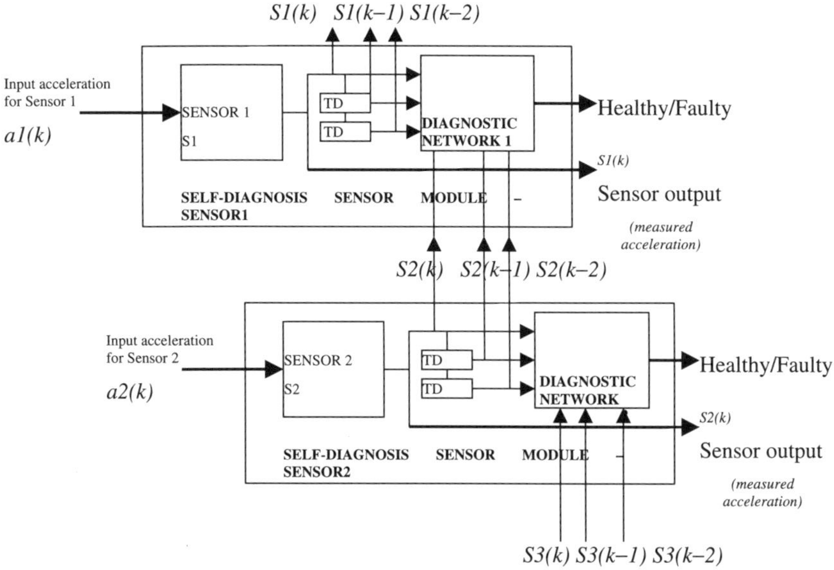

Figure 8.13: Configuration of the diagnostic Neural Network.

outputs form the input vector to the neural network. The network output represents the *Healthy/Faulty* condition of S1 at any instant of time (k). Since no feedback loop exists in the model, static error backpropagation (BKP) can be used to adjust the network parameters.

Based on these considerations, the electrical equivalent circuits in Figures 8.10 and 8.11 were simulated in SPICE in order to gather the input-output network training data. The ANN was trained and tested using Matlab. The best network performance, in terms of least false alarms and highest correct diagnosis rate, was obtained with a $6 \times 31 \times 17 \times 1$ network architecture.

The network performance on a test set, produced under the same conditions as the training set, is shown in Figure 8.15. The continuous line represents the correct diagnosis expected from the sensor, with $+0.99$ identifying the *Healthy* condition and -0.99 identifying the *Faulty* condition; the dotted line represents the actual DIAGNOSTIC NETWORK 1 response. A 'zero level' decision boundary would mean that the network correctly assesses the sensor's health, with two exception, both corresponding to the case where the sensor S1 is *Faulty* and S2 is *Healthy*.

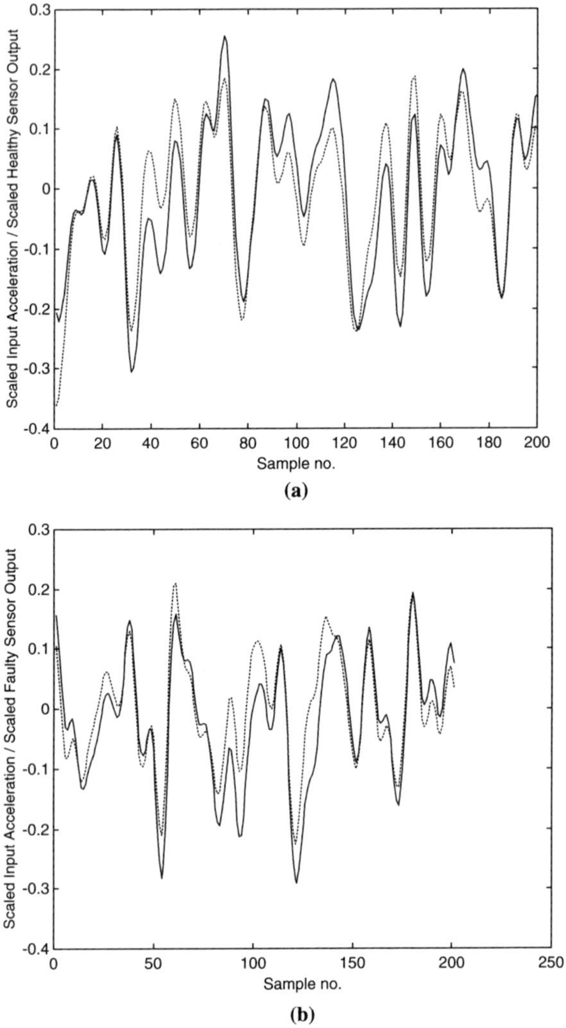

Figure 8.14: Scaled input acceleration for (a) a healthy sensor; and (b) a faulty sensor.

It has therefore been possible to design a working self-diagnosis module for the acceleration sensor considered, on the basis of its own output signal and a minimum of contextual information, non-application based, provided by one neighboring sensor. The results obtained encourage the continuation of this line of research towards multiple faults diagnosis. It should be noted

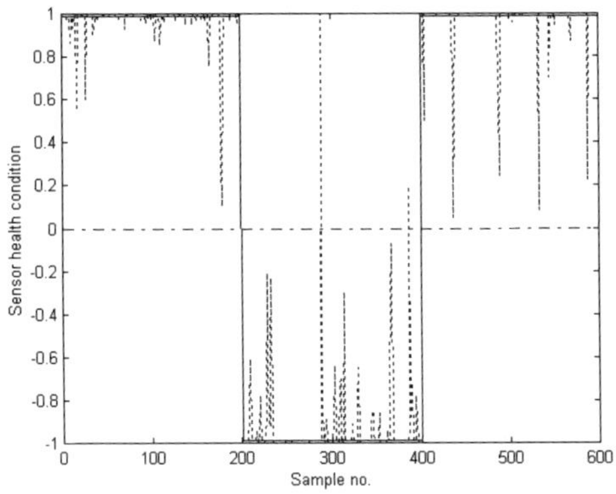

Figure 8.15: Diagnostic network performance.

here that a single neighbour providing context for the diagnosis module might not be sufficient for multiple faults identification and classification.

To summarise on the case study: the main thrust of the proposed method is the use of existing hardware resources in an array or network of sensors. The outmost merits of the method are:

- No redundancy in the array is necessary for the purpose of diagnosis alone;
- No central data mining is required at application level, but local, individual decision are taken by the sensors about their own state of health, which is a cogency characteristic;
- All, or at least most, of the information required for the detailed, application specific design and implementation of the method is gathered early in the sensing application design process;
- The communication level necessary for performing the diagnosis is low: at any one time, only two neighbouring sensors need to exchange data in order for the diagnosis decision on a particular sensor to be made;
- The method can be implemented in real-time; the diagnosis is performed based on a three steps back only history of the sensor readings;
- No *a priory* information is needed on the devices themselves as the sensor signatures are to be learnt and analysed by the NN;
- The method in itself is application independent and easily scalable; only the drawn rules for NN training are sensor type and application dependent.

Whilst the case study showed the feasibility of the proposed diagnosis method, the recipe is not complete and work needs carried out to extend and validate it both within a real application and from a generic viewpoint of a sensor network. In the context of this book, the case study should be read as supporting the advocated sensors and networks design choice of performant sensor based signal processing and network de-centrality.

8.7. Conclusion

By way of concluding this chapter, the following reasoning in support of AI techniques and tools is offered.

The basic purpose of sensors is measurement. Achieving high metrology performance is the primary design aim, which has been resolved through one or both of the following approaches: technological perfection (which is inherently expensive and difficult to achieve) or the application of structural or structural-algorithmic methods. Relaxing the requirement for technological perfection allows designers to achieve the same performance with lower cost, design effort and on shorter time-scales. The second approach not only allows for increased measurement accuracy but also allows the extension of functional capabilities of such systems. The concept of the cogent sensor extends this principle further. What is important is not so much the quality of measurement itself, but the quality of the information derived from it. With cogent sensors the information required by the application may be available, in the form required by the application directly from the sensor or network of sensors. The information may directly reflect the sensed data or it may have been deduced and sifted by the application of degrees of 'intelligence'.

In the stride to embed intelligence in microsystems, ANNs play an important role. However, although attempting to mimic the brain, present day neural networks are very far from exhausting the possibilities of a brain-inspired thinking architecture. There is much work still to be done in this area, particularly towards a better representation of information and knowledge.

Many examples of ANN/microsensor integration (sometimes algorithmic rather than physical integration) have been published, some of which have been surveyed here. Care was exercised to choose such examples that could, eventually, lead to the development of 'cogent', generic, application independent, autonomous sensors, although in their present state of development, the sensors in these examples might have only been classed as 'smart' or 'intelligent'. The suitability of bringing together MEMS and AI techniques has been proven many times over and, in the authors' opinion there is scope for great developments in this area.

Whilst when applied to isolated sensors neural networks' usefulness had to be strongly justified, when dealing with large numbers of sensors, their power is much more apparent. The more inputs and outputs are being processed, the more uncertain and noisy the overall data becomes and the more the need for non-deterministic procedures to bring together the best approach to extracting information. Several NN advantages in the context of sensor arrays were pointed out in this chapter: their application dramatically reduces the effects of individual sensor noise, accommodates sensor-to-sensor variations in arrays and networks by treating the variation as noise, and, most importantly, they are capable of multi sensor fusion and information inferring. All these, added to their general abilities of black box non-linear processing (much in need by the nonlinear sensors), objective optimisation, interpolation and extrapolation make NNs good candidates for a variety of functional enhancements in sensor systems. Another characteristic is their VLSI simple (particularly in current mode CMOS) and economical implementation together with their suitability to solve many of the problems a sensor system or a system of sensors might have, rather than be applied for a unique purpose (with 20–30 neurons per function they can calibrate, linearise, fault detect, fault classify, signal validate, signal restore, noise filter and compress, etc.). In this respect, the large, thousands of neurons on-chip trainable VLSI NNs, if integrated with MEMS could revolutionise sensing.

References

1. White, N. (2001) Smart move for intelligent sensors, *Sensor Review* **21**(1), Viewpoint.
2. Ko, W.H. and Fung, C.D. (1982) VLSI and intelligent transducers, *Sensors and Actuators* **2**, 239–250.
3. Varadan, V.K. (2001) MEMS- and NEMS-based complex adaptive smart devices and systems, *Proc. SPIE*, Complex Adaptive Structures, William B. Spilman (ed) **4512**, 25–45.
4. Ranky, P.G. (2002) Smart sensors, *Sensor Review* **22**(Issue 4), 301–311.

5. Gardner, *Microsensors, MEMS and Smart Devices.*
6. Frank, R., *Understanding Smart Sensors.*
7. Breckenridge, R.A. and Husson, C. (1978) Smart sensors in spacecraft — The impact and trends, *American Institute of Aeronautics and Astronautics and NASA, Conference on 'Smart' Sensors*, Hampton, VA., Novmber 14–16, AIAA, 6 p.
8. Yamasaki, H. (Ed.) (1996) *Intelligent Sensors*, Elsevier Science.
9. Corcoran, P. and Lowery, P. (1995) Neural network applications in multisensor systems, *Sensor Review* **15**(4), 15–18.
10. Roberts, S.J. and Penny, W. (1997) Neural networks: friends or foes? *Sensor Review* **17**(1), 64–70.
11. Dong, T.A. (2003) *Cascade Back-propagation Learning in Neural Networks*, http://www.nasatech.com/Briefs/May03/NPO19289.html.
12. Fang, W.C., Sheu, B.J. and Wall, J. (1998) VLSI neural processors based on optimisation neural networks, *NASA Briefs*, http://www.nasatech.com/Briefs/jan98/NPO19989.html.
13. Fang, W.-C. (2000) *Low-power Fast Machine Vision System on a Single IC Chip*, http://www.nasatech.com/Briefs/Feb00/NPO20449.html.
14. George, M.H. (1997) TESPAR paves the way for smart sensors, *Sensor Review* **17**(2), 131–137.
15. Goodwin, M. (1994) Neural processing set to boost sensor technology, *Sensor Review* **14**(3), 20–22.
16. Djahanshahi, H. *et al.* (2001) Quantization noise improvement in a hybrid distributed neuron ANN architecture, *IEEE Trans. Circuits and Systems II: Analogue and Digital Signal Processing* **48**(9), 842–846.
17. Faggin, F. and Mead, C. (1990) VLSI implementation of neural networks, In: Zorneter, S.F. (ed.) *An Introduction to Neural and Electronic Networks*, Academic Press Inc., Arlington, Virginia, USA, pp. 275–300.
18. Lindsey, C.S. (2002) *Neural Networks in Hardware: Architectures, Products and Applications*, Lecture notes, http://www.particle.kth.se/~lindsey/HardwareNNWCourse/home.html.
19. Yihua Liao, *Neural Networks in Hardware: A Survey*, http://ailab.das.ucdavis.edu/~yihua/research/NNhardware.pdf.
20. Pereira, D., Silva Girao, J.M. and Postolache, O. (2001) Fitting transducer characteristics to measured data, *IEEE Instrumentation and Measurement Mag.*, December, pp. 26–39.
21. Kraft, M. and Gaura, E. (2001) Intelligent control for a micromachined tunnelling accelerometer, *Proc. Int. MEMS Workshop (IMEMS)*, pp. 738–742, Singapore.
22. Patra, J.C., Van den Bos, A. and Kot, A.C. (2000) An ANN-based smart capacitive pressure sensor in dynamic environment, *Sensors and Actuators* **86**, 26–38.

23. Patra, J.C. and van der Bos, A. (2000) Auto-calibration and compensation of a capacitive pressure sensor using multilayer perceptrons, *ISA Trans.* **39**, 175–190.
24. Almodarresi, S. and White, N.M. (1999) Application of artificial neural networks to intelligent weighing systems, *IEE Proc.-Sci. Meas. Technol.* **146**(6), 265–269.
25. Narayanan, S. and Marks, R.J. (2002) Set constraint discovery: Missing sensor data restoration using auto-associative regression machines, *Proc. IEEE World Congress on Computational Intelligence*, Hawai, pp. 2872–2877.
26. Hines, J.W., Uhrig, R.E. and Wrest, D.J. (1997) Use of autoassociative neural networks for signal validation, *Proc. NEURAP97*, http://web.utk.edu/~hines/publications.html.
27. Hines, J.W. and Gribock, A.V. *et al.* (2000) Improved methods for on-line sensor calibration verification, *Proc. 8th Int. Conf. Nuclear Engineering*, USA, http://web.utk.edu/~hines/publications.html.
28. Mei, T., Ge, Y., Chen, Y., Ni, L., Liao, W.H., Xu, Y.S. and Li, W.J. (1999) Design and fabrication of an integrated three-dimensional tactile sensor for space robotic applications, *IEEE MEMS 99'*, Orlando, USA, pp. 112–117.
29. Lee, C., Kim, J., Babcock, D. and Goodman, R. (1997) Application of neural network to turbulence control for drag reduction, *Phys. Fluids* **9**, 1740–1747.
30. Koosh, V. and Babcock, D. *et al.* (1996) Active drag reduction using neural networks, *Proc. Int. Workshop on Neural Networks for Identification, Control, Robotics, and Signal/Image Processing*, http://www.rodgoodman.ws/Active%20Skin.htm.
31. Valentin, N. and Denoeux, T. (2001) A neural network-based software sensor for coagulation control in water treatment plant, *Intelligent Data Analysis* **5**, 23–39.
32. Gardner, J.W., Cole, M. and Udrea, F. (2002) CMOS gas sensors and smart devices, *Proc. IEEE Sensors 2002* **1**, 721–726.
33. Roppel, T., Wilson, D.M., Dunman, K., Becanovic, V. and Padgett, M.L. (1999) Design of a low-power, portable sensor system using embedded neural networks and hardware preprocessing, *Proc. Int. Joint Conference on Neural Networks IJCNN'99*, Washington, D.C., July 10–16.
34. Lee, D.S. *et al.* (2002) SnO_2 gas sensing array for combustible and explosive gas leakage recognition, *IEEE Sensors Journal* **2**(3), 140–149.
35. Colin, J. (2001) *Smart Sensors Extend Web Scale*, April, EETimes.
36. Hand, C. (2002) *ANNs for Organizing Sensor Webs*, http://www.nasatech.com/Briefs/july02/NPO30317.html.
37. Leopold, G. (2003) *US Military Deploys Neural Network Technology*, March, EETimes.
38. NASA (1998) *Neural Network Based Sensor Validation*, http://www.grc.nasa.gov/WWW/cdtb/projects/fdia/nnsv.html.

39. Xu, X., Hines, J.W. and Uhrig, R.E. (1999) Sensor validation and fault detection using neural networks, *Proc. Maintenance and Reliability Conference (MARCON 99)*, Gatlinburg, TN, http://web.utk.edu/~hines/publications.html.
40. NASA (1998) *Neural Network Based Sensor Validation*, http://www.grc.nasa.gov/WWW/cdtb/projects/fdia/nnsv.html.
41. Gaura, E.I. and Newman, R.M. (2003) Intelligent sensing: Neural network based health diagnosis for sensor arrays, *Proc. Advanced Intelligent Mechatronic*, Japan, pp. 36–365.
42. Ahmed, H. and Moussa, W.A. (2003) Optimising the performance of electrostatic comb-drive actuators using neural networks, *Proc. Int. Conf. MEMS, NANO and Smart Systems*, Alberta, Canada, pp. 62–68.
43. Liang, Y.C. *et al.* (2001) A neural-network-based method of model reduction for the dynamic simulation of MEMS, *J. Micromech. Microeng.* **11**, 226–233.
44. Zhou, N., Agogino, A. and Pister, K. (2002) Automated design synthesis for micro-electro-mechanical systems (MEMS), *Proc. DETC 2002: Design Automation*, Montreal, Canada, http://best.me.berkeley.edu/~aagogino/Papers.total.word.html.
45. Price, D., Scott, A., Edwards, G., Batten, A., Farmer, A., Hedley, M., Johnson, M., Lewis, C., Poulton, G., Prokopenko, M., Valencia, P. and Wang, P. (2003) *An Integrated Health Monitoring System for an Ageless Aerospace Vehicle, Structural Health Monitoring 2003: From Diagnostic & Prognosis to Structural Health Management*, Fu-Kuo Chang (ed.), DEStech Publications (Lancaster PA), pp. 310–318, www.ict.csiro.au/CISD/Publications/NDE/IWSHM4%20Paper%20031015.pdf.
46. Henry, M.P. (1994) Validating data from smart sensors, *J. Control Engineering* **41**(9), 63–66.

CHAPTER 9

SENSOR ARRAYS AND NETWORKS

by *Robert Newman and Elena Gaura*

9.1. Potential of Sensor Arrays

In many application areas for sensor systems there has been a move from systems relying on a few sensors to ones which make use of many sensors. It requires some understanding of the particular application domain to see why this is the case. One of the main considerations is that the over-provision of sensors can in some instances relax many of the constraints involved in the design of the sensing system, or allow sensor systems to be designed before the phenomenon under observation has been fully characterised. Consider an application in which the requirement is to measure the noise field emanating from a factory or airport. If the number of sensors is restricted it is necessary to know exactly the best location for each measurement to be taken so that a sensor can be located exactly on that spot. Planning the data gathering exercise requires a large amount of forward planning, and some kind of modelling, just to find out where the key locations for sensor placement are. Even then, it is quite possible that an unexpected sound peak, caused, for example, by constructive interference of sound reflections from buildings, will be completely missed. By contrast, if it is possible to use very many sensors the area of interest can be liberally covered with them, with the result that there is likely to be a sensor close to each optimum location and even unexpected phenomena will be measured.

The usefulness to the sensing system designer of large sensor arrays has been noted by several writers in the field including Culshaw [1], writing specifically about adaptive structures. Likewise, Varadan and Varadan [2] present a thorough exposition of the some of the many different sensor types that are amenable to use in what they call 'smart systems', (which are defined as 'a device or array of devices that can sense changes in its environment and makes a useful or optimal response by changing its material properties, geometry, or mechanical or electromagnetic response') and note the frequent requirement to use a multiplicity of sensors. The sensor types considered here include accelerometers, gyroscopes, acoustic ice sensors, acoustic crack sensors, and electronic tongues and noses. Actuators include pumps, valves and acoustic generators.

Because of the multifarious nature of sensing and actuating available, and the potential of multiple and many-sensor systems, a great deal of work has been directed towards the design of sensory systems which use very large number of sensors. Most of the so far suggested applications of this work only scratch the surface of the ultimate potential. As the technology matures many more innovative and useful applications will be developed.

Although these large sensor arrays have in some ways been seen as a solution to a range of metrological design problems actually realising them is very difficult. Their implementation poses a number of real problems:

- In many applications, the sheer size and weight of the sensor devices can be problematic. In some applications, such as monitoring of aircraft, excessive weight cannot be tolerated, and in many others, the mass of attached sensors can significantly alter the dynamic behaviour of the system, which is the intended subject of the measurement being made by the sensor array.
- In a conventionally designed system, each sensor would have to be connected individually to a multiple channel data acquisition system, which would be likely to be difficult to realise due to data bandwidth and processing power requirements.
- Connection of the sensors is also problematic. If we assume a passive, non-amplified sensor, then very good quality wiring must be used to connect individually each sensor to the data acquisition subsystem.
- The cost of the system can be very high. Typically the cost derives from two factors. Firstly, there is the cost of the devices themselves. As has been seen in Chapter 4, it is not just the manufacturing costs which can be high, the costs involved in calibration of many sensors can also be prohibitive. Secondly, in a system with very many sensors, the installation

costs can be high. Each sensor must be individually located and wired up, requiring large amounts of highly skilled labour.

Using conventional sensors to build a large sensor array is, therefore, not a practical proposition. MEMS sensors however, integrated with digital processing provide a solution to a number of the problems above.

The *size* and *weight* problem may be overcome because MEMS sensors are very small and light.

The *cost* problem may be ameliorated, since MEMS sensors are made using a mass manufacturing process, and therefore may be inexpensive if produced in sufficient volumes. If the sensors are self-calibrating and self-organising then the installation costs can be significantly reduced. As an extreme example of the possibilities raised by reductions in size and cost, in some suggested environmental monitoring applications (mainly military), the sensors are effectively thrown away, because installation can be a major expense or impracticable if the terrain to be monitored were occupied by hostile forces. One well publicised application of this type is 'smart dust' [3, 4]. Given a 'throw away' solution, sensors can be simply be deployed by dropping them from an aircraft. In non-military applications it helps if the sensors are small and light enough so as not to cause collateral damage if they land in unintended areas. Acoustic monitoring can be used to locate travelling objects emitting sound, with obvious applications in the battlefield arena, and less obvious ones in the civilian arena. Most of these applications are short term, rather than permanently fitted over an extended period of time.

The *connection* problem can be significantly simplified, for three reasons. Firstly, a sensor endowed with a digital processing capability may transmit its data on a multi-drop or wireless network, with considerable or total simplification of the wiring requirements. Secondly, a digital transmission medium can be considerably more robust than a low-level analogue one. Finally, the processing power may be used to process and reduce the data so there is less (and in some cases a lot less) data to be transmitted.

The building block for these systems is variously called by different writers the 'smart', 'intelligent' or 'cogent' sensor. Our definitions of these terms were discussed extensively in Chapter 8. The discussion in this chapter concerns mainly what we would define to be 'intelligent' sensors, which may be 'cogent' if the integrated processing power is harnessed for information transformation. However, the historical context of the design of these systems is important, because established practice has a habit of enduring into new situations, even to the stage where designs based on new practice may be easier to execute. Thus one can still find large sensor array systems based on sensors which are merely 'smart' by our definition. (Indeed, there is a complete standard for such 'smart' sensors, the IEEE 1451 standard, which is discussed extensively within this chapter).

This chapter will examine the hardware design (primarily digital electronic) of sensor array systems, starting with the design of an individual sensing element or 'node' in the array. The term 'node' comes from network usage. As shall be seen, generally, sensor arrays are implemented electronically as a network of sensing devices. When the term 'array' is used it refers to the replication of sensing devices used in a metrological system. The term 'network' refers to the mode of electronic connection of those sensing devices. Since sensing arrays are frequently connected as networks, the terms 'sensor array' and 'sensor network' are often, but not always, interchangeable.

The node design that will be introduced is grounded in the assumption that sensor arrays will generally be implemented using networks, and Section 9.3 traces the historical development of sensor array connection in order to establish that this assumption is well founded. This discussion continues to consider some of the latest connective technologies, namely wireless (generally radio) methods.

Section 9.4 outlines the broad design parameters and considerations for the next generation of sensor networks, based on the properties of sensory arrays which are emerging or likely to emerge in the near future. Section 9.5 discusses some of the more important of these considerations in detail, including networking network technologies and topologies, network discovery, 'locationing' (discovering the physical locality of nodes) and means of providing synchronisation or a common time standard between nodes. Software organisation within the nodes and the overall network is discussed in Chapter 10.

9.2. Node Design

Figure 9.1 shows a block diagram of a networked intelligent sensor, as might be used as a part of a networked 'intelligent sensor' based system, as

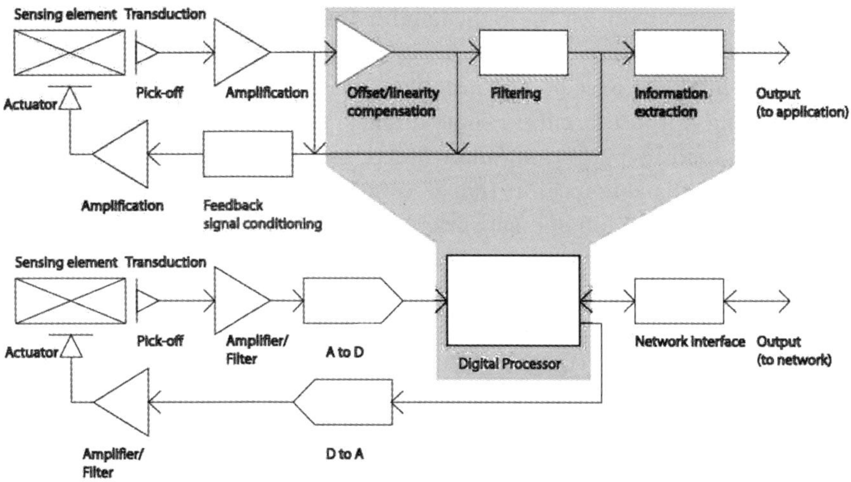

Figure 9.1: Networked intelligent sensor, compared with functional diagram used in Chapter 3 — Figure 3.1.

discussed above. The diagram also shows the functional block diagram for a sensor system, taken from Chapter 3. It is interesting to contrast the two. The functional diagram relates to the signal and information processing functions to be performed in a sensing system and there are many different possible hardware realisations of those functions. The block diagram in the lower part of Figure 9.1 is an architectural diagram of a piece of hardware, albeit a generically defined one. Within this architectural definition, there are many different variations of design detail that could be made. Nonetheless, several basic design decision have been set here, including the decision, discussed at length in Chapter 3, as to the stage in the signal processing chain to make the analogue to digital transformation. In this 'networked sensor', the compensation, filtering and information extraction functions take place in the digital domain, *at the sensor.* The 'output to application' is via a network, for which purpose an additional piece of hardware (and related software functions, which are the preserve of Chapter 10) – the network interface, appears in the diagram. The analogue to digital transfomation is indicated in this diagram, (which represents the hardware subsystems) by the analogue to digital converter (the reverse transformation, required by a 'hardware in the loop' sensor, occurs in the digital to analogue converter). On the left hand of the converters, the system functions in the analogue domain, on the right hand it functions in the digital domain. Filtering functions are shown as being implemented within the analogue domain (shown as part of the 'amplification/filtering function'), both for anti-aliasing purposes, and, in some designs, to relieve some of the digital processing load by performing filter functions using analogue filters.

This design is canonical in that it contains the essential subsystems to do every part of the job of a networked intelligent sensor. It has been introduced here because almost every hardware design for an intelligent sensing node follows this general architecture (indeed, the examples discussed below all conform to this design). There are some design variations on this theme. For a start, it is quite common to use a single processor to service several sensor devices. In this case, the analogue part will be replicated, once for each sensing devices. The converters may be replicated, or alternatively, an analogue multiplexer used to share one converter between several devices. Secondly, in systems using open-loop sensor devices, the lower chain will be omitted.

The design can be elaborated to improve performance in particular areas, but no new functions need to be added. The reason that this simple design can perform every necessary function is the flexibility of the digital microprocessor. Once the signal from the transducer has been digitised, suitable software on the processor can perform any required signal processing, so long as that signal processing is computable.

The canonical design in Figure 9.1 is a reasonably accurate representation of many sensor nodes being produced today. The most ubiquitous wireless intelligent sensor design to date is the Berkeley 'mote' design, which was originally published in 1998 [5] and has since been continuously refined, both in a commercial direction, and towards a single chip architecture. Table 9.1, produced from information from Falchi [6], King [7] and Crossbow [8] shows the resources available to several generations of mote. The first two are the original Berkeley designs, the others are the Crossbow and Intel developments of the mote. The devices themselves are illustrated in Figures 9.2 and 9.3.

The drivers for the Berkeley design has been the 'smart dust' vision of future ubiquitous sensing discussed in Chapter 1 and a plan of working towards this using available technologies. Currently, most of the feasibility work being done on 'mote' designs uses commercial, off the shelf (COTS) devices.

Table 9.1: Some contempory intelligent sensor designs.

Mote	Sensor device interface	Processor	Memory	Communications	Form factor
Renee	Mezzanine card	Atmel 8 bit 4 MHz	49 kB	916 MHz, software modulation	484 mm^2 rectangle
Mica 2	Mezzanine card (4 sensors) Analogue	Atmel 8 bit 8 MHz	644 kB	916/433 MHz hardware modulation 19.2 kbps	1800 mm^2 rectangle
Mica2Dot	Single sensor Analogue	Atmel 8 bit 4 MHz	644 kB	916/433 MHz hardware modulation 19.2 kbps	255 mm^2 disc
MicaZ	Mezzanine card (4 sensors) Analogue	Atmel 8 bit 8 MHz	644 kB	2.4 GHz ZigBee	1800 mm^2 rectangle
Intel mote	Digital interface	ARM 32-bit 12 MHz	586 kB	2.4 GHz Bluetooth	900 mm^2 rectangle

Using COTS technologies, the canonical device can be built using just three chips (2 electronic, 1 MEMS). For instance, if a single chip accelerometer with an integrated sigma-delta modulator, such as the one described in Chapter 5, is used, this will interface directly to the microprocessor, which in turn can, with suitable software, provide all the networking and communications facilities necessary (including the RF components, if wireless networking is required) to transmit the data to the host. The other electronic chip is for power supply conditioning.

We might ask, if designing sensor systems for realisation in the near future, if the COTS requirement is relaxed, and specialised MEMS and ASICS used, what is the likely scale of resource available in an intelligent sensor device? Taking 2004 technologies as a baseline, we can tabulate the chip real estate that will be required for the major subsystems for an intelligent sensor capable of cogent behaviour as defined in Chapter 8, assuming a requirement for 1 Mbyte memory.

Figure 9.2: Mica mote. From [8].

Figure 9.3: Intel motes. From [7].

Adding these together we come to a total chip area of around 49 mm^2, or 7 mm square (see Table 9.2) This is an easily feasible size for a commodity chip, and could include die space for MEMS components, if this was the integration route chosen, without the chip reaching an uneconomic size. Thus a single chip cogent MEMS sensor is easily possible using established technology.

One interesting point about this is that the processor is the smallest element here, even though this processor, a 200 MIP 32 bit RISC is far more powerful than those commonly found in current sensor motes. On the other hand, memory consumes a dominant share of the chip area. It is often

Table 9.2: Chip area for node subsystems.

Subsystem	Geometry	Technology	Chip area	Source	Ref.
2.4 GHz radio tranceiver	0.25u	BiCMOS	$8\,mm^2$	Rigge and Grewe	[9]
ARM 7EJS processor	0.18u	CMOS	$0.95\,mm^2$	ARM Ltd	[10]
Intel Static RAM cell	0.13u	CMOS	$5 \times 10^{-6}\,mm^2$ $= 40\,mm^2$ (1MB)	Yang *et al.*	[11]

the case that sensor node designs minimise processor size (for instance, the Berkeley mote designs use a very small processor) in an attempt to minimise the scale of hardware required by the node. However, a powerful processor does not consume a substantial amount of silicon real-estate, and could be a good investment, if the additional processing power can reduce the requirement for other specialist hardware. For instance, if the enhanced processor provides for additional digital signal processing, which in turn provides some type of data reduction, then this can save both memory space and communications bandwidth, both of which are critical resources in terms of the wider system design. In designing intelligent sensor devices, it is best to work from a systems point of view, and judge the design options on their merits.

9.3. An Architectural History of Sensor Arrays and Networks

As discussed above, there is currently a degree of convergence about the hardware architecture of a system composed of 'intelligent' or 'cogent' sensors. However, this convergence has not been arrived at arbitrarily, it is the culmination of a process of development, now enabled by the cost and size opportunities offered by MEMS and VLSI. To understand how this has been arrived at, it is best to take a historical perspective, since the starting point for designers of systems is generally existing practice, and many concepts and methods will carry through from what has gone before, sometimes beyond the point where they are valuable with current technologies. The natural starting point for examining the development of sensor arrays is industrial metrology.

9.3.1. *Where and how to Network*

As we have seen in Chapter 3, there is no clear-cut imperative to make the transformation from analogue to digital at a particular point in the signal chain. The decision depends on system level design considerations. The same thing is true of where in the signal chain to 'network', and hence where different system functions will be located in the context of an overall networked system. If we look once again at the processing chain introduced in Chapter 3, some parts will be replicated (at least the individual sensors and their associated amplification). Other parts will be shared between a number of processors.

Depending on the decision where to multiplex the data streams from multiple sensors to a single applications processor, different multiplexing and bussing technologies are appropriate. Figure 9.4 illustrates the options available at a functional level. Essentially the construction of a network entails the building of a single entity from a number of components. The position of the network connection which draws together the activity of many nodes in the functional chain determines which functions are distributed and which are centralised, and also the form of networking which must be used. The options depend on the 'intelligence' of the node. Thus

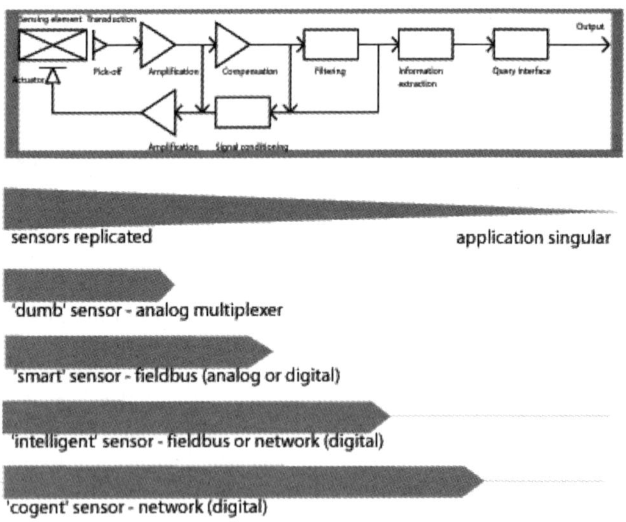

Figure 9.4: Processing chain and networking choices.

with a 'dumb' node (the top network option in Figure 9.4), the network must take the form of an analogue multiplexer. With a 'cogent' sensor (the bottom option) the network might be an ad-hoc, wireless digital network. Intermediate options (labelled here smart and intelligent sensors) allow a range of network choices, either analogue or digital. Very often, these choices have been directed by practical considerations such as the availability and cost of the required hardware to realise a particular choice.

As stated, with the old fashioned 'dumb' sensor, the choice is straight forward. Since the sensor has no associated processing functions, individual sensors are wired to a centralised data processing system, either using an analogue multiplexer, or multiple signal processing channels.

Early sensor array systems (and their corresponding 'networks') were small, composed of a few sensors. The sensor hardware and analogue signal processing was relatively cheap, while digital signal processing was expensive. The architectures employed minimised the digital part of the system. Prior to the development of integrated intelligent MEMS devices, this was the approach commonly used. Such systems were grouped together under the heading of SCADA (Supervisory Control and Data Acquisition) systems in an industrial context (here we discuss only the data acquisition role, not the control function of SCADA), and a specialist vocabulary built up around it. Typically a SCADA system would consist of a number of sensing devices connected to a Remote Telemetry Unit (RTU). In some systems the RTU would perform analogue to digital conversion, the sensors communicating using an analogue signal (typically a 4–20 mA current loop interface). The RTU (or several RTU's in a large system) are connected to a host computer system using a digital network. Sensors used in this way are already 'smart' under some uses of the terminology, since transducers do not produce a current loop output without (analogue) signal processing circuitry. Some such sensors have been produced using integrated MEMS technology. For instance Fenner, Kleefstra and Zdankiewicz [12] describe a single chip, MEMS vapour sensor with an integrated 4–20 mA current loop interface, a major advantage being proposed that it provides a very low cost means of interfacing with existing SCADA systems. The device itself is shown in Figure 9.5.

A SCADA system as described has a 'hierarchical' architecture, usually with two levels. At the top is the host, a computer or Programmable Logic Controller (PLC). The next layer is composed of one or more RTU's and at the lowest level are the sensors themselves.

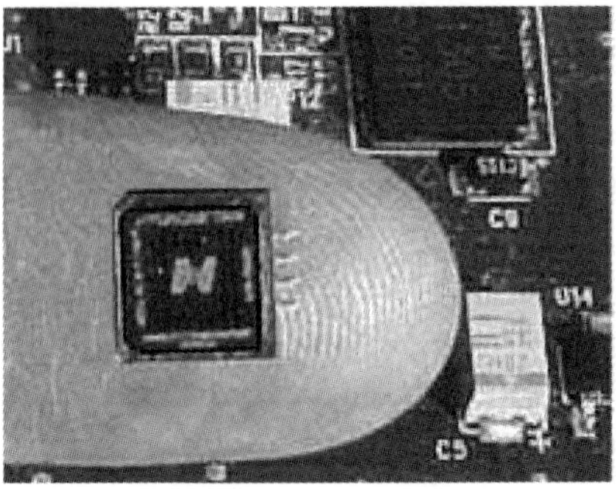

Figure 9.5: Hygrometrix HMX2000 micromachined relative humidity sensor. From [12].

A more sophisticated 'analogue' network is the IEEE 1451 family mixed signal interface standard [13], combining analogue signal transmission with digital control signals, either on shared or separate wire pairs, which is discussed in more detail later in this chapter.

The systems described above are a class of 'smart sensor', in which the 'network' which connects the sensors uses analogue signals. However, if the sensor is to be 'smart', there are many advantages to the use of this 'smartness' to allow a more simply wired and robust form of network than the analogue networks described. One major problem is that an analogue interface requires a pair of wires from every sensor to the RTU, making wiring complex and expensive. An immediate advantage could be had using a 'multi-drop' network, in which a single cable connects all of the devices and they share its use, usually using time-division multiplexing of some form. Although an analogue multi-drop network is conceivable, in practice they are much easier to design and build using digital technologies. Moreover, digital signals are much more robust and resistant to noise than analogue ones. Therefore multi-drop digital networks have become widely used in SCADA, where they are known collectively as 'Field Busses'.

There have been many different Field Bus systems, some proprietary (such as Profibus, Interbus, DeviceNet, Arcnet, Seriplex and others) and some standardised (Foundation Fieldbus, AS-I, IEC/ISA SP50, and others). For the purposes of this discussion, the features and precise function of the many different systems is not relevant. A summary can be found in Berge [14]. A sensor for use with a fieldbus architecture must inevitably include some digital circuitry, in order to connect to the fieldbus, and will generally need to integrate some sort of analogue to digital converter, to provide the digital data that a fieldbus requires. Thus a fieldbus sensor is an integrated mechanical/analogue/digital system. Most commonly, the majority of the digital functions will be implemented using some kind of microprocessor. An integrated fieldbus sensor is therefore a much more complex device than an analogue output 'smart' sensor. In practice, this is not a great constraint on the use of fieldbusses. With modern VLSI technologies, it is possible to integrate systems of this complexity onto a single chip, and allow space for the mechanical components. Compared with the traditional analogue output sensor a fieldbus integrated sensor has far superior 'ease of use' for systems integrators.

Organisation of fieldbus systems can follow the model of analogue connected systems, with each RTU using a fieldbus to interface several sensors. The RTU's are then networked to the central host using a digital network such as Ethernet. Alternatively, the RTU's may be omitted altogether, leaving the sensors to connect directly to the host, via the multi-drop network. For a large system this will require a network of higher bandwidth than the majority of fieldbusses, and for this reason industrialised versions of the Ethernet network have become increasingly popular as fieldbusses, allowing direct connection to a standard computer as host. It is single layer, all digital networking of the type exemplified by Ethernet based fieldbus systems that we might expect to become the norm for intelligent sensor networks, although as shall be seen in Section 9.5, such networks will not always be wired. Digital wired networks such as that described here form a second class of network hardware architecture.

The SCADA and fieldbus based view of sensor systems design is well understood by measurement engineers, and systems based on this type of architecture continue to be specified and built. Thus standards which facilitate their design and construction continue to be developed, even though, from the point of view of 'intelligent sensing systems', they may be something of an anachronism. The design principles for smart transducers operating within the SCADA/fieldbus view of system organisation are discussed and two smart transducer interface standards (IEEE 1451.2 STI and OMG STI) are compared by Elmenreich and Pitzek [15]. The 1451 standard set represents the state of the art in SCADA style sensor system architectures and is examined in some depth below.

9.3.2. *The IEEE 1451 Standards*

A description of the IEEE 1451 standards family is given by Allan [16]. Lee gives a full view of all of the standards from an updated "deployable" sensors view [17].

Five actual and proposed standards make up the IEEE 1451 family. They have been developed from 1994 under the direction of Kang Lee of the Sensor Development and Application Group at the National Institute of Standards and Technology, or NIST. The interoperable standards are explained and depicted below (Figure 9.6).

IEEE 1451.1 Standard:
Issued in 1999, defines a common object model for the components of a smart transducer and the networks that connect the transducers to the outside world. The model defines a network-capable application processor (NCAP).

IEEE 1451.2 Standard:
Issued in 1997, it specifies a transducer-to-microprocessor protocol and a transducer electronic data sheet (TEDS) format for digital point-to-point communications.

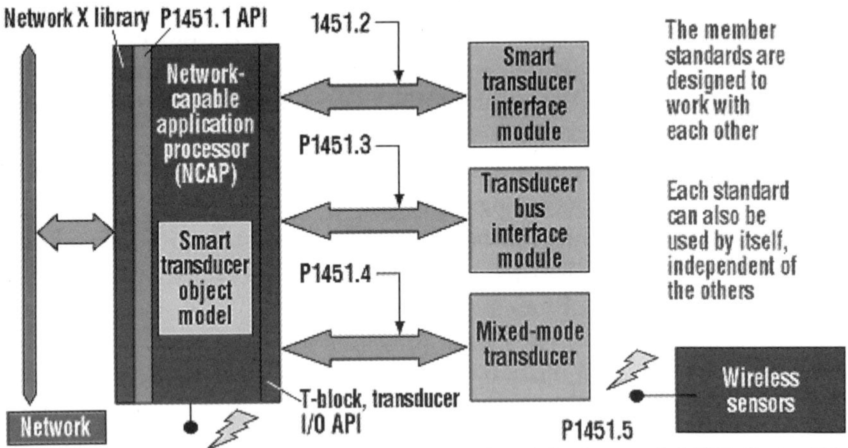

Source: "An Overview of IEEE 1451, Kang Lee," Sensors Expo & Conference, Fall 2002, Boston, MA.

Figure 9.6: IEEE 1451 standards family.

PIEEE 1451.3 Standard:
Defines a multidrop configuration for networking distributed transducers with TEDS.

PIEEE 1451.4 Proposed Standard:
Defines a mixed-mode interface that allows both digital signals (from the TEDS) and analogue signals (from the sensor) to share the same set of wires between the NCAP and the transducer. Balloting on this proposed standard should occur this year.

PIEEE 1451.5 Proposed Standard:
Addresses wireless communications between the NCAP and a transducer with a TEDS.

By comparison with the discussion of the canonical 'intelligent' sensor in Section 9.2, we see that the 1451 standards split the 'canonical' sensor into two parts, the NCAP and the transducers, and defines interfaces between the two. Thus, given that integrated intelligent sensors are already available, it would indeed appear to be a retrograde step (especially considering that the standard family is not complete). However, it does fit well with existing SCADA based practice, so there has been interest from manufactures in designing products around it.

One of the earliest interface products was the Bfoot NCAP from Agilent Technologies. The Bfoot thin Web server provided total plug-and-play Internet connectivity using standard interfaces. However, apparently it did not attract much market interest and was discontinued, but some other MEMS sensor manufacturers offer sensors that are customised to the standard.

These are 'smart' rather than 'intelligent' sensors under our terminology since they simply present a raw version of the transducer output, albeit packaged to a particular communications standard, and are exemplified by the Silicon Microstructures product shown in Figure 9.7.

In the context of SCADA based system design a two-level design approach arguably helps to reduce the overall complexity of a system by separating transducer-specific implementation issues from the interaction issues between different smart transducers and between transducers and the application. In this view, the transducer manufacturer will deal with instrumenting the local transducer and signal conditioning in order to export the transducer's service in a standardised way. Transducers manufacturers are therefore seen as liberated from interoperability issues between sensors,

Figure 9.7: Smart sensors designed for IEEE1451. From [18].

naming inconsistencies and network topology of the total system. The user of the smart transducer service will access its data via an abstract interface that hides the internal complexity of the transducers hardware and software.

The smart transducer interface and the communications interface are separated here (aligned to the ISO/OSI model, the first one relates to the application layer, whilst the second one relates to the physical layer mostly). The smart transducer interface gives access to the transducer features (measurement value, vendor ID numbers, diagnostic information, setup parameters) and provides services such as configuration, remote diagnosis, and real-time measurement. Kopetz proposes introducing distinct interfaces for various functional services [19], as follows:

- Real-time service interface (RS) — timely real time services during the operation of the system;
- Diagnosis and management Interface (DM) — opens a communication channel to the internals of the smart transducer (set parameters and retrieve information for diagnosis purposes for example); it is not time critical and is available without disturbing the real time service;
- Configuration and Planning interface (CP) — necessary in order to access configuration properties of a node; not time critical; during integration

this is used to provide the "glue" between various autonomous smart transducers in a network;

- The communications interface — defines the communication amongst the transducers in the network. It handles aspects such as communication baud rate, data encoding, flow control and message scheduling.

However, the IEEE 1451 appears to be struggling to maintain widespread success as a system architecture for large systems. Despite the efforts to develop standardised connectivity for smart transducers, to new or existing networks, by IEEE1451, the existence of "multivendor field-buses" and the appearance of powerful devices with an integrated Ethernet controller have made bridging necessary, thus increasing the cost of such transducers. These bridges if used could result in bottlenecks for the data streams, which means that the designer of a large system is almost inevitably faced with the design of a full-blown networked system, and in that context, the advantages offered by 1451 are not so clear.

9.3.3. *Networked Systems*

Industrial Ethernet systems have been relatively widely used in process and plant automation, for instance connecting together RTU's in a SCADA system. However, it is only recently that confidence in its 'unreliable'* channel control and new protocols supporting real-time traffic have been developed, and connecting sensors and actuators directly to such networks has become widely acceptable. Even so, although the prices of single-chip Ethernet interfaces are as low as the field-bus ones, the implementation of the required communication protocols involves the use of 'intelligent' sensors based upon powerful processors which have a considerable memory, in terms of the scale of processor that has been traditionally associated with small embedded systems. The feasibility of improved performance, low cost and added functionality (Internet visibility) was investigated and details of the hardware and available products given by Flamini *et al.* [20]. An assessment of cost and characteristics of existing sensors interconnected to Profibus-DP, CAN-bus2.0B and Ethernet802.3 was made. Upon evaluation

*The term 'unreliable' is used advisedly. Ethernet channel access control is probabilistic, thus any given level of performance cannot be absolutely guaranteed. Nonetheless, with enough spare network bandwidth, the network may offer a high reliability.

it was found that an Ethernet-capable sensor is comparable, as regards to cost, data throughput and complexity, to a field bus interface like Profibus-DP, or CANbus2.0B. As was stated in Section 9.3.2, Ethernet has gained popularity due to the fact that it is readily available. The feasibility of an Ethernet sensor with IP capability has been demonstrated at costs similar to the fieldbus interfaces mentioned above. Furthermore, to be simply managed by widespread software in the instrumentation area (like LabView), a proprietary smart and low cost solution based on UDP (User Datagram Protocol) has been realised.

Another class of network that is becoming of increasing importance is the wireless network. Again, there is a convergence between commercial computing practice, where wireless networks such as ISO 802.11b and g are in everyday use in offices, homes and coffee bars, and industrial practice, where RFID (Radio Frequency Identification Device) technology has introduced short range radio technologies into factories, plants and warehouses. Industrial sensing and metrology has tended to view wireless sensor networks as an extension of RFID technology, while systems engineers have tended to view the problems from a conventional networking angle, albeit using wireless networks. This can lead to a terminological inconsistency, which tends to mask the differences between these two related technologies.

The distinction between the two arises from a major difference between wired and wireless networks: nodes in wired networks can be powered from the network wires — this was a major consideration in the design of the original 4–20 mA analogue signalling standard and has been continued in many of the fieldbusses available. On the other hand, when designing a wireless network, the power available from the network (in the form of RF radiation) is zero, or very small. RFID has been designed to take advantage of the very small power available. RFID tags consist of an antenna, a capacitor (for short term power storage) and an RFID chip. The RF communications is triggered and powered by the interrogating device by rectifying the incoming signal and using it to charge the capacitor, so RFID tags are passive when not communicating with their host. In an RFID sensor device, the sensor and its associated circuitry must also be powered by the RF power from the host. As a result, RFID technology is incapable of continuous operation and must use very low power devices. As a consequence RFID based networks are inherently centralised and controlled by the interrogation device. However, at least at some level, it is still possible to conceive of an arrangement whereby there is a decentralised group of controller/interrogators and each such controller controls a group of RFID devices. A brief summary of the application of RFID to sensing is given by Want [21]. One such use is for sensing the temperature of food packages. A label on the package, an example by KSW Microtek [22] being shown in Figure 9.8, contains a humidity sensor connected to a very low power RF encoder. In response to the interrogating RF signal, the encoder can send back the sensed temperature, allowing the temperature of packages packed within a shipping container to be monitored. All control occurs in the data logger used to interrogate the RFID tags. The RFID sensor itself is at most 'smart' by our terminology.

By contrast, a wireless network node must be provided with considerable autonomy, and therefore require some source of power, and will be capable of continuous operation. In the design of such networks power conservation is often a key issue, and many ingenious designs for 'harvesting' power from the environment have been proposed, as were discussed in Chapter 4.

Such is the effect of the network on node design that many workers have believed it necessary to reject the many existing network standards and design new network systems specifically for sensor networks, in order to optimise the sensing performance of the system and minimise the resources

Figure 9.8: RFID temperature sensing label. From [22].

required by the individual node. Some of these specialist network designs are discussed in Section 9.3.6.

In summary, the hardware architecture of a sensor network system is primarily determined by the communications technology selected, since this places fundamental constraints on almost every aspect of the rest of the system design, from power usage to bandwidth available for data transmission. Whereas SCADA and RFID type networks require 'smart' sensors in order to present the sensor data in fieldbus or radio form, directly networked sensor systems require 'intelligent' sensors, so that they can participate in the protocols required to organise the network. Wireless sensors (apart from RFID sensors), in turn, require yet more self organisation, and thus intelligence from the sensor node.

9.3.4. *Network Organisation*

The discussion above has already touched on the issue of network organisation, from the point of view of making use of RFID sensors, which are not capable of independent activity, since they need at least power from an interrogating system to activate them. It was seen that an RFID based system must be organised with some kind of central controller. The simplest form of RFID based system is therefore a central controller, connected to a number of sensors. This type of network is known as a 'star' network.

A single controller may not have sufficient capacity to handle a large number of sensors simultaneously, in which case the solution is to network the controllers, turning the network into a 'tree'. Such organisations may be used with powered intelligent sensors, but are not the only possibility.

As the sensor node hardware becomes more powerful, which is almost an inevitable consequence of building larger networks, the options in terms of network organisation become wider. In particular it is possible to build a 'flat' network, in which all nodes are equivalent (i.e. there are no special nodes, or controllers). While these are in some ways more difficult to configure and design, there are a number of claimed advantages. Most of these are to do with the freedom from vulnerability associated with a special node, and with increased flexibility of configuration. Controller based networks will fail if the controller fails, whereas arbitrary failure of a node in a flat network can often be tolerated without compromising the performance of the network as a whole. When configuring a controller based network for a particular application, the location and organisation of the controller or controllers is a key decision. If the network changes or grows, the chosen configuration of controllers may not be suitable. A flat controllerless network may configure itself to suit any requirement, since it is not constrained by the need to accommodate a particular number of special nodes. Thus, flat controllerless networks are usually a preferred design option for *ad-hoc* or dynamic networks, particularly since they may be designed to configure themselves into a 'virtual' hierarchy in situations where this is advantageous.

9.3.5. *Wired Standards*

Wired sensor systems operate in a number of environments. Intelligent sensors will be connected according to a number of different wired protocols. We have discussed already the history of the development of plant level or SCADA sensor networks, from fieldbusses, through to industrial Ethernet. This development has been mirrored in several other fields of sensor deployment.

One of the earliest areas in which what would later be called sensor networks were deployed was for avionic systems use. The ARINC (Aeronautical Radio Inc) 429 and MIL-STD 1553 data busses have been in use since the 1970's. These provide a standardised, if somewhat primitive, means for avionic equipment to communicate. The organisation of such systems is similar to a SCADA system, the communication being between identified avionics equipment packages, rather than directly between sensors and the system using the data emanating from the sensor. The system allows a piece of equipment which gathers sensor data lo link with other pieces of equipment which may require use of that data.

Subsequently, as automotive designs became more sophisticated and reliant of sensing technologies, there has been a similar development in the automotive industry, where a networked intelligent sensor approach has become the norm. For automotive use the technology is called a 'controller area network' or CAN. A CAN is a high-integrity serial data communications bus for real-time applications. An example of a CAN system according to Freescale Semiconductor, one of the major suppliers of CAN oriented components, in an automotive context is shown in Figure 9.9 [23]. Typically, they operate at data rates of up to 1 Mb per second. Given the

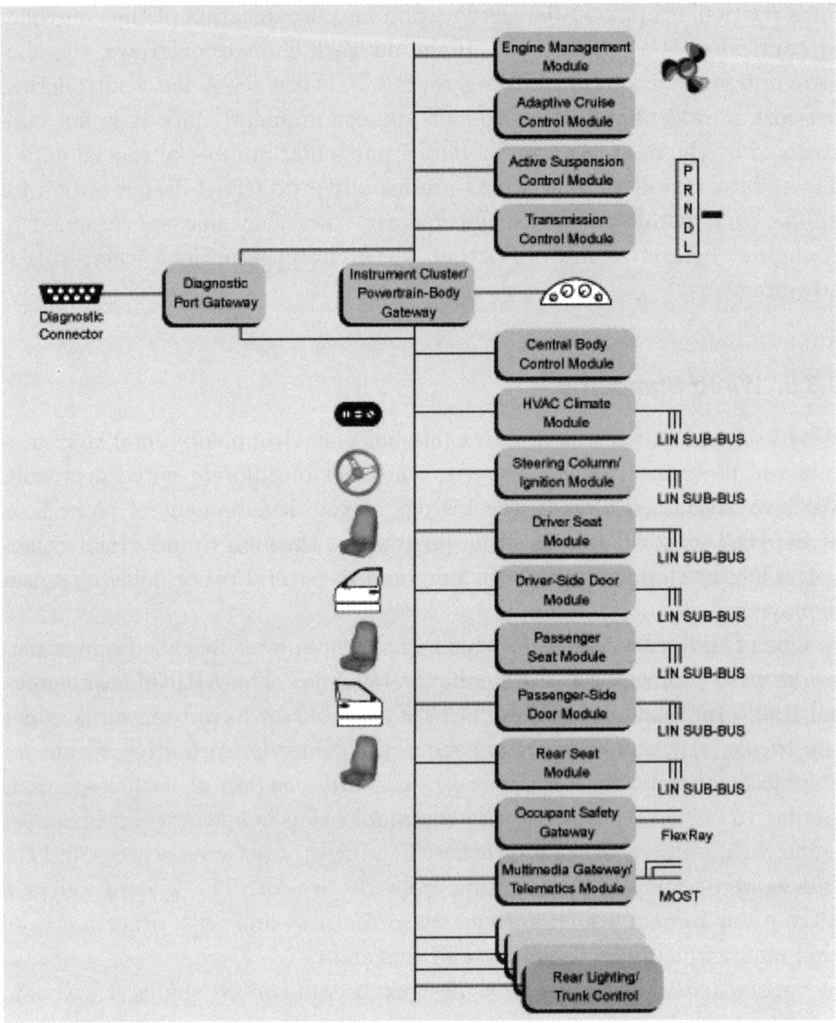

Figure 9.9: An Automative CAN application (Freescale Inc). From [23].

safety critical nature of the application, attention has been paid to the error detection and confinement capabilities. CANS for the automotive industry were originally specified by the Bosch company [24]. There are two ISO standards for CAN. The difference is in the physical layer, where ISO 11898 handles high speed applications up to 1 Mbit/second and ISO 11519 has an upper limit of 125 kbit/second. Available low cost technologies are often put to use in situations other than those originally intended and CAN networks have also been applied in industrial plant and laboratory contexts, as well as those for automotive use.

9.3.6. *Wireless Standards*

A very recent development is wireless connection of sensors, initially driven by frequent difficulties in particular situations in providing wiring to sensors. Initially, wireless sensing systems have been top down designs, driven by an application requirement and using available technology. Thus, a number of existing wireless communication standards have been pressed into service. A brief outline of some of these follows.

802.11 b and g

These are the standards typically used for wireless PC networking. In just the same way that the Ethernet wired network standards have become commonly used in industrial networks, largely supplanting previous specialised network standards, it is possible that these standards will be used for sensor networks. They bring with them a primary advantage of availability and low cost, on the back of huge production volumes for the commodity PC market. They also have very wide bandwidths (by sensor network standards) of 11 Mbit/s for 802.11 b and 54 Mbit/s for 802.11 g. Usable range is in the 10's of metres.

Bluetooth

This 'personal area networking' protocol was originally developed by Ericsson, the mobile phone company, as a means of connecting together peripherals to a computing device without wires. Subsequently it has been standardised by the IEEE as the standard, 802.15.1. Like the 802.11 protocols described in the previous paragraph, Bluetooth promises very low costs supported by huge consumer markets. However in several ways it is not well suited to sensor applications. Reliability and security can be low, the chance of interference from other Bluetooth networks is high, power usage can be high, the size of a network is severely limited and the range is very short (generally a few metres). Bluetooth has been used in the Intel development of the Berkeley mote.

ZigBee and 802.15.4

Recently a standard communication network system for wireless sensor networks has arisen, as the combination of two protocols. The result is the 'ZigBee' wireless networking system, which is built over the IEEE 802.15.4 Low-Rate Wireless Personal Area Network (WPAN) standard which defines the physical layer and medium access control. On top of this, ZigBee defines the network/security layers and the 'applications platform' or API. Both of these standards are designed with an emphasis on power saving, particularly for self powered nodes, expected to operate for long periods without recharging of their power source. Like most of the digital wireless networks, it operates in the 'unlicenced' bands, at 2.4 GHz, 915 and 868 MHz. The physical layer, 802.15.4, uses spread spectrum modulation methods which provide a degree of robustness in high noise environments. In view of the low power goals, transmission bandwidth is severely limited, compared with the other network standards, being 250 kbit/s in the 2.4 GHz band, but only 40 kbit/s in the 915 MHz and 20 kbit/s in the 868 MHz band. Range is 70 m. Although nodes have globally unique identifiers, only 64 k nodes can participate in a single network. 802.15.4 uses small, 128 bit packets. At the time of writing, the ZigBee 'applications platform' is still under development, but will include network configuration or 'logical topology establishment' into one of three configurations: a star, around special 'full function' network controller nodes, a fully connected mesh, or a tree of clusters. As well as these models it also supports point to point (a direct link between software processes within a node) and broadcast communication to all nodes in a network.

Zigbee seems destined to become the standard wireless interface of choice for wireless sensor nets. Crossbow Inc, commercial vendors of the Berkeley mote, have produced a development of it using ZigBee standards (see Table 9.1).

9.4. Systems Design Issues

From the above discussion we can make some observations, so as to draw out some requirements for the supporting systems architecture for a sensor network. These observations include the following:

- Arrays in the future are likely to contain very many sensors. Any network architecture adopted to link these arrays must be capable of scaling to include possibly thousands or millions of nodes. The design problem is closer to the 'world wide web' than it is to a local processor cluster.

- As discussed above, cost and power consumption of sensors will be a key consideration in many applications. The systems support will need to be implemented in a resource parsimonious way, avoiding computationally intensive tasks (as a part of the system support) and unnecessary communication. There is a complex trade-off that the system designer must make between power used for communications and that used for data processing. Sohabri *et al.* [25] estimate that it takes 3 J of energy to transmit 1 Kb of data a distance of 100 m. A general purpose microprocessor can execute 300 million instructions for the same amount of energy. While these estimates are based on year 2000 technology, it is likely that the ratio between the two will remain reasonably constant as technology improves, and thus it will always be worthwhile undertaking a considerable amount of data processing in order to reduce the amount of data being transmitted. Modern distributed computer systems do not often have the operational parameters to fine tune the partitioning of processing tasks between nodes so as to minimise communication.
- In the majority of applications proposed, communications between sensors is wireless. Thus, the architecture cannot be based on fixed topologies, or any particular topology. Wireless networks inevitably involve a level of transience, with nodes dropping in and out of communication, and the architecture must be capable of handling this. One example of such a network is the glacier monitoring system reported by Marshall [26] and Martinez [27]. Here wireless sensor nodes are placed on the bed of a glacier. They are designed to act like stones on the glacier bed, moving with the flow of the glacier. In such a network, the configuration will continuously change. In some applications sensors my join or leave the network, either because the sensor is faulty or for power conservation. The requirement for a system to adapt to node failures, by switching out faulty nodes, also means that the configuration of the network is liable to change as it operates. The support system will need to cater for this, preferably transparently to the applications designer. The software processes resident on the network need the property of *mobility*, that is they need to be relocated between physical nodes in such a way that their connectivity remains and any side-effects of the change in location are handled transparently.

- In some cases placement of the sensors is deliberate, in others it is arbitrary. The design of the architecture cannot assume a given physical placement. In fact, it must be possible for the system to discover its own physical topology on initialisation.
- Both homogenous and heterogeneous sensor systems are being proposed. This implies that the system will need to handle many different data types and volumes of data, and that these may required to be handled at the node level — the system needs to be viewable both as a collection of nodes (network connection points for one or more sensors) and as a collection of sensors (individual sensing devices).
- In many applications it will be advantageous for some applications level processing to be distributed to the sensor nodes. This becomes a parallel computing problem. Successful parallel computing has generally involved the hand crafting of codes to suit the particular environment, but in many of the cases discussed above, there is little that can be known for certain about the actual operational network, since placement and topology will be established on deployment and initialization. It will be necessary therefore for the system support both to provide support for parallel computing, but also to do so in a manner that allows automatic configuration and process allocation.

9.4.1. *Network Node Location*

As mentioned above, in some applications sensors are arranged into a predesigned geometrical arrangement, in some other, they are randomly placed.

Sensor networks may be broadly divided into two classes. The former we term *placed networks*, in which each of the sensor nodes is deliberately located at a particular position. The second class are *arbitrary* or *ad hoc networks*, in which some random process, such as dropping from an aircraft or mixing with a structural material, has been used for distribution. All array applications need to know the physical location of each sensor in the array for the signal processing to operate. The binding of the nodes' logical addresses to physical locations (which may be real, or a data point, where the data has been reconstructed using data fusion methods) is something that must ideally be handled transparently by the system. In a placed network it is necessary to bind node network addresses to physical locations before the network can be used. In an *ad-hoc* network there are two steps necessary before that can be done. The first is to identify the topology of the network called *network discovery*, the second to identify the relative physical positions of the nodes, called *locationing*.

9.5. Network Technology and Topology

Wired networks have many advantages in terms of bandwidth, electrical interference and power consumption but suffer from the crucial disadvantage of requiring that the wiring be installed and connected. This installation is likely to come at a considerable cost, and in applications such as smart structures it is possible that the wires may not be compatible with the structure in which they are installed. There could also be difficulties maintaining reliability of wired connections, particularly in some of the harsh environments found in structural and health monitoring applications. For these reasons, wireless networks are likely to be attractive even in situations in which the sensors are installed permanently in a fixed, static structure.

In a wired network, the designer must determine the arrangement or 'topology' of the wiring. Network textbooks tend to conflate the issues of the physical arrangement of the wiring (i.e. 'star', 'bus', 'ring') with the 'topology' from the point of view of the operating protocols. The latter point of view is the more important in most cases, and here the distinction should really be between passive 'broadcast' style networks, in which messages from a node reach all other nodes and switched networks in which messages are passed from node to node using a 'store and forward' approach.† Most networks used in sensor applications, including wireless networks and 'multi-drop' wired networks, such as the original Ethernet or CAN busses, fall into the former category. There are some exceptions, for instance the IEEE 1451 system, which, if viewed as a network, is of the active, switched variety.

However, whatever the 'real' underlying topology of a network system, there are reasons why another 'logical' topology may be imposed over it. There are two major reasons why this may be done.

†This issue has been further confused as technology evolves. Thus, although every textbook will tell you that the Ethernet is a passive, broadcast, bus network, in practice almost every large Ethernet installation is actually an active, switched, star network based around active message switches.

(i) Power saving

In a broadcast network, the transmitted power from a node must be sufficient to reach every node in the network. The radio frequency power received by one node from another in a wireless network follows the well known inverse square law, that is, the transmission power required to communicate between two nodes is proportional to the square of the distance between them. Consider the very simple three-node network shown in Figure 9.10. If node A wishes to communicate to node C, the power to make a direct communication is proportional to the square of the distance between them, i.e. $4d^2$. On the other hand the power required to transmit to node B is proportional to d^2. If node B then retransmits the message to node C, the total power used is $2d^2$, half the power required for the direct transmission. This suggests that an optimal configuration will be a fully connected mesh, such that each message navigates its way from sink to source via each intermediary, giving a transmission power for a n-node trip of n, rather than n^2, as would be the case for direct transmission.

(ii) Channel bandwidth

In a broadcast network, the 'channel' is available to only one node at a time. If the network is large, the available bandwidth may be limited severely. On the other hand, if the transmission power is limited so that the network can be separated into a number of non-overlapping 'cells' (in radio networks) or 'segments' (in wired networks), then nodes in different cells can use the channel simultaneously.

Topology design at this 'logical' level is a system issue, and the selected solution is dependent on many application level constraints. The choice will be governed by considerations of power, as discussed above; transmission speed and latency, since a single transmission can span a broadcast network,

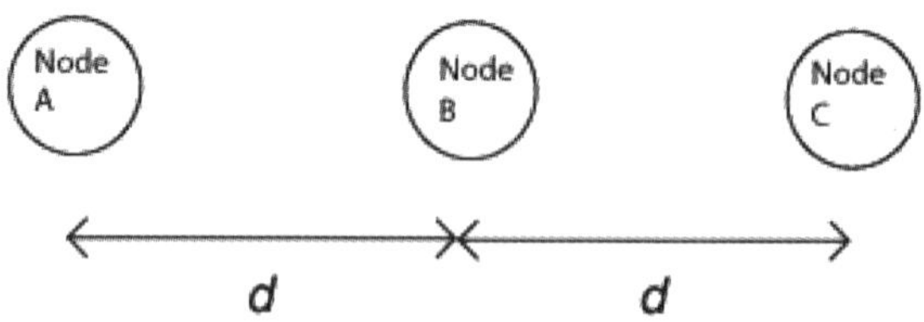

Figure 9.10: Store and forward transmission from A to B via C requires less power than direct transmission.

whereas a store and forward one will require several hops for a message to reach across it. The choice may also influence node complexity, since nodes in a store and forward network require sufficient processing power and memory capacity to be able to act as a 'store and forward'. Thus, it cannot be said that a particular topology is 'best', the choice will be determined by many application level or installation considerations. In wired networks, imposition of a logical topology will involve dividing the wire into segments, with switches between them, since the wire is too efficient a transmission medium to allow segmentation by means of control of transmission power. In wireless networks it is common for transmission power to be limited so as to impose a 'virtual topology' over the intrinsically flat broadcast network.

If the designer elects to use a 'store and forward' strategy, there still remains the choice between a 'flat' system, in which any node may act as a 'store and forward' or a hierarchical or clustered one, in which only some of the nodes, called clusterheads, store and forward.

In some cases, it will be necessary for a 'geographical', as well as a physical, topology to be established. For many data fusion and fault management procedures it is required to know the relative physical positions of the nodes. The distance between nodes can be discovered by measuring the time taken for messages to pass between them. Once the distance between each node and its neighbour is known, the physical topology of the whole system can be deduced by a repeated process of triangulation. In other reported work, nodes have contained a means for locating themselves, for instance by GPS.

After the process of network discovery the physical topology of the system is known. The virtual topology of the network must be established, according to the requirements discussed above.

9.5.1. *Sensor Network Protocols*

Researchers trying to provide the underlying system basis for distributed sensor networks view the problem either as the provision of a set of protocols, or as an operating architecture, or as a combination of both. This section deals with the various protocols which have been designed to address some or all of the concerns above, particularly in the context of sensor networks (more generalised distributed systems protocols are not covered here).

The consideration of power consumption drives the design of many sensor network protocols. Protocols designed for low power consumption build hierarchical, clustered networks, whereas those concerned with optimising or speeding data bandwidth through a network tend to build flat ones. As mentioned above, a clustered topology saves power by using clusterheads to forward messages, thus avoiding the transmission of messages across the whole network. The clusters form a hierarchy, and the number of levels in the hierarchy determines the tradeoff between power usage and transmission speed. In a radio network, this allows the transmitters to be turned down to a power which can just cover the local cluster. Several authors have considered the question of how far the power can be reduced [28, 29] by various optimisations of the topology specifically to minimise the transmission power necessary. Of course, each optimisation for one parameter will reduce the optimality of others. For instance, as the power is reduced the transmission error rate increases, requiring computation for error correction or sometimes retransmission. For a particular application, there will be an optimum point in terms of overall node and system power. It is to be expected that, as the field matures, generic standards will be adopted that are broadly optimal for a class of applications.

Another major tradeoff in protocol design is between 'stateless' and 'stateful' protocols. The 'state' referred to is a database of routes, or routing tables contained within the nodes. Obviously, routing tables for a large network can take significant memory resources within a node, and since the resources of sensor nodes tend to be limited, it is advantageous to use stateless protocols. Unfortunately, stateless protocols are very inefficient. Since nodes do not remember previously used routes, a route must be rediscovered for every single message that is sent. Since route discovery requires the sending of messages along paths which ultimately do not lead to the destination node, as well as the paths that do, this results in the sending many more messages than is required with stateful protocols. The most stateless protocol is the 'flooding protocol' in which a node, on reception of a message not addressed to itself, simply forwards that message to every other node that it knows about. Eventually, the message will be forwarded to the target node. However, in the process, every other node in the network will have forwarded the message, and the growth in traffic is exponential (since every message results in a number of forwards) and unending. Thus the flooding must be mediated if the network is to function. This is done by arranging for the messages to be identified (for instance by the address of the originating node and a serial number). Nodes keep a record of messages that they have recently received (thus the protocol is becoming 'stateful'). If a received message is a copy of one that has been recently received, then it is ignored. The network will still 'flood', but only to a 'depth' of one message. A further refinement towards a stateful protocol works as follows: on forwarding a message nodes stores the sender, intended recipient and the node the message was received from. When a node eventually receives a message intended for it, it transmits its reply to the node from which it received the forwarding transmission. That node forwards the reply to the node from which it received the original message, at the same time adding this routing step into its routing table. This step is repeated until the reply reaches the source of the original message. In the process, routing tables have been generated so that the next message from that source to the same destination can be routed directly along the same path. The most efficient 'stateful' protocol will maintain an optimum set of routing tables at each node to balance node resource usage and efficient routing. However, route tables are static information, and in a network in which the nodes are mobile, or subject to failure or entering and leaving the network for some other reason, stateful protocols are liable to fail.

9.5.2. *Non-hierarchical Protocols*

The quest in the design of non-hierarchical store and forward protocols is an optimum mix of stateful and stateless routing, using the stateless routing to 'discover' the connectivity of the network, and the stateful routing to convey the bulk of the traffic efficiently. Researchers also aim to optimise the routes, making sure that if multiple routes between nodes exist, the 'best' (according to whatever are the prioritised constraints for this particular design) one will be used. There follows a brief description of some of the sensor specialised protocols which have appeared in the literature.

The Self-Organising Medium Access Control for Sensor Networks (SMACS) [30] is a distributed, de-centralised protocol to allow the autonomous configuration of an *ad hoc* sensor network. The shared TDMA (Time-domain multiple access) communication channel is shared between *bootup* periods, in which nodes search for new nodes (stateless activity) to include within the network and communications periods, in which substantive communication takes place. SMACS addresses the *ad hoc* network discovery issues, and is teamed with a flat routing protocol, Sequential

Assignment Routing (SAR). Within the power constraints of a flat protocol, SAR tries to optimise power usage, using the energy resource and quality of service on each path to make routing decisions. The assumption is that nodes are communicating to a common destination, or sink. Multiple routing trees are built up from each node to the sink, avoiding nodes with low quality of service or energy reserves. Each sensor can then control which one-hop neighbour is used for relaying a message.

Directed Diffusion (DD) [31] attempts to mediate the overhead of flooding type protocols by directing the traffic along a 'gradient' which leads to the required destination. Nodes are named by attributes. The sensor names the type of data that it produces, sinks query by expressing an interest in data types. In propagation through the network a 'gradient map' is established reflecting the cost of transmission to a sink and the need for the data. Transmission of data from nodes is controlled by the gradient at each intermediate node.

The Minimum Cost Forwarding Algorithm for Large Sensor Networks [32] minimises the data resources needed in each node by eliminating routing tables for forwarding massages. Each node maintains an estimate of the lowest transmission cost from itself to the (in this case, single) sink or base node. On receiving a message, a node checks whether it is on the lowest cost path, and if so, retransmits the message. The cost estimates are established by the transmission by the base node of an 'advertisement' which is propagated through the network. Typically, nodes will receive copies of the advertisement through multiple paths through the network. On receipt of the advertisement by each node, the cost of transmission of that message replaces the current estimate, if it is lower.

Sensor Protocols for Information via Negotiation (SPIN) [33] assumes a non-specialised network, that is, all nodes are potentially sinks. In this case, route determination based on advertising (as with Minimum Cost Forwarding Algorithm for Large Sensor Networks) would flood the network with routing messages. Nodes with data to send, and those wishing to receive engage in local negotiations, and on successful negotiation, transfer the data. The protocol has variants, depending on the assumptions about the network medium, which changes the way in which negotiations are conducted. SPIN-PP assumes point-to point communications, such as a wired network, and negotiation is simple and assumed to be reliable. SPIN-EC adds an energy saving heuristic. SPIN-BC assumes broadcast channels, such as radio. Simultaneous responses to negotiation are eliminated by random time-outs. SPIN-RL refines this for the case of lossy channels.

Directional Source-Aware routing Protocol (DSAP) incorporates power considerations into routing tables. It is designed for fixed, rather than *ad hoc*, networks, and thus a dynamic locationing procedure is not included. Nodes are addressed with a 2D address, identifying the nodes place in the array. By comparing a destination node's address with its own, a transmitting node can determine in which direction to send a message.

These protocols are some of those more commonly cited in current sensor network research, and there are many others which have appeared but are not discussed here. In each case, the protocols design seeks to address the particular set of design problems which the authors feel to be paramount. Unfortunately, most if not all are based on theoretical studies as opposed to practical experience of large scale sensor networks, since there have been very few such networks built yet. Until such networks are built, it is very difficult to tell which of the theoretical design issues are actually pressing practical problems, and therefore which ones need special measures within the protocol stacks. The current research provides many answers, but until more practical experience is gained, it is not known which are the important questions.

9.5.3. *Hierarchical Routing Protocols*

Hierarchical protocols are intrinsically more 'efficient' than flat protocols, because the routing decisions are concentrated in a subset of the nodes, which may potentially collaborate together to produce efficient routing tables. The major questions in the design of such protocols is how the routing nodes or clusterheads are selected (or, by extension, how the clusters are formed) and how the collaboration between clusterheads operates. Some hierarchical protocols which have been reported are as follows.

The Low Energy Adaptive Clustering Hierarchy (LEACH) [34] assumes that all sensors are similarly constrained in terms of energy resources and the need to communicate. Its key feature is localised co-ordination of cluster set-up and operation. The operation is organised as temporal *rounds*, each starting with a set-up phase, followed by a data transfer phase. The set up phase starts with nodes randomly selecting themselves to be clusterheads. Other nodes then select which cluster to join, by measuring the signal to noise ratio for communication with the available clusterheads. The clusterheads then determine the transmission scheduling of their cluster members.

The Threshold sensitive Energy Efficient sensor Network protocol (TEEN) [35] is similar to LEACH, but designed for what the authors term *proactive* networks, which continuously monitor the environment, and therefore have data to send at a more or less continuous rate. TEEN reduces data transmission by requiring nodes to transmit data only when the variable being monitored by the sensor increases above a threshold. The Power-Efficient Gathering in Sensor Information Systems (PEGASIS) [36] protocol takes the cluster concept to its limit, with each node communicating only with its nearest neighbours, communication paths therefore being reduced to chains of nodes.

Other hierarchical routing protocols include MANET [37], TORA [38], AODV [39], DSR [40], INSIGNIA [41], RDMAR [42], STAR [43] and ODMRP [44]. The number of specialist protocols (and papers are still being written) for sensor networks reflects the fact that the large optimisation space allows many novel protocols to be generated, optimising for one factor or another. Again, only the acquisition of practical experience will tell which factors are of operational importance, then the solutions proposed above in the ongoing research will be applied within whichever protocols emerge as a result of that practical experience.

9.5.4. *Determining Location*

In the case of *ad hoc* networks, and other networks where the location of the nodes is not pre-determined, two tasks need to be undertaken with respect to the determination of the location of the nodes. The first is to determine the location of each node with respect to each other, while the second is the determination of the dimension space of the whole network. The second is vital, since applications require to know the location of the phenomena sensed by the sensor array.

The process of finding the location of nodes has become known as 'locationing'. In some cases sensor nodes are provided with an absolute means of location detection, such as an integrated Global Positioning System (GPS) receiver. Alternatively, in a system in which there is a base station, it may be possible for this station to locate the nodes of its network. However, in most cases it will be necessary to provide a mechanism whereby nodes can discover their location without recourse to the facilities. Ultimately all location mechanisms depend on distance measurement and triangulation (or, more accurately, trilateration).

Some suitable methods for locationing in sensor networks have been surveyed by Savarese and Rabaey [45]. The first requirement is to discover the range (distance) between two communicating nodes.

This can be derived from measurement of physical parameters of the received signals, and is discussed in some detail by Bulusu *et al.* [46]. Possible methods are as follows:

- Assuming a constant transmission strength, the received signal strength indicator (RSSI) is proportional to the inverse square of the distance between the transmitting and receiving nodes.
- If the transmission distance is long enough to make transmission time measurable, and the time of the transmission is known, then the travel time of the message can be discovered from the time of arrival, and the distance determined, the signal velocity being constant (the speed of light for radio transmissions). To get some measure of the quantities involved, the speed of light is $3 \times 10^8\,\mathrm{ms}^{-1}$, so a 300 MHz clock allows measurement of distance to a resolution of 1 m. Greater resolution can be gained by basing the estimate on the transmission time of a sequence of messages alternated between the two nodes, so long as the delay between messages remains constant.
- The transmitted signal used for ranging does not have to be electromagnetic. Particularly, acoustic signals may be attractive for ranging, since they travel slower, producing more easily measured time delays. Girod and Estrin describe a system using both acoustic and electromagnetic ranging [47].
- Another way to provide the required information for a locationing scheme based on trilateration or, in this case, triangulation, is to compute the angle of each reference point with respect to the mobile node in some reference frame. The position of the mobile node can then be computed using triangulation methods. This technique requires a steerable antenna, which, in micro-systems, could well make use of MEMS actuators.

Once the distance between pairs of nodes is known, their relative locations may be found by triangulation. If the location x_i, y_i, z_i and range r_i of neighbouring nodes is known, then the location (x, y, z) of the node under discussion can be calculated. Triangulation produces a set of simultaneous equations.

$$\begin{aligned}(x_1 - x)^2 + (y_1 - y)^2 + (z_1 - z)^2 &= r_1^2\\ (x_2 - x)^2 + (y_2 - y)^2 + (z_2 - z)^2 &= r_2^2 .\\ &\cdots\\ (x_n - x)^2 + (y_n - y)^2 + (z_n - z)^2 &= r_n^2\end{aligned}$$

Obviously, to solve for x, y and z requires three equations (and therefore the location and range of three neigbours). Additional equations can be used to refine the estimate of location and mitigate effects of range measurement errors by taking a least-mean squares value for all of the estimates available.

This method produces a relative position of a single node with respect to neighbours. To produce the position of all sensors in the network using this method requires that at least three sensors be located initially. This is done using an *Assumption Based Co-ordinates* (ABC) algorithm, which places the first node at the co-ordinate origin, the second co-linear on the x-axis and the third co-planar in the xy plane. The assumption is made that an (arbitrary) node is placed at co-ordinates $(0, 0, 0)$. The second node is placed at $(r_{01}, 0, 0)$ where r_{01} is the range between the two nodes. The third node is placed at

$$\begin{aligned}x_2 &= r_{01}^2 + r_{02}^2 + r_{12}^2/2r_{01},\\ y_2 &= (r_{02}^2 + x_2^2)^{1/2},\\ z^2 &= 0.\end{aligned}$$

The fourth node is placed at

$$\begin{aligned}x_3 &= r_{01}^2 + r_{03}^2 + r_{13}^2/2r_{01}\\ y_3 &= r_{03}^2 - r_{23}^2 + x_2^2 + y_2^2 - 2x_2x_3/2y_2\\ z^3 &= (r_{03}^2 - x_3^2 - y_3^2)^{1/2}.\end{aligned}$$

When four nodes have been so located, the remainder can be found using triangulation.

However, it is very unlikely that the range estimates will be very accurate. Errors in range measurement will cause inconsistencies in the overall location map of the network, since there will be errors in the individual distance measurements from which it was built. Having made an initial estimate of the overall topology of the network, using the ABC algorithm and iterative triangulation, this estimate can be improved by using various optimisation algorithms to adjust the estimates of the ranges between nodes until a consistent overall topology is established. This process is called *Global Balancing.* In *Global Topology Discovery,* the ABC algorithm is applied to locate a set of 'anchoring nodes'. This information is then transmitted to neighbouring nodes, which transform their own co-ordinate space to match that of the anchoring nodes, and then so constrain their own neighbours, until a uniform co-ordinate system has spread through the network.

Nagpal *et al.* [48] discuss the requirements for the process of co-ordinate system optimisation, and calculate a minimum size system within which an accurate result can be obtained. Within a set of constraints, an estimate of 15 nodes is obtained. Nicolescu and Nath describe an *Ad Hoc* Positioning System (APS) based on triangulation techniques. Particular attention is given to minimising transmission power, while still retaining sufficient duplication of measurements to provide good accuracy [49]. Iyengar and Sikdar describe a similar trilateration based system [50]. Savvides *et al.* discuss some Kalman filter based methods for refinement of the location grid, and estimate the processing power required for this task, producing a figure of 3.7 Mflops, considerably above the processing power found in most intelligent sensor designs. Using the triangulation principle, Nasipuri and Li have produced a locationing system based on fixed beacons which provide directionally rotating transmissions from which nodes can calculate their relative position [51].

There are also methods of locationing which do not make use of triangulation methods. One such localisation technique is the proprietary Location Pattern Matching technology, used in US Wireless Corporation's Radio-Camera system [52]. This relies on signal structure characteristics. By combining the multipath pattern with other signal characteristics, a signature unique to a given location is created. The system includes a signal signature database for a location grid of a service area. This database is generated by the driving of a vehicle through the coverage area. The vehicle transmits signals to a monitoring site, as would a sensor node from the vehicle's location. The incoming signals are analysed (with respect to signal strength, group delay, multipath reception and similar characteristics) and the unique signature for each square in the location derived. To determine the position of a transmitter, the system matches the transmitter's signal signature to an entry in the database, allowing location to be determined using the signal from a single transmitter.

Beacon based schemes provide a number of transmitters or 'beacons' at known locations. If the reception areas for a number of beacons overlap, a node can locate itself from the set of beacons that can be received. A locationing scheme based on beacons is described by Bulushu *et al.* [53]. He *et al.* [54] describe a similar, range free locationing system. Rather than using beacons, this system effectively treats every node as a beacon, obtaining a location map by analysing the connection sets of all nodes in the network, using a protocol called the 'centroid algorithm'. This is one of a class of algorithms known as 'proximity' or 'connectivity' locationing algorithms. Patawi *et al.* describe how they may be combined with trilateration algorithms based on received signal strength to obtain a system which shows a standard deviation range error some 50% better than a straight-forward trilateration system [55].

9.5.5. *Synchronisation between Nodes*

Sensing is inherently a time based activity, and therefore it is necessary that in an array there be some concept of 'array time'. Generally, each node will be equipped with a local clock source, but these are unlikely to be accurate enough to ensure uniform time throughout the network. It is necessary therefore for the protocols to include some means of time synchronisation between nodes. The protocols discussed in Sections 9.5.2 and 9.5.3 do not include this function, presuming a uniform time through the network. A complete operating system for a sensor network must therefore include synchronisation functions. Romer discusses some of the issues involved, from a viewpoint of assessing the suitability of solutions adopted in traditional distributed systems for the more specialist case of sensor networks [56]. The discussion starts with some observations on the characteristics and requirements of sensor networks. Firstly it is observed that routing, and consequently message delay time in *ad hoc* networks is not predictable in the same way as many 'traditional systems', due to the variety of different logical network topologies that might be adopted. In particular, the process of incremental network discovery (or more accurately re-discovery), needed as membership of the system changes, can take a long time. A period of 5 seconds is quoted for the Bluetooth network (of maximum size 255 nodes). In addition sensor networks have particular requirements for energy efficiency, robustness and scalability. As real-time systems, they also have a requirement that is called 'immediacy', the notification of an event detected by the sensor as soon as possible after its occurrence.

Four approaches to network synchronisation that have been used in 'traditional' *ad hoc* systems are discussed here.

Delaying techniques [57–59] assume that there is a small upper bound D on the message delay in the network, i.e., it takes at most time D to send a message from any node in the network to any other node. Messages m are sorted by their time stamps $m \times t$ into a list in ascending temporal order. When a new message arrives at the receiver, it is inserted into this list at the right position. The first element m_0 of the list is removed and used as soon as the node's clock shows a value greater than $m_0 \times t + D$. By this time, all messages with time stamps smaller than $m_0 \times t$ must have been received and inserted into the list due to the upper bound D on the message delay. This approach does not require any extra message exchanges for achieving temporal message ordering, but does not satisfy the immediacy requirement. Using delaying techniques, if the value of D is smaller than the maximum network delay then messages may arrive out of temporal order. If a large value, such as the 5s that would be required for Bluetooth, is used, this precludes immediacy, since evaluation of all messages is artificially delayed by D at the receiver.

Heartbeat protocols such as the one proposed in [60] assume FIFO channels between all nodes. Again, the receiver maintains an ordered list of messages. The first message m_0 is delivered as soon as the receiver has received a message m_i, with $m_i \times t > m_0.\ t$ from each node i in the network. The FIFO property ensures that the receiver has received all messages with time stamps earlier than $m_0 \times t$ at this point in time. In order to prevent starvation, all nodes send timestamped heartbeat messages to the receiver at regular intervals Δ. A small value of Δ results in a large message overhead, since each node will send a heartbeat message after Δ, which is neither scalable nor energy efficient. A large value of Δ precludes immediacy, since evaluation of all messages is artificially delayed by Δ at the receiver in the worst case.

Logical time [61, 62] defines partial orders between pairs of events, the ordering operators being before ($\rightarrow$), after ($\leftarrow$), or unrelated ($||$). Logical time does not provide a measure for elapsed real-time. Many applications for sensor networks will require absolute times for events occurring at different times in the network. Logical time is not therefore sufficient in the WSN domain.

Causal message ordering is similar to temporal message ordering and uses logical time. It ensures that messages are delivered in the sequence of their logical timestamps, so long as the temporal relation between them is not undefined ($||$). Causal message ordering has been an active research topic in many areas such as distributed databases, real-time systems, and fault tolerant systems. Examples include solutions for distributed systems in general [63, 64], for mobile computing systems [65, 66], and as part of total ordering multicast protocols with support for causal order delivery of messages from multiple sources [67–70]. However, as a logical time scheme, causal ordering cannot give absolute time references required in sensor networks.

These synchronisation and ordering methods have been used, both singly and in combination, to produce a number of practical synchronisation methods, described below.

Romer [71] describes a Temporal Message Ordering Scheme (TMOS). This scheme uses duplicate routing of timestamped messages between a local group of sensors, organised as a logical ring, the messages being sent both ways round the ring. When a node receives a duplicate of a message already received, it knows that all messages with an earlier timestamp in the ring have been delivered, and the message can be delivered to the application. It is claimed that TMOS is both energy efficient and scalable.

Girod *et al.* [72] describe a scheme which attempts to provide local, synchronised clocks throughout the network. The system uses Reference Broadcast Synchronisation (RBS), which synchronises a set of receivers with each other, as opposed to synchronising receivers with transmitters. It is claimed that this results in significantly better error distribution than algorithms which attempt to measure a round-trip delay. RBS daemons act as both senders of broadcast time reference messages, every 10s, and receivers, timestamping the messages and reporting back to the senders. Sending daemons collect the returns and collate them into a table of clock conversion parameters between pairs of nodes, which are broadcast back to the local nodes. The data is used to adjust the nodes clocks to achieve synchronisation.

Contrarily to Girod's work, Ganeriwal *et al.* [73] argue that pairwise syncronisation performs better than RBS. They describe a Timing-Sync Protocol for Sensor Networks (TPSN) using this approach. It is claimed that in this type of network, messages can be timestamped at the instant of transmission, this avoiding the indeterminacy that Girod *et al.* cite as the major failing of pairwise schemes. TPSN follows (but is not restricted to) an 'always-on' model, in which the whole network attempts to synchronise itself with a reference node.

Elson and Estrin [74] have described a *post facto* synchronisation algorithm. This avoids the need to keep local clocks continually synchronised with each other, which would result in a continuous protocol overhead and also prevent nodes form selectively powering down to save power. Nodes are normally unsynchronised, they only synchronise when an event occurs which requires a common time reference between nodes. Each node records the time of the event according to its local clock. Another node, acting as a third party then transmits a synchronisation message, giving its local time. All nodes then normalise their record of the event timing with respect to the synchronisation message.

RBS and pairwise schemes such as TPSN can use *post facto* synchronisation to provi synchronisation between clusters of already synchronised nodes.

Karp *et al.* [75] provide an extended analysis of the mathematical theory behind synchronisation, and derive from this a synchronisation protocol which can use any message as a synchronisation signal. It involves data being 'blocked' into regular messages which contain the data for the preceding period, as well as current estimated timestamps and timestamps for the message data, for messages sent during the period and for messages received during the period. Nodes can use the timestamps to update their own estimates of the timestamps. In addition, at a longer interval, special messages are sent to allow correction of timing skew between nodes.

Similarly to the network discovery and locationing problems, the theoretical research in network synchronisation is running ahead of practical experience. Until such experience is gained, and real application systems are built, it is difficult to say with any authority the degree and precision of synchronisation will actually be required, and therefore the extents to which synchronisation schemes need to be elaborated. Several of the schemes described here will provide a single, network wide time frame, down to some resolution governed by the speed of the processing available to execute the synchronisation algorithms and the latency and bandwidth of the network medium. However they do so at the cost of additional protocol complexity and network traffic. The issue that has still to be resolved is what is the level of synchronisation actually required, and what is the affordable cost in terms of complexity and additional traffic. It is likely that the optimum synchronisation scheme will depend on the precise parameters of the sensing application.

9.6. Conclusion

As we have seen, in this chapter, protocols and networking arrangements for sensor networks, particularly of the wireless variety, provide a wealth of research topics, and as a consequence, there is an enormous amount of current research being undertaken. While this is a good state of affairs for the researcher, it is more perplexing for the designer of operational systems, who will be looking for the kind of stability offered by a secure standardised networking system.

The standardisation activity is inevitably dragging somewhat behind the research. As a consequence, the largest standards activity in the field, IEEE1451 really relates to the previous generation of multiple sensor systems, based on SCADA architectures. For the current developments in *ad hoc* wireless, intelligent sensor networks, the standards for network level protocols are beginning to emerge, in the form of ZigBee and related protocol sets. The remaining major issues which are to do with network discovery, topology, synchronisation and fault management are still far away from the stability and consensus required for establishment of operational standards.

As a result, systems designers are confronted with two choices. The first is to abandon hope of using these new-fangled autonomous network architectures and fall back on older, well understood and standardised systems architectures such as SCADA and IEEE1451. The second choice is to build systems using the latest research, but accept that every system is likely to be a 'one-off', quickly superseded by the following research results.

The authors believe that there is another way that avoids either of these two choices. That is to build the system around a framework which allows the inclusion, in a structured way, of new research results as they become available. The key to ordered development of these systems depends on the adoption of suitably structured 'component based'[‡] design at the next higher

[‡]We acknowledge that 'component based' is a confusing term for the electronics (hardware) engineer, for whom all designs are 'component based'. The term here refers to software construction, undertaken in such a way that the software is organised as 'components' which can be simply interchanged. The hardware engineer knows that this is indeed a strained metaphor, because if you update your hardware design by simply swapping components, chances are that the result will be hot and expensive.

levels, (the 'presentation' and 'applications' levels for those familiar with OSI terminology), so that new 'components' can be developed and swapped in as the opportunity arises . This forms the topic of the next chapter.

References

1. Culshaw, B. (2001) Complex adaptive structures: design considerations, *Proceedings of SPIE* **4512**.
2. Varajan, V. K. and Varajan, V. V. (2000) Microsensors, microelectromechanical systems (MEMS) and electronics for smart systems and structures, *Smart Materials Structures* **9**, 953–972, IOP Publishing Ltd.
3. Doherty, L., Warneke, B. A., Boser, B. E. and Pister, K. S. J. (2001) Energy and performance considerations for smart dust, *International Journal of Parallel and Distributed Systems and Networks* **4**(3).
4. Warneke, B. A. and Pister, K. S. J. (2002) Exploring the limits of system integration with smart dust, *Proc. IMECE 2002*, ASME.
5. Hsu, V., Kahn, J. M. and Pister, K. S. J. (1998) *Wireless Communications for Smart Dust*, Electronics Research Laboratory Technical Memorandum Number M98/2.
6. Anastasi, G., Falchi, A., Passarella, A., Conti, M. and Gregori, E. (2004) Performance measurements of motes sensor networks, *Proceedings of the 7th ACM International Symposium on Modeling, Analysis and Simulation of Wireless and Mobile Systems.*
7. King, R. *Intel Mote*, presentation at Berkeley Wireless Research Center, http://bwrc.eecs.berkeley.edu/Seminars/King-6.27.03/Intel%20Seminar.ppt.
8. http://www.xbow.com/Products/Wireless_Sensor_Networks.htm.
9. Rigge, L. and Grewe, T. (2003) *Tradeoffs Shape Multimode WLAN Silicon Designs*, URL: http://www.commsdesign.com/showArticle.jhtml?articleID=16500948, Agere Systems.
10. ARM Ltd., *ARM Technical Support FAQs*, http://www.arm.com/support/faqip/3735.html.
11. Yang, S., Bai, P., Bramblett, T., Crew, B., Hussein, M., Jacob, P., Kenyon, C., Mcintyre, B., Moon, P., Sivakumar, S., Tufts, B., Thompson, S., Tyagi, S. and Bohr, M. (2000) *A 130 nm Generation Logic Technology Featuring 70 nm Transistors, Dual VT Transistors and 6 Layers of Cu Interconnects*, International Electron Device Meeting, San Francisco.
12. Fenner, R. L., Kleefstra, M. and Zdankiewicz, E. (2002) A micromachined water vapor sensor for home appliances (Sensor Technology and Design), *Sensors Magazine.*
13. The IEEE Family of Transducer Interface Standards (2003) Roger Allan, (ed.) Online ID #2990, http://www.elecdesign.com/Articles/ArticleID/2990/2990.html.

14. Berge, J. (2001) *Fieldbuses for Process Control: Engineering, Operation and Maintenance*, ISA.
15. Elmenreich, W. and Pitzek, S. (2003) Smart transducers — principles, communications and configuration, *Proceedings of the 7th IEEE International Conference on Intelligent Engineering Systems (INES)*, pp. 510–515.
16. The IEEE Family of Transducer Interface Standards (2003) Roger Allan (ed.) Online ID #2990, http://www.elecdesign.com/Articles/ArticleID/2990/2990.html.
17. Lee, K. (2003) The smart transducer interface standard, *NIST Workshop on Data Exchange Standards at the Construction Site*, USA, May 29, http://www.bfrl.nist.gov/861/CMAG/CMAG_workshop/lee.pdf.
18. Silicon Microstructures Inc, http://www.si-micro.com/.
19. Kopetz, H., Holzmann, M. and Elmenreich, W. (2001) A universal smart transducer interface: TTP/A, *International Journal of Computer System Science & Engineering* **16**(2), 71–77.
20. Flamini, A., Ferrari, P., Sisinni, E., Marioli, D. and Taroni, A. (2002) Sensor interfaces: from field-bus to Ethernet and Internet, *Sensors and Actuators A* **101**, 194–202.
21. Want, R. (2004) *Enabling Ubiquitous Sensing with RFID*, Computer, pp. 84–86, IEEE Computer Society Publications.
22. KSW Mikrotek Gmbh, http://www.ksw-microtec.de/www/startseite_de.php.
23. Freescale Semiconductor Inc, http://www.freescale.com/webapp/sps/site/application.jsp?nodeId=02WcbfNZnLNnms&tid=tAhl.
24. Robert Bosch gmbh (1991) CAN specification Version 2.
25. Sohabri, K., Gao, J., Ailawadhi, V. and Pottie, G. J. (2000) Protocols for self-organisation of a wireless sensor network, *IEEE Personal Communications* **7**(5), 16–27.
26. Marshall, W., Roadknight, C., Wokoma, I. and Sacks, L. (2003) Self-organizing sensor networks, *UbiNet 2003*, London, UK.
27. Martinez, K., Ong, R., Hart, J. K. and Stefanov, J. (2004) GLACSWEB — a sensor web for glaciers. *Adjunct Proc. EWSN 2004*, Berlin, Germany.
28. Pursley, M. B., Russell, H. B. and Wysokarski, J. S. (1999) Energy efficient routing in frequency hop networks with adaptive transmission, *Proc. IEEE Military Communications ConferenceMILCOM.*
29. Raghaven, A. R., Baum, C. W. and Russell, H. B. (1999) A distance-vector routing protocol with consistency checking for mobile distributed direct-sequence packet radio networks, *IEEE Military Communications ConferenceMILCOM.*
30. Sohabri, *et al. op. cit.*
31. Estrin, D., Govindan, R., Heidemann, J. and Kumar, S. (1999) Next century challenges: scalable co-ordination in sensor networks, *Proc. 5th Annual ACM/IEEE International Conference on Mobile Computing and Networking*, pp. 263–270.

32. Ye, F., Chen, A., Liu, S. and Zhang, L. (2001) A scalable solution to minimum cost forwarding in large sensor networks, *Proc 10th International Conference on Computer Communications and Networks*, pp. 304–309.
33. Heinzelman, W., Kulik, J. and Balakrishnan, H. (1999) Adaptive protocols for information dissemination in wireless sensor networks, *Proc 5th Annual ACM/IEEE International Conference on Mobile Computing and Networking.*
34. Heinzelman, W., Chandrakasan, A. and Balakrishnan, H. (2000) Energy efficient communication protocol for wireless micro sensor networks, *Proc 33rd Annual Hawaii International Conference on System Sciences*, pp. 3005–3014.
35. Manjeshwar, A. and Agrawal, D. P. (2001) TEEN: a routing protocol forenhanced efficiency in wireless sensor networks, *International Proceedings of 15th Parallel and Distributed Processing Symposium*, pp. 2009–2015.
36. Lindsey, S. and Raghavendra, C. S. (2001) PEGASIS: Power efficient gathering in sensor information systems, *Proc. International Conference on Communications.*
37. Corson, S. and Macker, J. (1999) *RFC-2501 Mobile ad hoc Networking (MANET): Routing Protocol Performance Issues and Evaluation Considerations*, http://www.rfc-editor.org/rfc/rfc2501.txt.
38. Park, V. and Corson, S. (1999) *Temporally-Ordered Routing Algorithm (TORA) Version 1 Functional Specification,* http://ftp.sunet.se/pub/Internet-documents/internet-drafts/draft-ietf-manet-aodv-05.txt.
39. Perkins, C. and Royer, E. (2000) *Ad Hoc On Demand Distance Vector Routing*, http://ftp.sunet.se/pub/Internet-documents/internet-drafts/draft-ietf-manet-aodv-05.txt.
40. Broch, J., *et al.* (1999) *The Dynamic Source Routing Protocol for Mobile Ad-Hoc Networks*, http://ftp.sunet.se/pub/Internet-documents/internet-drafts/draft-ietf-manet-dsr-03.txt.
41. Ahn, G.-S., *et al. INSIGNIA*, http://ftp.sunet.se/pub/Internet-documents/internet-drafts/draft-ietf-insignia01.txt.
42. Aggelou, G. and Tafazolli, R. (1999) *Relative Distance Micro-discovery Ad Hoc Routing (RDMAR) Protocol*, http://ftp.sunet.se/pub/Internet-documents/internet-drafts/draft-ietf-manet-rdmar-00.txt.
43. Garcia-Luna-Aceves, J. J. (1999) Source Tree Adaptive Routing (STAR) Protocol, http://ftp.sunet.se/pub/Internet-documents/internet-drafts/draft-ietf-manet-star-00.txt.
44. Lee, S.-J., *et al.* (2000) *On-Demand Multicast Routing Protocol (ODMRP) for Ad-Hoc Networks*, http://ftp.sunet.se/pub/Internet-documents/internet-drafts/draft-ietf-manet-odmrp-02.txt.
45. Savarese, C. and Rabaey, J. (2001) Locationing in distributed *ad hoc* wireless sensor networks, *IEEE Proceedings on Acoustics, Speech and Signal Processing*, pp. 2037–2040.
46. Bulusu, N., Heidemann, J. and Estrin, D. (2000) GPS-less low-cost outdoor localization for very small devices, *IEEE Personal Communication.*

47. Girod, L. and Estrin, D. (2001) Robust range estimation using acoustic and multimodal sensing, *Intelligent Robots and Systems, Proceedings 2001 IEEE/RSJ International Conference* **3**(29).
48. Nagpal, R., Shrobe, H. and Bachrach, J. (2003) Organizing a global coordinate system from local information on an *ad hoc* sensor network, *2nd International Workshop on Information Processing in Sensor Networks (IPSN 03)*, Palo Alto, CA.
49. Niculescu, D. and Nath, B. (2001) *Ad Hoc* Positioning System (APS), *Global Telecommunications Conference, 2001*, GLOBECOM '01. IEEE **5**, 25–29.
50. Iyengar, R. and Sikdar, B. (2003) Scalable and distributed GPS free positioning for sensor networks, *Communications*, Volume 1.
51. Nasipuri, A. and Li, K. (2002) *A Directionality Based Location Discovery Scheme for Wireless Sensor Networks, WSNA'02*, Atlanta, Georgia, USA.
52. http://www.uswcorp.com/USWCmainpages/our.htm.
53. Bulusu, N., Heidemann, J. and D. Estrin (2001) *Adaptive Beacon Placement. Distributed Computing Systems.*
54. He, T., Huang, C., Blum, B. M., Stankovic, J. A. and Abdelzaher, T. (2003) *Range-Free Localization Schemes for Large Scale Sensor Networks, MobiCom'03*, San Diego, CA, USA.
55. Patwari, N. and Hero, A. O. III, *Using Proximity and Quantized RSS for Sensor Localization in Wireless Networks, WSNA'03*, San Diego, CA, USA.
56. Romer, K. (2001) Time synchronization in *ad hoc* networks, *Proc. 2001 ACM Int. Symp. Mobile Ad Hoc Networking and Computing (MobiHoc'01)*, Long Beach, CA.
57. Mansouri-Samani, M. and Sloman, M. (1997) GEM — a generalised event monitoring language for distributed systems, *IEE/IOP/BCS Distributed Systems Engineering Journal* **4**(25).
58. Nelson, G. J. (1998) *Context-Aware and Location Systems*, PhD thesis, University of Cambridge.
59. Shim, Y. C. and Ramamoorthy, C. V. (1990) Monitoring and control of distributed systems, *First Int. Conference of Systems Integration*, Morristown, USA, pp. 672–681.
60. Hayton, R. (1996) *OASIS: An Open Architecture for Secure Interworking Services*, PhD thesis, University of Cambridge.
61. Lamport, L. (1978) Time, clocks, and the ordering of events in a distributed system, *Communications of the ACM* **21**(4), 558–565.
62. Mattern, F. (1998) Virtual time and global states in distributed systems, *Workshop on Parallel and Distributed Algorithms*, Chateau de Bonas.
63. Kearns, J. P. and Koodalattupuram, B. (1989) Immediate ordered service in distributed systems, *9th Int. Conf. Distributed Computing Systems (ICDCS 89)*, Newport Beach, USA, pp. 611–618.
64. Schiper, A., Eggli, J. and Sandoz, A. (1989) A new algorithm to implement causal ordering, *Workshop on Distributed Algorithms*, Nice, France, pp. 219–232.

65. Quaireau, S. and Laumay, P. (2001) Ensuring applicative causal ordering in autonomous mobile computing, *Workshop on Middleware for Mobile Computing (Middleware 2001)*, Heidelberg, Germany.
66. Skawratananond, C., Mittal, N. and Garg, V. K. (1998) *A Lightweight Algorithm for Causal Message Ordering in Mobile Computing Systems*, Technical Report TR-PDS-1998-11, Parallel and Distributed Systems Group, University of Texas at Austin.
67. Birman, K. P. and Joseph, T. A. (1987) Reliable communication in the presence of failures, *ACM Transactions on Computer Systems* **5**(1), 47–76.
68. Garcia-Molina, H. and Spauster, A. (1991) Ordered and reliable multicast communication, *ACM Transactions on Computer Systems* **9**(3), 242–271.
69. Jia, X. (1995) A total ordering multicast protocol using propagation trees, *IEEE Transactions on Parallel and Distributed Systems* **6**(6), 617–627.
70. Ng, T. P. (1991) Ordered broadcasts for large applications, *IEEE Int. Conf. Distributed Computing Systems.*
71. Elson, J. and Romer, K. (2002) Wireless sensor networks: a new regime for time synchronization, *Proceedings of the First Workshop on Hot Topics in Networks (HotNets)*, Princeton, New Jersey, USA.
72. Girod, L., Bychkovskiy, V., Elson, J. and Estrin, D. (2002) Locating tiny sensors in time and space: a case study, *Proceedings of the International Conference on Computer Design (ICCD 2002)*, Freiburg, Germany.
73. Ganeriwal, S., Kumar, R. and Srivastava, M. B. (2003) Timing-sync protocol for sensor networks. SenSys'03, Los Angeles, CA, USA.
74. Elson, J. and Estrin, D. (2001) Time sychronisation for wireless sensor networks, *International Proceedings of 15th Parallel and Distributed Processing Symposium*, pp. 1965–1970.
75. Karp, R., Elson, J., Estrin, D. and Shenker, S. (2003) *Optimal and Global time Synchronization in Sensornets*, CENS Technical Report 0012.

CHAPTER 10

WIRELESS AND AD HOC SENSOR NETWORKS

by *Robert Newman*

The previous chapter considered the design concerns, to do with arrays of intelligent sensors, and the design of intelligent sensor nodes themselves. This chapter looks at the same components, but from the point of view of the system into which they are integrated, and the concerns of designing a viable system as a whole. This further elaboration is required since large, *ad hoc* networks are extremely complex. Chapter 9 covered *what* the network system had to do, but producing a system organisation that is actually capable of performing the tasks described requires some overall structuring abstractions, in order to tackle the complexity of the overall design. The question of what precisely these structuring abstractions should be is an active research topic in its own right, and in this chapter two different approaches to the matter are presented.

At this level, the concern is with the connectivity of the nodes, from a system point of view, rather than their position in space (although there must be a way to discover this). Rather than considering the sensor system as an array, the topic of interest is the behaviour of the system as a *network*, which is the hardware and software realisation of the array. The chapter starts by examining some of the proposed applications for such networks, in order to provide some context for the following discussion. System design abstractions fulfil a role as a 'working metaphor' for the system designers, and must therefore be accessible and useful for the design team. Next is an examination of the role and composition of likely design teams for very large sensor network systems. The preliminary part of the chapter is completed with a set of design assumptions for distributed systems and a 'philosophy' for their design. The network functions put forward in Chapter 9 are then revisited in the context of these assumptions and philosophy and a 'layered model' proposed as the first structural abstraction. One of the abstractions which is often used, and which has a place in the proposed model, is an 'operating system', or a set of software providing a uniform program interface for applications programs. A brief overview of some operating systems designed for use with networked sensor systems is presented, followed by a detailed examination of three of them, which represent the current state of the art in sensor operating systems. A major concern that the set of structuring abstractions adopted must deal with is retrieval of information from the sensor network. Two alternative, though not mutually exclusive models are discussed: agent systems and applicative query systems. The chapter finishes with the proposal of a sensor support system architecture. The aim of this architecture is to define the interface points between the layers of software which handle the many functions which may be included in a large sensor network. This approach allows the definition of a set of standardised interfaces to the various parts of the software system, interfaces which are appropriate and useful to the people fulfilling different roles in the overall system design team.

10.1. Sensor Network Applications

In order to derive suitable models to structure the design of a complex entity such as a very large sensor network, it is necessary to know what the system will be used for, and also to know the kind of structures that have been found useful. Below a number of existing and proposed applications of sensor networks are reviewed in this context.

10.1.1. *Seismic or Vibration Monitoring for Military, Seismic, Structural Monitoring Purposes*

Sensor arrays are frequently proposed as solutions to various vibration monitoring systems, be they for seismic, military or structural monitoring purposes. Most of those which have actually been built so far make use of simple sensor devices connected to central intelligence, often making use of wireless technology to obviate the physical problems of connection.

For instance, Varadan and Varadan [1] put forward a structural monitoring system using integrated sensors, described by the authors as 'smart MEMS sensors'. In the light of the discussion in Chapter 8, it is important to understand how the authors use the term 'smart' — in this case it is limited to a frequency based wireless interrogation of the sensors in the array — all data processing and reduction occurs outside the array in a separate processor. Another system is proposed by Krantz *et al.* [2]. This is a wireless system for remote query and powering of an array of sensors embedded in a composite structure. Here again, the sensors are essentially passive devices. Brotherton and Johnson [3] and Wang and Chang [4] both describe an advanced structural monitoring system in which information derived from a large array of vibration sensors is used to deduce and localize structural anomalies using neural network techniques. While both of these systems use centralised analysis, this type of analysis mechanism is readily partitioned into a parallel implementation suitable for distribution around a network of processors, as would be found in an intelligent sensor network. Barai and Pandey [5] have characterised the use of neural network techniques to diagnose and localise structural damage based on vibration data from an array of sensors. Similar NN based diagnosis systems have been described elsewhere [6–8].

It can be seen that all of these systems could have been structured as a network of cogent sensors, if the wireless sensor nodes were endowed with sufficient processing power, allowing them to execute information extraction software (which in these applications was centralised).

10.1.2. *Acoustic and Environmental Monitoring*

Another application area present in the literature is pressure field monitoring. Nagel [9] discusses several examples of intelligent multi-sensor clusters, including meteorological applications, in which the cluster has temperature, pressure and humidity sensors, along with a rain guage. It is also noted that the addition of an 'electronic nose' sensor would allow atmospheric chemical sensing, and therefore pollution monitoring. Acoustic sensor arrays may be used for 'beamforming' [10], in which the array of microphones is used to locate the source of sound to a high degree of accuracy by modeling the beam of sound based on the magnitude and phase of the excitation of the microphones in the array. Ruffin [11] surveys a range of military applications of sensor arrays, including vibration, environmental, chemical, optical and RF sensors. In each case the use of an array allows sensing and localisation of targets to a high degree of precision.

One of the most ambitious proposals related to the Berkeley 'smart dust' concept, and one of the most audacious proposals for a large sensor array is the Global Environmental MEMS Sensors (GEMS) application (Figure 10.1) [12]. The idea behind this is to provide a global environmental monitoring system based on the deployment of millions of multi-function wireless MEMS intelligent sensors in the global atmosphere. The result is a sensor network of staggering proportions. Currently, an extended feasibility study is being undertaken to determine whether this ambitious MEMS application offers better cost/benefit characteristics than more conventional weather data collection methods. The feasibility study has included mathematical modelling of ways of distributing the sensors (from drone aircraft), the time it will take for sensors, on average, to fall to earth (this is the useful lifetime of a sensor), the way in which sensors will be distributed by winds and other weather patterns and outlines of an overall systems architecture for data gathering from these enormous mobile sensor networks. GEMS is a sensor array which cannot conceivably be realised except as a wireless sensor network, and therefore makes a very good case that the ability to build

Figure 10.1: GEMS — a three dimensional sensing system. From [12].

networks of cogent sensors, wirelessly connected, is an enabling technology which will allow the design of completely new types of sensory system.

A more down to earth example of such a network is the glacier monitoring system reported by Marshall [13]. Here wireless sensor nodes are placed on the bed of a glacier. They are designed to act like stones on the glacier bed, moving with the flow of the glacier. Whereas GEMS is clearly an array application, the aim being to use a multitude of sensors to measure field phenomena, the glacier monitoring system focuses on the readings from an individual sensor, the wireless networking aspect being one of implementation convenience — it being difficult to provide wires to a sensor embedded in a glacier.

Similar considerations dictate the use of a wireless sensor network for observing the breeding behaviour of a bird (the Leach's Storm Petrel) [14] on Great Duck Island, Maine, USA. The sensor network provides a way of monitoring their behaviour without causing the disturbance that would occur as a by-product of human observation. Sensor nodes are installed inside the birds' burrows and on the surface. Nodes can measure humidity, pressure, temperature, and ambient light level. Burrow nodes are equipped with infrared sensors to detect the presence of the birds. The burrows occur in clusters and the sensor nodes form a multihop *ad hoc* network. Each network cluster contains a sensor node with a long-range directional antenna that connects the cluster to a central base station computer. The base station computer is connected to a database back-end system via a satellite link. Sensor nodes sample their sensors about once a minute and send their readings directly to the database back-end system.

Zebranet is a wireless sensor network in operation [6] at the Mpala Research Center in Kenya for monitoring individual large animals. Of particular interest is the behaviour of individual animals, interactions within a species, interactions among different species and the impact of human development on the species. The observation area may be as large as hundreds or even thousands of square kilometres. Animals are equipped with sensor nodes, shown in Figure 10.2. An integrated GPS receiver is used to obtain estimates of their position and speed of movement. Light sensors are used to give an indication of the current environment, a simple deduction being made from ambient light level to likely location. Each node logs readings from its sensors every three minutes. Whenever a node enters the communication range of another node, the sensor readings and the identities of

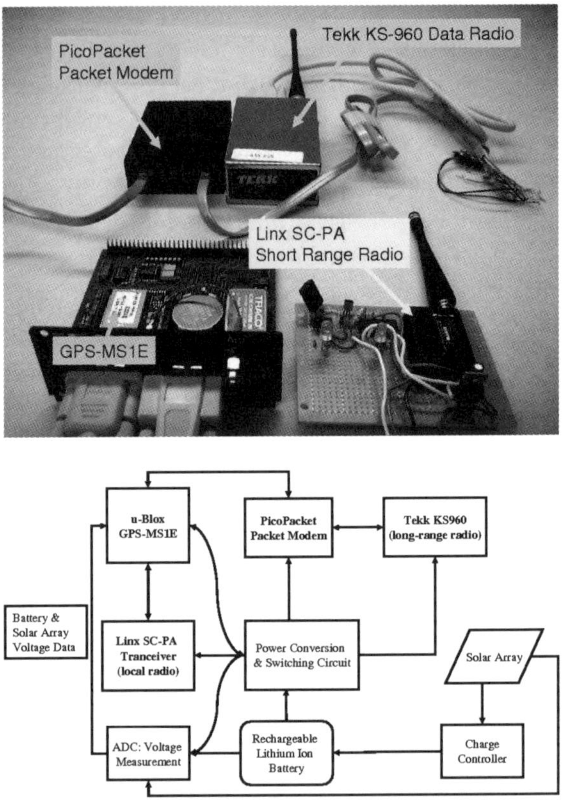

Figure 10.2: Zebranet sensor node. From [6].

the sensor nodes are exchanged (i.e. data is flooded across network partitions). At regular intervals, a mobile base station (e.g. a car or a plane) moves through the observation area and collects the recorded data from the animals it passes.

A survey of some of these and several other wireless sensor applications is given by Romer and Mattern [15].

10.1.3. *Features of the Applications*

The proposed applications exhibit a range of ambition, in terms of system organisation. The most straightforward are the structural monitoring applications, which tend to use fixed sensor arrays and, maybe as a consequence of this, use little embedded intelligence within the sensor node — what intelligence there is, being limited to serving remote queries to the sensors. By contrast the environmental monitoring systems generally make use of *ad hoc*, mobile networks, with a range of sophisticated node behaviour, including interpretation of sensor data (making them fully cogent sensors), complex interaction (with speculative information forwarding in the case of Zebranet) and advanced data analysis techniques.

10.1.4. *The Sensor Network as a Field Sensor*

In many of the cases above, particularly in the structural monitoring examples, the phenomenon being observed takes the form of a field with a complex pattern in space (usually two dimensional, although one can envisage three dimensional monitoring systems using MEMS technology). The positioning of many sensors in space allows the pattern of the field to be determined, albeit often with complex computation. Since it is the pattern of the field that is the unknown, prior decisions cannot be made as to where to position sensors. The only solution is to deploy an array of sensors, with a sufficient spatial density so as to construct a good 'map' of the field. The number of sensors depends on the 'resolution' that the application demands. It should be noted that what is being observed here is a field in an area of a two (or three) dimensional space, so the number of data points (and therefore sensors) increases as the square (or cube) of the dimension of the space. This can lead to very large numbers of sensors being required to provide even a moderate 'map' of the field. If a degree of redundancy for 'fail safe' operation is included, the required number of sensors increases further. Stasweski *et al.* [16] have demonstrated the use of genetic algorithm based placement methods to optimise such redundant arrays, but if the sensors are sufficiently cheap, or funding sufficiently generous, it is possible to use an over-provison of sensors and use a stochastic distribution mechanism, (such as the aforementioned distribution by means of throwing them out of the back of an aircraft, as in the smart dust proposals).

Another scenario in which arbitrary placement may be found in the future is the construction of 'smart materials'. If the 'smart' characteristics of a material depends on the distribution of sensors through the material, then intelligent MEMS sensors could provide a means of making the sensors sufficiently cheap to make such materials viable. The structural monitoring applications discussed above have all used sensors deliberately located, either in a regular pattern or at key data points on the structure. However, the cost of placing individual sensors over the material could be high. An alternative would be to mix the sensor chips into the 'matrix' of a composite material, which would lead to them being distributed arbitrarily through the material as it is layed up. For this to be viable, the sensors must be small and light enough not to make a significant alteration to the structural properties of the material into which they are embedded, cheap enough to use in huge quantities, and have a robust wireless communications mechanism which does not cause interference or radiation problems. The computation involved with deriving information from the sensor data must be sophisticated enough to make sense from the arbitrarily placed array.

Several of the proposed and operational system discussed in Section 10.1.1 are mobile in nature. It is this mobility of the wireless sensor, and its ability to travel with the object being observed (be it wind, glacier or zebra) that makes wireless sensor networks such a powerful tool in environmental sensing applications. Some of the issues posed by the design of mobile systems (which are intrinsically *ad hoc*), such as network discovery, location finding and network synchronisation were discussed in Chapter 9. In this chapter, the requirement for mobility is simply one of the constraints that will shape our attitude to system design. It is one of the harder design issues to deal with, and any design approach that lays claim to being generic across wireless sensor networks must certainly handle the problem of mobility.

10.2. System Designers' Role

Design of these huge, complex systems is not something that will be undertaken by a single engineer, or even a small single disciplinary team. Rather, the development will require the collaboration of large, multi-disciplinary teams, of designers and specialists from many different backgrounds. We can envisage the type of teams likely to be involved by using the GEMS proposal as a working example.

As a starting point, the design team will need to be headed by application specialists, in this case meteorologists. The basic operational parameters of the system will be determined by one or more system engineers. This will inevitably be a 'helical' process and evaluating proposals for different operational parameters of the overall system, using iterative evaluations of 'what if scenarios' to move towards a feasible design. The scenario evaluation is informed by preliminary design work from the other specialists in the team, to be identified below. Essentially, the scenario calls for distribution of microscopic, wireless 'weather stations' into the atmosphere. These are designed with an inbuilt 'parachute', and so over a period of time, float down to the ground. To keep the atmosphere 'seeded' the sensors need continual replenishment, replacing sensors which have fallen to the ground, or otherwise been destroyed. The overall economics of the operation depends on the rate of replenishment, which in turn depends on the time they take to float down and the density of deployment.

Extending the discussion to the next stage of concern, the outline design of components on which to base further detailing of the scenario, these parameters depend on the required precision of the data field measurement, the transmission range of the sensors and their weight and detailed aerodynamic design. Once again, discovering these is likely to be an iterative process. The initial, speculative feasibility studies may be based on 'guesstimates', but ultimately the feasibility will depend on the availability of accurate design estimates. To produce these includes adding to the team. Firstly, we need aerodynamic and micro-mechanical design expertise, to outline a design and estimate the characteristics of an individual sensor. Whether or not these designs are feasible depends on the detail sensor and electronic design, so we need MEMS, electronic, RF and VLSI design skills available. With these skills, the basic, physical feasibility issues can be addressed, but there remains the question of whether the software system functionality is feasible, which in turn depends on the architectures and the software technologies adopted. Merely establishing feasibility is a considerable multidisciplinary operation, and the specification and detail design is obviously more complex still.

Multi disciplinary design teams are not always easy to construct or manage, even in environments where there is a long history behind them. Essentially, the limiting effects are human, as opposed to technical. They include:

Territorialism:
Specialists tend to protect their own area of specialisation against intruders, who in the case of multidisciplinary projects, includes people from outside

their specialism trying to impose requirements or constraints. Thus, people will hang on to their favoured technical solution to the death, even if it is entirely inappropriate.

Isolationism:
Specialists tend to have limited understanding of, or sympathy with other disciplines, even if they are adjacent. Under this mind set, non-communication becomes a goal, rather than a problem, particularly if it prevents territorial encroachments (see above).

A little knowledge is a dangerous thing:
This is an affliction which affects specialists from adjacent disciplines. If I understand the basic principles of the next door discipline I obviously know all about it, so why is that specialist making it all seem so difficult? This attitude is often observed from electronic engineers with respect to programmers. Programmers suffer from it too, but, as they rarely know anything about electronic engineering, they save their contempt for systems analysts.

Localised perfectionism:
The goal becomes to achieve a perfect result, according to local metrics, without regard to the global aims. Thus, a MEMS device designer might strive to attain optimum performance when the imperfections might be more easily eliminated using signal processing.

The language problem:
One big issue is that adjacent disciplines often use the same terminology to denote subtly (or completely) different concepts. The result of this can be extended mis-communications which would be amusing if they weren't so damaging.

The management solution to all of these problems depends on finding a way for all team members to understand the global objectives, and their own role in achieving them. This chapter is not about management skills, but about technical solutions. However, a technical solution may be used to address a management problem, and it is this technical solution which we now address. Building a team around a shared and understood set of objectives is facilitated by adoption of a design philosophy to which they can subscribe to a shared and well understood architectural framework, against which participants can gauge their own roles and benefit from clearly designed interfaces with the rest of the team.

Thus, architectures for complex systems such as those in Sections 10.1.1 and 10.1.2 are not merely about technical considerations. The design of the architecture must be technically sound, but an important indicator of merit is the extent to which the architecture defines and clarifies the interface between the people working on the development, as well as the software and hardware components. To an extent, the technical and human aspects of the design are concerned with same thing — clarity of interfaces and well defined functions within components. However, one also has to consider the role and skill set of the people working with that component. It would, for example, be unreasonable for an application domain expert (perhaps a zoologist working with Zebranet) to be expected to master the detail of device level programming in order to fulfil their role in the team.

The discussion in the rest of this chapter focuses on the architectural options for a major part of the system, the 'systems middleware', the layers of software and hardware which provides the interface between the application designer and the sensor hardware designer. In doing this task, the 'middleware' provide a standardised set of systems requirements against which the sensor hardware and its signal processing can be specified, producing an abstracted view of the sensor hardware and its support systems, which allows the decoupling of applications design from the lower levels of design.

10.3. Design Assumptions for *Ad hoc* Networks

Many of the applications discussed in Section 10.1 come into the category of '*ad hoc*' networks, in that the disposition and configuration of the network cannot be determined at design time, and there is no opportunity for human intervention to configure the network at deployment time. Several of them are in addition 'mobile' networks, in that the nodes may change location or enter and leave the network while it is operating. *Ad hoc* networks are an important research topic, not just in the field of sensor network design, but on account of the widespread increasing interest in wireless communications. Wireless sensor networks are an important thread in the wider field of *ad hoc* and mobile networks, and probably account for the majority of sensor network research currently being undertaken. Therefore, despite the fact that not all of the applications of sensor networks are wireless, or mobile, the emphasis here will be on this type of network. A design philosophy which can cope with the wireless, mobile case can certainly cope with wired, fixed networks, although it may not be as efficient or optimised for those situations.

As was seen in Chapter 9, 'traditional' sensor networks, developed from the SCADA heritage, would have a central controller, and would also have local network switches, or RTU's. This design philosophy has been carried forward to the IEEE 1451 sensor bus standard. While such a design can be (and is at times) easily translated to wireless communication, simply by replacing the sensor/RTU and RTU/host links with wireless ones, 'wireless sensor networks' of this variety will not be considered further here. Although extremely useful in a limited class of applications, their design does not pose any new problems, at least from the systems configuration point of view.

On the other hand, *ad hoc* networks do pose considerable design problems, on a par with the most complex of those found in large-scale computer systems. The added complication that is the problems must be solved using a minimum of communication, processing and power resources. This is one reason why wireless sensor network research is one of the most fertile areas of computer and operating systems research today. It simply provides an interesting and demanding context for workers in those fields.

In this spirit, we will make a number of assumptions about the basic design parameters of wireless sensor networks for the remainder of this discussion. While they may not apply in their entirety to any given design, most other choices can be seen as 'relaxations' of these cases. The assumptions are:

(i) The network will be organised on a decentralised or 'controllerless' basis. This is not to say that some nodes may not assume, from time to time or even permanently an organising role amongst their peers — in fact most decentralised systems work this way. It rather means that there will be no special purpose (in a hardware or software sense) nodes within the network.

(ii) The network may be heterogeneous both in terms of the sensor load that they carry and other information resources. It is, however homogeneous in terms of nodes ability to participate in the basic operational protocols of the network — this is a corollary of the 'no special nodes' decision in point 1.

(iii) Communication will be wireless, and essentially broadcast, at least in a locality. That is, any node may communicate with a set of nodes in its own locality.

(iv) Nodes are self contained, and power constrained.
(v) The nodes are in some way uniquely identified, but have no other built-in or intrinsic knowledge of the system configuration.
(vi) Nodes are mobile, and may enter and leave the network at any time in the network's life. This also accounts for nodes that are not 100% reliable — a failing node is seen simply as leaving the network.
(vii) There is one or more application system using the sensor network. These application systems may connect to the network arbitrarily. One way of looking at this is to say that 'queries' (i.e. information requests) may be injected into the network at any point.

Together, these assumptions constrain the system design in a particular direction, at practically every turn, removing the easy options.

10.4. Distributed System Design Philosophy

The individual design problems that have to be surmounted in the building of a large, complex, *ad hoc* network system are sufficiently complex that it would be unreasonable to expect to introduce the same design issues for each application system design. To do so would entail each designer of a new application solving again some of the most difficult problems of systems design.

Unfortunately, most of the work described to date in the literature addresses specific concerns of a specific class of application, even when generality is claimed. For instance Mitchell *et al.* [17] describe a 'Distributed Computing and Sensing Architecture'. Clusters of sensors are served by local processors which are connected in a wireless network. Data analysis includes the use of fast Fourier transform (FFT). Calculation of the required processing and data transmission resources for a 1000 sensor, 200 cluster systems have been made — A centralized processor using 10.7 Mflops of computing power, 20 Mbytes of memory and 10 Mbytes of transmitted data may be replaced by a distributed system requiring 53 500 flops and 100 kbytes of memory in each cluster and 20 kbytes total transmitted data. This system demonstrates some of the advantages to be gained using a fully distributed environment. Distribution has reduced both the processing load and the overall data transmission requirement. However, it requires to be adapted from top to bottom for each new application.

By contrast, however, in the general programming of computer applications, programmers have become used to the idea that the designers of operating systems will have 'abstracted away' the complexities of the underlying systems. Hence, in the same way, the essential concerns of how the nodes collaborate and are organised need to be addressed in a standardised manner, by means of a general purpose applications programming interface (API). This API must allow the programmer 'transparency' of their own programming goals, unhindered by the low-level requirements of organising and maintaining operation of the sensor network. The question remains, what level of 'transparency' is needed to allow an applications designer to work efficiently? For instance, is it reasonable to expect every application using a sensor net to be responsible for determining the location of each node in that net, given that in most sensory applications physical location is a vital piece of information, or should this be handled by the 'system', and if so, how? At practically every stage in the application design, system level design questions are raised. Rather than looking for an individual answer to each of these questions, we will seek an answer from the literature in distributed system research.

Requirements for transparency* have long been a goal of distributed system design, and have been classified by Coulouris *et al.* [18]. Seven 'transparencies' have been defined, all of relevance to sensor networks, as follows:

Location transparency:
Services in a distributed system should operate independent of their location, and the location of the service should not be apparent or important to the user. In the context of a sensor network location transparency requires that all nodes, particularly in an *ad hoc* system, be essentially equivalent, that there be no nodes of a special type that require a particular process

*It has been commented by some reviewers that these 'transparencies' should more accurately be called 'opacities', since a system that supports them will inevitably obscure the detail of what's really going on at a hardware level. While the reviewers have a point, this usage is not ours. 'Transparency', used in this way, is reasonably widely accepted within a particular community, therefore we have not changed its use here. In a way, the point reflects two different design viewpoints. For someone who is an electronics engineer, achievement of these 'transparencies' obscures the view of the parts of the system that are their concern. For someone who is an applications software designer, the 'transparencies' clear the view of the application software concerns, abstracting away the clutter that is the complex underlying structure of the system.

located on that node. The system support must handle communications between processes independently of their location.

Concurrency transparency:
Processes may operate concurrently and share resources without interference between them. This requires that all sensors be equipped with properly concurrent operating system kernels.

Replication transparency:
Multiple instances of resources may be used to increase reliability and performance without knowledge of the replicas by users or applications programmers. This transparency is required to enable fault management techniques based on redundancy within the network.

Failure transparency:
Enables the concealment of faults, allowing applications to complete their tasks despite the failure of hardware or software components. So far as sensor networks are concerned, this transparency is essentially a statement of the requirement to handle sensor and node faults.

Mobility transparency:
Allows the movement of resources and clients within a system without affecting the operation of users or programs. As well as location transparency, discussed above, the requirement here is that a process be relocated to a different node without modifying its state.

Performance transparency:
Allows the system to be reconfigured to improve performance as loads vary. This is essential if the full computational resources of the network are to be used effectively, or overall power usage within the network is to be minimized by optimum partitioning and location of processes.

Scaling transparency:
Allows the system and applications to expand in scale without change to the system structure or the applications algorithms. This is a basic requirement for a generic way of interfacing to sensor networks.

We add one further transparency criterion, since sensor systems are real-time systems.

Temporal transparency:
The time domain behaviour of the system, in terms of event capture and synchronization, must remain the same regardless of how computational tasks and resources are distributed in the network.

The difficulties involved in producing systems which possess all of these transparencies may be gauged by the fact that no modern, commonly used distributed operating system achieves all of them. The latest releases of Microsoft Windows achieve three or four of them (depending on how pedantic you are), and therefore could not be considered a suitable basis for distributed sensor networks. On the other hand, fifteen years ago, research systems were demonstrated offering all of them, for instance, Plan9 from Bell laboratories [19] which was usable in 1989.† Thus it is tenable to expect an operating environment for sensor system to similarly achieve all of the transparencies. The key to the achievement of the transparencies is the adoption of a suitably layered architecture, in which each layer serves to abstract away a certain level of specific concerns, translating them to generalised, conceptualised objects. At first sight, the definition of such an architecture is trivial: simply draw up some layers and give them names. In practice, things are more difficult. Definition of too many layers causes inefficiency, as does allocation of the wrong functions to different layers, since communication between system functions ends up occurring through several layers of the system. In severe cases, in a distributed system, communications between functions residing on the same node can end up occurring via the network, rather than directly between the processes serving those functions. This occurs because (if the layered system is inappropriately designed) the routing 'layer' does not know which processes are 'local' and which are 'remote' and therefore uses the network to communicate between them all. Obviously, the inefficiencies of network usage caused by messages between software components in the same node being routed via the network is not permissible in systems such as sensor networks, and neither are the many other inefficiencies that can occur as a result of poorly designed layered systems.

†The fact that the market leader under-performs when against systems which have languished in relative obscurity for years has more to do with the nature of the market than with technical excellence.

10.5. Network Design Considerations

In designing the layers of system software that will control the network, it is necessary to take all of the transparency requirements into account as the various software components are designed. It is only if the detail design of each part of the software is carried out in such a way so as not to compromise the transparency requirements that they will be present in the final product. Here we look at some of the key concerns of operating a distributed sensor network form the point of view of the eight transparencies.

10.5.1. *Network Discovery*

Network discovery refers to the process of establishing an operating connection topology for an *ad hoc* network. The process was discussed in Chapter 9. Here we are interested not so much in the details of the network discovery process, but the ramifications on the design of the systems support middleware.

The network discovery function is driven by several of the transparency requirements:

- Mobility transparency dictates that the connectivity of the system be able to change as the nodes move, in a way transparent to the applications designer. Failure transparency determines that it is necessary to reconfigure the connectivity of the network in case of node failure.
- The established topology also affects both location transparency and scaling transparency. The distribution abstraction layer must provide topology management services, according to the needs of the specific installation and the application.
- Location transparency and our initial assumptions demand that there be no special nodes, far less a central controller, and in its absence, the job must be done in a collaborative, distributed way.

On initialisation, the first task is to establish communications with neighbouring nodes, the second to deduce from the information gained the connectivity of the network, the third to impose a suitable routing topology, either static or dynamic. Whereas a static routing scheme conserves power, since the discovery of routes is not needlessly repeated, it is also not sufficiently adaptable for *ad hoc* networks, in which the membership and topology changes. Thus, *ad hoc* networks require one set of solutions to network discovery, and mobile networks a further set. Distributed algorithms for network discovery and maintenance have been discussed by several authors [20–25]. There are a number of subtle factors in their design. In Chapter 9 we observed how the communications load in distributed search algorithms of the type required here, such as flood routing, can quickly grow exponentially, as every node tries to contact every other node. In any circumstance this is a bad idea, but in applications that are power limited, as many intelligent sensor applications will be, this is a particular problem. Generally, in the design of sensor network systems one cannot simply apply solutions from other areas of networking. The wider area provides a useful source of prior art, but in the context of sensor systems every part of the system must be carefully evaluated against the particular constraints, of which the dominating one is the need for power conservation and therefore the resultant requirement for elimination of unnecessary network usage and selection of protocols which are particularly efficient in terms of their bandwidth usage prevails.

10.5.2. *Node Locationing*

Given that sensor networks exist to sense physical phenomena at a particular set of locations, the location of the sensors must be known before the network is useful. Locationing algorithms were discussed in Chapter 9. In terms of the distributed computing transparencies, locationing is tied up with mobility transparency, scaling transparency, replication transparency and failure transparency. The link with mobility transparency is obvious — mobile nodes and changing membership of the network necessitate every application program to continually check location unless this task is handled transparently by the system software layers. Scaling transparency simply demands that whatever locationing mechanism is used, it scales with the network. It is also linked with replication and failure transparency, our next topic, since failure of a node at a particular location will require to be 'made good' in a way transparent to the application programmer. If replication (i.e. redundancy) at a location is used to make good node failures, then that must also be transparent at the application level.

10.5.3. *Fault Management*

Should the node at a particular location fail, unless the system has replication transparency and failure transparency, an application programmer is responsible for handling the consequences of that fault. As discussed in Chapters 8 and 9, fault management is an essential function in any sensor network. Essentially, fault management strategies depend on the existence of some redundancy in a network, so that failed sensors can be switched out and the information for which they were responsible gained or generated from some other source, such as other sensors in the locality. The actual mechanism for this must be transparent to the application programmer.

10.5.4. *Information Extraction*

The network exists to yield information, yet the application designer has no knowledge of the deployment of the sources of the data from which that information is derived within the nodes of an *ad hoc*, mobile network. Unless each application is to include its own mechanisms for mapping data sources to the specified information requirements, these mechanisms must be handled within the system middleware. Inevitably, this means that the application software will have little direct control over the deployment or use of the sensor resources (in any case, this is inevitable, given our initial assumption that there could be multiple applications using the network simultaneously). Ideally, the components which locate and return the required information must posses all eight transparencies, since those producing the software to do with top-level information processing are unlikely to have either the expertise or the desire to deal with network specific issues such as locationing, fault tolerance, synchronisation and so-on. The topic of information extraction from the network is dealt with more fully in Section 10.8.

10.5.5. *Task Migration*

The next problem concerns how the application code which needs to be distributed over the sensor network is handled — the various forms this may take are discussed in Section 10.8. For the network to be used at its optimum, it is required that various application level computations occur distributed through the network. This is particularly the case in networks that are limited by power or data bandwidth, because it can be shown that in many cases the power consumed by reducing data to the specific information required by the application is less than that required to transmit the data to a central processor.

At the same time, location, mobility, replication and performance transparency must be maintained. The system design should not burden the applications designer with knowing where to locate specific parts of the application, or where particular network capabilities or services reside.

The practical problem with distributing application code is that the network is defined in a very 'soft' way. The precise topology of a network, and the member nodes, may not be known until the network is deployed, and moreover, it may well change as the network operates. Thus, it is impossible for the application designer to hard code the distribution of the application. Rather, constraints and requirements must be set and the system must take care of distribution to satisfy those constraints and requirements.

10.5.6. *Network Synchronisation*

The observation of the temporal behaviour of the phenomena that are being sensed by the sensor network is of the essence of signal analysis. We have discussed in Chapter 9 the problems of synchronisation within an *ad hoc* network, particularly in the context of ensuring the maintenance of a correct ordering of events across the network. However, the application requirements are likely to be stricter. It is natural for the applications which do this analysis to deal with events in a common time-frame, meaning that events must be correctly timed, rather than simply ordered. Unless they are to be burdened with mapping a reference clock to the ordering of events in a distributed system, to maintain temporal transparency entails the maintenance of such a common time reference by the system middleware.

10.5.7. *Achieving the Transparency Requirements*

The consequences of the transparency requirements mean that the system as a whole is best seen as an assembly of autonomous, self organising, co-operative nodes. Any other organisation makes it hard to achieve one or more of the transparencies. With co-operative autonomous nodes it is possible, but still conceptually difficult.

As was stated in the conclusion to Chapter 9, many of these design issues of *ad hoc* sensor systems are still the topic of active research, and are far from being resolved. No solutions have emerged as the established orthodoxy. Thus we cannot construct a middleware system which commits to one published solution or another for any of the basic functions such as locationing or synchronisation. Rather, we are looking for a system of components, with defined interfaces between them, which can be substituted for different components embodying alternative solutions as the research progresses. This evolution of the middleware should proceed in such a way that the function of the application based on this middleware should be unchanged (or possibly enhanced) as the middleware changes. Although it avoids the problem of having to design all of the components (and solve all of the deep research problems) now, it is still difficult to conceive how such a flexible component based middleware will work. Once again, the conceptualisation of the problem is eased by the adoption of a structural abstraction.

10.6. Layered Model

In this section we propose a layered model for mobile, *ad hoc* sensor network systems, which would also be applicable to sensor networks which contain at least some of the design problems inherent in the mobile, *ad hoc* type of network. The purpose of the model is similar to that which has emerged for the Open Systems Interconnect (OSI) model for computing network systems. While OSI has not produced the ability to openly interconnect systems, as was originally envisaged by the authors of the standard, it has at least provided a reference point against which proposed architectures can be compared and assessed. Through the provision of a reference terminology, comparative evaluation of alternatives becomes simpler. A worthwhile question would be — why cannot OSI be applied directly to sensor systems? The answer is that it can, however it was originally designed to define heterogeneous computing networks, at a time when circuit switched communication (as opposed to packet switching) was the norm. It is something of a stretch to apply it to a homogeneous (at least, as far as purpose of the network is concerned) *ad hoc*, mobile network, for which efficient operation is a priority. One of the problems with OSI is that it is over elaborated as a model (with seven defined layers), leading to over complex and inefficient partitioning of protocol stacks.

Thus, a model adopted for mobile *ad hoc* sensor systems needs to be as simple as it can be, while still separating out different concerns into cleanly separated entities (Figure 10.3). The one proposed here consists of just four layers. Conceptually, the lower two reside on each node, and provide a standardised way of interfacing to the hardware and a standardised node to node interface. The upper two layers are distributed around the network, and provide support for the system wide functions. At the top is the API itself, supported by the top-most layer.

The lowest levels are the per-sensor node system software (or firmware) that deals with the interface and support of the hardware and processor functions. In line with modern operating system terminology this is called the 'hardware abstraction layer' or HAL. In particular, it includes device drivers for the sensor interface (usually A to D), timing functions for acquisition and communication control and for the network communications functions themselves.

The next layer includes resource and process management at the node level and is called the Node Abstraction Layer (NAL). The HAL and NAL functions are those commonly provided by a concurrent embedded operating system, and it is envisaged that an 'off the shelf' operating system will provide these layers. Options are discussed in Section 10.7.

The next layer is the per-sensor node system layer which deals with the sensor system considerations such as the mapping of application level

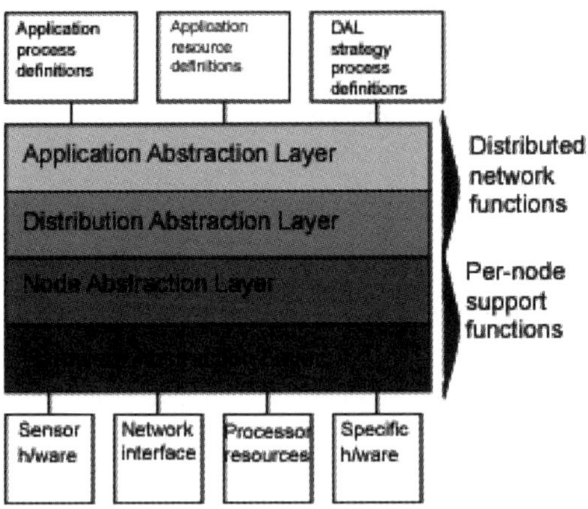

Figure 10.3: Four layer model of a sensor network.

concepts (e.g. 'location' and 'time') to sensor level concepts (e.g. 'node address', 'localised network topology', 'sensor clock' and 'event ordering'). This level is called the 'distribution abstraction layer' or 'DAL'. The DAL function would be provided by a number of processes which run continuously on each sensor node, and together dictate the generalised strategy for performing these mappings. It is the DAL processes which are liable to be updated as research progresses and improved strategies become available. For example, one of the DAL processes is likely to implement the algorithms that locate the nodes in real space and, if the nodes are mobile, keeps that location map current. So far as the application designer is concerned, all that is required is that the location for which information is required is specified. If that information is location specific, the DAL deals with the specifics of how that physical location is mapped to the actual array of sensor devices. As the DAL locationing process is improved, the application level code need not change, but the estimates of location available from the DAL should improve, or track changes better, or become more resistant to failures within the network — whatever the result of the improvements in locationing algorithms produced by the research turns out to be.

The DAL processes are defined by process objects accessible at the top, API level, and can therefore be updated or adapted for different classes of application. This standard API, which is the interface that the applications programmer uses, is termed the Applications Abstraction Layer, AAL. Since, in an *ad hoc* network, the applications programmer can have no advance information on the number or deployment of the nodes, the AAL supports an interface to the network as a whole. In practice, this will mean communication with an arbitrary node, or set of nodes. In a mobile network, the relative location of any particular node and the 'user' (whether the user is a person or a computer system) — who will need to be tapped into the network at some level — is liable to change, since the user is presumably fixed but all or any of the nodes may move. Therefore the node 'selected' to communicate with the user is also liable to change.

The design and even specification of the components which fit at the higher levels of this model, the DAL and AAL, are still unclear. Some thoughts and likely directions are put forward in Section 10.8. The lower levels can, however, be sourced 'off the shelf' as has been stated. The next section examines some possible candidates for these lower levels.

10.7. Sensor Network Operating Environments

The lower two levels of the model are generally realised using an operating system, which executes on each and every node in the network. The operating system provides a support environment for programs which run on the node, the support environment from the programmers point of view is expressed as an Applications Programming Interface (API). Specifically, an operating system provides functionality to furnish memory, process and concurrency management for the Node Abstraction Layer and device management for the Hardware Abstraction Layer. As intelligent sensor networks are realized, investigators are designing specialised operating systems and systems support for them.

There is a rich seam of work on wireless sensor operating systems environments available in the literature. Some of the most prominent ones will be discussed in Section 10.7.1. Much of this work demonstrates the potential in performance terms of intelligent sensor networks.

However, most sensor network operating systems so far proposed do not offer support at the level or sophistication necessary to address all of the transparency concerns discussed in Section 10.4, yet alone in a generic way. Design of a system or architecture that does address them involves considerations at a completely different level of abstraction from those obtained by a bottom up design starting from the requirements of the immediate sensor and its associated circuitry. It is in this context that we start our survey of some of the 'likely candidates' for components which would fit the two bottom layers of our model.

10.7.1. *Comparative Study of Operating Systems*

In this section we compare the design approach of three operating systems which have been either designed specifically for wireless sensor networks (in the case of Tiny-OS and the EYES system), or, in the case of the third, is a member of the class of general, real-time embedded operating systems that could form a suitable base for the software of an integrated intelligent sensor. This particular system has been selected because it provides a very simple starting point for those developing their own distributed sensor systems, since it is has openly and freely available source code, is modular and configurable to a particular requirement and has about the lowest memory and processing resource requirement of all but specialised minimal operating systems.

Hence, the three systems are Tiny-OS, the DCOS/AmbientRT operating system for the EYES (EnergY Efficient Sensor networks) project and eCOS (embedded Configurable Operating System). All three operating systems use a component based architecture, which means that rather than being constructed as a single, monolithic whole, they are constructed as a set of software 'components' which can be configured to provide for the requirements of a wide variety of applications, without the inclusion of non-standard or application specific code into the operating system.

Tiny-OS has been developed specifically as a part of the Berkeley 'Smart Dust' research [26]. The design focuses directly on the needs and characteristics of the intelligent sensor nodes, or Motes, envisaged by that project. These are:

- Small physical size and low power consumption, which implies limited physical parallelism and controller hierarchy and a direct-to-device interface for input/output. Typically, the resources of a mote are small. The Berkeley 'Mica' mote has available 640 kB of code memory and 4 kB of data memory;
- Concurrency-intensive operation, being able to handle multiple inputs and outputs simultaneously;
- Diversity of design and usage. Individual mote designs will be application specific, not general purpose and there will be huge device variation;
- Largely unattended and numerous motes, requiring robust operation and 'narrow' interfaces.

The 'mission' of TinyOS is consequently to 'enable rapid innovation and implementation while minimising code size as required by the severe memory constraints inherent in sensor networks' [27].

The EYES project is a European Union funded project comprising of The Centre for Telematics and Information Technology (CTIT) — University of Twente, the Netherlands; Nedap N.V., the Netherlands; Consorzio Nazionale Interuniversitario per le Telecomunicazioni (CNIT), Italy; Rome University La Sapienza, Italy; Technical University of Berlin, Germany and Infineon Technologies, Austria. The DCOS/AmbientRT operating system [28] is the node operating system used to support the EYES mote.

The design parameters are similar to those for Tiny-OS, the mote having similar resource levels. The design goals are motivated as follows:

> *DCOS is a Real-Time Operating System (RTOS) for embedded devices with very limited memory, processing, and energy resources. Despite these limitations, DCOS/AmbientRT has powerful features like, real-time scheduling, online reconfiguration, and support for a modular data driven architecture [28].*

The key difference from TinyOS here is the 'online reconfiguration' — TinyOS is configurable only at system build time. There are other fundamental differences of design philosophy, as we shall see.

The third operating system discussed here is eCOS, a product developed by Cygnus Solutions, later acquired by Red Hat (the well known Linux company), as a configurable concurrent operating system. eCos is an open-source product, distributed under the Gnu public license (GPL) and now has the IPR rights vested in the Open Software Foundation. It sits alongside the well known embedded Linux operating systems, but eCOS is designed to be smaller, more flexible and energy efficient than Linux, to suit smaller processors, without memory managers. However, it cannot be installed on systems as tiny as those served by TinyOS and AmbientRT, as, at its smallest, it occupies 3 K of ROM and 1 K of RAM, for a kernel that includes a scheduler, memory management, RTC support, several threads and interrupt handlers. eCOS has been ported to many different hardware platforms, mostly using 32-bit processors. eCOS is fully 'component' based, allowing a tailored operating system to be configured to suit a specific set of requirements from a set of predefined 'components' shown in Figure 10.4. Versions can be assembled with a choice of API's: the well known POSIX standard Unix API, which allows compatibility with a range of Unix software, or the uITRON API, a generic API aimed particularly at embedded systems. Full details of eCOS are available on line in the eCOS Reference Manual [29].

In the comparison of the three operating systems below, the factors which contribute specifically to the NAL and HAL in the four layer model will be considered. These are the functions concerning hardware abstraction, device management, memory management, task abstraction, synchronisation and scheduling and communication.

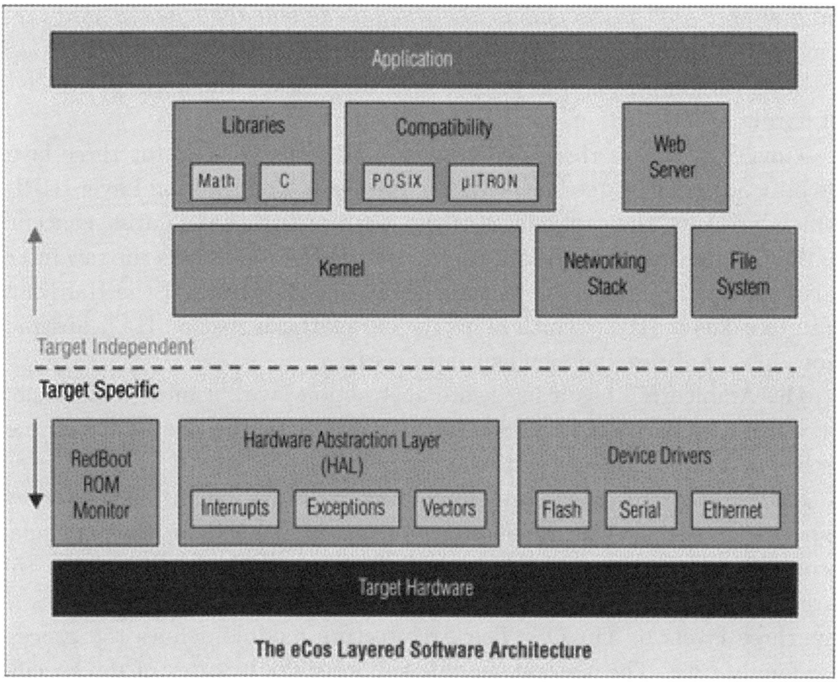

Figure 10.4: The layers of the eCOS operating system. From [29]. The 'Target Specific' section corresponds with our HAL, while the 'Target Independent' Layer corresponds with our DAL. The 'Application' in this case will be the components which constitute the NAL and AAL as defined in this chapter.

Hardware abstraction

Hardware abstraction is the process of provision of the variable hardware base (entailed with the use of different hardware designs and components) with a consistent programming interface. On this, software can be based, which implements a variety of applications services. Often this is a package or layer of software called the hardware abstraction layer, or HAL — this is a similar, but has sometimes slightly different usage, to the HAL in our four layer model. In particular, in some systems the HAL doesn't necessarily abstract away the hardware level concerns from programs working in higher level layers, but does at least present the specific characteristics of that hardware to a consistent format — this means that the applications programmer still has to change the program to suit the specific hardware, but those changes are at least localised and well ordered.

In all three systems, the programmer may choose the appropriate level of hardware abstraction.

TinyOS 2.0 has a three tier hardware abstraction layer (or three layer architecture). The three layers are the Hardware Presentation Layer (HPL) which 'presents' the capabilities of the hardware using the native concepts of the OS, the Hardware Adaptation Layer (HAL) which uses the raw interfaces provided by the HPL to build useful abstractions and the Hardware Interface Layer (HIL) which converts the platform-specific HAL abstractions into hardware-independent interfaces.

The AmbientRT has a hardware abstraction layer, a minimum version of which can be placed in 3800 bytes, along with the rest of the kernel (scheduler, data manager and memory manager).

eCOS has a HAL which presents a fully hardware independent interface (apart from specific executable object code requirements for the particular processor in use). The HAL is typically built up with three modules: the Architecture, the Variant, and the Platform module, with similar roles to the three layers of TinyOS. The Architecture module defines the processor family type. The Variant module supports the features of the specific processor in the family. The Platform module extends the HAL support to tightly coupled peripherals like interrupt controllers and timer devices.

In all three cases, the level of HAL support provided is sufficient to allow abstraction of node specific hardware details away from higher level programs, although the eCOS arrangement is perhaps the cleanest and most flexible.

Device management

In terms of our four layer model, device management is also part of the HAL. As 'tiny' operating systems both TinyOS and AmbientRT employ a minimalist approach to device management, with device registers essentially being directly programmed. As they are both component based, the basic component entities are the same as their elemental program building blocks (in AmbientRT this is called a DCE, in Tiny OS a 'component'). Effectively, a component dedicated to handling a device serves as a device driver, without a dedicated device-driving subsystem within the operating system. The provision of mutual exclusion between tasks using a single device is not explicitly dealt with within the device driver, as it would be in some other systems (including eCOS). In both operating systems the mutual exclusion mechanisms provided by the operating system, described below, provide the means for dealing with this.

By contrast, the eCOS device management system follows the form of the traditional Unix style operating system. It is designed as a general purpose framework for supporting device drivers, which can include all classes of drivers from simple serial to networking stacks. Components of the I/O package, such as device drivers, are configured into the system in the same way as other components and end users may add their own drivers to this set. Each device in the system has a unique name, such as '/dev/console' and '/dev/serial0', where the '/dev/' prefix indicates that this is the name of a device, which allows for generic, named devices, as well as more flexibility. Basic functions are provided to send data to and receive data from a device. Additional functions are provided to manipulate the state of the driver and/or the actual device. These functions are, by design, quite specific to the actual driver. The driver model supports layering; in other words, a device may actually be created 'on top' of another device.

In this respect, TinyOS and AmbientRT fall shy of the needs of the four layer model, as put forward in Section 10.6, in particular allowing hardware specific concerns through the HAL into the NAL. eCOS, which includes modular device drivers, allows the inclusion of a set of device drivers which can abstract away the details of the devices from higher levels of software.

Memory management

Tiny OS adopts a strict component based model for its memory management. Each 'component' is associated with a separate stack frame which contains all of its data. There is no heap (dynamically allocatable memory). To allow this simple model, the components (which may be tasks, events or handlers) are strictly layered, with calling only allowed to lower level components, thus there can be no cycles in the call chain.

AmbientRT, like TinyOS, is an explicitly data-centric operating system. Thus, the memory management model is based around data abstractions, in this case called Data Centric Entities (DCE). Data is a generalisation of memory objects and events, where an event is for example the occurrence of a hardware interrupt. Each distinct memory object or event is called a Data Type (DT). The DCEs are used in a publish/subscribe system that allows them to react to DTs produced by others. New configurations can be achieved by altering the set of active DCEs and modifying their subscriptions. A task — or data set can grow or shrink in size. Dynamic memory allocation is employed to reserve and free memory space of arbitrary size in a dedicated heap area.

eCOS memory management follows a more traditional OS model, with each process or task being allocated a stack frame, which offers a running process the usual resources of stack and heap. How these resources are used is up to the application designer.

Given the likely scale of the software required for the DAL and AAL, it is likely that the larger scale memory management provided by eCOS will be more capable of servicing the storage requirements of those layers.

Task abstraction

All three systems provide concurrency, although the models vary. The underlying task model in AmbientRT and eCOS are similar, namely quasi-concurrent time sharing threads of execution. Whereas eCOS couples this with the Unix like thread model, the abstraction employed in AmbientRT is linked to the data-centric architecture. In the data-centric model employed, data is a generalisation both of memory objects and events. In turn a task is seen as a subprogram of an application that performs an action, in response to some event. Whereas eCOS provides tasks with individual stack space, in AmbientRT all tasks share a stack, simplifying task switching. TinyOS operates an event-driven task model. It executes only one program consisting of selected system components and custom components needed for a single application. There are two threads of execution: tasks and hardware event handlers.

All three operating systems provide abstractions for the concurrent execution of programs and the handling of non-determinate events, as will be required for WINS applications.

Synchronisation, mutual exclusion and scheduling

TinyOS tasks are functions whose execution is deferred. Once scheduled, they run to completion and do not preempt one another. Hardware event handlers are executed in response to a hardware interrupt and also run to completion, but may preempt the execution of a task or other hardware event handlers. The scheduling model is simple, but makes no guarantees of schedulability.

By contrast, AmbientRT operates a sophisticated scheduler, which builds on research in real-time scheduling. An 'earliest deadline first' (EDF) scheme operates. The real-time scheduler in AmbientRT uses dynamic priorities. This means that the priority of a task relative to the priorities of other tasks changes over time. The scheduler uses the absolute deadline as the priority of a task. The task with the earliest absolute deadline has the highest priority. EDF belongs to a small set of scheduling algorithms which can guarantee to find a schedule, should a feasible one exist. The programmer of AmbientRT tasks is required to provide the scheduling parameters of a task, so that the schedule can be determined.

eCOS does not include a scheduler as a built in component, but can be configured with a custom scheduler. Two types of schedulers are available in standard eCos[‡]: a multilevel queue scheduler and a bitmap scheduler. The multilevel queue scheduler allows threads to be assigned with a priority level from 0 to 31, with 0 being the highest priority and 31 being the lowest. At each level multiple threads can execute and pre-emption between levels is supported, allowing higher level threads to execute while lower level threads are halted. Time-slicing is also supported within a priority level and across priority levels, allowing each thread a predetermined execution time before relinquishing resource to the next thread. The multilevel queue scheduler supports Symmetric Multi-Processing (SMP). The bitmap scheduler is a simpler, and more efficient, scheduler. It also allows 32 priority levels similar to the multilevel queue scheduler, but only one thread is allowed to execute on each level. Preemption is supported with this scheduler, but time-slicing between threads is not, and is not needed because each level only allows a single thread. Thus, the standard eCOS falls somewhere between the simple, priority-less scheduler of TinyOS and the dynamic priority scheme of AmbientRT.

Once concurrency is provided within an operating environment, measures must be taken to synchronise between tasks, where a certain ordering

[‡]As a fully configurable system, there is nothing to stop a completely new scheduler being written and included, if that is what is required.

of events is a priority, and to ensure mutual exclusion of tasks between commonly used resources. eCOS includes support for a full set of synchronisation and mutual exclusion primitives. All of the following synchronisation methods are supported in eCOS: mutexes, counting semaphores, flags, spinlocks, condition variables, and mailboxes. Mutexes or mutual exclusion objects allow threads to share resources serially. Each thread in turn locks the resource, uses it, and then unlocks it for next thread to use. Counting semaphores use a count associated with each resource to track its availability. If the resource is not in use, the first thread to request it will be granted the resource. If however, several threads are waiting on the resource the highest priority thread will be given access first. Flags are 32-bit words where each bit in the word can represent a resource condition. Threads can execute when a single condition is met or when a specific combination of conditions are represented by the flag. eCOS supports low level synchronisation with spinlocks. As a thread needs a resource it checks a status flag, if the resource is in use the thread 'spins' in a tight loop polling the flag until the resource is free. Condition variables allow a thread that is using the resource to signal another thread(s) when the resource becomes ready.

Mutual exclusion in the AmbientRT kernel is obtained through the scheduler. It provides automatic synchronisation of shared resources. The scheduler compares on initialisation the resource usage lists of every task and generates per task, on the basis of the relative deadlines, a threshold value. Comparing this threshold value of a task to the deadline of another indicates whether they share a resource. When the scheduler determines which task is allowed to run it will not only compare the dynamic priorities of the tasks, but also the threshold of the running task to the deadline of the candidate task. If they share a resource, the scheduler will first let the running task finish even if the candidate task has an earlier deadline. In this last situation the candidate task is blocked by the running task, and, the added delay because of this, is called the blocking time.

Because tasks and hardware event handlers may in TinyOS be preempted by other asynchronous code, programs are susceptible to synchronisation and mutual exclusion problems. The avoidance of these conditions is left to the programmer to resolve, either by accessing shared data exclusively within tasks, or by having all accesses within atomic statements. The compiler reports potential data races to the programmer at compile-time.

All of the systems provide sufficient synchronisation and inter-process communications facilities for our purposes, although eCOS allows a variety of approaches, as distinct from the imposition of a particular model of task abstraction.

The eCOS and TinyOS scheduling policies are inadequate for serious real-time systems, which we would expect our sensor networks to be. AmbientRT, on the other hand, includes a genuine real-time scheduler. The component based structure of eCOS would, however, allow the inclusion of a genuine real-time scheduler.

Communication

Tiny OS includes support for application level messaging between nodes. The messaging follows an 'active messaging paradigm'. The motivation for this is an assertion that legacy communication such as TCP/IP, sockets, and routing protocols such as OSPF cannot be used, being bandwidth intensive and centred on 'stop and wait' semantics. Active messaging is designed to have real time constraints and low processing overhead. It provides a distributed eventing model, in which networked nodes send events to each other.

Messaging in AmbientRT is also a directly kernel supported activity. The communication follows the data-centric design of the OS, rendering, it is claimed, conventional inter-process communication unnecessary.

eCOS, as a general purpose but still component based operating system, allows different IPC mechanisms to be used. Standard components are available to implement the TCP/IP protocol stack, and Unix style sockets between processes (reflecting that one of the target applications for eCOS is lightweight IP routers).

Review of design philosophy

Interestingly, while eCOS, a 'small' operating system, confines its concerns to the lower two layers of our four layer model — the hardware and node abstraction layers, both TinyOS and AmbientRT, which are both 'tiny' systems, venture into the distribution abstraction layer as well, providing 'data-centric' facilities designed to achieve some level of location, concurrency and replication transparency. In addition, AmbientRT addresses the issue of mobility transparency by providing a direct means for migration of processes, and some of the concerns of temporal transparency, by way of its sophisticated real-time scheduler. By contrast, none of these concerns are dealt with as an integral part of eCOS, which is defined as a 'traditional' real-time operating system. This reflects the design motivation of the various designers. eCOS was indeed conceived as a general purpose real-time OS, for which sensor networks were but one of very many possible applications. TinyOS and AmbientRT were specifically aimed at wireless sensor networks. Their design was motivated by the manifest design issues of these networks, of which location, concurrency and replication transparency are at the forefront. Thus it was natural that their designs include some measures specifically targeted at these requirements. Another driving concern for TinyOS and AmbientRT was an understanding that in many wireless sensor networks the node level computational and storage resources would be minimal. For this reason both were specified to operate on very small processors, thus compromising their ability to deal completely with the transparency issues. As was seen in Chapter 9, there are likely to be few cases in which it will be necessary to specify nodes with a truly tiny processing capability for reasons of cost, space or power usage, and thus it should be possible to provide sufficient computational and storage resources in the network to provide most if not all of the transparencies. If this is the case, an OS such as eCOS may be a better starting point, since it provides a straightforward API at the NAL level, while not predetermining the strategies required to achieve all of the transparencies.

10.8. Application Services

There has been some research published that relates to the functions that have been defined here as the DAL and AAL, although the lower level concerns tend to get included in the mix as well. Below is a brief summary of some that have appeared in the literature. So far as the user of a sensor network is concerned, the network exists as a source of information. The major role of the Application Abstraction Layer (AAL) is therefore to provide a simple and consistent means of extracting information from the network. There have been several pieces of research reported which address specifically this problem, which are outlined below.

The Sensor Information Networking Architecture (SINA) [30] defines a middleware layer that allows the querying and monitoring of a network of wireless sensors. Lower levels of the architecture allow nodes to autonomously self-organise in a hierarchy of clusters, in order to optimise transmission power usage. SINA includes an attribute naming scheme, which decouples node queries from their physical network addresses.

Peti *et al.* [31] discuss the design of an architecture which addresses the requirements for data exchange, diagnostic services, system integration, constraint checking and dynamic reconfiguration, viewing these as *distributed systems* issues. Their approach is to use the CORBA (the Common Object Request Broker Architecture) distributed programming interface, which tackle some of the problems of integration and reconfigurability, but not those of mobility. The architecture provides a consistent and location independent programming interface, but not a systematic way of handling the system management requirements specific to sensor networks. These require that the function of the system be considered as a whole. Fundamentally, this is a *top-down* view, requiring that some basic architectural decisions to be made at a system level, before their effect can be considered on the design of an individual intelligent sensor node — thus it is unlikely to be the basis for a generic system.

In order to begin to produce a manageable strategy for the way in which applications tasks are distributed around the network, it is necessary first to adopt some underlying model of the role and interaction of these tasks. Both of the systems above apply a particular model of application task distribution. In SINA the tasks are 'queries', in Peti *et al.* work, using CORBA the tasks are distributed 'objects' or agents. These two represent the two common approaches to the information extraction problem in *ad hoc* systems, the agent and query bases methods.

The intelligent agent model has become popular in recent years. An 'agent' is a self contained, mobile, self-organising piece of software, dedicated to performing some set task. The underlying idea is that the system functions are embodied in such agents, which migrate around the network and collaborate to obtain some optimum configuration for the task in hand (although some schemes may involve an agent which centralises or mediates allocation between a group of agents). The attraction of the agent model is that it naturally supports a 'self-organising' view of the network, a great advantage when the exact configuration or distribution of the network is not known at design time. Rather than attempting to describe all the configurations possible, the designer designs agents, whose behaviour is to seek out optimum organisations, according to the parameters of the particular application.

Whereas CORBA is a general purpose agent architecture, designed for 'big systems', many agent based protocols have been designed specifically for sensor systems, including SPAM [32] which resolve conflicts between agents using distributed negotiation. The conflicts that arise are of the nature of resource allocation, the resource in sensor networks being the sensor and the associated processor. Ortiiz *et al.* [33] describe a set of inter-agent negotiation protocols for *non-additive* domains — environments in which the domains of different sensors are not independent, they may overlap, and where tasks may include interaction between neighbouring sensors. These protocols are *centre-based* algorithms, in which the negotiation is mediated by one *centre agent.* A number of variations of these protocols were evaluated, using a simulation of the sensor network challenge problem. Soh *et al.* [34] describe protocols for formation of coalitions of agents using machine learning techniques, particularly case based reasoning (CBR). Horling *et al.* [35] describe a complete design for a distributed wireless sensor network, based on an agent model, using the SPAM protocol.

One issue that has prevented agent based systems gaining widespread acceptance for reliable systems concerns the lack of certainty that an agent will succed in it appointed task. Some experimental validation is required to gain some confidence that agent based systems will indeed produce reliable results. Lesser *et al.* pose a 'sensor network challenge problem' [36] which serves as a good case study for the design of co-operative sensor application problems, including agent based approaches. The problem involves the distributed allocation of sensors within a network. Each sensor consists of three heads, each of which produces a 120 degree scanning region, so as to cover a complete 360 degree circle. Measurements from each sensor can only be taken from one head at a time. Communication between sensors is restricted, using a low speed, unreliable RF system, spread over eight channels, and capable only of reception or transmission at a time. The sensors cannot share a channel. Each sensor has a CPU, which is capable of supporting one or more processes. The sensors themselves are radar sensors, and the purpose of the array is to track moving targets (actually model trains). The sensors must organise themselves in such a way as to be able to co-operatively determine the targets position using triangulation, while making optimum use of power and the very constrained communication resources of the network. This challenge problem provides an application capable of simulation, against which different agent protocols can be measured, although few, if any, of the proposed protocols have yet been measured in this way.

A second way of approaching the information extraction problem is the query based model. Essentially this is a declarative model, whereas the agent based scheme is an imperative one. Both can be seen as ways of distributing programs through the network which will control the gathering of data from it. Yao and Gerkhe [37] propose a query mechanism called 'Cougar'. It allows the interrogation of sensor networks using declarative queries, which shields users from the physical characteristics of the network. Efficiency is gained by use of a query optimiser, which, it is claimed, vastly reduces resource usage.

Sadagopan *et al.* describe ACQUIRE, another declarative query mechanism [38]. ACQUIRE seeks to minimise energy usage in resolving queries by an efficient, incremental mode of query resolution, which involves the response being returned to the querying node by the node which finally resolves the query.

Other self-organising models can also be used to structure an *ad hoc* system. The 'cellular autonoma' model, based on a biological metaphor, is one. Whereas agents are based on ideas of human interaction, whereby 'negotiation' achieves results through co-operation, celluar models tend to be driven by models of reproduction and fitness for survival. Future systems might include mixtures of both models.

10.9. Proposed Sensor Support System Architecture

The four layer model put forward in Section 10.6 is a structure within which a number of different architectures could be fitted.[§] For the authors' work, it helps define the components that will be needed, as well as the place they fit in the overall scheme of things, and the nature of the interfaces that will be neded. The complete definition of an architecture using the model is an

[§]The word 'architecture' has been used with some abandon in this section, mostly quoted from other authors. We distinguish 'model' as a framework which essentially provides a set of names for system functions and provides some information about the relationship between those. An 'architecture' is a refinement of a model which identifies some or all of the components which provide the functions, and starts to detail their inter-operation. Thus a model says there will be, say, a hardware abstraction layer. An architecture defines how that layer is built.

extended task, which has hardly been started. Here some of the components which will operate in each layer of the model, and on which development is underway, will be presented. The design aim is to specify components which are needed to resolve the transparency concerns discussed in Section 10.4, but to perform this in a way that is itself transparent — that is we can update or modify the components below a particular inter-layer interface without affecting the components above. This additional transparency is necessary because much of the research on which the specification of such an architecture depends is still on-going. However, it is timely to start to think about actual implementation of some of the 'dream applications' put forward in Chapter 1, and at the beginning of this chapter. Thus, freedom is needed to update components and evaluate new technologies with as little influence on the complex task of defining the application system itself — the layer interfaces provide the buffer between different parts of the development team envisaged in Section 10.2.

10.9.1. *The Hardware and Node Abstraction Layers*

Essentially, these two layers may be provided by an existing embedded operating system, and given the amount of research that is still required to define the rest of the architecture, there seems to be little point in re-inventing the wheel. From the discussion in Section 10.7, it will be seen that neither TinyOS nor AmbientRT quite fits the four layer model, and are both restricted in terms of the scale of software that can be hosted on a node. The intention is, therefore to use an appropriately configured eCOS system to provide the HAL and NAL. This decision entails the use of a somewhat larger node (in processing and memory usage terms) than would be represented by the Berkely Mote or its derivatives. However, given that this extra power need not actually increase the cost or size of the node, and that many of the 'dream applications' are beyond the capabilities of the current generation of motes, this does not seem to be an unreasonable constraint to bear.

10.9.2. *Defining the Applications Abstraction Layer*

As discussed in Section 10.8, the choice for the AAL is essentially between an agent based interface or a query interface. Essentially, they are equivalent. Both are means of 'tasking' the network, the salient difference being that a 'query' is declarative, whereas an 'agent' is imperative, the advantage of declarative methods is that they can be proved to select the required information, if it is there whereas imperative solutions appear to have an advantage of efficiency and execution speed. Our proposal for the AAL API is a query langauge called ASQue [39]. ASQue follows the same general plan as the query languages described in Section 10.8, being a predicate based, incrementally resolved in a distributed manner. ASQue is formally defined, the proposal being that efficiency of operation can be gained by semantically checking the queries at design time, this obviating the need for layers of checking software. If this proves feasible, it will provide a declarative way of phrasing sensor network queries, but with the efficiency of imperative systems (at least at run time).

10.9.3. *The Distribution Abstraction Layer*

The remaining layer is the per-sensor node system layer which deals with the sensor system functionss discussed above (network discovery, process migration, inter process communication, fault management, locationing, synchronisation, message routing), 'distribution abstraction layer' (DAL). This is in the form a number of software processes which run continuously on each sensor node, and together dictate the generalised strategy towards the systems functions mentioned. These processes are defined by process objects accessible at the top, API level, and can therefore be updated or adapted for different classes of application. The simple interprocess communication methods supported by eCOS will make it possible to decouple the design and development of these componets, so that they can be 'mixed and matched' as they are developed. They are as follows: Network Discovery, Locationing Synchronisation, Message routing, Process migration, Inter-node process communication, Fault detection and Fault Management. Sufficient basic research exists in the literature to provide at least an ouline for an initial example of each of these processes, much of the work being presented in Section 10.5 being immediately relevant. Thus, the production of a 'first shot' at the DAL is primarily a development issue. As the components listed above are used in real-world systems it will become apparent which have adequate levels of performance in the field, and which will require further research and development.

10.10. Conclusions

Research in wireless sensor networks is a busy area of activity, reflecting the number of important and exciting applications scenarios which are on the horizon. Much progress has been made in the design of systems support for *ad hoc* sensor networks. Several approaches have been surveyed here, which deal with the structure and organisation of the system and issues such as self-organisation and mobility have been addressed. There is a sound basis for the development of small scale sensor networks with two dedicated operating systems which have been developed for this purpose: TinyOS and AmbientRT.

However realising the very large scale scenarios will require large, multidisciplinary teams. For these teams to prosecute the development efficiently, it is necessary to adopt a view of systems design which clearly defines interfaces between the different parties developing the system, and provides the means of insulating those working in one field of research from knock-on effects of the developing research of those working in other fields — unless, of course those effects are beneficial, when the benefits need to be distributed as efficiently as possible. To enable this state of affairs, we have proposed a four layer systems model, which defines boundaries between different concerns, and outlined an architecture based on that model, an architecture which could be seen as an 'operating system' for intelligent sensor networks.

The design of the proposed intelligent sensor network 'operating system' is not a straightforward task. If the applications designer is to be freed from the necessity to cover day to day network and array management issues in each new design, then the systems support must take care of higher level issues than does a straightforward operating system. Indeed, the system proposed here is layered over a classical operating system, which resides on each and every node in the sensor network. At the network level, the system must provide functions, such as those of process migration and fault management, which are still not found in the distributed operating systems in use today. Thus there are many research issues remaining to be solved if the use of large arrays of intelligent sensors is to become a readily available tool for the solution of leading edge engineering problems, as has been proposed.

References

1. Varajan, V. K. and Varajan, V. V. (2000) Conformal and embedded IDT micro sensors for health monitoring of structures, *Smart Structures and Materials 2000: Smart Electronics and MEMS, Proceedings of SPIE* **3990**.
2. Krantz, D., Belk, J., Biermann, P. J. and Troyk, P. (2000) Project summary: applied research on remotely queried embedded microsensors, *Smart Structures and Materials 2000: Smart Electronics and MEMS, Proceedings of SPIE* **3990**.
3. Brotherton, T. and Johnson, T. (2001) Anomaly detection for advanced military aircraft using neural networks, *Proc. IEEE Aerospace Conference.*
4. Wang, C. S. and Chang, F.-K. (2000) Diagnosis of impact damage in composite structures with built in piezoelectrics network, *Smart Structures and Materials 2000: Smart Electronics and MEMS, Proceedings of SPIE* **3990**.
5. Barai, S. V. and Pandey, P. C. (1997) Time-delay neural networks in damage detection of railway bridges, *Advances in Engineering Software* **28**, 1–10, Elsevier Science Limited.
6. Kim, S.-H. and Yoon, C. (2000) Structural monitoring system based on sensitivity analysis and a neural network, *Computer-Aided Civil and Infrastructure Engineering* **15**, 309–318, Blackwell Publishers, Malden., MA.
7. Loh, C.-H. and Yeh, S.-C. (2000) Application of neural networks to health monitoring of bridge structures, *Nondestructive Evaluation of Highways, Utilities and Pipelines IV.*, Aktan., E. and Gosselin., S. R. (eds.), *Proceedings of SPIE* **3995**.
8. Yun, C.-B. and Bahng, E. Y. (2000) Substructural identification using neural networks, *Computers and Structures* **77**, 410–452, Elsevier Science.
9. Nagel, D. J. (2002) Microsensor clusters., Microelectronics journal **33**, 107–119 Elsevier Science.
10. Chowdhury, S., Ahmadi, M. and Miller, W. C. (2002) Design of a MEMS acoustical beamforming sensor microarray, *IEEE Sensors Journal* **2**(6), 617–627.
11. Ruffin, P. B. (2002) MEMS based sensor arrays for military applications, *Smart Electronics, MEMS and Nanotechnology, Proc. SPIE* **4700**.
12. Manobianco, J., Evans, R. J., Pister, K. S. J. and Manobianco, D. M. (2004) *GEMS: A Revolutionary System for Environmental Monitoring, Proc. Nanotech 04*, NSTI, pp. 422–425.
13. Martinez, K. G. *Presentation to Nextwave Technologies*, available at http://www.nextwave-interface.org/docs/pdfs/e-031030-22.pdf.
14. Mainwaring, A., Polastre, J., Szewczyk, R., Culler, D. and Anderson, J. (2002) *Wireless Sensor Networks for Habitat Monitoring, WSNA*, Atlanta, USA.
15. Römer, K. and Mattern, F. (2004) The design space of wireless sensor networks, *IEEE Wireless Communications* **11**(6), 54–61.

16. Staszewski, W. J., Worden, K., Wardle, R. and Tomlinson, G. R. (2000) Fail-safe sensor distributions for impact detection in composite materials, *Smart Material Structures*, IOP Publishing, **9**, 298–303.
17. Mitchell, K., Sana, S., Liu, P., Cingirikonda, K., Rao, S. and Pottinger, H. J. (2000) Distributed computing and sensing for structural health monitoring systems, *Smart Structures and Materials 2000: Smart Electronics and MEMS.*, Varadan, V. K. (ed.), *Proc. SPIE* **3990**.
18. Coulouris, G. F., Dollimore, J. and Kindberg, R. (2001) *Distributed Systems: Concepts and Design*, 3rd edition, Pearson Press.
19. Pike, R., Presotto, D., Dorward, S., Flandrena, B., Thompson, K., Trickey, H. and Winterbottom, P. (2002) Plan 9 From Bell Labs, http://plan9.bell-labs.com/sys/doc/9.pdf.
20. Kasetkasem, T. and Varshney, P. K. (2001) Communication structure planning for multisensor detection systems, *Proceedings of the IEE Conference on Radar, Sonar and Navigation* **148**, 2–8.
21. Zou, Y. and Chakrabarty, K. (2004) Uncertainty-aware and coverage-oriented deployment for sensor networks, *Journal of Parallel and Distributed Computing* **64**(7), 788–798.
22. McGlynn, M. J. and Borbash, S. A. (2001) Birthday protocols for low energy deployment and flexible neighbor discovery in *ad hoc* wireless networks, *MobiHOC'01*, Long Beach, CA, USA.
23. Inanc, M., Magdon-Ismail, M. and Yener, B. (2003) *Power Optimal Connectivity and Coverage in Wireless Sensor Networks*, RPI Computer Science Technical Report, TR # 03-06.
24. Cerpa, A. and Estrin, D. (2002) ASCENT: adaptive self-configuring sensor networks topologies, INFOCOM 2002. *21st Annual Joint Conference of the IEEE Computer and Communications Societies, Proceedings* **3**.
25. Chen, B., Jamieson, K., Balakrishnan, H. and Morris, R. (2002) Span: an energy-efficient coordination algorithm for topology maintenance in *ad hoc* wireless networks, *ACM Wireless Networks Journal* **8**(5).
26. Details of TinyOS available at http://www.tinyos.net/.
27. http://www.tinyos.net/special/mission.
28. Hofmeijer, T., Dulman, S., Jansen, P. G. and Havinga, P. J. M. (2004) DCOS, a real-time light-weight data centric operating system, In Sahni, S. (ed.), *IASTED Int. Conf. on Advances in Computer Science and Technology (ACST)*, St. Thomas, Virgin Islands, USA, ACTA Press, Calgary, Canada, pp. 259–264.
29. Red Hat, Inc., Garnett, N., Larmour, J., Lunn, A., Thomas, G. and Veer, B. *eCOS Reference Manual*, http://ecos.sourceware.org/docs-2.0/.
30. Srisathapornphat, C., Jaikaeo, C. and Shen, C.-C. (2000) Sensor information networking architecture, *International Workshops on Parallel Processing*, pp. 23–30.
31. Peti, P., Obermaisser, R., Eimenreich, W. and Losert, T. (2003) An architecture supporting monitoring and configuration in real-time smart

transducer networks, *Proc. IEEE International Conference on Computational Cybernetics.*

32. Mailler, R., Vincent, R., Leser, V., Shen, J. and Middlekoop, T. (2001) Soft real-time, co-operative negotiation for distributed resource allocation, *Proc. 2001 AAAI Fall Symposium on Negotiation.*
33. Ortiz, C. L., Rauenbach, T. W., Hsu, E. and Vincent, R., Dynamic resource bounded negotiation in non-additive domains, pp. 61–107.
34. Soh, L.-K., Tsatsoulis, C. and Sevay, H., A satisficing, negotiated, and learning coalition formation architecture, pp. 110–138.
35. Mailler, R., Horling, B., Lesser, V. and Regis, V. (2004) The control, coordination, and organizational design of a distributed sensor network, *IEEE Transactions on Systems, Man, and Cybernetics: Part A.*
36. Lesser, V., Ortitz, C. L. and Tambe, M. (2003) Distributed sensor networks: introduction to a multiagent perspective, In Lesser, Orlitz and Tambe (eds.), *Distributed SensorNetworks: A Multiagent Perspective.*, Kluwer Academic Publishers, Norwell, MA, ISBN 1-4020-7499-9, pp. 1–8.
37. Yao, Y. and Gehrke, J. (2002) The cougar approach to in-network query processing in sensor networks, *SIGMOD Record*, **31**(1).
38. Sadagopan, N., Krishnamachari, B. and Helmy, A. (2003) The ACQUIRE mechanism for efficient querying in sensor networks, *IEEE International Workshop on Sensor Network Protocols and Applications (SNPA'03)*, held in conjunction with the *IEEE International Conference on Communications (ICC 2003)*, Anchorage, Alaska.

CHAPTER 11

REALISING THE DREAM — A CASE STUDY

by *Elena Gaura and Robert Newman*

11.1. Introduction

As a conclusion to this book, it is appropriate to look at new areas of technology, research and exploration that the intelligent MEMS sensor technologies that form the subject of the book make possible. This will be done by means of an outline feasibility and design for a speculative exploration mission, made possible by the combination of MEMS and cogent sensor technologies. The prospective mission plan will be fleshed out using information drawn from the chapters of the book, to show how each of the topics that has been discussed contributes to the overall feasibility of a 'dream application, such as the one that will be proposed. The original purpose of this study was to provide an attractive scenario for the promotion of the importance of pervasive microsensing technologies and to inform a continuing research agenda in this area. The scenerio has provided case studies which have helped the development and refinement of many of the ideas put forward in this work. As such it provides an excellent vehicle to draw those ideas together, and place them in the context of their joint and several contribution to the new sensing technologies, which will enable completely new ways of approaching the solution of some hitherto very difficult sensing problems.

The work on this scenario started shortly after the failure of the Beagle II Mars mission. This might be unfamiliar to non-British readers. For a summary of the mission itself the reader is referred to the mission website [1]. Suffice it to say here that the Beagle 2 Mars mission was an initiative to build an economic lander as part of the ESA Mars Express expedition. The scenario development produced an alternative mission profile, based on the use of a host of MEMS based intelligent multi-sensor motes. As the scenario became more detailed, it became apparent that, at least in hardware terms, it was distinctly feasible using current technology. The mission plan envisaged is itself aligned with the 'smart dust' view of sensor systems which was discussed in Chapters 1, 9 and 10, and has now been developed in sufficient detail to allow it to be used to evaluate the way in which the techniques and research put forward in this book could be utilised to enable it.

This concluding chapter will examine the scenario, and discuss the way in which it explores the subject matter of the rest of the book. First the envisaged mission, and the way that it could make use of MEMS and cogent sensor technology will be discussed. Following that, in Section 11.3 and outline design study for an intelligent sensor node which could be used in this mission is presented, drawing from subject matter within the previous chapters of the book. Section 11.4 examines the various sensor technologies that have been presented in earlier chapters, and explores how they might be incorporated into the sensor for this mission. Section 11.5 puts forward a scenario for the early part of the mission, how the sensors would be deployed and initialised and Section 11.6 looks at the cogency within each node, the level of processing power required to support it and how it will support the operation of the sensor and its part of the overall mission. Finally, Section 11.7 concludes the chapter and the book.

11.2. The Mission

Imagine that MEMS sensing technologies could be used to build a probe with dimensions of 6 mm in diameter and 40 mm long. If made of solid titanium, such a probe would weigh 5 grams. This provides an upper limit for its mass, since in practice parts of it will be constructed using materials lighter than titanium (such as silicon, for instance) and the probe is unlikely to be completely solid. The Beagle 2 probe weighed 33.2 kg and was designed to have the highest scientific payload ratio of any lander built so far. For the mass of the Beagle over 3000 of the micro probes could be landed. The NASA Spirit and Opportunity rovers which are at the time of writing still exploring the Martian surface each weighed 185 kg [2], the equivalent of 37 000 micro probes. These figures suggest that the rocket technology exists to land a very large sensor array, onto the surface of another planet.

The questions that will be addressed in the rest of the chapter concern the feasibility of such an array itself, whether a probe of such size could carry any useful instrumentation, how would they be deployed and how would the outcomes of the exploration be extracted from such an array?

The micro-lander is called a 'Daisy' (for Distributed Artificially Intelligent SensorY). One possible design for it, which will be developed in this chapter, bears some resemblance to an earthly Daisy, being shaped like a flower with petals (which are actually an active optical antenna — see the illustration in Figure 11.1). Since the authors are not planetary scientists, assumptions have had to be made about the experiments to be carried. In the end, a few sensor types from recent planetary missions have been borrowed. The experiments that these probes will carry will allow scientists to investigate a number of different planetary phenomena, as follows:

- The visual environment — most surface probes will be equipped with cameras so that the immediate environment can be observed visually. This mission will be no exception.
- Atmospheric chemical composition. The gaseous composition of the atmosphere seems to be of interest, particularly those gases which either support or are by-products of life.
- Soil chemical composition. For similar reasons, soil composition is of interest.

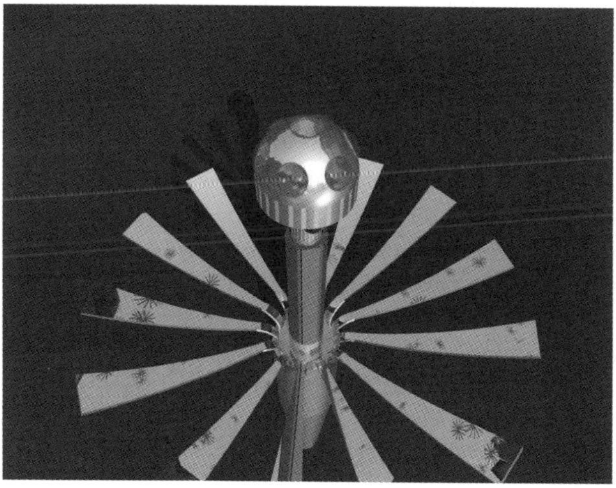

Figure 11.1: The Mars DAISY. Illustration by Jim Tabor.

- Wind speed and direction. Planetary meteorology is a persistent interest of planetary scientists.
- Ambient temperature will also be measured, again for meteorological purposes.
- Atmospheric pressure will be measured, for altitude estimation and meteorological purposes.
- Seismic events are of interest because they indicate the activity of the geology of the planet.
- Magnetic field sensing, to allow mapping of the planet's magnet field and location of large concentrations of magnetic materials.

Genuine planetary scientists would probably produce a different set of experiments, and the resultant probe would be very different. As was discussed in Chapter 10, real development teams for this kind of project are large and multidisciplinary. The authors here are conducting a thought experiment, based on what they think such a team might come up with — it must be remembered that what we are producing is a scenario not a product — any shortcomings in the 'design' demonstrates the point made in Chapter 10, that to really design complex systems such as this requires a wide range of expertise and advanced design methodologies.

The advantage of mobile explorers or 'rovers' is that they allow a wide area to be explored. How wide an area of exploration would the microlander approach allow? As was seen in Chapter 9, sensor arrays allow field observations in two dimensions to be made, covering a wide area. Again, some working assumptions must be made, as well as some simplifications to provide some rough and ready estimates. As a starting point, it is assumed that the sensors can communicate over a range of 100 m (a more realistic figure will be derived later). This figure dictates the density of the coverage. The likely deployment pattern will be random, but for this first estimation a square array is assumed. If 37 000 sensor probes are available (the equivalent to one of the NASA rovers) and if these probes were arranged in a square at 100 m intervals then the sides of the square would be $\sqrt{37\,000} \times 100$ m, or 19.2 km. In their 19 months (at the time of writing) the NASA rovers have travelled about 3.5 km, and so have explored a 3.5 km strip of the planet. Thus it would seem that the sensor array approach will allow exploration of a much greater area. Of course, there are trade-offs to be made between the density of coverage and the detail of observation required, so it may be that the planetary scientists decide on a denser packing of the array than proposed here. Nonetheless, the area covered by the sensor array seems to be far larger than the equivalent mass of rover, and moreover, the sensor array observes continuously at every point in the array since there is a continual sensor presence everywhere in the array, whereas the rover can only view around only one point at a time: its current location. This first estimate of the capability of the sensor array seems to be positive. To find out whether it is as attractive as this first sight would appear, some questions need to be answered:

- Is it possible to pack all of the experiments into a 40 mm long microsensor?
- How will the sensors be powered?
- Will it be possible for the sensors to communicate their results back to earth? how would such a network be organised and maintained?
- What level of 'intelligence' will be required in each sensor probe, how will it be implemented?
- How will the array be placed on the planet?
- How will the network be devised?
- What might be learned from the information gained from the array?

These questions will be addressed by reference to the relevant chapters in the book, but first, some rough estimates of the operating parameters of the micro-probe and mission plan will need to be made.

11.3. Initial Rough Design

As discussed in Chapter 9, Section 9.4, when designing an array there is a choice between a homogeneous array, in which all probes are the same, or a heterogeneous one in which there are different types of probe involved. The assumption made is that a homogeneous array is selected, for a number of reasons:

- Special nodes make a system vulnerable. Planetary exploration is a high risk business. Imagine that the system was designed with some special master communication nodes, if one of these failed to make it to the surface, a whole area would be unable to communicate. With a homogenous

Figure 11.2: A cloud of sensors approaching the surface. Illustration by Jim Tabor.

array, all nodes have the same capability, so the system is not vulnerable to the loss of a few.

- A single type of node reduces the design effort, at least of the hardware itself, since there is only one type of node to be designed.
- A homogeneous array is arguably more flexible. Past planetary probes have been called on to make investigations not forseen in the original project plan, and it is to be expected that the sensor array would not be an exception to this. A homogeneous array has full measurement capacity at every point in the array, and so is able to be adapted to any application which is possible within the basic sensory package.

11.3.1. *Communications*

A consequence of a homogeneous array is that every node must be capable of sending information back to the explorers on earth. It would appear unlikely that a 40 mm mote would have sufficient transmission capability to reach the Earth from Mars, so there will need to be an intermediary satellite (a common technique for communication with surface probes). Even to communicate with a relatively local satellite, the transmission from the mote will be required to be highly directional.

To achieve such directional transmission, optical communication is necessary, in order to allow an antenna with sufficient beam forming power to fit within the 4 cm size and allow communication to the host satellite. Given the predominantly red colouration of the planet, blue has been selected as the colour of the laser, to ensure maximum contrast against the background.

11.3.2. *Power Budget*

For long term term operation on the surface of Mars, some of the techniques of power harvesting, discussed in Chapter 3, Section 3.7 will need to be used. Photovoltaic solar power generation is the most likely candidate, this being the only easily accessed and predictable power source available on another planet. There follow some rough calculations to estimate the amount of power that might be available to the probe.

The Martian solar irradiance at noon is 590 Wm^{-2} [3]. A 20 mm diameter solar collector with a fill-factor of 0.5, which could be incorporated within the 40 mm Daisy has an area of 0.000157 m^2, and thus the power collected (at noon) is 93 mW. Allowing for night, and the variations of illumination over the day, we might expect an average value of just 1/10th of this, giving an averaged budget of 10 mW. Spectrolab, Inc. has demonstrated conversion efficiency for a silicon photovoltaic cell of 36.9% [4]. Using the more efficient GaAs technology, an efficiency of 40% is not unreasonable. Our total electrical power budget is therefore 4 mW. Since the available power input varies according to the time of the Martian day and the ambient conditions, battery storage will be required to even out the power supply, storing power gathered during periods of high incident light to periods of high power usage. The bulk of the mass for the Daisy needs to be at the front (in the spike) so that it is aerodynamically stable on descent, and falls point down. This mass could be formed from the power storage battery (since we want all mass to be usefully employed, as it costs a spectacularly large amount of money to get that mass to Mars). The spike will have a volume of 130 mm^3, given the speculative dimensions in the working assumptions that have been made. Lithium ion cells have a power density of 300 W-hours/l, so the capacity of a battery filling the volume of the spike is 39 mW/hours — at the 4 mW budget, enough to power the mote for 10 hours.

To achieve a probe which operates within this power budget, use will need to be made of the power conservation and management techniques discussed in Chapter 3, Section 3.7. The design of the probe will require to be power aware, and in its operation, care must be taken to minimise operations which consume large amounts of power.

The most power hungry communication is likely to be with the host satellite, so an estimate is needed of how much power this will consume. We can assume that the satellite will be equipped with a large and efficient optical antenna system. The required incident power is determined by the acceptable bit error rate (BER), the noise floor of the photodetector and the size of the reflector system [5]. For a BER of 10^{-10} (as required by SONET, Synchronous Optical Network) the required signal to noise ratio is 12.732 [5]. The required optical modulation amplitude (OMA) is given by:

$$\frac{i_n SNR}{\rho},$$

where i_n is the input referred noise figure for the front end amplifier, ρ is the responsivity flux of the photodetector. The average optical input power P_{AVG} is

$$\frac{OMA(r_e + 1)}{2(r_e - 1)},$$

where r_e is the 'extinction ratio' — i.e. the power ratio between a logic '0' and a logic '1'.

We can estimate this figure using figures for appropriate 'off the shelf' parts (custom designed components would probably be better, but the commercially available parts allow a quick working estimate to be made). A Centonics OSD1-5T has $\rho = 0.2\,\mathrm{A/W}$ and a MAX3266 transconductance amplifier an i_{in} of 200 nA. Thus, in this example the OMA is

$$2 \times 10^{-7} \times 12.732/0.2 = 12.732\,\mu\mathrm{W}.$$

P_{AVG} is $12.732 \times 5/6 = 10.6\,\mu\mathrm{W}$, assuming an r_e of 4. Thus a $1\,\mathrm{m}^2$ beam gathering reflector on the satellite is assumed, the Daisy must be equipped to transmit a power of $10\,\mu\mathrm{Wm}^{-2}$ at the orbital height of the satellite. The height of a geostationary orbit of Mars is 1.71×10^7 m.

The minimum beam angle for a 20 mm radius reflector operating at a 405 nm wavelength (as formed by the Daisy petals) is $1.22\lambda/a = 1.22 \times 405 \times 10^{-9}/0.02 = 2.47 \times 10^{-5}$ rad. At the orbit height the beam diameter is 422 m, its area around $140\,000\,\mathrm{m}^2$, and therefore the power required in the beam is 1.4 W. Assuming a laser diode efficiency of 20% [6], then 7 W of input power is required for communication with the satellite. This power usage is sustainable within the overall package, but will obviously require to be used very sparingly. As discussed in Chapter 9, Section 9.5, the logical organisation of the network will have to be such as to minimise transmission power. This will dictate a clustered organisation of motes, whereby Daisy to Daisy communication with store and forward transmission of messages across the array is the norm (to recap from Chapter 9, the power required for store and forward communication rises in proportion to the distance, whereas that for direct communication rises as the square of the distance). An estimate is needed of the power required for node to node communication. If we allocate a quarter of the power budget to such communications (1 mW into the laser), then, assuming the same receiver sensitivity figures, and an effective receiver and transmitter area of $10^{-4}\,\mathrm{m}^2$, to allow for the fill factor and beam steering geometry, the required irradiation at the receiver is $10 \times 10^{-6}/10^{-4} = 10\,\mathrm{mW\,m}^{-2}$. The minimum beam angle is $1.22 \times 405 \times 10^{-9}/0.01 = 5 \times 10^{-5}$ rad. This gives a communications range from sensor to sensor of about 70 m, which could be exceeded where necessary by using the full available power of the laser. Much of this communications strategy requires the ability to provide a sophisticated, steerable, active antenna, the design of which will be discussed in the next section.

11.3.3. *The Active Antenna*

In order to achieve the different modes of transmission (into the sky to reach the satellite, and horizontally to communicate with a neighbour) it is necessary to have a highly steerable antenna system. Since the communication is optical, what is required is a combination of adaptive MEMS optics, as described in Chapter 6. To achieve the required characteristics and steerability it is likely that a number of adaptive mirrors will be required. The Daisy profile shown assumes an upward firing photodiode. A membrane deformable mirror or mirrors (see Chapter 6, Section 6.2) reflects the beam onto the main focussing mirror. Membrane deformable mirrors can be made in a large enough size (20 mm diameter) but such a component would be vulnerable in descent, and would not pack efficiently. The design sketched for the Daisy, which may test current technology beyond its capabilities, is a multi petal design, in which each petal is deformable to produce the required shape of beam forming mirror to steer the beam in the required direction. The shape of the reflector must be varied from a straightforward parabolic bowl, when the target is overhead, to a complex, reflex configuration when the target is on the same lateral plane. One possible technology is PVDF piezo film, which can certainly produce the scale of deflections required when used in a bimorph configuration [7]. Advantages are the simplicity of fabrication, the complex forms can be obtained by suitable patterning and intelligent control of the electrodes printed on the surface of a PVDF bimorph. Disadvantages include poor high temperature capability (135°C, which means the daisies will need to be re-entry protected) and the requirement for high drive voltages. These will require a special high voltage controller chip, to which the petals could be bonded. This chip could double as the general power management for the mote.

Probe mass can be saved if parts of the design serve several purposes. The 'petals' of the Daisy can provide other functions as well as forming part of the antenna. Firstly they can focus incident solar radiation onto the photovoltaic/optical reception cell, Also they could have a role in controlling the descent of the Daisy, by altering the aerodynamic drag as it falls. Using the onboard accelerometers which are part of the instrument package, it could be possible for the mote to navigate as it descends.

11.4. Sensor Technology

In this section, the way in which the sensing requirements listed in Section 11.2 could be satisfied using the technology discussed in this book will be discussed.

11.4.1. *Acceleration Sensor*

Accelerometers are an important part of the instrument package on the Daisy, reflecting their ubiquity in the overall field of sensing, which causes them to be a very well developed class of sensor. Thus, the designers of this part of the sensor package could take their pick from the fabrication techniques described in Chapter 2, and ensure that appropriate pick-off electronics are designed as described in Chapter 3. High performance capacitive pick-off accelerometers can be integrated 'side-by-side' with adequate ancillary electronics using a CMOS process. The performance can be enhanced using additional signal processing within the integrated electronics utilising either the closed-loop control techniques described in Chapter 5, or the Artificial Neural Network linearisation methods elaborated in Chapter 7, or a combination of both. A major advantage of these techniques, critical in an application such as this, in which thousands of probes will be used, is that they provide consistent performance despite manufacturing spreads. Along with the automatic calibration and compensation techniques discussed in Chapter 4, this will drastically reduce the cost of the mission. The economies will be made by decreasing the proportion of sensor chips that must be rejected as being out of specification, and reducing the amount of skilled work required to characterise individually each sensor.

11.4.2. *Atmospheric Sensor*

The chemical sensing technology to sense the atmosphere and soil composition require a further design choice. Electronic 'noses' are an established MEMS technology, but generally need to be tailored for the detection of a specific chemical or class of chemicals. For scientific investigation a more generally capable chemical sensor is needed. One type of general purpose sensor is a gas chromatograph. A MEMS based Fabry-Perot chromatograph [8] has been described by Crocombe and is illustrated in Figure 11.3. This type of sensor includes a Fabry Perot interferometer (essentially two parallel semi reflective plates). An actuator can vary the distance between the plates, thus changing the frequency of transmission of the spectrometer,

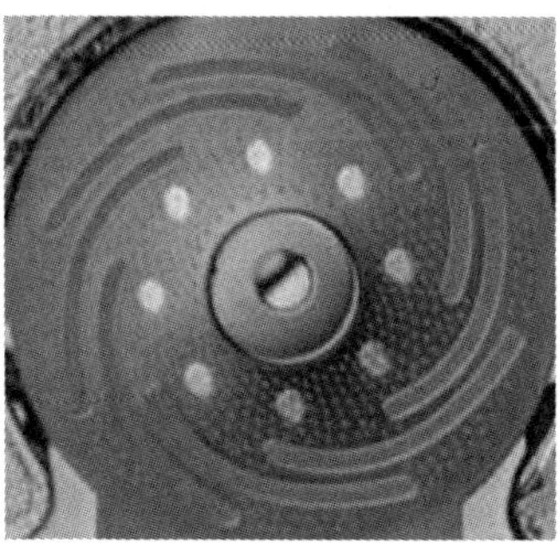

Figure 11.3: Fabry-Perot Interferometer chip. From [8].

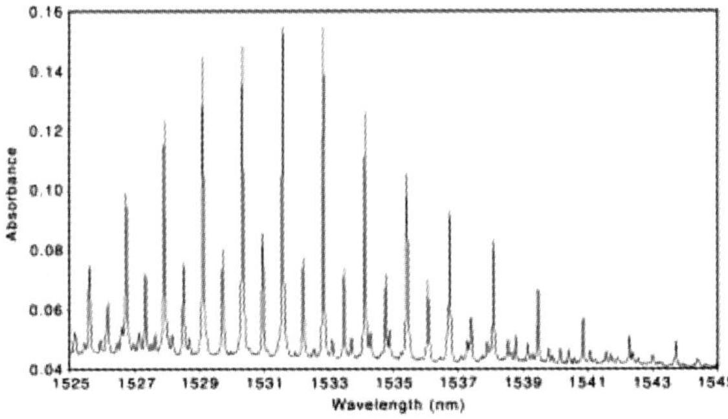

Figure 11.4: Acetylene spectrum from F-P interferometer chip. From [8].

allowing it to scan optical frequencies to produce a characteristic transmission spectrum of a sample through which an illuminating beam passes. The Fabry-Perot chemical sensor is a further example of the use of the active optical MEMS techniques described in Chapter 6, and could be fabricated using the nitride film techniques described there. The type of output is illustrated in Figure 11.4, and is the spectrum of acetylene gas, scanned by the MEMS spectrometer shown. The nature of this data, which has been produced by the type of MEMS sensor envisaged in this scenario, will inform the discussion on the processing resources required for the node.

11.4.3. *Pressure Sensing*

Examples of pressure sensors were shown in Chapter 2, using capacitive or piezo-electronic pick off techniques. Alternately, a very sensitive pressure sensor can be fabricated using an F-P cell, as described in the previous section. In this application the pressure causes displacement of one of the plates which in turn modulates the intensity of light passing through the cell.

11.4.4. *Soil Chemical Sensor*

The difficulty in chemical sensing soil in an environment without water (assuming that Martial soil has no water) is conveying the constituent chemicals to the analyser. A simple solution would be to use the transmission laser to vapourise the sample. The 7 W of power available should be sufficient, so long as it can be focussed onto the sample to be vaporised.

This provides a further application (and design challenge) for the adaptive optical MEMS discussed in Chapter 6. The gaseous sample needs to be conveyed to an FP gas sensor for analysis. The sample could be collected on landing by providing a bore through the spike, so that soil is forced up into it on impact. Physical arrangements for routing the laser beam and conveying the gaseous sample to the sensor are not straightforward, and require detailed design of the probe to verify that there is indeed a practical solution.

11.4.5. *Magnetometer*

Several magnetometers have been described in the literature. They typically require a magnetically active element which moves according to the prevalent magnetic field, the movement being sensed using any of the established MEMS pick-off techniques (capacitive, piezo or tunnelling). The magnetically active element can be provided using either a permanent magnet, which requires specialist materials within the MEMS, or by using electromagnetism, which consumes current. The determination of which would be the best choice in this application requires detailed design in order to precisely evaluate the options and trade-offs. The feasibility of the technology is established, a mechanical MEMS magnetometer has been described by Moreland *et al.* [9], and a similar device, with optical pick off (suitable for integration with the optical pick-off Fabry-Perot sensors) patented [10].

11.4.6. *Thermal Sensing*

As discussed in Chapter 3, temperature sensors can use thermocouples or semiconductor junctions. The latter have the advantage of being readily available as components within standard semiconductor technologies, without the need for specialised metal layers, as are needed to fabricate thermocouples. Chapter 3, Section 3.3 presented an example of a thermal wind speed and direction sensor, which could be used to provide the wind sensing needs of the instrument package.

11.4.7. *Image Sensor*

This subsystem is straight forward, being a silicon CCD or CMOS image sensor, as is used in a modern digital camera. The resources that need to be integrated are an estimated 500 pixel array for the pick-off and image sensors for imaging the surroundings. The design goal is a 360% image circle. This can be provided with four 2000 pixel image patches and 90% image optics for each. This requires an 8500 pixel array, on a 5 mm square chip, easily within current technological capabilities.

11.4.8. *The Sensing Package*

Clearly from the above, the technologies that have been explored in Chapters 2 to 6 of this book can potentially provide all of the sensing capabilities required for the probe. The precise packaging and format of the various sensors would not be known until a detailed design had been done, which is outside the scope of this 'thought experiment' — however it is possible to speculate about possible configurations. In terms of technologies, the sensors can be classified as follows:

- optical sensors — the image sensor and pick-off for the Fabry-Perot chromatograph;
- active optical MEMS — the beam steering mirrors, the Fabry-Perot chromatograph, the laser beam focussing optics for soil vapourisation (and pressure sensor, if this option is selected);
- mechanical sensors using capacitive, piezo or tunnelling pick-off — accelerometers and magnetometers (and pressure sensors, if this option is selected);
- thermal sensors for temperature and wind detection.

It is possible that sensors using these different technologies could be integrated together on a single chip to form a 'lab on a chip'. However, as was discussed in Chapters 2 and 3, the level of integration achievable depends on the features of available VLSI technologies. A technology that could integrate all of the above onto a single chip would need to be specially developed for this application, and would therefore be likely to be prohibitively expensive. Therefore, a more probable scenario is that separate chips would be needed to provide some of the more specialist sensing types, the precise allocation of sensing devices to chips being a detail design decision. These chips would need to be integrated into a single sensing package using the 'side by side' or 'vertical' integration techniques described in Chapter 3. Given that there are four sensing techniques to be used, a package of four chips seems to be a possible scenario. Four chip 'side by side' packages are reasonably common. However, the form factor of the probe described would seem to dictate a 'vertical' arrangement since the geometry of some of the sensing devices using these techniques that have been published in the literature suggests that it would be possible to integrate the suggested sensors within 5 mm chips, which would fit with the speculative profile of the Daisy probe if arranged vertically. Integration of four chips in a stack would be an advance on the current state of the art.

11.5. Deployment

The proposed method of deployment for the sensor array is as follows. The daisies would be packed into a small re-entry vehicle, illustrated in Figure 11.5, which protects them from the heat of entry into the Martian atmosphere. During transit they could be attached to a tape, which would provide electrical connections to charge their batteries and provide initial programming of their processors (which will be discussed in Section 11.8). After atmospheric entry, the daisies would be ejected from the re-entry vehicle, and would descend individually to the planet's surface, where their

Figure 11.5: The re-entry capsule, containing the package of Daisies.

velocity would be sufficient to embed their spikes into the surface and effectively fix them in place. The distribution could be entirely stochastic, but it could be advantageous for an individual sensor to be able to steer itself towards a desired lending spot, as proposed in Section 11.3. It is to be expected that a proportion of the probes will fail during the descent, or will fail to embed properly in the soil. Thus some attrition rate must be allowed for, but the overall risk to the mission is still far smaller than it would be for a single probe, such as the ill-fated Beagle 2. This resilience is an essential advantage of sensor arrays, as was discussed in Chapter 9.

The detailed planning of the manufacture and deployment of the sensors must take into account the calibration and compensation of all of the sensors in each and every probe to be deployed. Although planetary exploration is not a 'low price' market, the costs of calibrating manually thousands of sensors individually is likely to be high. Since each of the micro-probes will have digital processing capabilities, it will be possible to use the techniques of auto-calibration and compensation described in Chapter 4. Each sensor will require a reference against which it can be calibrated. It is likely that the journey that the probes must undertake will provide a variety of environmental changes which will exercise the sensors, allowing the possibility

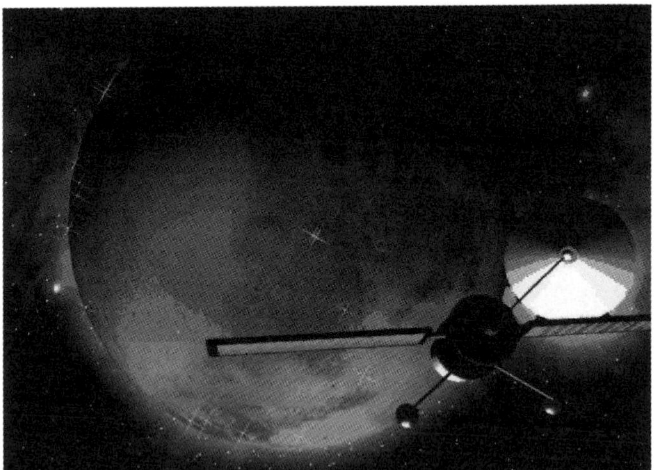

Figure 11.6: Daisies communicating with the host satellite. Each patch of light is a clusterhed, forwarding information to the satellite.

to use this excitation for calibration. Calibrated, reference sensors against which the values detected by the micro-probes' sensors can be compared, will be required. If these are placed in the re-entry vehicle, and that vehicle is provided with sufficient processing capability to communicate the values to its load of probes, then the calibration can be delayed until the probes have been through the most extreme of the environmental changes that they will face during the journey. By performing this final calibration as late as possible, any changes in the operation of the sensors caused by the stress of the journey can be compensated out.

11.6. Operation, Control and Communication

Once the probes are deployed they must start operations, going through the various procedures described in Chapter 9. The first step is network discovery. Each probe must scan for neighbours and a distributed map of the network built up. This is likely to be a lengthy process, since it will be some time before the array is physically complete, as probes land at different times. The network discovery process will therefore share some of the characteristics of a mobile system, although these motes, as described would not be fully mobile.* There are some other factors which will make the design of this particular network discovery task individual. Firstly, due to the directional nature of the antenna, the discovery process will require 'scanned' communication, as opposed to 'broadcast' communication. This would seem to reduce the chances of any two motes making contact within a given period of time, and is therefore likely to extend the discovery process. Against this, the daisies are highly power limited and therefore the discovery process is required to be efficient. Because of these considerations, the design of a network discovery process for this particular application is likely to be complex and specialised.

The directional nature of the antenna, and its ability to 'scan', also means that once in operation, the network of sensors is likely to communicate in a dramatically different manner from one in which the nodes are capable

*At least, as described here. This kind of scenario building exercise often results in the generation of original ideas. One member of one of the discussions which built this scenario is sure that the Daisy petals could be used to allow the Daisy to self-right, if it falls at a bad angle, or even to make it 'walk', in which case it becomes a micro-rover.

of simultaneous omni-directional transmission and reception. The network protocols at all levels are likely to reflect this, and to be completely different from those designed for a more normal network. The details of a protocol stack which would fit the requirements of a steerable, directional antenna system remain to be worked out. However, from the point of view of the ease of design of the higher levels of software in the network (which are likely in themselves to be complex) it is important that the detail and side effects of the software required to service unusual communications protocols be prevented from affecting the design of the software which handles the information coming from the sensors. Otherwise, those who design that software will need to become conversant with the protocol design.

It is considerations such as these that dictate that software architectures adopted, as discussed in Chapter 10, must be 'component based' in design and appropriately layered, allowing specialised versions of support functions such as network discovery and communications to be tailored for a specialised need, without impacting other parts of the system design. It is unlikely that such complex applications as the Mars Daisy could be successfully built without the adoption of sufficiently advanced systems software and operating system, such as those discussed in Chapter 10.

Once the network is operational, it must be maintained. While the sensors in this application are not mobile, they are located in a hostile environment, and a failure rate, even after successful deployment, must be anticipated. In order to maintain the operation of the array fault detection techniques, such as those discussed in Chapter 8, will be needed. In highly energy constrained applications, such as this, the fault detection techniques to be used must be carefully selected. The Artificial Neural Network method decribed in Chapter 8 has the particular advantage in this application of the minimisation of communication required for reference data, and the use of a low level of computation power (at least, relative to some of the other methods which have been put forward).

11.7. Querying the Array

One of the most attractive features of a large sensor array, particularly one implemented using a network of cogent sensors, such as the one in this thought experiment, is its versatility. The array itself is capable of gathering huge amounts of data. In this application, the sensors are multi-modal, and so the different types of data emanating from the array can be used to synthesise many different pieces of information. The specification of type of information required is the preserve of the planetary scientists. As was discussed in Chapter 10, a means must be developed to allow applications specialists to pose queries which can be sent to the sensor network in order to extract the required information. The imagination and expertise of planetary scientists will allow the use of the array in ways not even though of by the present authors, so here we use as an example the form of planetary data gathering which gathers most attention in the popular media from past missions and see how that data could be provided using the speculative sensor array — image data. Previous missions have placed cameras on 'rovers' which can be moved to explore a region in detail. Using an array of image sensors the task changes somewhat, a sensor must be located close to the required viewpoint, and the image data taken from that sensor, and transmitted back to the viewer. If a viewpoint is required as seen from a location between sensors, the task is somewhat more complicated. Images from adjacent sensors must be used to build up a three dimensional model of the view, using stereoscopic methods (a technique commonly used in the field of computer vision), and that model visualised from the desired viewpoint.

Large amounts of data can be potentially collected, but with such expensive communication resources, substantial data reduction needs to take place. Images cannot be sent across the array in anticipation of there being a need for them. Instead, the query must diffuse into the network, and data transmitted only in response to a specific request.

11.8. A Cogent Sensor

In Chapter 8 a terminology was developed, distinguishing between 'smart', intelligent' and 'cogent' sensors. Comparing with the definitions given there, it will be seen that the sensor package described here fits fully into the category of cogent sensors. Not only is its whole task to measure and react to the environment, but in order to make the data collection feasible, the sensor device must also interpret and reduce data, producing refined information to be transmitted to the planetary scientists. To produce that information the sensors must be able to react to complex queries. Their response will often be collaborative, using the data from several or many sensors to produce the information required. That collaboration and information extraction must occur at the sensor node itself, since communication is simply too expensive, in terms of power usage.

This kind of data reduction, transformation and decision making envisaged requires a substantial processing and memory capability. We might wonder whether a powerful enough processor is feasible within the constraints of the scenario. As an example, a suitable processor core, the ARM7TDMI-S, implemented using a $0.13\,\mu m$ technology requires a die area of $0.32\,mm^2$, and has a power consumption of $0.11\,mW/MHz$ [11].

Since there is $25\,mm^2$ chip area available if the 5 mm square chip profile discussed in Section 11.5 is used, the rest of the processing chip (if a single chip processor is specified) can be devoted to storage. FLASH densities (at $0.13\,\mu$) are $0.052 \times 10^{-6}\,mm^2/bit$ [12].

SRAM densities currently are $2.43 \times 10^{-6}\,mm^2/bit$ [13]. Thus, there is room for 500 Mbit flash or 10 Mbit SRAM. Obviously, flash memory will form the bulk of the 'storage memory'. If we allocate 0.5 Mbyte (4 Mbit) RAM, then there is room for 37.5 Mbyte flash, to provide program memory and longer term data storage. Looking at the power usage, allocating 2 mW of the budget to the processing allows the processor to operate at an average rate of 18 MHz. Although the figure is low in comparison to current desktop clock rates, it is sufficient to undertake appreciable amounts of processing, if the whole software system is designed to be efficient. In any case, it is substantially more processing power than has been available to the majority of planetary probes in the past.

What could be done with a node of this capability? Below is a thought experiment within a thought experiment, to explore the sequence of operations that a single Daisy might be engaged in. Whilst the Daisy array, as described, can provide planetary scientists with a host of useful (for them) information, what the common person thinks of as space exploration is walking on another planet. Can the Daisy field be used, if not to actually allow people to walk on Mars, to provide the information which will allow them to have the experience using virtual reality and telepresence techniques? In essence, virtual reality is a grand scale computer game, which will provide the virtual explorer with the sensory input needed to maintain the illusion of the real walk, based on a mathematical model, commonly called a 'physics engine'. The physics engine needs to be built up using information derived from the Daisy network. The first requirement is for a relief map of the area through which the walk is to be taken. After locationing, each Daisy knows its position relative to others in the field. The communications satellite can provide absolute location for the Daisy field, since it has a very large antenna with a very narrow beam, and can therefore locate the sensors with which it is communicating very accurately. Once this is done, the Daisies have possession of the required relief map, distributed between them. However, the map is in the form of a 'cloud of points', with the precise location of each Daisy as a three dimensional co-ordinate. A better form, for the physics engine, would be as a NURBS (Non-Uniform Rational B-Spline) surface, and such a form would also be more economical to transmit back to the surface. Derivation of a good NURBS surface from a cloud of points is a current computer graphics research topic. Should a good algorithm for making the conversion be found, it could be distributed between the Daises, which would then be able to transmit back a NURBS relief map of the surface of the planet.

The next stage in building a convincing visual illusion of the planet is to 'texture map' and 'bump map' the NURBS surface to give it the appearance of the actual surface. The required colour images will obviously be derived from the image sensors within the daisies. However, in their raw form they are unsuitable for this purpose. What is required is a large image, from a viewpoint directly above the surface. What is available is a number of small images, from a viewpoint of each Daisy. Not only is the viewpoint incorrect, but the fields of view are likely to overlap, resulting in uneccesary duplication of data. Furthermore, the transmission of every single image will result in large communication overheads. Again, the cogency withing the Daisies must be used, to derive a single top-view image of the field from the multiplicity of individual images. The reader will begin to appreciate at this point that a general purpose query mechanism for this type of application must be very sophisticated.

Once a convincing visual illusion is generated, the other sensory information available from the Daisies can be used to produce a full sensory environment. Wind speed and direction, temperature, chemical composition of the atmosphere (smell) and seismic data are all available, and can be simulated in the explorer's virtual environment, should the stimulatory technology be available.[†] All of these quantities are best provided as three dimensional field data, maybe using analogues of the graphical cloud of points mapping techniques.

The virtual world must be maintained, as the real Martian environment changes. Once again, low communication bandwidths dicate that the complete data from each sensor not be transmitted. As the sun sets, rather than the illumination at every sensor, only the general change in level and colour of illumination needs to be transmitted, except where the change is abnormal, due to shadow or reflection effects. Only the changed position of the sun in the sky should be sent back home, rather than a new image of the surroundings of each Daisy.

After sunset, the temperature will drop. The daisies need to track and transmit the advancing pattern of cooling, instead of sending complete new temperature maps with each infinitesimal change.

Thus, to maintain the telepresence illusion, each Daisy is engaged in a complex sequence of data gathering, information extraction, collaboration and communication. This is fully cogent behaviour, and to enable it will require sophisticated and ingenious programming within the individual node, probably using artificial intelligence techniques.

11.9. A World of Applications

The application above is perhaps 'science fiction' in nature, but is very feasible within the bounds of the technologies and research described in this book. With only a little imagination, many other applications can be seen for the technologies which will indeed make profound changes to our world. These applications, some of which were introduced in Chapter 1, will change the way in which environmental monitoring is performed, potentially allowing for provision of information on the detail of our environment to a level never before achieved. For structures, plants and transport systems health, safety and usage monitoring will be available at a level of precison completeness and continuity of coverage never before envisaged. The consequences of this will be more efficient, safer and more dependable

[†]One would hope that the stimulatory equipment will translate the actual environment into one which can be tolerated by the explorer, otherwise he or she will be frozen and asphyxiated.

buildings, transport systems and industries. Together with MEMS actuation technologies, the sensing technologies allow for new adaptive systems, providing more efficient aerodynamics and hydrodynamics. However, none of the technologies described in this book is capable of producing these results on its own — what is required is an integration of them all. MEMS sensors have already proven to be a potent technology in terms of size and cost reduction of everyday sensors. When integrated with processing capability they can handle a range of further applications, in particular being more readily deployed and adapted to a specific application problem. Networked sensors, particularly wireless ones, make it practical to use large arrays of sensors which can provide precise and detailed data from environments not previously amenable to sensing. Together these elements form a potent whole, Wireless Intelligent Networked Sensors. However, to release this potency advanced distributed software design techniques are required. This book has covered a range of the topics to show how the different fields of research and study that have contributed to the whole combined technology, and placed each in context against the others. The field as a whole advanced some time ago from a specialism in itself to a multidisciplinary enterprise in which defined specialisms are required and distinguished. As with any complex technological undertaking, providing team members with the context and understanding to appreciate theirs and others role in the team, and the contribution to the whole undertaking of the collected set of skills and knowledge, is a vital precondition to success. As a survey of the emergent field, with a breadth of coverage of the whole of that field, the authors hope that this book will be at least one contribution to the work of the researchers and developer from many disciplines who will make some of the 'dream applications' a reality.

References

1. http://www.beagle2.com/index.htm.
2. NASA Jet Propulsion Laboratory, Maers Exploration Rover Mission, http://marsrovers.jpl.nasa.gov/home/.
3. Williams, D. R. Mars/Earth Comparison, http://www-k12.atmos.washington.edu/k12/resources/mars_data-information/mars_earth_comp_NSSDC_and.html.
4. Spectrolab Inc. press release, http://www.spectrolab.com/com/news/news-detail.asp?id=152, July 25, 2003.

5. Maxim Inc., application note, HFAN-4.0.3, December 2000.
6. Johnson, N. M., Nurmikko, A. V. and DenBaars, S. P. (2000) Blue Diode Lasers, *Physics Today*.
7. Measurement Specialities Inc., Piezo Film Sensors Technical Manual, www.msiusa.com.
8. Crocombe, R. MEMS technology moves process spectroscopy into a new dimension, *Spectroscopy*, Europe.
9. Moreland, J., Jander, A., Beall, J. A., Pavel, K. and Russek, S. E. (2001) Micromechanical torque magnetometer for *in situ* thin film measurements, *IEEE Transactions on Magnetics* **37**(4).
10. US Patent 5,998,995 Microelectromechanical (MEMS)-Based Magnetostrictive Magnetometer.
11. ARM INC., product data, http://www.arm.com/products/CPUs/ARM7TDMIS.html.
12. Hitachi Ltd., product announcement, http://www.electronicstalk.com/news/hit/hit127.html.
13. TSMC Inc., product announcement, http://www.tsmc.com/chinese/technology/t0101.htm.

INDEX